High Voltage Vacuum Insulation

BASIC CONCEPTS AND TECHNOLOGICAL PRACTICE

This book is dedicated to the memory of Arthur Maitland, a driving force behind the 1960s' flowering of the subject and a valued contributor thereafter. Sadly, he was not spared long enough to witness the publication of his contribution to the present text.

High Voltage Vacuum Insulation

BASIC CONCEPTS AND TECHNOLOGICAL PRACTICE

Edited by

R V Latham

Department of Electronic Engineering and Applied Physics
Aston University, Birmingham, UK

ACADEMIC PRESS

Harcourt Brace & Company, Publishers

London   San Diego   New York
Boston   Sydney   Tokyo   Toronto

ACADEMIC PRESS LIMITED
24-28 Oval Road
LONDON NW1 7DX

U.S. Edition Published by
ACADEMIC PRESS INC.
San Diego, CA 92101

This book is printed on acid free paper

A catalogue record for this book is available from the British Library

ISBN 0-12-437175-2

Printed and bound in Great Britain by
TJ Press Ltd, Padstow, Cornwall

Contents

Part II- Technological Practice

Author Profiles

Guy Blaise

Department of Physics
Orsay University
France

Guy Blaise is Professor of Physics at the Orsay University, where he also conducted studies of the secondary emission of positive ions for his PhD thesis at the Laboratoire de Physique des Solides. In this early work, he related the positive emission to the virtual bound states of the electronic structure of 3d transition metal alloys, through the formation of auto-ionising states on the outgoing atoms. He then carried out research on various aspects of the application of ion beams for quantitative analysis using a mass spectrometry technique based on the thermalisation and post-ionisation of sputtered particles. The major theme of his recent work since 1990 concerns the problem of charge trapping in insulators and the subject dielectric polarisation/relaxation processes in to the flashover breakdown, or phenomenon.

Alain J Durand

Thomson Tubes Electroniques
2 Rue Latecoere, BP23
78141 Velizy-Villacoublay Cedex
France

Alain J Durand, graduated from the Ecole National Supéireure des Télécommunications in 1976. He is now Project Manager of R & D in the High Frequency Department (TDH) of Thomson Tubes Electroniques.

Claude le Gressus

Commissariat a L'Énergie Atomique
Centre D'Études de Bruyeres-le-Chatel
B.P 12, 91680 Bruyeres-le-Chatel
France

Claude le Gressus is working at the Commissariat à l' Energie Atomique, Centre d'Etudes de Bruyéres-le-Chatel, France. After his Diploma d'Ingenieur de l'Institut des Sciences Appliquees de Lyon, he did his PhD at Lyon University on the adsorption of reactive gas on insulators. Since 1970 he has been working on insulator charging and related phenomena. He has developed a laboratory for Scanning Electron Microscopy, Scanning Auger Microscopy and other electron Spectroscopy Technics. Since 1980 he has worked on the investigation of breakdown phenomena mechanisms and on the origin of the analogy between mechanical and electrical properties of insulators. He is co-operating with a multidisciplinary scientific community. He is active in organising International Conferences and he belongs to the boards of various Scientific Societies.

Burkhard G A Jüttner

Max-Planck-Institut Für Plasmaphysik
Mohrenstr. 40/41 (Hausvogteiplatz 5-7)
D-10117 Berlin
Germany

Burkhard Jüttner was born in Prague, Czechoslovakia, in 1938, and studied physics at the Humboldt- University, Berlin (PhD Thesis 1971). In 1985 he was made Professor at the Academy of Sciences, Berlin. Since 1991 he has been with the Max-Planck-Institute for Plasma Physics, Berlin and the Humboldt-University Berlin (the latter since January, 1994). His scientific interests centre on basic research in to high-current gas discharges, especially electrode problems of vacuum breakdown and vacuum arcs. These studies particularly involve field electron emission, surface physics, and plasma-wall interactions in devices for controlled thermonuclear fusion.

Rod Latham

Department of Electronic Engineering & Applied Physics
Aston University
Aston Triangle
Birmingham
B4 7ET
United Kingdom

Professor Rod Latham followed both his undergraduate and postgraduate education at Queen Mary College, University of London, and was awarded his B.Sc Honours Degree in Physics in 1958, and his Ph.D for work on Ferroelectrics in 1963: in 1984, London University also awarded him a D.Sc for his contribution to the field of Surface physics. He is a Fellow of both the Institutions of Electrical Engineers and Physics, and is a Chartered Engineer. He is Professor of Applied Physics in the Department of Electronic Engineering and Applied Physics at Aston University, having previously served as its Head: more recently he has held the post of Pro-Vice-Chancellor of the University with special responsibilities for research. His own research interests centre on how solid surfaces in vacuum respond to high electric fields, with particular reference to the phenomenon of "cold" or field-induced electron emission. From a technological perspective, this interest translates, on the one hand, into the development of new types of electron sources, and on the other, to improving the voltage hold-off performance of vacuum-insulated electrode configurations employed in electronic devices and systems. Space applications feature prominently in his work, and in this context he has received extensive funding from the US through their SDI Innovative Science and Technology programme. Professor Latham has previously written a specialist reference book on this subject and published over 100 scientific papers. He is a member of the permanent scientific committees of two international conferences. In the past, he has held senior visiting appointments at the Ecole Superieure d'Electricité in Paris, the University of Geneva, and CERN. He was the 1988 IEEE Whitehead Memorial Lecturer at the IEEE's Conference on "Electrical Insulation and Dielectric Phenomena".

Arthur Maitland

Lately of Department of Physics & Astronomy
University of St Andrews
Scotland
United Kingdom

Arthur Maitland was born in Blackburn, England in 1928. He studied physics as a part-time student for a London University External Degree and graduated with a BSc in 1956. In 1972 he was awarded a DSc degree by St Andrews University on the basis of papers on ionised gases published while working for a number of electrical engineering companies and St Andrews University. He established laser research at St Andrews in 1964, and he came to lead a group of some 14 physicists there, working in the areas of lasers and their applications, gas discharge tubes, ionised gases, optical methods of signal processing, and applications of ionised gases to fast switching. He became Professor of Physics in St Andrews in 1993. Professor Maitland was a Fellow of the Institute of Physics, published over 40 research papers on lasers, ionised gases and fast switching, and was co-author of Laser Physics (North Holland) and Vacuum as an Insulator (Chapman and Hall). He also held about 40 patents and was a consultant for the gas tubes division of EEV Co, Chelmsford, England. He died of cancer on 30 June 1994.

Boleslaw Mazurek

Department of Electrical Engineering
Technical University
Wroclaw
Wroclaw -51
Poland

Boleslaw Mazurek is Professor of High Voltage Technology and Electrical Materials at the Technical University of Wroclaw. He received the PhD and DSc from TU of Wroclaw in 1975 and 1984, respectively. In 1985-86 he was a Visiting Professor in the HV group at University of Waterloo and VARIAN in Canada. In 1970-85 he was preoccupied with cryelectronics and applied superconductivity. He was also engaged in investigations of the breakdown mechanisms in vacuum, gases and solid dielectrics. He has written over 60 papers.

H Craig Miller

616 Ruskin Road
Clearwater
Florida 34625
USA

H. Craig Miller was born in Northampton, Pennsylvania USA 1932. He received his BA and MS degrees in Physics from Leigh University in 1954 and 1955, and the his PhD degree in Physics from Pennsylvania State University in 1960. He has worked in the field of electrical discharges in vacuum for over thirty years (mostly with General Electric Co.), with special interests in vacuum arcs and surface flashover of insulators. HC Miller is a member of the AAS, the American and European Physical Societies, the IEEE and the American Vacuum Society. He is also treasurer of the Permanent International Scientific Committee which is responsible for the continuing series of International Symposia on Discharges and Electrical Insulation in Vacuum. Dr Miller has over 40 technical publications, including a comprehensive bibliography on electrical discharges in vacuum.

Hasan Padamsee

Floyd R Newman Laboratory of Nuclear Science
Cornell University
Ithaca
NY 14853-5001
USA

Dr Hasan Padamsee is an adjunct Professor at Cornell University and the Project Leader of the Superconducting Radio Frequency Group pushing the advancement of the technology for particle physics accelerators at the high energy and high current frontiers. At Cornell he has been conducting research and development on superconducting radio frequency (SRF) cavities for over 20 years. One of these years he spent at CERN in Geneva, Switzerland, working with the SRF group there to develop superconducting cavity technology, now being utilised to upgrade the Large Electron Positron (LEP) collider. He collaborated with CEBAF in Newport, News, Virginia an accelerator for precision nuclear physics constructed from superconducting cavities of the type developed at Cornell. Recently he has been active in promoting the application of superconducting cavities to the next generation linear collider, collaborating with the TESLA project at DESY laboratory in Hamburg, Germany. He was recently elected Fellow of the American Physical Society for his contribution in the field.

M Frank Rose

Space Power Institute
231 Leach Science Centre
Auburn University
Auburn
Alabama 36849
USA

M. Frank Rose is Director of the Space Power Institute and Professor of Electrical Engineering at Auburn University, Auburn Alabama. He has been active in researching and defining the conditions under which high power electrical systems can be made to operate reliably in the space environment. Under his direction, the Institute has contributed substantially to the SDIO SPEAR series of space experiments which were designed to determine the operating limits of high voltage space systems. In addition, he designed and built a hypervelocity impact facility which effectively duplicates a substantial portion of the space debris spectrum and is unique in the world. He has been an active contributor to the analysis and definition of the impact phenomena on IDED. His personal research interests are centred around electrical breakdown in the space environment, hypervelocity impact phenomena, and thermal management associated with high power space systems. Dr Rose holds 14 patents and is the author of 104 technical papers on various aspects of space environmental effects hypervelocity impact phenomena, energy storage and conversion technology. He received his PhD in Solid State Science from the Pennsylvania State University in 1966 and a BA in Physics from the University of Virginia in 1961. He is a fellow of the IEEE, Senior member of the AIAA and a member of Sigma Xi.

Arvind M Shroff

Thomson Tubes Electroniques
2 Rue Latecoere, BP23
78141 Velizy-Villacoublay Cedex, France

Arvind Shroff graduated from Ecole Spéciale de Mécanique et d' Electricité in 1953, and was awarded his Doctorate in Physics in 1965. He was manager of the Cathode and Ceramique R & D Laboratory of the High Frequency Department (TDH) of Thomson Tubes Electroniques, until his retirement in December 1993.

Ningsheng Xu

Department of Electronic Engineering & Applied Physics
Aston University
Aston Triangle
Birmingham
B4 7ET
United Kingdom

Ningsheng Xu was awarded a BSc degree in Semiconductor Physics in 1982 from Zhongshan (Sun Yatsen) University, China, and subsequently appointed Assistant Lecturer in the Department of Physics at the same University. From 1983 to 1986 he followed his postgraduate research training at Aston University (Birmingham, UK) sponsored by the scholarships from both the Chinese and UK governments. He was awarded a PhD degree in 1986 for his work on field-induced electron emission from microstructures on cathode surfaces. Since then, he has been a Research Fellow in the Surface Science Group of Aston University. During the early phase of this period, he expanded his research interests to include fast atom and ion surface interactions, time-of-flight mass spectrometry and the development of both pulsed fast atom and ion beam sources. More recently, he has concentrated on the fundamental mechanisms responsible for "cold" electron emission phenomena that are relevant to the insulation of high voltage by vacuum and solid insulators. He is also interested in the development of new types of cold cathode electron sources. Dr Xu is a member of the Institute of Physics, is a Chartered Physicist, and has published over 40 scientific research papers.

Preface

The past decade has seen a sustained growth in the diversity and complexity of device applications where vacuum is required to support either high voltages or high electric fields. This is particularly true in the space industry, and more specifically for the development of space-based pulse power systems. A primary aim of this book is therefore to provide a specialist update of the technological advances that have accompanied this quiet revolution. In part, the text is designed to be a replacement for the Editor's earlier book *High Voltage Vacuum Insulation: The Physical Basis*, Academic Press, 1981. However, by including an expanded treatment of the practical aspect of the subject, through a selection of specialist contributions from internationally recognised workers in field, this new book also provides a modern "working manual" for this specialist technology that is generic to such a wide range of practical applications. Equally, this mixed format makes the text suitable for use as a basis for "special topic" lecture courses at either Undergraduate or Postgraduate level within Electrical/Electronic Engineering or Applied Physics Departments of Universities and Polytechnics.

To meet the needs of the reader who is unfamiliar with the subject, the material has been presented in two broad sections. Part I is basically introductory, and concerned with fundamental physical concepts. For clarity, it has been divided into two sub-sections, with Chapters 2-7 being devoted to the properties of the "pure" vacuum gap, and Chapters 8 and 9 to a consideration of the growingly important question of what are the physical processes that influence the behaviour of a "bridged" vacuum gap where a solid insulator has to support a high voltage (or field) in a vacuum environment. In contrast, Part II focuses on particular applications of high voltage technology, and is therefore primarily targeted at practising technologists. Within this framework, each chapter is written as a self-contained entity with inclusive referencing.

In the opening chapter of Part I, the reader is given a general perspective of the electrical insulating properties of the vacuum gap, together with a brief historical survey of the main developments in the identification of the fundamental physical processes that ultimately determine its performance. This is followed in Chapter 2 by a detailed account of the design criteria for practical high voltage electrodes and how their operational characteristics are influenced by such basic parameters as their material, geometry, separation and surface preparation. This "technological" description of a vacuum gap is followed in Chapter 3 by an account of a very important series of experimental investigations involving the development of specialised in situ diagnostic techniques for dynamically identifying the physical processes that actually occur in normal operational gaps, and are responsible for the i.e. externally measured characteristics. To complement these experimental findings, Chapter 4 presents a review of the theoretical models that have been developed to explain the physical origin of the "pinhole" phenomenon that is responsible for parasitic prebreakdown electron emission. In Chapters 5, the discussion moves forward to explain how breakdown, the ultimate practical disaster, can be initiated by this "cold" microscopically-localised "cold" electron emission process, whilst Chapter 6 presents an account of how controlled simulation studies using the special case of a point-plane diode geometry have been used to verify a range of theoretical predictions. To conclude this group of chapters dealing with the "pure" vacuum gap, an account is given in Chapter 7 of another type of breakdown mechanism that results from the presence of charged microparticles, and how this mechanism has again been evaluated using laboratory simulation techniques.

Although the flashover of solid insulators has always been an important technological phenomenon, it is only relatively recently that it has assumed a heightened practical importance. This is due in part to the fact that technological advances have led to major improvements in the insulating performance of "pure" vacuum gap, so that the solid insulator has become the "weak link" of many device applications. Also, however, the "bridged" gap as assumed a growing importance in the "active" role of a fast-acting, high-current vacuum switch. To provide a comprehensive review of the current status of this subject, Chapters 8 and 9 presents two contrasting viewpoints that are currently being debated in the literature. Thus, Chapter 8 reviews the measured performance characteristics of insulators, and interprets these in terms of a long-standing electron-hopping surface breakdown. By comparison, Chapter 9 presents an alternative explanation of these same characteristics which is based on an entirely new type of physical model where the trapping and de-trapping of charge in the bulk of the insulator plays a crucial role.

The six chapters comprising Part II of the book aim to present a "technological" perspective of the subject. This challenge has been approached by selecting authors who are practising scientists or engineers in key technologies, and asking them to review the subject of high voltage vacuum insulation, as it is viewed and practised in their industry. For example, special considerations are involved in the design of a "mixed" insulating system consisting of vacuum-insulated metal electrodes in combination with their associated ceramic support structures. In many applications, there are also a range of problems associated with the application of non-stationary voltages. Indeed, it is this latter consideration of how typical insulating structures perform under pulsed field conditions that forms the subject of Chapter 10, the first "key" technology" of Part II. This is followed in Chapter 11 by a discussion of the special problems involved with "active" insulating systems where the internal environment is being continuously modified by evolved products; i.e. such as occurs with electron tube devices employing dispenser cathodes. In contrast, the superconducting RF cavity structures discussed in Chapter 12 represent an ideal "single-electrode", ultra-clean metallic system that, not perhaps surprisingly, has achieved some of the highest reported operating fields. The discussion is again broadened out in Chapter 13 to the important question of how the performance of HV vacuum devices and systems is influenced by the harsh conditions existing in space. This is followed in Chapter 14 by a consideration of the related topic of vacuum insulation under cryogenic conditions. The book is concluded in Chapter 15 with a review of the mechanisms thought to be responsible for initiating the vacuum arc as it occurs in such practical devices as vacuum switches: this chapter is also designed to act as a "bridge" between the present book and the literature devoted to the wider subject of plasma and arc physics.

Whilst most of the cited references have been taken from well known scientific periodicals, a considerable number have also been drawn from the proceedings of international conferences devoted to particular applications of HV technology. There is however, one particular symposium that takes a broader generic view of the subject, and has consequently been cited widely throughout the text. This is the Proceedings of the International Symposium on "Discharges and Electrical Insulation in Vacuum", subsequently abbreviated as Proc. ISDEIV, that was initiated in 1964 at the Massachusetts Institute of Technology (Proc. I-ISDEIV) with subsequent two-yearly symposia being held at various other centres throughout the world. The two other important conference series that have been extensively referred to throughout the text are the Pulse Power Conference (PPC) and the Conference on Electrical Insulation and Dielectric Phenomena (CEIDP); both of which are held biennially in the USA. Since these proceedings have provided a regular up-dating of the major developments across the field, they are an invaluable source of detailed information for the reader who wishes to pursue the subject beyond the scope of this volume. More generally, a representative selection of review articles are cited at the end of Chapter 1 which provide an alternative perspective on most of the topics covered in the present text.

Acknowledgements

I should firstly like to extend my warmest thanks to those valued colleagues who have contributed specialist chapters to the book. Without their support, the project would not have been possible.

I am also grateful to those who have given their permission for the reproduction of Figures and Tables. In all instances, I have endeavoured to make full acknowledgement, and apologise for any accidental omissions.

My particular thanks are due to Rachael Taylor for her tireless devotion in producing the camera-ready text, and to Andrew Abbot for preparing and processing all the figure work appearing in the text.

Finally, I affectionately acknowledge the constant support and encouragement I have received from my family whilst undertaking this project.

Rod Latham
Editor

Editor's Notes

1.   To avoid interrupting the intellectual development of the text, only the author's name appears at the beginning of each chapter. However, full details of their applications and professional experience can be found in the "Author's Profile" located on pages vii - xi.

2.   Although SI units have been used consistently throughout this book, it will be noted that "pressure" has been expressed in terms of the millibar, where 1 mbar $\equiv 10^2$ Pa. This choice is based on the continuing policy of some vacuum equipment manufacturers who calibrate their pressure monitoring systems in the mbar, a unit that is similar in magnitude to the more familiar "torr" (i.e. 1 torr = 133 Pa ~ 1 mbar).

3.   As already stated in the Preface, reference has frequently been made to papers appearing in the two-yearly Proceedings of the International Symposium on Discharges and Electrical Insulation in Vacuum. For simplicity, this source has been abbreviated to Proc. I, II XVI-ISDEIV, where I (1964) and II (1966) were held at M.I.T in the U.S.A, III (1968) in Paris, IV (1970) at Waterloo, Canada, V (1972) in Poznon, Poland, VI (1974) in Swansea, UK, VII (1976) in Novosibirsk, USSR, VIII (1978) in Albuquerque, USA, IX (1980) in Eindhoven, The Netherlands, X (1982) in Columbia, USA, XI (1984) in Berlin, GDR, XII (1986) in Shoresh Israel, XIII (1988) in Paris, France, XIV (1990) Sante Fe, USA, XV, (1992) Darmstadt, Germany, XVI, (1994) Moscow - St Petersburg, Russia, 1994.

Reference is also frequently made to papers taken from the Proceedings (IEEE Reports) on the biennial Conference on Electrical Insulation and Dielectric Phenomena (abbreviated CEIDP) held every even-year in the USA and the biennial Pulse Power Conference (abbreviated PPC) held every odd year in the USA.

Symbols and Abbreviations

A_e	effective emitting area
α_v	pulse over-voltage coefficient
β	geometric field enhancement factor
C	specific heat
C_c	Cranberg constant
γ	electron penetration depth/relativistic constant
D	diffusion constant/electric displacement/microcrater diameter
D(w)	transmission coefficient
d	electrode separation (i.e. electrode gap)
δ_e	secondary electron yield coefficient
E	macroscopic electric field
E_b	macroscopic DC breakdown field
E_b'	macroscopic pulsed breakdown field
E_c	critical microscopic breakdown field
E_g	microscopic electric field between a microparticle and an electrode
E_ℓ	microscopic field defining the onset of space-charge limited emission
E_m	local microscopic field
E_o	operational macroscopic gap field
e	electronic charge
ε	dielectric constant
ε^*	high frequency dielectric constant
ε_o	permittivity of free space
ε_r	relative dielectric constant
ε_e	total electron energy
ε_f	fermi energy

F_d microparticle detachment force

F_{vw} Van der Waals adhesive force

f spectral half width

h height of a microprotrusion

I total prebreakdown current

I_i ion current

I_s sample current

i_{EE} "explosive" electron emission current

i_F field electron emission current from a single emitter

i_{OF} low-temperature field electron emission current from a single emitter

i_{TF} high-temperature field electron emission current from a single emitter

i_p photomultiplier current

j_F low-temperature field electron emission current density

j_{TF} high-temperature field electron emission current density

K Thermal conductivity

k Boltzmann's constant

M_p microparticle mass

μ_o Fermi energy at $T = O$

$\mu(T)$ chemical potential

n_e electron density

n_i ion density

N_d donor concentration

N_t trap concentration

$N(W,\varepsilon_e)$ total energy supply function

P polarisation

$P(W)$ normal energy distribution

$P(\varepsilon_e)$ total energy distribution

ρ density

ρ_t tunnelling resistivity

Q_p	microparticle charge
Q_T	total charge transfer
q_i	impact ionisation charge
R_a	radius of anode "hot-spot"
R_g	"Gaussian" radius of electron beam cross-section
R_p	microparticle radius
r	tip radius of microprotrusion
r_c	arc crater radius
r_p	plasma polaron radius
S	spectral shift
s	thickness of ambient oxide film
σ	surface charge density
σ_y	yield strength of electrode material
T	temperature
T_b	temperature at the base of a microemitter
T_e	electron temperature
T_i	Nottingham inversion temperature
T_o	ambient temperature
T_p	phonon temperature
T_r	temperature at the tip of a microemitter
$T(r,x)$	temperature distribution in an anode "hot-spot"
T_{oo}	temperature at the centre of an anode "hot-spot"
t	time
t_b	total breakdown time
t_c	contact time of a bouncing microparticle/commutation time
t_d	breakdown delay time
t_p	pulse length
t_t	transit time of a microparticle
τ	charge relaxation time

τ_A thermal response time of an anode "hot-spot"

τ_b arc burning time

τ_c thermal response time of a cathode microprotrusion

τ_α heat spread parameter in an anode "hot-spot"

τ_s characteristic tip-sharpening time

τ_t charge exchange time of a bouncing microparticle

U_k kinetic energy of a microparticle

V externally applied DC gap voltage

V_{ac} externally applied AC gap voltage

V_b DC breakdown voltage of a gap

V_b' pulse breakdown voltage of a gap

V_{ip} externally applied impulse voltage

V_o operational voltage of a gap

V_p potential of a charged microparticle relative to an electrode

v microparticle velocity

v_a anode plasma expansion velocity

v_c critical microparticle impact velocity/cathode plasma expansion velocity

v_f cathode flare velocity

v_i impact velocity of a microparticle

W total power

w power density

ϕ_m work function of metal

ϕ_I work function of insulator

χ electron affinity/dielectric susceptibility

Ψ wave function of electron

$\Omega(T)$ electrical resistivity

ω angular frequency

Z atomic number

ABBREVIATIONS - Scientific

CW	Continuous wave
EEE	Explosive electron emission
EHT	Extra high tension
FEE	Field electron emission
FL	Fermi level
F-N	Fowler-Nordheim
FWHM	Full width at half maximum
HF	High frequency
HV	High voltage
I-V	Current-voltage
LEO	Low earth orbit
M-FEE	Metallic field electron emission
PF	Pulsed Field
RF	Radio frequency
SEM	Scanning electron microscopy
STEM	Scanning tunnelling electron microscopy
T-F	Thermally assisted field emission
TWT	Travelling wave tube
UHV	Ultra high vacuum

ABBREVIATIONS - General

CEDP	Conference on Electrical Insulation and Dielectric Phenomena
DEIV	Discharges and Electrical Insulation in Vacuum
PPC	Pulse Power Conference

1

Introduction

RV Latham

1.1   The Breakdown Phenomenon

Although vacuum is used extensively for the insulation of high voltages in such devices as X-ray tubes, electron microscopes, power vacuum switches, particle accelerators and separators etc., the reliability of its performance is limited by the operational risk of an unpredictable "sparking" or "arcing" between the high voltage electrodes, when the insulating capability of the vacuum gap is suddenly lost and "electrical breakdown" is said to have occurred. For some devices, such as sealed-off high voltage vacuum diodes with oxide cathodes, a breakdown event is likely to be an irreversible process and catastrophic from both the operational and financial points of view since, not only will the cathode probably be damaged, but the high voltage gap is subsequently likely to break down at a very much lower voltage: at all events, the future performance of the device will almost certainly be permanently impaired.

Fig 1.1   A high speed video image illustrating the "explosive" nature of a vacuum breakdown event.

In other applications however, such as vacuum switches and particle separators, occasional breakdown events, although undesirable, can generally be tolerated without disastrous consequences: that is, provided special precautions are taken to control the energy that is available in the external circuitry for dissipation in a gap during a breakdown event. To illustrate the dramatic nature of such breakdown events, Fig 1.1 is a highspeed photograph, recorded in the author's laboratory, of a discharge event between two plane-parallel electrodes. It shows the "explosion" of an isolated prebreakdown electron emission site, or "pin-hole", captured with the aid of the Transparent Anode technique to be described in Chapter 3.

Not surprisingly, this practical limitation of the insulating capability of vacuum has had a profound influence on the design of high voltage vacuum equipment: in the short term, the problem is conventionally tackled by such empirical and often expensive procedures as maximising the dimensions of vacuum gaps, using special electrode materials and surface preparation techniques, or incorporating sophisticated electronic protection circuitry. For the long term, there has been a considerable research investment directed towards obtaining an understanding of the fundamental physical processes that give rise to breakdown, so that by taking informed precautions the insulating capability of high voltage vacuum gaps can be improved. It is with the details of this research programme and the technological implications of its findings that this book is principally concerned.

The most serious hazard in the early applications of vacuum insulation was the presence of excessive residual gas, which manifested itself either directly as a relatively high ambient pressure in the gap, probably little better than $\sim 10^{-6}$-10^{-5} mbar, or indirectly as transient pressure bursts caused by the thermal desorption of gas from the electrodes and vacuum chamber walls by localised electron or ion bombardment processes. Thus, if the local pressure p in the gap approaches a value where the mean free path of electrons becomes less than the dimension of the electrode gap d, the necessary conditions will be created for avalanche ionisation and the spontaneous establishment of an arc between the electrodes. This situation has been quantified as the well known Paschen Law [1], which defines the sparking or breakdown potential V_b in terms of a function of the product pd whose detailed form can be found, for example, in the writings of Von Engel [2], Morgan [3] and Llewellyn-Jones [4] on the electrical properties of low pressure ionised gases. However, as a result of the many advances that have taken place in high vacuum technology, where it is now standard practice to use baked-out vacuum systems in which ambient pressures of $<10^{-7}$ mbar can be guaranteed, this mechanism of electrical breakdown is no longer regarded as a threat and, accordingly, will receive no further treatment in this text.

Thus, whilst the use of relatively gas-free ultra high vacuum (UHV) conditions greatly improves the high-voltage insulating capability of a vacuum gap, its electrical breakdown is still ultimately initiated by some form of discharge process arising from the creation of an ionisable medium in the gap. Since this can now only be derived from an increase in the local metal vapour pressure, it follows that any physical explanation of this form of breakdown must be based upon electrode surface processes that lead to the vaporisation of electrode material.

1. 2   Historical Background

The breakdown phenomenon was first investigated scientifically by Wood [5] in 1897 and somewhat later by Earhart [6], Hobbs [7] and Millikan and Sawyer [8]. From these early studies it was established that, even prior to breakdown, a vacuum gap has a small but finite conductivity as evidenced by the flow of "prebreakdown" currents whose magnitude increased rapidly with increasing gap voltage until breakdown occurred. For mm gap separations, it was found that the corresponding breakdown field was typically $\sim 10^8$ Vm^{-1}, although it depended somewhat on the electrode material. A further observation of considerable practical significance was that the breakdown voltage was independent of pressure in the range 10^{-8}-10^{-5} mbar. Millikan and Sawyer [8] also discovered that the voltage hold-off capability of a given gap can be significantly improved if it is subjected to an initial "conditioning" procedure (see Section 2.3) whereby the voltage is increased in small steps such that all major prebreakdown current instabilities are allowed to decay before the next voltage increment is applied. In 1920, Millikan and his subsequent co-workers [9-11] embarked on a decade of study into the source of the noisy but reversible component of the prebreakdown currents that flow between a pair of broad-area high voltage electrodes. They established that they were electronic and originated from a cold emission process (now known as field electron emission - FEE) at isolated points on the surface of the cathode which gave rise to complementary fluorescent spots on the anode.

At this time, these emission sites were assumed to be localised regions of the electrode surfaces where there was either an "effective" reduction in the work function through the Schottky effect [12] at field-enhancing microfeatures associated with the intrinsic microscopic roughness of electrode surfaces, or with a "real" reduction in the work function due to the presence of isolated chemical impurities. The noisy nature of the currents was attributed to the back-sputtering of ions produced by electron collision processes in the gap. Millikan and Lauritsen [11] also established that this prebreakdown current I had a well-defined empirical dependence on the gap field E such that a graph of log I versus 1/E gave a reversible straight line, i.e. I = A exp (-B/A) where A and B are constants. It was found however that the slope B and intercept A of such plots were very sensitive to the electrode surface conditions. Other early workers of note whose principal concern was with the origin and role of these highly localised cold emission processes included Hull and Burger [13], Snoddy [14], Beams [15] and Ahearn [16]. The initial conclusion to emerge from their studies was that breakdown was due to the intense localised heating and consequent vaporisation of the anode by the bombardment of electrons emitted from these point sources. However, Ahearn [16] extended the understanding of the phenomenon by considering the possibility of breakdown being cathode initiated following the field-induced rupture of current emitting projections.

One of the first really comprehensive investigations into how the operational breakdown voltage of a gap depends on such practical parameters as the electrode material, surface preparation and gap spacing etc. was undertaken by Anderson [17] in 1935. Although the general practical conclusions to emerge from this and many subsequent studies of its kind will be reviewed in Chapter 2, special mention should be made in this historical context of Anderson's identification of the "total voltage effect". This is associated with large cm-gap regimes supporting hundreds as opposed to tens of kilovolts, where it is found that breakdown tends to be voltage rather than field-dependent, and not apparently related to the prebreakdown electron emission currents which are frequently absent or negligible in such regimes, i.e. where the macroscopic gap fields are significantly lower than those existing in the earlier mm-gap studies of cold emission processes. To illustrate this distinction, it can be pointed out that, whereas it is possible for a 0.5 mm vacuum gap to support ~20 kV without breaking down, a 10 cm gap will support less than 1 MV.

Another important observation associated with this type of large-gap breakdown event was that electrode material is transported across the gap, apparently arbitrarily from either the cathode or anode. In these early experiments, this phenomenon was investigated by using dissimilar electrodes, say copper and aluminium, so that optical spectroscopy techniques could be used to identify the presence of traces of a "foreign" element on a given electrode. Although it was concluded from these findings that there was evidently some additional electrode surface process which was common to both cathode and anode, no satisfactory physical explanation for it emerged until 1952 when Cranberg [18] proposed his "clump" hypothesis. This breakdown model introduced for the first time the concept that loosely adhering microscopic particles ("clumps", or nowadays more frequently referred to as "microparticles") may be torn from an electrode surface by the applied field and, because of their charge, accelerated across the gap to impact on the opposite electrode as high velocity microparticles, causing localised fusion and vaporisation of electrode material that is sufficient to trigger the breakdown of the gap. The immediate appeal of this simple mechanism was that it provided an explanation of why breakdown events are unheralded, independent of the prebreakdown current and can result in material transfer between electrodes.

There have been many subsequent refinements and developments of the original model: for example, there is the "trigger discharge" model of

Olendzkaya [19], whereby a charged microparticle can give rise to a local breakdown-initiating discharge between itself and an electrode at very close distances of approach. As a result, the concept of microparticle-initiated breakdown is now firmly established and will be discussed at length in Chapter 7. Apart from these theoretical considerations, there have also been extensive laboratory simulation studies using artificially generated microparticles to verify many of the theoretical predictions: these will also be fully reviewed in Chapter 7.

At about the same time that Cranberg proposed his microparticle-initiated breakdown mechanism, Dyke and his co-workers in America [20-22] made the next major contribution to the subject with their publication of the first quantitative results from a controlled laboratory study on the electron emission-induced breakdown mechanism identified by earlier workers. The approach followed by this group, which will be fully described in Chapter 4, was to simulate the field emitting microprotrusions thought to be present on broad-area electrodes by laboratory-etched micropoint emitters similar to those used in the field emission microscope pioneered by Muller [23]. With the much improved vacuum conditions available at that time, they were able to use such emitters to study the breakdown characteristics of a well defined point-plane diode and establish that there is a critical emission current density of about 1×10 Am^{-2} at which the emitting surface becomes thermally unstable, cathode material is vaporised and breakdown is consequently initiated.

The relevance of these studies to the understanding of the breakdown mechanisms operating in a broad-area gap situation was further emphasised by the electron optical profile imaging of the surfaces of miniature electrodes by Bogdanovskii [24] in 1959, and later for broad-area electrodes by Little and Smith [25] and other workers, that positively confirmed the existence of microprotrusions having a sufficiently "sharp" geometry to produce the required geometrical field enhancement for FEE to occur. Based on information of this kind, Tomaschke and Alpert [26] were then able to show that the measured macroscopic breakdown field of a thoroughly out-gassed broad-area tungsten gap corresponded to a microscopic breakdown field of $\sim 6^{-7} \times 10$ Vm^{-1} at the tip of a tungsten microprotrusion; i.e. in agreement with the value quoted earlier that was measured directly using control micropoint emitters.

By the mid 1960s it had been recognised that a given field emitting microprotrusion would also give rise to an associated anode "hot-spot" where the electrode surface becomes locally heated by the bombardment of the fine pencil of electrons emitted by the cathode protrusion. Thus, if the local power density of the beam reaches some critical value, it is possible for anode material to be vaporised and hence the initiation of an arc between the electrodes. Such a mechanism is conventionally termed "anode-initiated" breakdown, as opposed to "cathode-initiated" breakdown when the microprotrusion itself becomes thermally unstable. In order to be able to predict which of these initiating mechanisms is most likely to precipitate the breakdown of a gap, unified theories were developed firstly by Chatterton [27] and Slivkov [28] in 1966 and in a somewhat more complete form by Charbonnier et al [29] in 1967, which established the critical initiating criteria for the two mechanisms in terms of the gap field, the electron emitting properties of the microprotrusion, the cathode and anode thermal diffusion characteristics and finally, the material constants of the electrode material.

The general qualitative conclusion to emerge from these theories is that a sharp thin microprotrusion will give rise to "cathode-initiated" breakdown whilst a more blunted geometry will favour the "anode-initiated" mechanism: for a quantitative evaluation of the criteria, the reader is referred to Chapter 5. It must however be pointed out at this early stage that, in their original idealised form, these breakdown initiation models take no account of a number of other physical processes that must be assumed to occur in practical high voltage gap regimes. Thus, more recent analyses of gap behaviour have attempted to incorporate, for example, the effects of ion desorption, ion bombardment (sputtering), surface diffusion of contaminants, and the more dramatic changes in the microtopography of an electrode surface that can result from localised fusion processes.

To complement the development of these models for electron emission and microparticle-based breakdown, there has also been a considerable effort to obtain experimental evidence supporting their validity. For example, electron optical imaging techniques have been developed for both identifying the position of cathode emitting sites and measuring the dimensions of their associated anode hot-spots; optical and X-ray techniques have been used to measure the temperature of "anode spots"; mass spectrometry has been employed to monitor both the desorption of gas and metal vaporisation due to thermal processes at both the cathode and anode; elemental analysis by optical spectroscopy and neutron activation techniques have confirmed that material is transferred between electrodes, presumably by microparticle processes; finally, scanning electron microscopy has played a major role in identifying significant features associated with the microtopography of electrodes. The main findings of these and other techniques will be reviewed in Chapters 3 and 7 when the reader has become fully acquainted with the detailed theories of electrical breakdown.

1.3   Practical Considerations

In parallel with these fundamental studies of gap phenomena, including both prebreakdown and breakdown-initiating mechanisms, there has been an on-going development programme in the technological practice of high voltage vacuum insulation, whose basic aim has been to improve the voltage hold-off capability of HV gaps. Referring to Fig 1.2, there are in fact two basic types of gap configuration. The first, where the space between two metal electrodes is entirely vacuum, is referred to as an MVM gap, whilst the second is a "mixed" situation where the gap is bridged by a solid insulator, and is referred to as an MIM gap. In performance terms, it would generally be assumed that for a given electrode separation, the voltage hold-off of an MIM gap would be lower than that of an MVM gap. Although these two regimes are intimately related at the practical level, they have traditionally been assumed to be controlled by entirely different sets of physical processes and, as a result, have been treated as largely independent entities. This perspective is in fact currently under review, but for the purpose of this text we have followed the standard practice and treated the MVM and MIM regimes independently in Parts IA and IB respectively.

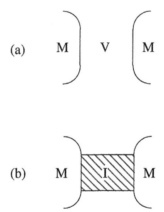

Fig 1.2   Schematic illustration of (a) the "pure" and (b) the "bridged" vacuum gap.

MVM Gaps   The experimental philosophy for characterising the operational performance of this type of regime has generally been to follow the "destructive" procedure of directly measuring the breakdown voltage V_b of a given gap regime. This approach suffers from the disadvantage that it gives rise to irreversible electrode damage, and a consequent lack of reproducibility, or "scatter", in breakdown data: a better approach, which at present remains a future objective, would be to "non-destructively" predict V_b from the prebreakdown characteristics of a gap. The considerations that have featured most prominently in these breakdown studies have been the choice of electrode material, the detailed design of electrode profiles and the final preparation of their surfaces prior to assembly. When under vacuum, it then becomes equally important to consider the external circuit parameters and the available in-situ processing procedures for gas desorption and electrode "conditioning". For some applications, it is also possible to consider the influence of the operating pressure of the gap and the residual gas species. The inescapable conclusion to emerge from most of these investigations is that it is very difficult to achieve a high level of performance reproducibility among a set of apparently identical experimental regimes. In practical terms, this has therefore meant that large "safety factors" have to be included in the design of commercial devices, such as adopting the often expensive procedure of increasing the separation and hence dimensions of electrodes. Whilst recognising this limitation, it is nevertheless possible to identify reproducible trends in behaviour among obviously different electrode regimes, and it is these that have converged to form the basis of modern high voltage engineering practice.

Although the details of such trends will be reviewed in the following chapters, some indication of their more important practical implications may be given at this stage. Thus, it is now common practice for commercial high voltage electrodes to be machined from stainless steel, molybdenum or titanium alloys, although there are many applications where other considerations make it desirable to use copper. To avoid undesirable "edge effects", electrodes are carefully profiled to avoid the local enhancement of the applied field. Prior to assembly, the high-field surfaces of electrodes have to be subjected to some form of polishing procedure. Traditionally, this used to involve them being mechanically polished to a sub-micron "mirror" finish followed by a final ultrasonic cleaning procedure; the so called "commercial" polishing procedure. However, as described in Chapters 3 and 12, it has recently been demonstrated that the main "enemy" of HV electrodes is contamination, and so the tendency has been towards chemical, or electrochemical polishing procedures in conjunction with clean-room rinsing and assembly procedures.

The electrode separation d is set by the voltage hold-off requirements. For *short gaps* of $d \lesssim 2$ mm, where instabilities predominantly stem from cathode-initiated electron emission processes, breakdown is field-dependent, with the minimum d-spacing being determined by a critical breakdown field that is typically $\sim 3\text{-}5 \times 10^7$ Vm^{-1}: in practical terms, this corresponds to 20 kV being safely supported across a 1 mm gap. On the other hand, for *large gaps* of $d \gtrsim 5$ mm, where anode-initiated processes are thought to take over the dominant role, breakdown becomes voltage-dependent so that the minimum d-spacing has to be determined from a power relation of the form $V_b \propto d_m$, where m typically takes a value in the range 0.4 to 0.7 [30]. Thus, whereas with the short-gap field-dependent behaviour pattern, a 5 mm gap might be expected to support ~ 150 kV, in practice it will only have a voltage hold-off capability of perhaps ~ 100 kV.

The choice of in-situ electrode conditioning treatment depends very much upon the function of the high voltage device. For the more common "passive" type of application, where the principal aim is to avoid arcing between electrodes, a prebreakdown "current" conditioning procedure is generally followed. However, for "active" applications, such as the contacts of power frequency vacuum circuit breakers, where intermittent arcing has to be tolerated, "spark" conditioning is more appropriate. In all applications, however, it is essential to minimise any capacitive energy stored in the external circuit that can be dumped across the gap during a breakdown; wherever possible, it is also desirable to have large current-limiting resistors (~ $M\Omega$) in series with high voltage electrodes.

Apart from being the ultimate manifestation of electrical breakdown, the vacuum arc, as intimated above, is also of vital technological significance as a transient condition that severely limits the operational performance of high voltage vacuum switches [31-33]. It is not surprising therefore that the phenomenon has been the subject of an intense independent research effort to establish the underlying physical mechanisms that control it; in particular, how the regenerative processes operating at the complementary anode and cathode "spots" sustain its burning, and what parameters ultimately determine how rapidly it can be extinguished. A detailed discussion of the properties and applications of the vacuum arc is in Chapter 15 of this book. For the interested reader, there are also comprehensive review articles [34-36] or more extended texts [37,38], that would provide an ideal introduction to the subject.

MIM Gaps   To reflect the growing technological demand for an understanding of the physical mechanisms responsible for the flashover of solid insulators (see Fig 1.1), Chapters 8 and 9 forming Part IB of this book are respectively dedicated to reviews of a) the practical performance of MIM gaps and b) new theoretical models of the breakdown process. In this context, there are broadly two types of operating regime that have to be addressed: those where an insulator is only required to support a high electric field under "passive" vacuum conditions, and those where the insulator has the additional demand placed on it of having to operate in an "active" environment that can progressively modify its material and electrical properties.

The design of insulator structures for most "passive" applications is generally well within the capability of contemporary technology. Thus, the type of data presented in Chapter 8 should be sufficient for determining the optimum shape, dimensions, bulk material requirements, and surface treatment for a particular application. In contrast, the situation is very much more complicated with "active" applications where surfaces can be subjected to ionising radiation or particulate contamination. Examples of this latter category are devices that have to perform in the harsh space environment, electrode spacers in high energy particle accelerators, and the support structures in sealed-off electron tube devices that employ dispenser cathodes. In all of these instances, insulators are found to be significantly more vulnerable to random flashover events that generally constitute a major threat to the subsequent stable operation of the device or system. It is to the physical cause of this externally-induced behaviour that Chapter 9 is principally directed. This discussion will highlight the important role played by trapped charges, and how the associated relaxation processes could provide an explanation for many of the observed breakdown phenomena.

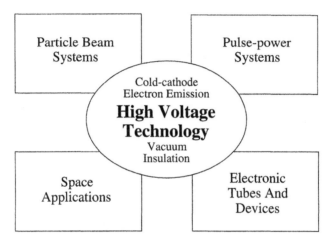

Fig 1.3   Illustrating the generic character of high voltage vacuum technology.

1. 4   Technological Applications

As illustrated in Fig 1.3, the range and diversity of technological applications where the insulation of high voltages plays a crucial role is extremely diverse. Indeed, to use modern terminology, it can truly be said to be a *"generic"* technology. At one end of the spectrum, there are small millimetre-sized devices, such as vacuum switches, and at the other there are such "large technology" applications as MeV electrostatic particle accelerators and laser confinement fusion systems. In general, each of these latter large-system applications has its own research community, infrastructure, conference arrangements, and associated published proceedings where full details of the relevant technologies can be found. For example, the electrostatic particle accelerator community have a regular International Conference on "Electrostatic Accelerator Technology" whose proceedings would provide the interested reader with all the necessary technical information about all aspects

11

of these very complex high voltage systems. For example, they contain detailed information about the design and operation of ion sources, vacuum columns, outer pressure vessels, the characteristics of the graded ceramic accelerator electrode assemblies, initial conditioning procedures, the threat to long-term stable operation resulting from the degrading effect of ionising radiation, and all other aspects of the peripheral high voltage control systems.

Likewise, other research communities employing specialist high voltage technology have their own international networks for disseminating their research findings. To provide the reader with a broader appreciation of the scope of this field, the chapters forming Part II of this book present contemporary "snap-shot" reviews of six such technology areas. Apart from offering an alternative perspective of the basic concepts described in the nine chapters of Part I, these specialist chapters also provide detailed information about how these same basic concepts have been translated into technological practice across a range of HV engineering applications. Whilst each of these chapters is intended to be a free-standing review, the new reader is nevertheless encouraged to become broadly familiar with the fundamental material in Part I before approaching the practical domain.

Part II opens with a consideration of the very important and rapidly expanding subject of pulse power technology. For those requiring a "working manual" of this topic, there is no better source of information than the Proceedings of the biennial IEEE-sponsored Pulse Power Conference held annually at differing locations in the United States. To complement this source of detailed information, Chapter 10 reviews some of the key developments and trends that have taken place in pulsed power technology over the past decade or so, including both CW and single-shot systems. Thus, because of the strategic military importance of this subject, the major effort has been largely focused on advanced microwave communication systems of various sorts, and particle beam-based weapon systems. With regard to the former category, a premium has been placed on improving the performance of existing systems, particularly in terms of their reliability and power output. This has resulted in a great deal of effort being expended in developing new procedures for minimising the destructive effects of surface flashover and multipactoring: however, in order to handle the transmission of these higher power levels, a corresponding investment has had to be made in improving the design of microwave windows. For single-shot pulse power delivery systems, the emphasis has been on the development of faster and more robust switches, including the triggered vacuum gap and solid state photoconductive devices. There has also been a major effort to develop new types of pulsed electron and X-ray sources for both military and civilian use; an example of the latter being the application of X-ray lithography to the microelectronics industry. In both of these areas of applications, there has been an over-riding requirement to ensure that devices are compatible for performing reliably in the harsh space environment, particularly in low earth orbit (LEO).

This discussion is followed in Chapter 11 with a consideration of the particular problems that are involved in encapsulating an electron tube device in a sealed-off vacuum envelope, and what, if any, special precautions have to be taken when the device employs a dispenser-cathode that constitutes a potential source of in-situ particulate contamination. These issues will be addressed, either directly or indirectly, through a short "case history" of the design, commissioning and operation of the electron gun assembly of a commercial travelling wave tube (TWT).

RF cavities are widely used in particle accelerating and storage systems. Conventionally, these been made from copper: however, with the growing demands from nuclear physicists for ever higher particle energies, there has recently been a major effort to enhance their performance by operating them under superconducting conditions with significantly higher accelerating field gradients. To achieve this objective, it has been necessary to invest an enormous technological effort over a ten-year period to develop specialised procedures for suppressing the onset of parasitic field emission, which would otherwise catastrophically degrade their performance. As will be described in Chapter 12, the key stages in the processing of a cavity are a) its initial chemical polishing, rinsing, and assembly under ultra clean conditions, b) in-situ high temperature thermal cycling under UHV conditions, and c) a final gas conditioning treatment. To illustrate the degree of success achieved by this type of processing sequence, it is sufficient to cite the three-fold up-grading of the peak accelerating field that is routinely achieved in the 300 MHz cavities now installed and operational in the LEP collider at CERN in Geneva. This review also includes an account of how sophisticated diagnostic techniques have been specially developed for studying the fundamental physical processes that ultimately limit the performance of RF cavities.

The harsh space environment presents a new and very challenging set of problems to the reliable long-term operation of HV devices and systems. Of particular significance are the threats posed by the low density plasma, including atomic oxygen, that are faced in low-earth-orbit situations, together with the effects of ionising radiation and microparticle impact that occur under all circumstances. In Chapter 13, an account is given of how the space technologist has firstly to quantitatively characterise the conditions prevailing in the space environment, and then to develop mitigating techniques to cope with the potential hazards faced by spacecraft. A notable feature of this review is the inclusion of in-situ data directly recorded in space from rocket-borne experiments.

Since the discovery of high Tc materials, there has been a renewed technological urgency to exploit their potential of providing low-loss power transmission systems. To reflect these developments, Chapter 14 discusses the physical and practical issues involved with the insulation of high voltages under cryogenic conditions for both MVM and MIM regimes. Thus, whilst the basic mechanism responsible for parasitic "cold" emission processes exhibits no special dependence on the superconducting state, there are nevertheless sound physical reasons, and convincing evidence, to conclude that the mechanisms responsible for initiating breakdown are significantly quenched by maintaining electrodes at low temperatures. In the case of a superconducting power transmission cable, for example, this could translate into an increase of between 20% to 50% in voltage hold-off performance compared to its value under ambient conditions at 300K.

To conclude this group of "technological" chapters, and indeed to round off this text, Chapter 15 presents a review of basic vacuum arc phenomena, and a survey of the types of switching devices in which vacuum arcs play a crucial role. The first part of the chapter is devoted to a detailed account of the diagnostic techniques that have been developed for studying the evolution process of an arc, starting with the formative stage of the cathode spot. From extensive data on how the key arc and spot parameters of size, material composition, life time, and movement depend on the gap conditions, a physical picture, or model, is built up as to how an arc is firstly initiated and then sustained. By way of providing a bridge to wider reading, the chapter concludes with an account of how vacuum arcs may be used to generate electron and ion beams.

1.5   Reviews and Bibliographies

For the reader who is interested in the historical development of this subject, there have been a number of specialist review articles that trace our growing understanding of the mechanisms thought to be responsible for initiating vacuum breakdown. The most notable of these are by Alpert, Lee, Lyman and Tomaschke [39], Slivkov, Mikhailov, Sidorov and Nastyukha [40], Rakhovskii [41], Davies [42] and Latham [43]. In addition, there are the reviews of Chatterton and Farrall that appear respectively in references [37] and [38], which are edited collections of writings on the broader topic of electrical breakdown of gases. For the reader who requires a more concise initial introduction to this subject than is offered by the present text, there are also several contemporary review articles by the present author and others [44-51].

The first detailed bibliography devoted to the phenomena associated with the initiation of breakdown was compiled by Hawley and Maitland [52] in 1967. Whilst this was valued by research workers at that time, it has been successively superseded by the bibliographies of Bacon, Grissom, Mangan, McClure and Newton [53] in 1972, by Farrall [54-56] in 1972, 1974 and 1976, and most recently by that of Miller [57-59] respectively in 1978, 1980 and 1990. Of these, the up-dated Farrall bibliography is undoubtedly the most comprehensive, since it covers not only prebreakdown and arc phenomena, but is also cross-referenced under a large number of "categories of interest". However the more recent bibliography of Miller would be a particularly valuable adjunct for the research worker who requires the most recent collation of information on the topics covered in this book.

1. 6 References

1.   Paschen, F., *Weid. Ann.*, **37**, 69-75, 1889.
2.   Von Engel, A., "Ionised Gases", Oxford University Press, 1955.
3.   Morgan, CG., Fundamentals of Electrical Discharges in Gases. *In* "Handbook of Vacuum Physics, (Vol.2 - Physical Electronics Electronics"), Pergamon, New York, 1965.
4.   Llewellyn-Jones, F., "Ionisation and Breakdown in Gases", Methuen, 1966.
5.   Wood, RW., *Phys. Rev. Ser.* I, **5**, 1-10, 1897.
6.   Earhart, RF., *Phil. Mag.*, **1**, 147-59, 1901.
7.   Hobbs, GM., *Phil. Mag.*, **10**, 617-31, 1905.
8.   Millikan, RA. and Sawyer, RA., *Phys. Rev.*, **12**, 167-70, 1918.
9.   Millikan, RA. and Shackelford, BE., *Phys. Rev.*, **15**, 239-240, 1920.
10.   Millikan, RA. and Eyring, CF., *Phys. Rev.*, **27**, 51-67, 1926.
11.   Millikan, RA. and Lauritsen, CC., *Proc. Nat. Acad. Sci.* (US), **14**, 45-49, 1928.
12.   Schottky, W., *Z. Physik*, **14**, 63-106, 1923.
13.   Hull, AW. and Burger, EE., *Phys. Rev.*, **31**, 1121-8, 1928.
14.   Snoddy, B., *Phys. Rev.*, **37**, 1678-85, 1931.
15.   Beams, JW., *Phys. Rev.*, **44**, 803-7, 1933.
16.   Ahearn, AJ., *Phys. Rev.*, **50**, 238-53, 1936.
17.   Anderson, HW., Elec. Eng., **54**, 1315-20, 1935.
18.   Cranberg, L., *J. Appl. Phys.*, **23**, 518-22, 1952.
19.   Olendzkaya, NS., *Rad. Eng. Electron Phys.*, **8**, 423-9, 1963.
20.   Dyke, W.P. and Trolan, J.K., *Phys. Rev.*, **89**, 799-808, 1953.
21.   Dyke, WP., Trolan, JK., Dolan, W.W. and Barnes, G., *J. Appl. Phys.*, **24**, 570-6, 1953.
22.   Dyke, WP., Trolan, JK., Martin, EE. and Barbour, JP., *Phys. Rev.*, **91**, 1043-54, 1953
23.   Muller, E.W., *Z. Physik*, **106**, 541-50, 1937.
24.   Bogdanovskii, GA., *Sov. Phys.- Solid State*, **1**, 1171-6, 1959.
25.   Little, RP. and Smith, ST., *J.Appl. Phys.*, **36**, 1502-4, 1965.
26.   Tomaschke, HE. and Alpert, D., *J. Vac. Sci. Technol.*, **4**, 192-8, 1967.

27. Chatterton, PA., *Proc. Phys. Soc.* (London), **88**, 231-45, 1966.
28. Slivkov, IN., *Sov. Phys. Tech. Phys.*, **11**, 249-53, 1966.
29. Charbonnier, FM., Bennette, CJ. and Swanson, LW., *J. Appl. Phys.*, **38**, 627-33, 1967.
30. Maitland, A., *J. Appl. Phys.* , **32**, 2399-2407, 1961.
31. Selzer, A., *IEEE Trans. Ind. Appl.*, **1A-8**, 707-22, 1972.
32. Althoff, FD., Elektrotech, ZA. **92**, 538-43, 1971.
33. Farrall, GA., *IEEE Trans.*, Parts, Hybrids and Packaging, **11**, 134-8, 1975.
34 Farrall, GA., *Proc. IEEE*, **41**, 1113-36, 1973.
35. Rakhovskii, V.I., *IEEE Trans. Plasma Science*, **PS-4**, 81-102,1976.
36. Miller, HC., *IEEE Trans. Plasma Science,* **PS-5**, 181-96, 1977.
37. JM. Meek and JD. Craggs (eds), "Electrical Breakdown of Gases", John Wiley, New York and Chichester, 1978.
38. JM. Lafferty (ed.), "Vacuum Arcs", John Wiley, New York and Chichester, 1980.

Historical Review Articles

39. Alpert, D., Lee, DA., Lyman, EM. and Tomaschke, HE., *J. Vac. Sci. Technol.*, **1**, 35-50, 1964.
40. Slivkov, IN., Mikhailov, VI., Sidorov, NI. and Nastyukha, AI., NTIS Report AD745471, 1972.
41. Rakhovskii, VI., NTIS Report AD773868 1973.
42. Davies, DK., *J. Vac. Sci. Tech.*, **10**, 115-21, 1973.
43. Latham, RV., *Phys. Technol.*, **9**, 20-25, 1978.
44. Toepfer, AF., *Adv. Electron Phys.,* **53**, 1-45, 1980.

Contemporary Review Articles

45. Noer, RJ, Appl. Phys., **A28**, 1-12, 1982.
46. Nelson, JK., *Eng. Dielectrics,* **II-A**, 445-570, 1988.
47. Mesyats, GA and Proskurovsky, D.I., "Pulsed Electrical Discharges in Vacuum", Springer, Berlin, 1989.
48. Miller, HC., *IEEE Trans. Elec. Insul.,* **24**, 765-86, 1989.
49. Eckertova, L., *Int. J. Electronics*, **69**, 65-78, 1990.
50. Latham, RV. and Xu, NS, *Vacuum*, **42**, 1173-81, 1991
51. Le Gressus, C., and Blaise, G., Conf. on Elec. Insul. and Dielec. Phenom., pp.574-90, 1993.

Bibliographies

52.	Hawley, R. and Maitland, A., "Vacuum as an Insulator". Chapman and Hall, London 1967.
53.	Bacon, F.M., Grisson, J.T., Mangan, DL., McClure, G.W. and Newton, J.C., "A Bibliography on Arc Discharges", Sandia Labs. Report SC-B-72 3233, 1972.
54.	Farrall, GA., "A Cross-Referenced File on Vacuum Arc and Vacuum Breakdown Phenomena". General Electric CRD Report 72CRD169, 1972.
55.	Farrall, GA., First up-date of [54] above; General Electric CRD Report 74CRD178, 1974.
56.	Farrall, GA., Second up-date of [54] above; General Electric CRD Report 76CRD052, 1976.
57.	Miller, HC., "A Bibliography and Author Index for Electrical Discharges in Vacuum" (1897-1976), NTIS Report No. GEPP-366, 1978.
58.	Miller, HC., First up-date of [57] above (1897-1978), NTIS Report No. GEPP-366a, 1980.
59.	Miller, HC., Second up-date of [57] above (1877-1979), *IEEE Trans. Elec. Insul.*, **25**, 765-860, 1990.

2

The Operational Characteristics of Practical HV Gaps

RV Latham

2. 1   General Considerations

From the introduction to this text, the reader will already have begun to appreciate the physical complexity of a high voltage vacuum gap. Not only does its behaviour depend on a large number of disposable macroscopic parameters, e.g. the electrode material, their geometry and separation, the residual gas pressure etc., but also upon the ill defined microscopic properties of electrode surfaces as determined for example by their preparation procedures and the operational "history" of the gap. Equally, there is a wide range of physical phenomena, such as electron emission, microparticle, ionic, electrode heating and thermal diffusion processes, that can operate within such a complex regime and significantly influence the performance of a gap. Thus, although there is an extensive literature describing detailed investigations into the behaviour of both "experimental" and "practical" gap regimes, it is not surprising that unspecifiable microscopic differences frequently result in a large "scatter" between the findings reported from apparently identical systems. Nevertheless, it is possible to identify certain underlying and reproducible dependencies of the gap behaviour on its macroscopic parameters that may be used as design criteria by high voltage engineers. The aim of the present chapter is therefore to provide a survey of these general trends, with particular reference to the uniform field, plane-parallel vacuum gap since this has been the most widely studied electrode geometry. The initial focus of the discussion will be on the DC insulating properties of gaps, leaving a consideration of the more complex AC and pulsed field regimes until the end of the chapter.

2. 2   The Plane-Parallel DC Gap

In order to identify the key physical processes that control the performance of an HV gap, it is desirable to choose the simplest and most reproducible experimental regime. For this reason, the plane-parallel gap, with its built-in uniform field condition, has invariably been used as the "test bed" for most fundamental investigations. Indeed, the literature of the 1960s and 70s, when this subject was at its height, abounds with studies of this regime; often, it must be said, at the expense of the more practically relevant non-uniform geometries.

2.2.1   Experimental Regime

The basic experimental facilities required for studying the external behaviour of a high voltage vacuum gap under controlled laboratory conditions are illustrated schematically in Fig 2.1. In the case of a plane-parallel geometry, the electrodes consist of two identical circular discs mounted on electrically insulated feed-throughs to a UHV chamber which should be a demountable and bakeable stainless steel system capable of reaching a pressure of $\leq 10^{-9}$ mbar: all components, particularly the electrical insulators, must therefore be unaffected by temperature cycling up to ~ 250°C. One of the electrodes is generally designated the "high voltage" (HV) electrode (either positive or negative polarity) and accordingly has a specially designed mounting insulator capable of holding off the maximum voltage that is to be applied across the gap. The mounting for the other "low voltage" or (LV) "earthy" electrode incorporates some form of bellows linked translational manipulator for controlling the electrode separation d: it is sometimes desirable for this assembly to include an additional "tilt" manipulating facility for ensuring the accurate parallelicity of the electrodes. For further practical details of this type of basic experimental investigation, the reader is firstly referred to a representative selection of standard papers on the subject [1-5]; however, other examples may be found in the bibliographies of Farrall and Miller cited in the Introduction.

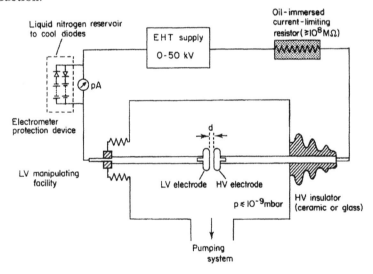

Fig 2.1   A schematic representation of the type of experimental facility required for studying the current-voltage characteristic of a high voltage vacuum gap.

To avoid the possibility of undesirable "edge effects", such as would arise from the local enhancement of the uniform gap field at a highly radiussed sharp edge of an electrode [6], it is essential to machine the outer edges of plane-parallel electrodes so that they have a rounded profile as shown in Fig 2.1. The effect is illustrated in more detail in Fig 2.2 which contrasts the distribution of electric field lines associated with pairs of unprofiled and profiled electrodes. In its ideal form, this profile must ensure that the electrode surface field never exceeds its uniform mid-gap value. This generalised requirement has been quantified by detailed field computations as the Rogowski [7], Bruce [8], or Harrison [9] profiles which are defined in terms of the electrode separation d, such that short gaps require a more highly radiussed profile than large. As a result of the proven operational advantages to be gained from this precaution, electrode profiling has now been adopted as standard practice in high voltage engineering. It should also be noted at this stage that the electrode surface preparation is of great importance in determining the performance of a gap; however, a detailed consideration of this topic will be delayed until Section 2.5.5.

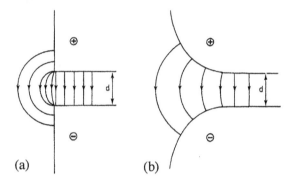

Fig 2.2   The electric field lines associated with a planar electrode gap formed from (a) *unprofiled electrodes*, where the sharp edges result in a local enhancement of the gap field, and (b) *profiled electrodes*, that give a uniform surface field.

Since, as will be discussed in subsequent sections, the performance of a high voltage gap is extremely sensitive to the presence of residual gas and other contaminants, such as vacuum pump oil, it is essential to choose a suitable evacuating system. Thus, getter-ion or turbomolecular pumping systems are in common use since they employ no intermediate pumping medium that could potentially contaminate the experimental environment: getter-ion pumps do, however, have the disadvantage of having low pumping speeds, whilst turbo molecular pumps are generally more expensive. On the other hand, oil diffusion pumped systems are relatively cheap, fast, and if adequately trapped, can achieve the required performance: a carefully designed system employing a liquid nitrogen trapped polyphenol-ether oil filled diffusion pump, a supplementary titanium sublimation pumping facility, and a trapped rotary roughing pump, can reliably achieve operating pressures of $\lesssim 5 \times 10^{-10}$ mbar.

The associated external electrical circuitry required for DC measurements is also shown in Fig 2.1 and is seen to consist of a simple series combination of a conventional low current ($\lesssim$ 1mA) EHT module having a continuously variable and internally monitored output voltage, an oil-immersed current-limiting resistor (typically $\sim 10^8$ Ω), and a picoammeter for measuring the circuit current. For an experimental field range $1 \lesssim E \lesssim 20$ MV m^{-1}, where E = V/d for the plane-parallel electrode gap being considered here, it follows that the specifications of the EHT unit will depend upon the d-spacing of the electrodes. Thus, a 0-50 kV unit would be adequate for d $\lesssim$ 2 mm, but the maximum voltage requirement would rise steeply to $\sim$ 250 kV for a d $\sim$ 5 mm and to $\sim$ 0.5 MV for d $\sim$ 1 cm [2]. A particularly desirable feature of such units is a "fine" control of the output voltage through a continuously variable "helipot" arrangement. To protect the picoammeter against a flashover of the gap, it is desirable to shunt it by some form of protection device such as a pair of biased-off zener diodes connected back-to-back as shown in Fig 2.1. If these components are also cooled to liquid nitrogen temperature it is possible to increase the residual resistance appearing across the combination to $\lesssim 10^{12}$ Ω. Under these circumstances, it will be greater than the input resistance of the picoammeter, so that only a negligible fraction of the total gap current will be shunted past the meter: this procedure also has the advantage of reducing the electronic noise generated in these components. For similar reasons, it is important to ensure that the resistance to earth of the low voltage electrode should also exceed 10^{12} Ω.

2.2.2   Current-Voltage Characteristic

It will be assumed in the following discussion that the vacuum chamber pressure is < 10^{-8} mbar which implies that the system has been baked to $\sim$ 250°C and that the "test" electrodes have therefore been partially out-gassed. To observe the general form of the current-voltage response, the applied voltage V derived from the EHT supply shown in Fig 2.1 is slowly increased until the picoammeter records a small but continuous current of $\lesssim 10^{-10}$ A. i.e. as opposed to the transient displacement current ($\propto$dV/dt) associated with the electrode capacitance. On further increasing V, Fig 2.3 illustrates how this "prebreakdown" current I will rise rapidly and become progressively unstable (see insert) until, without warning, the gap will break down; i.e. an arc is struck between the electrodes so that the gap resistance falls to zero and the circuit current is only limited by the external resistance. The properties of such arcs, and the condition required to "strike" and sustain them will be the subject of Chapter 15; at this stage, it is sufficient to note that the value of V at which this occurs is termed the breakdown voltage V_b of the gap. Although this general behaviour is broadly common to all high voltage vacuum gaps, the detailed nature of the response, particularly the level of the prebreakdown currents and value of V_b, is very sensitive to the macroscopic parameters of the gap, the electrode surface preparation procedure and its operational "history".

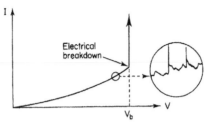

Fig 2.3   The typical form of the non-ohmic current-voltage characteristic of a high voltage vacuum gap.

For the special case of a pair of freshly prepared "virgin" electrodes, the current-voltage (I-V) or current-field (I-E), response exhibits several unique features. Thus, referring to curve 1 of Fig 2.4, there is firstly a threshold field in the range 5-15 MV/m associated with the initial switch-on of the prebreakdown current; i.e. where the prebreakdown current jumps discontinuously from an effective zero-value to ~ 1-10. Subsequently, it increases smoothly with field, although it may exhibit one or more further switching processes, as manifested by step-like increases in the magnitude of the current. Eventually, the gap will spontaneously break down at fields that are generally in excess of 20 MV/m. If however the field is reversed before breakdown occurs, the hysteresis effect shown dotted in Fig 2.4 is observed; i.e. where the current decays smoothly to zero, but over an extended field range. If the field is now re-cycled, the prebreakdown current follows the dotted curve to exhibit a smooth reversible I-V characteristic with no switching discontinuities: such a gap is said to be "conditioned".

The effect of increasing the ambient pressure can be somewhat arbitrary as illustrated by curve 2 of Fig 2.4, but it generally results in a lowering of the switch-on field, higher current magnitude, higher current noise levels, and a lower breakdown field. In fact, at pressures in excess of 10^{-5} mbar, there have also been reports of an "ignition" effect (see Section 2.4.3), in which the current can switch between two states with an associated hysteresis between them.

As shown in the insert to Fig 2.3, and also in Fig 2.4, prebreakdown currents are not only very "noisy", but they also frequently exhibit with the superposition of current spikes known as "microdischarges". The duration of these pulses of current can vary from 50-1000 µs, whilst their magnitude can exceed the mean prebreakdown current by perhaps two orders; in general, this latter phenomenon is most pronounced at higher pressures and higher voltages, i.e. as $V \rightarrow V_b$. The physical origin of these microdischarge pulses has been considered by Arnal [10] and a number of other early authors [1,11-

13], with several mechanisms offering a plausible explanation for their occurrence. Firstly, they could be manifestations of sub-critical instances of the "explosive" electron emission or microparticle based breakdown mechanisms to be discussed respectively in Chapters 5 and 7. Secondly, they could stem from regenerative ionisation process resulting from the local desorption of "pockets" of gas remaining in the electrode surface after the initial bake-out.

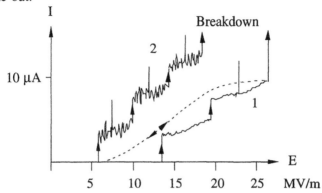

Fig 2.4   Illustrating the effect of pressure on (a) the DC current-voltage characteristic of a typical HV gap. Curve 1: $P \geq 10^{-9}$ mbar. Curve 2: $P \leq 10^{-6}$ mbar.

Thirdly, there is the type of ion-exchange mechanism analysed by Gerasimenko [14] in which positive and negative ions are ejected from contaminant films on electrode surfaces. This latter mechanism is assumed to be initiated by say a random positive ion created in the gap that is then accelerated by the field to generate further negative ions on impact with the cathode, which subsequently generate more positive ions on impact with the anode etc. Thus, if the ion multiplication factor at the two electrodes is greater than unity, the process will develop in an avalanche fashion to give the observed microdischarge.

The question of whether a microdischarge can lead to a spark (i.e. a breakdown) has been investigated by Berland et al. [15], who concluded that whilst there is a definite correlation between the two processes at low gap voltages, the same is not true at higher voltages. More recently, McKibben [16] has considered the practical significance of microdischarges in the context of the design and performance of accelerator tubes where electrodes tend to have large gaps (> 1 cm) which is the type of regime principally associated with the occurrence of microdischarges. He notes that the phenomenon can be partially eliminated by using electrodes made of titanium or its alloys, since these materials rank as having a "low ion exchange loading", especially in clean vacuum conditions where organic contamination is minimised. The operation of tubes containing an inert gas such as helium at a pressure of

~ 10^{-5} mbar is also beneficial (see Section 2.4.2). Even with these advantageous design features, it is still important to take further precautions to control the remaining microdischarge events that will inevitably occur during the initial voltage application; otherwise, a virgin gap will almost certainly break down at a voltage that is significantly lower than that to be anticipated from its macroscopic specifications, and will thereafter be unlikely to ever achieve its maximum insulating capability. To avoid such an irreversible and operationally disastrous occurrence in practical applications, it is customary to subject virgin electrodes to one of the carefully controlled in-situ "conditioning" treatments to be described in Section 2.3. These result in a gap that has not only a stable and reversible current-voltage characteristic as shown by curve 1 in Fig 2.4, but, more important, has the maximum possible voltage hold-off capability. However, as will be discussed in Section 2.3, the effectiveness of "conditioning" in stabilising prebreakdown currents and raising V_b varies considerably among electrode materials. For the most favourable regimes, it can reasonably be anticipated to achieve breakdown fields of $\gtrsim$ 30 MVm^{-1} although this is still over two orders of magnitude below the theoretical value of ~ 6.5×10^9 Vm^{-1} for a "perfect" vacuum gap [2].

2.2.3 Fowler-Nordheim Representation of the Prebreakdown I-V Characteristic

Following the recognition that stable prebreakdown currents originate from "cold" electron emission processes at perhaps one or two points on an electrode surface (see Chapter 3), it was a natural progression to look for an interpretation of them in terms of the well established Fowler-Nordheim quantum mechanical tunnelling theory of field emission from metal surfaces. As explained in the Introduction, it was this line of reasoning that led to the concept of there being isolated microprotrusions on an electrode surface where the uniform gap field E is geometrically enhanced by a factor β to a microscopic value βE (see Fig 4.3) that is above the threshold of ~ 3×10^9 Vm^{-1} necessary for "metallic" field electron emission to occur. Although the theoretical development of this model will be delayed until Chapter 4, it is possible to anticipate the general form of the current-voltage characteristic. Thus, from the detailed Fowler-Nordheim theory, the dependence of the low-temperature emission current density j_{OF} on the local surface field E, given by equation 4.2 can be acceptably simplified by letting the functions t(y) and v(y) $\rightarrow$ 1, so that

$$j_{OF} = \left(B_1 E^2 / \phi \right) \exp\left(-B_2 \phi^{\frac{3}{2}} / E \right) \qquad 2.1$$

26

where ϕ is the work function of the electrode material and B_1 and B_2 are known fundamental physical constants (see Section 4.2.1). With $E = \beta E = \beta(V/d)$ in the present case of a plane-parallel electrode geometry, and assuming there to be just one dominant site having an emitting area A_e, the total field emission prebreakdown current I will be given by

$$I = A_e j_{OF} = \left(A_e \, B_1/\phi \right)\left(\beta V/d\right)^2 \, exp\left(-dB_2 f^{\frac{3}{2}} bV\right)$$

or in logarithmic form

$$ln\left(I/V^2\right) = \left(dB_2 \, \phi^{\frac{3}{2}}/\beta\right)\left(I/V\right) + ln\left(A_e B_1 \beta^2 / d\phi\right) \qquad 2.2$$

It follows therefore that a plot of log I/V^2 against $1/V$, known as a Fowler-Nordheim or F-N plot, should give a straight line if this type of field emission mechanism is responsible for the prebreakdown currents. As illustrated in Fig 2.5, this is indeed confirmed experimentally for a wide range of applied fields, and accordingly led to the general acceptance of the above explanation of the non-ohmic conductivity of an HV gap. It should be added however, that F-N plots do frequently depart from linearity at very high fields (as shown dotted in Fig 2.5), but this behaviour can generally be attributed to subsidiary physical processes, such as the onset of space charge effects (see Chapter 6) rather than indicating false assumptions in the basic emission model.

The linearity of the F-N plot has the further advantage of providing information about both the local field enhancement β at an emission site and its apparent emitting area A_e. Thus, it will be seen from equation 2.2 that with B_1 and B_2 being known, and assuming an appropriate value for ϕ (which can be taken as an effective constant of ~ 4.5 eV for most electrode materials), β can be found from the slope of the F-N plot, and then used to find A_e from the intercept. It also follows that since β is inversely proportional to the slope, low-β "blunt" emitters (i.e. those that are likely to be stable in a practical situation) will be identified by their giving a steep-sloped F-N characteristic. Estimates of β and A_e taken from the extensive published literature on the behaviour of vacuum gaps reveal that they can assume widely varying and apparently arbitrary values within the typical ranges of $100 \lesssim \beta \lesssim 1000$ and $10^{-16} \lesssim A_e \lesssim 10^{-12}$ m^2. It should be added, however, that there have also been occasional reports of much higher predicted values of A_e, even approaching the unlikely dimension of 10 m^2 [17].

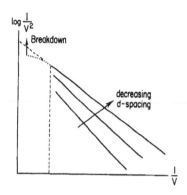

Fig 2.5   An illustration of how the Fowler-Nordheim (F-N) plots of the current-voltage characteristic of a gap are influenced by the electrode separation.

Whilst the lower values of these parameters, say $\beta \lesssim 150$ and $A_e \lesssim 10^{-15} \text{ m}^2$, can readily be interpreted in terms of the above model, it becomes increasingly difficult to reconcile the higher β and A_e, values with field emitting microprotrusions having realistic physical dimensions. In fact, this apparent anomaly will be considered further in the concluding chapter of this book in the light of recent experimental investigations that have provided further insight into the electron emission mechanisms operating on broad-area electrodes.

2.2.4   Electrode Separation

With the type of laboratory regime illustrated in Fig 2.1, it can readily be demonstrated how the d-spacing of the electrodes has a strong controlling influence over the prebreakdown I-V characteristic of the gap. Experimentally, this is observed as a rapid fall in I with increasing d at constant V, whilst in terms of the associated F-N plot, the effect results in a systematic shift of the characteristic as illustrated in Fig 2.5. This behaviour can be readily understood from equation 2.2, which shows how the slope of an F-N plot is directly proportional to d, whilst the intercept varies inversely with d. In practice, however, it is found that the measured variation in the F-N plots with d cannot be entirely accounted for in the above terms: it also appears that the field enhancement factor β appearing in equation 2.2 depends on d [1,2,18,19]. This effect is highlighted more dramatically by the experimental observation that the macroscopic gap field required to maintain a given prebreakdown current *decreases* with increasing d. Initially, this behaviour was attributed to the geometrical influence of the electrode edge profile [2]; however, using an anode probe hole technique to study individual sites in the centre of electrodes, Jüttner [18,20] was subsequently able to show that the phenomenon was in fact related to the voltage across the electrode gap and accordingly termed it the Total Voltage effect. Assuming prebreakdown currents to originate from microprotusions, he interpreted the effect in terms of the equilibrium that exists on an emitting surface between the sputtering of adsorbed oxygen by positive ions and the recovery of the surface by the migration of oxygen atoms. On this picture therefore, higher voltages would correspond to higher sputtering rates, and hence emission currents, since an oxygen-free surface has a lower work function. However, following a more recent model of Beukema [21], to be discussed in a different context in the following section, it is also possible that a voltage-dependent increase in emission could stem from the enhanced field-growth by surface migration of field emitting microprotrusions; i.e. such as could result from a rise in the cathode surface temperature following its bombardment by increasingly energetic positive ions. In this context, the published data [1,19,22] suggest that this effect is most marked with those electrode materials that give unstable gaps, e.g. copper and aluminium. As will be discussed in Chapter 4, these and other effects are open to an alternative explanation in terms of a more recent solid-state model of the emission process.

The important practical consideration of how the breakdown voltage V_b (see Figs 2.3 and 2.4) depends on the electrode separation d will be dealt with formally in Section 2.5.3. However, it can be said in summary here that at *small gaps* (d $\lesssim$ 0.5 mm), where breakdown is assumed to be initiated by an electron emission based mechanism, V_b is proportional to d, whereas at *large gaps* (d $\gtrsim$ 2 mm), V_b shows a less steep dependence on d.

2.2.5   Electrode Polarity

On the basis of both the traditional microprotrusion and more recent solid-state explanations of the reversible prebreakdown current-voltage characteristic (see Chapter 4), the behaviour of a given gap will be determined almost entirely by the microscopic detail of the electrode surface chosen to be the cathode. It follows therefore that since it is virtually impossible to prepare microscopically identical electrode surfaces, any gap will have two independent and frequently very different I-V characteristics, according to the alternative polarity arrangement of the electrode [23,24].

The operational stability of these two modes may be evaluated in terms of the respective emitting properties of the two surfaces as given by the β-values obtained from the associated pair of F-N plots: thus, the arrangement having the steeper sloped (low-β) plot can reasonably be assumed to have the higher voltage hold-off capability.

Beukema [21] has reported on another polarity effect that can occur with the present type of plane-parallel experimental gap regime, where one electrode is designed to carry the high voltage with the other being effectively earthed through some form of low-impedance current recording device (see

29

Fig 2.1). An analysis of the I-V characteristics of such systems reveals that, at any given operating pressure, there are consistently higher emission current levels, and consequently lower hold-off voltages, where the "high voltage" electrode is made the cathode. This effect was attributed by Beukema [21] to a three-stage mechanism involving (a) the electron desorption and ionisation of gas from the anode, (b) a subsequent increase in the local temperature of the cathode by impinging ions, followed by (c) the field-activated surface diffusion process described by Shrednik et al. [25,26] and Cavaille and Drechsler [27] which leads to the build-up of field-enhancing features that then

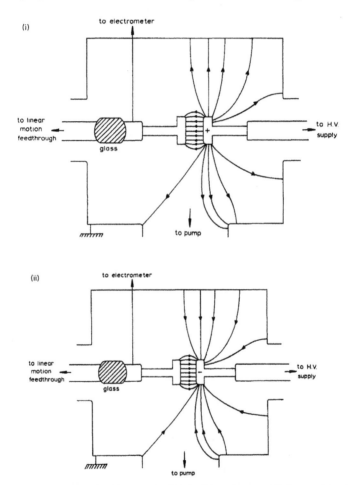

Fig 2.6   The electric field lines associated with a plane-parallel gap when the "high voltage" electrode is (i) the anode (normal operating polarity) and (ii) the cathode. In case (ii), the chamber is effectively part of the anode and can therefore provide an additional source of positive ions through electron desorption process. (From Beukema [21], with permission.)

act as new high-β electron emission centres. To understand how the influence of this mechanism can be sensitive to the electrode polarity, it is necessary to compare the field distribution for the two cases.

Thus referring to Fig 2.6, it is apparent that with arrangement (ii) i.e. the cathode at HV, the walls of the experimental chamber will also act as an anode, so that compared with arrangement (i), there will be a vastly increased surface area from which gas can be desorbed by electrons emitted from sites on the edges of the cathode. Although the practical significance of this effect has not been quantified, the evidence points to it being a potentially important consideration in the design of high voltage devices.

2.2.6 Electrode Temperature

This parameter is not generally treated as a variable quantity in practical high voltage devices. There are however practical situations where a device is subjected to temperature cycling, such as in space applications. It has therefore become increasingly important to investigate how the I-V characteristic of broad-area gaps varies with temperature. Considering first the effect of *increasing* temperatures it will be seen from Fig 2.7 how the emissivity of a 1 cm^2 Cu cathode shows a marked increase with its bulk temperature [28]. As will be demonstrated in Chapter 3, this effect arises from two physical processes: firstly, the emission mechanism itself is stimulated by an increase in temperature, and secondly, there is an increase in the number of emission sites contributing to the total prebreakdown current. Both of these important findings are difficult to explain in terms of a "metallic" emission model, and will be the subject of a detailed analysis in Chapter 4.

Turning to the effect of *decreasing* the electrode temperature, most studies of this kind have concentrated on investigating how the emission process is influenced by the onset of the superconducting state. Not surprisingly therefore, early investigations were devoted to low-T_c superconducting regimes [24,29-35]; more recently, however, comparable studies have been made with high-T_c regimes [36-38]. Indeed, such is the importance of this topic, that Chapter 12 of this text has been devoted to a detailed discussion of the theoretical and practical implication of field emission regimes under superconducting conditions. In addition, Chapter 12 is devoted to a discussion of the design and operation of superconducting RF accelerating cavities.

It would however, be appropriate at this stage to make a number of summary observations. Thus, if it is assumed that gap currents stem predominantly from field electron emission processes at isolated metallic microprotrusions, i.e. as traditionally assumed, it could be anticipated on theoretical grounds (see Chapter 4) that in the low temperature range

0 < T < 300K, the I-V characteristic should be virtually independent of temperature. Such a behaviour has indeed been confirmed for example by Septier and his co-workers [31,32] for low-T_C tantalum electrodes, where it was found that the F-N plots of the I-V characteristic at 4.2K, 11.7K and 77.3K were indistinguishable from one another. Similar findings have been reported by Cobourne and Williams [34,35] for low-T_C niobium, vanadium, tantalum and aluminium electrodes. To these must be added the more recent findings of Shkuratov [36,37] and Mazurek et al. [38], namely, that high-T_C materials also show no apparent transition in their field emission behaviour as an emitting cathode is cooled through its T_C. The physical implications of this important finding will be further considered in Chapter 4 when discussing a "non-metallic" emission mechanism.

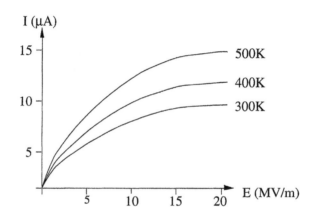

Fig 2.7   The influence of electrode heating on the prebreakdown emission current.

2.3   Electrode Conditioning

As indicated in the previous section, this type of in-situ treatment is generally an essential prerequisite if a HV gap is to achieve its full insulating capability. The basic concept behind any "conditioning" process is to safely quench as many as possible of the sources of prebreakdown current and "primary" microparticle events, so that the total number of potential hazards to the stability of the gap is significantly reduced. Since this objective is clearly of immense practical significance, there has been a considerable research effort directed towards developing techniques that can be relied upon to improve the voltage hold-off performance of gaps. As a result, four alternative procedures have been evolved.

2.3.1 "Current" Conditioning

This is both the simplest and probably the most widely used technique, and is particularly suitable for the initial treatment of "virgin" electrodes that are to be used in "passive" applications. The procedure involves using the same basic circuitry illustrated in Fig 2.1, whereby the applied voltage is increased in small discrete steps such that the prebreakdown current is allowed to stabilise at each stage before progressing. This sequence is illustrated schematically in Fig 2.8, and highlights how there is a general fall in both the noise level of the continuous prebreakdown current and the frequency of random current "spikes" or "microdischarges": the time constant of this stabilisation process is typically ~ 30 min.

Another common feature of "current" conditioning is the sudden and irreversible fall in the mean level of the prebreakdown current such as follows a microdischarge in the V_3 conditioning stage. In this technique, the current-limiting series resistor plays the important role of a negative feedback element, in that if there is a random surge in the prebreakdown current, the voltage across the resistor will correspondingly increase and result in a lowering of the gap voltage, thus controlling the mechanism responsible for the surge. If the conditioning process is arrested at say the voltage V_4, where the stable prebreakdown current is i_4, the gap will have a stable and reversible current-voltage characteristic up to the limiting co-ordinates (I_4, V_4); i.e. similar to that shown by the smooth curve of Fig 2.3. In practical applications, this conditioning procedure is of course continued until the intended operating voltage is reached, or more frequently exceeded by an acceptable safety margin of ~ 25%. There is clearly a limit to the effectiveness of this treatment since the current drawn from the residual "stable" emitting sites will eventually reach the critical level for the onset of thermal instability and the consequent breakdown of the gap. However, with an optimum electrode design, coupled with the correct choice of electrode material and surface preparation (see the following section), it can be reasonably anticipated for example, that hold-off voltages of 10 kV, 100 kV and 1 MV can be reliably achieved for gaps with d values of 0.5, 5 and 50 mm respectively. It should also be added that if the polarity of the gap is changed (or if it is intended for AC operation), the above conditioning procedure has to be repeated for the other polarity connection.

There has been considerable speculation as to the physical processes that give rise to the stabilising effects of "current conditioning" [39]. Traditionally, this discussion has been mainly in terms of the influence of high fields on electrode microprotrusions, loosely adhering microparticles and microreservoirs of occluded gas, since these were thought to be the most important agencies responsible for the various components of the prebreakdown current. Thus, the sharpest field-emitting microprotrusions are assumed to be either thermally blunted following excessive electron emission or, if mechanically unstable, are removed altogether from the electrode surface by the strong electromagnetic forces associated with the applied field: such an event is clearly suggested by the type of sudden fall in the prebreakdown current that can occur during a conditioning "stage" (see Fig 2.8). As will be discussed later in Section 2.3.3 and Chapter 4, current conditioning may also be explained in "electronic" terms if a "non-metallic" emission mechanism is assumed.

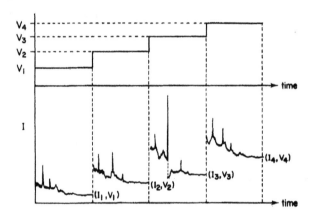

Fig 2.8   A schematic illustration of the "current" conditioning technique.

Consideration must also be given to the role of loosely adhering microparticles, since it has been demonstrated experimentally [40] that, with a slowly increasing conditioning voltage, these will be detached at the lowest possible field and hence acquire only a minimum charge; as a result, their single-transit kinetic energies will be too low for a breakdown-initiating explosive impact. Instead, they will either be harmlessly removed from the gap by repeatedly bouncing between the electrodes as described in Section 7.4.5, or become permanently welded to an electrode surface by the modified trigger discharge mechanism of Beukema [40] discussed in Section 7.2.1. The disappearance of random charge transients or "microdischarges" (see Fig 2.8) from the prebreakdown current during its stabilisation is frequently cited as evidence in support of the suppression of microparticle activity in a gap. Similarly, the field-desorption of residual gas trapped in grain boundaries etc. will be more controlled under a slowly increasing field, so that the risk of a large gas burst sufficient to initiate a breakdown is minimised. The existence of such a gradual desorption process is directly confirmed by mass spectrometer measurements described in Section 3.6.3, or indirectly as the general reduction in the "noise" on the prebreakdown current as conditioning proceeds. The involvement of gas in the conditioning process is further supported by the observation that the time constant associated with the decay of the prebreakdown current at a given voltage illustrated in Fig 2.8 is pressure-dependent, even down to $\sim 10^{-10}$ mbar. A detailed discussion of the effects related to the presence of residual gas will be given below in Section 2.4.

2.3.2 "Glow-discharge" Conditioning

In this approach, the primary aim is to use the sputtering action of low-energy gas ions to remove contaminants from electrode surfaces and thereby, as discussed in Section 2.2.2, minimise the incidence of microdischarges and possible breakdown events: a secondary improvement could also result from an increase of emitter work function following its coverage by inert gas atoms. The experimental procedure involves raising the pressure in the vacuum chamber to $\sim 10^{-3}$-10^{-2} mbar via a suitable gas handling system so that a low-voltage AC glow discharge can be struck between the electrodes. This low-current discharge can be conveniently derived from a 50 Hz source and, for optimum results, is typically run at $\sim$ 30 mA for up to $\sim$ 1 h: longer periods of ion sputtering can often lead to a rapid degradation of the performance of a pair of electrodes. At the conclusion of this treatment, the electrode chamber is of course re-evacuated to $\lesssim 10^{-8}$ mbar before making any high voltage measurements. Investigations into the effectiveness of this technique have shown it to be markedly dependent upon the residual gas species used to promote the glow discharge. Thus, from among the gases studied (H_2, D_2, He, Ar, N_2, SF_6 and "dry" air), helium, followed by nitrogen, consistently give the best results; i.e. as measured by the improvement in the hold-off voltage of the gap and the increase in the slope of the F-N plot of the current-voltage characteristic [41-43]. Considerable advantages are often claimed for this treatment over "current" conditioning, and so it is not surprising that it is common practice to subject such high voltage devices as particle-beam separators to an initial "glow-conditioning" treatment with helium at a pressure of $\sim 10^{-3}$ mbar [42].

2.3.3 "Gas" Conditioning

Apart from the adverse effects of running a gap at an elevated pressure, namely an increase in the incidence of microdischarges and the potential risk of breakdown events (see Fig 2.4), there can also be a beneficial effect, as manifested by the well known and widely used gas conditioning process. As illustrated in Fig 2.9, this technique involves running an HV gap at a progressively higher field in a helium atmosphere at currents of a few microamps and a pressure of around -10^{-5} mbar, and allowing the emission currents to quench over a 20 minute period to their asymptotic limit. This

35

procedure is repeated at incrementally increasing field levels until the final operating field of the device is achieved.

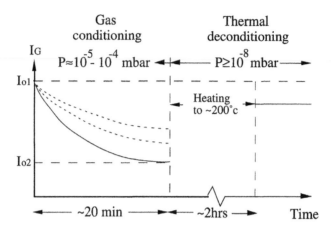

Fig 2.9   The general properties of the gas conditioning phenomenon, highlighting the deconditioning "Temperature" effect.

The gas normally used for this process is helium, and the phenomenon has been traditionally interpreted in terms of the blunting of metallic emitters by the sputtering action of high-energy gas ions [44-46]. Indeed, there is recent evidence [47] indicating that there are certain experiential regimes, particularly those involving small-area cathodes, where this sputtering mechanism dominates the conditioning process.

However, with the emergence of new experimental evidence which questioned the validity of the F-N "metallic" whisker model (see Chapters 3 and 4), it was necessary to reconsider the sputtering interpretation of the gas conditioning phenomenon. In addition, there was no satisfactory explanation of why helium is more effective than other gases. To clarify some of these issues, a detailed investigation has recently been undertaken to determine how both the gas species and temperature influence the conditioning process [48].

The first important finding to emerge from this investigation was that all the gases studied did in fact exhibit a positive conditioning effect. However, as illustrated in Fig 2.10, there is a built in "voltage effect", whereby each gas species has a characteristic voltage range over which it is effective as a conditioning gas. From a technological standpoint, this finding indicates that the optimum conditioning procedure for a device might well involve a change of gas as the voltage is progressively raised. The second important finding concerns the permanency of the conditioning, since recent studies have also shown [48] that the process can be significantly reversed by low-temperature thermal cycling in the range 200-300°C. Indeed, as indicated in Fig 2.9, up to 80% of the original current can be retrieved by such a procedure. Furthermore, the dashed lines on this figure illustrate how the effectiveness of the process falls off with increasing electrode temperature.

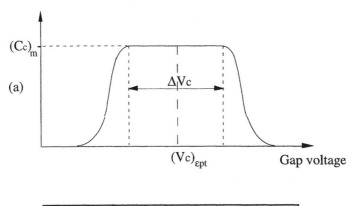

Fig 2.10   The general properties of the gas conditioning phenomenon, highlighting in a) the "temperature" effect, and in b) the "voltage" effect (After [48].)

Thus, it is important that gas conditioning is the final treatment process of electrodes, and that the procedure is not used where a device is required to tolerate temperature fluctuations in excess of 100°C. A concluding observation on these findings is that the well known "current" conditioning technique discussed in Section 2.3.1 could be a manifestation of the voltage effect, whereby the various residual gases progressively have a conditioning effect as the gap voltage is gradually raised. As will be discussed in Chapter 4, these findings may be explained in terms of a reversible "electronic" model involving a "non-metallic" emission mechanism that is promoted by metal-insulator microstructures.

2.3.4   "Spark" Conditioning

In a situation where a previously "conditioned" high voltage gap has become "deconditioned", such as following an inadvertent breakdown, it is usually possible to retrieve a large measure of its original performance by subsequently subjecting it to repeated sparkings: equally, this same procedure

37

can be applied to condition virgin polished electrodes up to their optimum insulating capability. This technique is also known as "spot-knocking", after the concept that its dominant aim is to remove localised high-β cathode emission sites or "spots" by arc erosion [49]; however, it is also efficient in eliminating microdischarges and diminishes the incidence of microparticle processes. For these reasons, it is widely used commercially as a method of maintaining the rating of high voltage vacuum devices. The electrical circuitry required for over-stressing the gap need be no more complicated than that shown in Fig 2.1, although it might be necessary to reduce the series resistor to ~ 100 kΩ in order to increase the rate of dissipation of energy in the spark. Alternatively, the effect can be achieved by an over-voltage pulse technique such as described by Miller and Farrall [49]. It is found that the number of sparkings required to achieve the maximum regeneration of a gap is unpredictable but usually varies between five and ten for common electrode materials such as copper [46,50], stainless steel [46] and molybdenum [51,52]; however, this number can be much larger for severely arced electrodes. With copper, for example, Cox [50] showed that the breakdown field increased with the number (up to 5) of low-current (< 1 A) conditioning arcs, and interpreted his findings in terms of a local hardening of the surface in the vicinity of emitting regions.

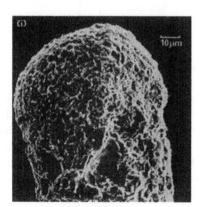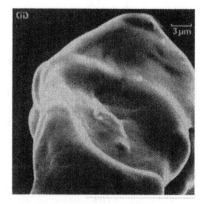

Fig 2.11   Illustrating the electrode "polishing" effect of nanosecond discharges: (i) on an initially eroded copper cathode, and (ii) the same cathode after "polishing" with 2 ns discharges. (From Jüttner [55], with permission.)

In order to apply this conditioning technique effectively, it is important to note that the external capacitance associated with the gap should be minimised so that only a limited amount of energy (≲ 10 J) is dissipated in the gap during sparking [53]. From studies of the effectiveness of this type of conditioning in the important practical regime of the high voltage vacuum switch, Farrall [54] has shown that if a spark discharge is struck between the contacts during their closing sequence, they should in turn be opened with low-current arcing if the maximum voltage hold-off capability is to be subsequently achieved: with a "cold" opening (i.e. in which no arc current is drawn), the spark discharge is likely to have a deconditioning effect.

Using a more sophisticated form of spark conditioning, Jüttner [55] has recently shown that it is possible to achieve an electrode "polishing" effect using nanosecond discharges under UHV conditions, and that this procedure can lead to a dramatic improvement in the voltage hold-off capability of a gap. The change in the microtopography of an electrode surface brought about by this "polishing" procedure is vividly illustrated by the "before" and "after" scanning electron micrographs of Fig 2.11.

It was also shown that a similar "polishing" effect can be achieved with pulses as long as 500 ns provided the current rise rate is kept below a critical value of 0.5-1 Ans^{-1}. From such observations it would appear that this approach to conditioning has considerable potentiality and will be further considered by Jüttner in Chapter 15.

2.4   Residual Gas Pressure

From our earlier discussion of the gas conditioning effect (see Section 2.3.3), it is clear that the nature of the residual gas environment can have a major influence on the prebreakdown characteristics of an HV gap. Equally, we know that a breakdown event generally stems from a prebreakdown emission process (see Figs 3.2 and 5.2). It could therefore be reasonably anticipated that the ultimate insulating properties of a vacuum gap, i.e. its voltage hold-off capability, would also be influenced by the residual gas environment. In this section, we shall review the findings obtained from studies aimed at identifying the key dependencies of the breakdown voltage V_b.

2.4.1   Dry Air

It has long been known from practical experience, particularly with accelerator tubes [16], that there is a "pressure effect" associated with the insulating properties of a high voltage gap [56-64] that is particularly marked at large d-spacings [60-65]. This manifests itself most obviously by the existence of a characteristic pressure p* at which the breakdown voltage V_b of a gap reaches a maximum value V_b* that can be as much as 50% higher than its mean hold-off voltage at normal operating pressures. The value of p* will vary according to the electrode material, their geometry, the gap separation and the residual gas species, although it will typically be in the range $(1-150) \times 10^{-4}$ mbar: among the inert gases however, for which the effect is generally the most marked, it has been reported that there is no great dependency on the particular species [59].

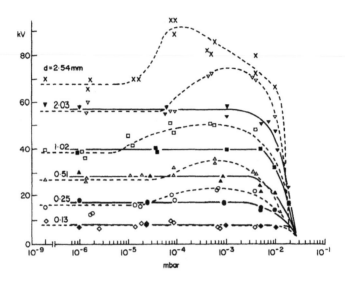

Fig 2.12   The pressure effect, illustrated by a family of plots showing how the DC (dashed lines and open symbols) and AC (solid lines and closed symbols) breakdown voltages of a pair of plane-parallel Cu-electrodes depend on the pressure and gap separation. (From Hackham and Altcheh [66], with permission.)

This later work has identified a number of general trends which are illustrated in Fig 2.12 for the case of "commercially" polished plane-parallel copper electrodes having air as the residual gas medium. From this set of breakdown data, it will be seen that all of the plots are characterised by having a low-pressure plateau region ($V_b = V_b$), a peak in the voltage hold-off capability ($V_b = V_b^*$) and a final high-pressure region where the onset of collision ionisation processes results in a rapid fall in V_b. It is also apparent that whilst there is a general increase of V_b with gap separation, the optimum operating pressure p^* corresponding to V_b^* remains nearly constant for all but the largest gap separation plots. There are, however, bound to be significant variations of p^* among electrode materials as is shown in Table 2.1 which collates the results obtained from this type of measurement. From this sample of data, it will be seen that both the value of V_b^* for a given gap separation (2 mm for the examples given), and its percentage enhancement through the "pressure effect" over the constant low-pressure value V_b, show a significant dependence on the electrode material. Among this group of electrode materials, it is evident that copper and niobium are most susceptible to this effect, which suggest that they have surfaces that are particularly "active" to the gas species present in air, i.e. principally oxygen and nitrogen. Conversely, there have also been reports of electrode materials such as sterling-silver that do not exhibit a pressure effect [67]; in such cases, the electrode surfaces must be physically and chemically inert to the ambient gas species.

Table 2.1

Material	p* (mbar)	V_b *	(V_b* - V_b)/V_b
Aluminium	1.5×10^{-4}	48	11
OFHC Copper	1.0×10^{-3}	75	32
Nickel	2.0×10^{-6}	97	7
Niobium	5.0×10^{-4}	90	25

Another significant manifestation of the "pressure effect" is a general lowering of the prebreakdown current levels at any given gap field, and it is this observation that has been focused upon as providing some insight into the fundamental phenomena that are responsible for the effect. However, because of the physical complexity of the electrode surface processes that are likely to occur in the pressure range 10^{-4}-10^{-3} mbar, it has to be recognised at the outset of this discussion that a complete interpretation of the observed behaviour must inevitably be very difficult and probably inconclusive. One explanation that has received wide support is due to Alpert et al. [44], and has been referred to already in Section 2.3.3. This is based on the concept that in the pressure range being considered, it is possible for there to be a reduction in the local field enhancement at emitters, and hence a lowering of their β-factors (see equation 2.2), through the sputtering action of low-energy ions formed by electron collision processes in the vicinity of the tips of field emitting microprotrusions. Apart from obtaining verification of the essential features of this mechanism from simulation studies on single micropoint emitters, these same authors showed that the mechanism could also be used very effectively as the basis of the in-situ "gas" conditioning technique previously described in Section 2.3.3.

There is however one feature of the pressure effect that cannot be explained solely in terms of this model, viz. its reversibility as the pressure is cycled, since this requires there to be either a re-sharpening of previously blunted emitters or the creation of new ones. Two further physical processes that have respectively been considered in this context are the surface diffusion mechanisms discussed by Shrednik et al. [25,26], Cavaille and Drechsler [27]

41

and Litvinov et al. [68], and the welding impact of cathode-bound microparticles proposed by Beukema [40]; both of which will be referred to again in greater detail in the following two chapters. In an alternative approach, Jüttner et al. [69] and Hackam [64] have attributed the "pressure effect" to an increase in the work function of an emitter following the physisorption of gas at the higher pressure. This interpretation is particularly attractive since it also provides an apparently plausible explanation for the reversibility of the phenomenon; viz. that gas previously adsorbed at high pressures will be field-desorbed at low pressures and thereby return the emission currents to their previous higher levels. However, this model must remain speculative at this stage since very little information exists on precisely how gas is adsorbed in the complex regime of a broad-area electrode surface: it is only for the very much simpler regime of small area emitters where, for example, Jüttner et al. [69] have been able to quantify how the emission mechanism is influenced by the adsorption of gas (see Section 3.4.2).

From the above discussion, it is evident that the "pressure effect" should more specifically be associated with the properties of the cathode surface since this is the source of electron emission and hence determines the level of the prebreakdown current at any given field. It follows therefore that the breakdown characteristics of a gap having dissimilar electrode materials will show a pronounced polarity dependence: for example, Table 2.1 indicates that an aluminium-nickel combination would be markedly more stable with the aluminium electrode as the anode. Equally, for non-planar electrode geometries, where there is a large disparity between the electric fields acting on the two electrode surfaces, there will be a similar polarity effect.

Thus, in experiments using a point-plane geometry [43], the breakdown strength of the gap is generally over 30% higher when operating with a positive-point polarity, i.e. where the planar cathode surfaces are subjected to a much reduced field. However, it is important to note that with this latter geometry the effect is observed at much smaller d-spacings than with planar electrodes.

2.4.2   "Foreign" Gas Species

Investigations into the pressure-dependence of V_b have also been extended to other controlled gas environments [43,53], including hydrogen, helium, argon, nitrogen, oxygen, carbon monoxide and sulphur hexafluoride. These studies have shown that, with the possible exception of hydrogen, a "pressure effect" similar to that illustrated in Fig 2.12 for dry air is observed for the remaining gas species when used in conjunction with a wide range of electrode materials. There is however, a considerable variation in the magnitude of the effect, although the evidence tends to suggest that most of these isolated gas species provide a more favourable insulating medium than dry air, with the relatively inert gases of N_2, He and SF_6 generally giving the highest voltage hold-off characteristics. As discussed in the previous section for the case of dry air, the physical interpretation of these pressure effects is likely to be very complex, probably involving several physical processes. However, by referring to selected studies of how the residual gas species influences the insulating properties of high voltage gaps, it is possible to give a reasonably consistent explanation of the experimental data on the basis that an improved insulating capability stems predominantly from an increase in the emitter work function [43]. Thus for systems in which chemisorption processes are likely to occur, such as those involving O_2 and CO on copper, nickel, molybdenum, tungsten and stainless steel, it has been shown that ϕ can typically increase in the range 1-2 ev. Similarly, from the literature on ion bombardment phenomena [70, 71], it is clear that physi-sorption processes can give rise to comparable increases in ϕ. He ions, for example, are known to be adsorbed more efficiently than Ar ions [70,71], and it has been suggested [43] that this could offer an explanation of why He gas gives a higher breakdown voltage than argon. It must however be emphasised that the actual changes in the work function associated with any given combination of electrode material and gas species can show a large scatter depending critically on the initial electrode preparation procedure and any subsequent in-situ treatments: for this reason, it is frequently very difficult to reproduce published data.

2.4.3   The "Ignition" Phenomenon

This effect is associated with the pressure range 10^{-5}-10^{-4} mbar and is observed experimentally at some arbitrary voltage V_i as a sudden increase in the prebreakdown current by several orders of magnitude to a new "ignited" mode where the gap current becomes very much less dependent on the applied voltage. The phenomenon is illustrated schematically in Fig 2.13, from which it is also seen that, when the gap voltage is cycled, there is an associated "hysteresis" of the I-V characteristic similar to that found with a low-pressure glow discharge where there are distinct "striking" and "extinction" voltages.

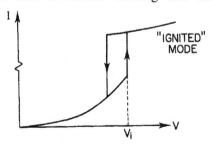

Fig 2.13   A schematic illustration of how the "ignition" phenomenon manifests itself on the current-voltage characteristic of a high voltage gap.

43

Although most reports of the effect have been related to systems having a residual air medium [72-74] there is also evidence to suggest that it may be common to most gas species. Physical explanations of the phenomenon tend to be somewhat vague but are either based upon some form of localised regenerative ionisation process that stems from the selective desorption of gas at an anode "hot-spot", or in terms of the onset at a critical field of an enhanced electron emission resulting from resonance tunnelling via virtual energy states associated with adsorbed contaminant atoms. In the context of this discussion on step-like discontinuities in the I-V characteristic of a high voltage gap, mention should also be made of an observation by Cox [75] that it is also possible to observe a type of irreversible hysteresis effect under UHV conditions, where regenerative ionisation processes are unlikely to occur. Instead their origin has been associated with a "switching" phenomenon that is a characteristic feature of a more complex type of emission process to be discussed in Chapter 4.

2. 5   Design Criteria for High Voltage Electric Assemblies

For most practical device applications where a high voltage, or field, has to be supported by a vacuum gap, the design engineer is faced with the problem of evolving a compromise between two independent sets of design criteria. On the one hand he has to ensure that the operational principles of the device are satisfied (e.g. the electric field requirements) and on the other, he has to have regard for the constraints imposed by the fundamental physical considerations that limit the insulating capability of the vacuum gap. In this section, we shall attempt to indicate how the disposable parameters of the system may be "tuned" to achieve this balance.

2.5.1   General Considerations

It will almost invariably be the case that the operational principle of the device will dictate the general form of the electrode geometry. Beyond this, however, there remains the choice of the electrode material, and surface preparation procedure, together with the optimisation of the edge profile and effective area of the electrodes. The relative importance of these factors has been investigated under controlled laboratory conditions in numerous "development" studies that have broadly followed one of two approaches. In the first, which includes the majority of such studies, the electrode geometry is reduced to its simplest symmetrical configuration, viz. the plane-parallel, uniform-field gap, so that the influence of the electrode material, surface preparation procedure, conditioning treatment and residual gas pressure on the I-V characteristic of the gap can more readily be assessed. The aim of the second approach is to more closely simulate the type of electrode geometries found in practical devices, so that it is possible to study the effects associated with non-uniform gap fields, such as would exist, for example, with sphere-sphere, sphere-plane, point-plane and concentric cylinder electrode combinations.

2.5.2   Electrode Geometry

Whilst the general form of a gap geometry is automatically decided by the function of the device involved, the final detailed design of the electrode profiles is determined by the physical operating requirements of the gap; e.g. a specific operational field E_o for the case of a particle separating module or a particular voltage hold-off capability V_o with say an HV vacuum switch. In practice, the design usually involves optimising the d-spacing of the electrodes (discussed in the following section) in conjunction with their effective area A_E, since both of these parameters have a strong but opposing influence on V (and hence E_o): thus, an increase in d at constant A_E will lower the gap field at a given applied voltage and hence result in a higher V_o, whereas an increase in A_E at constant d will generally lead to a lower V_o since more potential hazards in the form of emitting sites and microparticles will become exposed to the constant gap field [76].   This area-effect has also been shown to account for discrepancies in the published breakdown behaviour of plane-parallel electrode gaps of a given material, such as for example with stainless steel [77]. It must however also be appreciated that, because of the electrode edge-profiling requirements discussed in Section 2.2.1, an increase in d will correspondingly require an increase in the radius of the edge profile, so that, unless the overall dimensions of the electrodes are proportionally increased, there could be a sufficient fall off in the uniformity of the gap field to impair the operation of the device. This observation therefore explains why there has to be a general scaling-up of electrode dimensions in applications requiring say, a megavolt hold-off capability. It is also possible that the violation of this requirement in experiments where the d-spacing of fixed geometry electrodes is varied, could give rise to a spurious d-dependence of the field enhancement factor β.

In the rare situation that the electrode geometry is not constrained by some specific operational requirement, and that the sole aim is to achieve the maximum hold-off voltage for a given gap spacing, laboratory experiments have shown that an impressive improvement can be derived from using a positive point-plane geometry [21,78,79].   In fact, from comparative measurements with stainless steel electrodes, Hackam [43] has shown that a three to four fold gain in the voltage hold-off capability can be achieved by this approach. Pivovar et al. [79] have also shown that there is still a small but significant gain when the above geometry "expands" to a sphere-plane configuration. One very plausible explanation for the favourable voltage hold-off characteristics of this type of regime is that, for any given applied voltage, the field on the planar cathode surface will be significantly lower than with a pair of similarly spaced plane-parallel electrodes. For example, a simple calculation using equations 2.3 to 2.5 below shows that for a geometry having a tip of radius ~ 0.01 mm and a d-spacing of 0.1 mm this difference will be nearly 20% [43]. Accordingly, there will be lower levels of electron emission from any potential sites and hence a much reduced risk of breakdown occurring. Beukema [21], on the other hand, has discussed an alternative explanation for the phenomenon in terms of how gas desorbed from the anode can augment field emission processes on the cathode surface: with such a model, small anode surface areas (as in the present case of a "point" anode) would therefore give rise to only a minimal effect. Also, with the combination of a low cathode field and the very small anode area associated with a "point" electrode, there will be less chance of microparticle processes occurring. It must however be emphasised that the above behaviour only applies to short gap regimes. At large gaps, microparticle processes become dominant, and there are plausible arguments given in Section 7.4.5 which suggest that any polarity effect with non-uniform gap geometries will be reversed; i.e. it becomes more favourable to make the high field electrode the cathode.

In the course of designing gaps with non-uniform electrode geometries, it will normally be necessary to calculate the macroscopic field at the surface of an electrode to ensure that it does not exceed the critical breakdown value ~ 6-8 × 10^7 Vm^{-1} given in Section 2.5.3. Although such calculations are frequently very complicated, involving specialised mathematical techniques [80-84], some geometries can be treated analytically. Thus, in the case of a hemisphere-plane configuration, for example, the axial field is given to a good approximation [80] by

$$E_{oy} = 2dV \left[d^2 (f+1) + y^2 (f-1) \right] / \left[d^2 (f+1) - y^2 (f-1) \right]^2 \quad 2.3$$

where y is the axial co-ordinate, d the gap length, V the applied voltage and the function f is given by

$$f = \frac{1}{4} \left\{ (2p-1) + \left[(2p-1)2 + 8 \right]^{\frac{1}{2}} \right\} \quad 2.4$$

where $\quad p = 1 + d/R \quad$ 2.5

with R being the radius of the hemispherical electrode.

46

2.5.3   Electrode Separation

Although the electrode separation d is a crucial gap parameter, it is also very difficult to fix since its influence cannot be specified in absolute terms. This is because the voltage hold-off capability of any given electrode geometry is not a simple function of d, but rather depends upon a number of parameters, some of which cannot be physically quantified. Under these circumstances, it is only possible to have a meaningful discussion of the dependence of V_b on d among gap regimes of a given electrode material and that have been prepared according to some standard procedure involving "optimum" surface polishing and in-situ conditioning treatments (see respectively Sections 2.5.6 and 2.3). However, even with a strict adherence to such recommended guide-lines, there will still be a significant "scatter" in the insulating performance of identically prepared gaps. For this reason, it is important to incorporate "safety factors" of at least 25% when fixing the d-spacing of gaps in commercial devices.

As a consequence of this ill-defined behaviour of vacuum gaps, the design procedure for any specific application must almost inevitably involve a series of trial-and-error experiments with laboratory mock-ups to establish the correct gap spacing. Some initial guidance can however often be derived from published performance data or related experimental regimes, and to this end, the reader is referred to the extensive bibliographies cited in the introduction to this text, which contain reports on numerous and diverse electrode combinations. The main unifying condition to emerge from such studies is that for *small gaps* (d $\lesssim$ 0.5 mm), breakdown tends to be field-dependent, whereas at large gaps (d $\gtrsim$ 2 mm) it becomes voltage-dependent. In physical terms, the "small-gap" behaviour corresponds to breakdown being initiated by one of the electron emission based mechanisms discussed in Chapter 4, whilst the "large-gap" behaviour becomes dominated by the microparticle processes discussed in Chapter 7.

For the special case of plane-parallel gaps that have been prepared according to optimum specifications, the typical dependence of V_b on d will take the form illustrated in Fig 2.14, where for d $\lesssim$ 0.5 mm the linear relation

$$V_b \propto d = E_b d$$

$$2.6$$

holds, and for d $\gtrsim$ 2.0 mm, i.e. beyond the transition region 0.5 $\lesssim$ d $\lesssim$ 2.0 mm, the power relation

$$V_b \propto d^\alpha = K d^\alpha$$

$$2.7$$

takes over [1,22,53,77,85-93]. Under favourable conditions, the low-gap macroscopic breakdown field E_b will be in the range 6-8 $\times$ 10^7 Vm^{-1} [1], whilst in the high field case, the constant of proportionality K and power α can vary respectively between ~ 40-45 kVm^{-1} and 0.4-0.6, depending on the experimental conditions of the cited investigations.

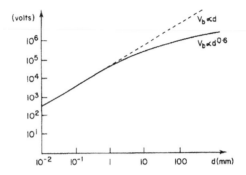

Fig 2.14   Illustrating how the breakdown voltage V_b of a plane parallel high voltage gap typically depends on the electrode separation d.

2.5.4   Electrode Material

It has long been recognised that the correct choice of this parameter is of vital importance if a high insulating capability is to be achieved for a given gap design. Nowadays, this is facilitated by drawing on the past experience of numerous laboratory and practical studies that have led to the establishment of an approximate hierarchy in the suitability of materials for use as high voltage electrodes. Thus, although there are many instances of "overlap" in the published data on the relative merits of materials, it is generally agreed that at one extreme there are materials such as copper and nickel, whose electrodes are particularly subject to instabilities, whilst at the other, there are stainless steel and titanium that consistently give stable electrodes and hence have come to be the "accepted" materials of high voltage engineering: molybdenum must also be included in this latter group since it too is becoming more accepted for use in practical devices. A widely used type of stainless steel is AISI Spec. 304 (18/8 chrome/nickel alloy), whilst in the case of titanium there is the IMI alloy Ti-318 (~ 89% titanium, 5-7% aluminium, 3-4.5% vanadium and 0.3% iron). In attempting to identify the physical characteristics that single out these materials as being particularly suitable for high voltage applications, it can first be noted that they are all "physically hard" with numbers > 300 on the Vickers Scale.

This property alone is not however a sufficient requirement, since there are other refractory materials such as tungsten that have never been found to give consistently stable electrodes. A further significant property that is shared particularly by stainless steel and titanium is their resistance to corrosion effects such as oxidation, which, in part, is thought to be attributable to the fact that the surfaces of these materials are both characterised by having a strong insulating ambient oxide film ~ 5 nm thick [94]; i.e. as opposed to the semiconducting ambient oxide films that are commonly found on metals such as copper [95]. This type of surface property is thought to be a particularly important prerequisite for high voltage electrodes and will be considered in greater detail in Section 7.4 when discussing the behaviour of microparticles in HV gaps. Another important property of these "accepted" high voltage materials is that they are equally the "accepted" materials of UHV technology; i.e. they have the property of being non-gassy. Finally it should be noted that current technological experience indicates that there are other material properties, such as the distribution and composition of surface impurities, that are of crucial importance in determining the behaviour of high voltage electrodes. In this context, the reader is particularly referred to Chapter 12.

2.5.5 Electrode Surface Preparation

According to the established models for breakdown initiation, the "ideal" electrode surface would be one that is free of both and superficially adhering microparticles. This therefore indicates the need for an electrode "finishing" treatment in which the operational surface is first subjected to some sort of microscopic polishing procedure and then to a final cleaning treatment for removing all traces of superficial debris. Attempts have therefore been made to evaluate the relative effectiveness of standard metallurgical polishing techniques viz. those based on mechanical, chemical and electrochemical processes. Significantly, it was found by Williams and Williams [96] that the latter two techniques could occasionally produce very stable high voltage electrodes. Nevertheless, the mechanical polishing approach was for many years generally accepted as the most reliable, and was consequently adopted as standard practice in high voltage engineering. Electrodes whose preparation is based on this technique are said to be "commercially" polished.

The first stage of this polishing procedure is to take the freshly machined electrodes and further smooth their operational surfaces by grinding them with progressively finer carborundum grits. After a thorough washing, they are then polished to a 0.25 μm "mirror" finish using a succession of alumina-based diamond pastes. The final and very vital stage in their preparation is to immerse them for long periods in an ultrasonic bath successively filled with trichloroethylene, acetone, methanol and distilled water to ensure that all traces of grease and polishing material are removed from their surface. Great care also has to be taken to avoid any re-contamination during the handling procedures involved in their final assembly: it is therefore particularly important that these latter operations are conducted in a "clean room" environment so that, for example, the risk of dust particles collecting on their surface is minimised. Despite these precautions, an electrode surface prepared in this way will still have a very complex microscopic structure, as illustrated by the scanning electron micrograph of Fig 2.15 which shows a typical (25 μm^2) area of a virgin, mechanically polished copper electrode. Apart from its evident non-planar microtopography, it is also likely to have non-uniform micro-mechanical properties due to varying degrees of work-hardening within the highly disturbed upper surface layer.

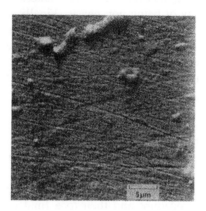

Fig 2.15   A scanning electron micrograph showing the typical microtopography of a 0.25 μm "commercially" polished planar copper electrode.

Similarly, there will be wide variations in the local levels of oxidation and contamination (e.g. decomposed hydrocarbons), and this will result in the surface also having non-uniform micro-electric properties. In addition, there will be many isolated non-metallic micro-inclusions, derived for example from the polishing treatment, that represent potential electron emission sites (see Chapter 3 and 4), and hence a hazard to the stability of a gap. To overcome these disadvantages electrodes are nowadays routinely prepared by using either a "chemical" or "electro-chemical" polishing process: the choice between these two approaches being determined by the constraints of the particular application.

In the case of superconducting Nb cavities to be discussed in Chapter 12 it has been found preferable to use a chemical polishing technique [27]. This finding highlights the fact that the performance of an electrode is not necessarily related to the "smoothness" of its surface microtopography. Thus, although electro-polishing produces a smoother Nb surface, a chemical polishing procedure generally results in a more reliable operational surface. It is also essential to recognise that both polishing processes can give rise to "self-contamination", and that consequently it is necessary to build in filtration procedures for removing suspended particulate material from both the gas and liquid media.

An alternative approach to the problem of producing an "ideal" electrode surface is to use surface coating techniques, since these would apparently offer the immediate advantages of "smoothing-out" the microtopography and "burying" loosely adhering particles. Accordingly, there have been investigations into the effectiveness of not only evaporated metal films [97] but also dielectric films [89,98-100], since these would have the further advantage of lowering the electric field over an electrode surface. However, the general conclusion of these studies is that whilst there can be initial advantages in the form of lower and more stable prebreakdown currents, the ultimate voltage hold-off capability of such gaps is not significantly increased, and, more important, that after a breakdown it is virtually impossible to apply standard conditioning procedures to recover the insulating performance of the gap. This type of irreversible change is attributed to a rupturing of the film by the strong electro-mechanical forces associated with the gap-field, where they locally overcome the non-uniform adhesive forces between the film and electrode. More likely, however, these changes arise from the explosive switch-on of local emission centres (see Chapter 4). Such an event would not only give rise to a highly unstable microtopography, but it would also inject "foreign" debris into the gap that could either act as breakdown initiating microparticles or become attached to an electrode surface to form field emission centres: as will be discussed in Chapter 4, this latter possibility is particularly likely in the case of dielectric material. It is possible however that with more sophisticated metal film deposition techniques where, for example, greater care is taken to achieve a uniform expitaxial growth of the deposited film over the substrate electrode, this approach may ultimately prove to be very profitable.

2.5.6   External Circuitry

The breakdown of a gap is always accompanied by the rapid dissipation of capacitive energy stored in both the gap itself and the associated external circuitry [101], where normally this latter contribution is much the larger since gap capacitances rarely exceed a few pF, particularly at large d-spacings. In fact, it is this energy that plays a vital role in "feeding" the growth of the initiating arc and hence determining the extent of the surface damage ultimately sustained by electrodes during a breakdown event. To avoid this process leading to an irretrievable loss in the voltage hold-off capability of a gap, it is imperative to take precautionary measures, firstly to limit the available energy, and secondly, to control its rate of dissipation [53]. In practical terms, this means on the one hand ensuring that all stray capacitance associated with the electrode mountings and connecting leads is reduced to a minimum, and on the other, inserting a current-limiting resistor in the HV line (such as shown in Fig 2.1) to protect the gap against the dissipation of the reservoir of capacitive energy associated with the EHT generator. Experimental investigations into optimum spark conditioning procedures discussed in Section 2.3.3 have indicated in fact that provided the value of this series resistor exceeds some limiting value, which is typically in the range 1-3 kΩ, its influence on the DC breakdown voltage of the gap falls off very significantly [43,88,102]. Physically, this requirement appears to correspond to the situation where the time constant of the external circuit, which controls the rate of dissipation of the capacitive energy, should be greater than ~ 10 ns which is the characteristic time constant associated with the current growth during the initiation of a vacuum arc (see Chapters 6 and 15). For most DC and low-frequency AC applications, this condition presents no problem; in fact, to limit any discharge current and hence the resulting electrode damage, it is conventional to increase the external resistance to ~ MΩs. It is only in the case of high frequency AC ($\gtrsim$ 100 MHz) or nanosecond pulse applications that the external circuit becomes critical.

2.6   Applied Voltage Waveform

There are many practical devices, such as magnetrons, klystrons, vacuum switches and high voltage electronic rectifiers [103-107], where it is necessary to use the insulating capability of vacuum for holding off time-dependent voltages. Probably the simplest example of such a waveform would be the 50 Hz sinusoidal voltage derived from a mains-driven high voltage transformer; others would be RF signals, square wave pulse trains and impulse voltages. It is therefore clearly of immense technological importance to know in detail how vacuum gaps respond to these types of voltage waveform; accordingly the subject has been made the specific focus of Chapter 10. The following brief survey will therefore be inevitably limited in scope.

2.6.1   Sinusoidal Voltages

From comparative measurements of the low-frequency ($\lesssim$ 250 Hz) AC and DC breakdown characteristics of plane-parallel "conditioned" gaps of a given electrode material and gap spacing [53,66,88,108,109], it can generally be concluded that the peak AC breakdown voltage tends to be somewhat higher than the corresponding DC value. However, the effect is not very marked and frequently only involves a difference of ~ 10% in the two breakdown voltages; the magnitude of the discrepancy also appears to be independent of frequency up to ~ 250 Hz. These observations should be further qualified by noting that a more lengthy preparatory procedure is required for gaps that are to be used for a experiment, since in this case both electrodes have to be conditioned; e.g. a larger number will be required to achieve a comparable level of conditioning [53]. In common with the DC behaviour of gaps, these studies show that the AC response is similarly sensitive to the choice of electrode material, and that from the limited sample of data, there appears to be a common hierarchy of electrode materials and a common optimum design specification for both DC and AC applications.

The dependence of the peak AC breakdown voltage $(V_b)_{AC}$ on the d-spacing of plane-parallel gaps has been investigated by Hackam and Altcheh [53,66] for a range of materials, including aluminium, copper, nickel, niobium and sterling silver. It was found that, as with the DC experiments, there is a clear transition between the small-gap and large-gap behaviour, being particularly marked with sterling silver, for example, which followed the empirical dependence

$$(V_b)_{AC} = 45d^{0.85} \qquad \text{for } 0.05 \lesssim d \lesssim 1.5 \text{ mm} \qquad 2.8$$

and

$$(V_b)_{AC} = 57d^{0.3} \qquad \text{for } 1.5 \lesssim d \lesssim 2.8 \text{ mm} \qquad 2.9$$

with $(V_b)_{AC}$ being given in kV; i.e. differing significantly from the DC behaviour defined by equations 2.6 and 2.7. Since however the other members of this small sample of materials had a nearly linear small-gap dependence coupled with a large-gap power dependence that more closely approximated to the DC behaviour, it has to be concluded that there is insufficient evidence to establish the existence of a distinct AC breakdown characteristic. The same authors also investigated whether there is an AC "pressure effect" [66] comparable to that illustrated in Fig 2.12 for DC gaps. Their findings indicate that whilst a similar phenomenon does occur, it is less general among electrode materials, where, from a sample set of measurements with aluminium, copper, nickel and niobium electrodes, a "pressure effect" was only observed with the latter two materials. There also appears to be no correlation between the optimum pressure values associated with the DC and AC operation for a given electrode material, so that in the case of nickel, the "optimum" DC pressure is $\sim 3 \times 10^{-6}$, whilst for the AC case it is $\sim 2 \times 10^{-3}$ mbar: in fact, the available data indicate that any "pressure effects" with AC regimes only occur above $\sim 10^{-5}$ mbar.

There is very little published information about the AC breakdown behaviour of gaps at high frequencies. Experiments at RF frequencies [108,109] have shown that gap performances are indistinguishable from their low-frequency characteristics. However, at MHz frequencies [110], it has been found that the gap-dependence of the breakdown voltage (in kV) takes the form $V_b = 31d^{0.7}$ over the range $1 \lesssim d \lesssim 3.5$ mm, which begins to show some departure from the low-frequency behaviour defined by equations 2.8 and 2.9.

This trend is further reinforced by the findings of Halpern et al. [111] who showed that the breakdown voltage of a 5 mm gap at a frequency of ~ 2800 MHz was significantly higher than its equivalent DC value.

2.6.2   Impulse Voltages

Farrall [112] has considered the breakdown behaviour associated with this type of ramp waveform from both a theoretical and experimental point of view. His analysis, which applies to large gap situations (d $\gtrsim$ 1 mm), where breakdown is assumed to be initiated by the Cranberg [85] microparticle based mechanism, distinguishes between experimental regimes using constant *rise rate* and constant *rise time* pulses, and shows that the gap-dependence of $(V_b)_{ip}$ for these respective regimes should take the form

$$(V_b)_{ip} \propto d^{5/6} \qquad\qquad 2.10$$

and

$$(V_b)_{ip} \propto d^{5/2} \qquad\qquad 2.11$$

From a review of the available data from pulsed over-voltage experiments [88,113], he was able to tentatively confirm that these types of breakdown event followed the above relations and were therefore very probably the result of microparticle processes. Subsequently Lloyd and co-workers [114,115] have made a detailed study of the behaviour of vacuum switches under impulse voltage testing. They concluded that fully conditioned switch contacts generally exhibit a distinct "small" and "large" gap behaviour, where for d $\lesssim$ 1.5 mm

$$\begin{aligned} (V_b)_{ip} &\propto d \\ &\simeq 40d \end{aligned} \qquad\qquad 2.12$$

and for d $\gtrsim$ 2

$$\begin{aligned} (V_b)_{ip} &\propto d^{1/2} \\ &\simeq 64d^{1/2} \end{aligned} \qquad\qquad 2.13$$

with $(V_b)_{ip}$ given in kilovolts. Whilst recognising that such a transition has traditionally been attributed to a change in the breakdown mechanism from field-emission-initiated at "small" gaps (d $\gtrsim$ 1 mm) to microparticle-initiated at "large" gaps (d $\lesssim$ 1 mm), they make the important point that the time scale of large-gap breakdown events is too short to involve any mechanism requiring the particle to make a complete transit of the gap. Accordingly, they suggest that large-gap events could result from the trigger discharge mechanism described in Section 7.2.1(ii), which is associated with the *launching* of a

54

microparticle. However, in the light of more recent findings to be discussed in Chapter 4, it is possible that such events could also have resulted from a thermal instability following the rapid "switching-on" (i.e. within tens of nanoseconds), of a more complex type of non-metallic electron emission mechanism associated with insulating micro-inclusions.

2.6.3   Pulse Voltages

The response of a high voltage gap to a fast-rise "square" voltage pulse has been extensively studied. In particular, microsecond and nanosecond pulse techniques have been used to investigate the fundamental physical processes responsible for initiating a high voltage vacuum arc. However, in view of the importance of this topic, a full discussion of these experiments will be delayed until Chapters 6 and 10 for the respective cases of a point-plane electrode geometry and a range of device applications.

2. 7   References

1.   Pivover, LJ. and Gordienko, VI., *Sov. Phys-Tech. Phys.*, **I**. 908-12, 1963.
2.   Alpert, D., Lee, DA., Lyman, EM. and Tomaschke, J., *J. Vac. Sci. Technol.*, **1**, 35-50, 1964
3.   Bennette, CJ., Swanson, LW. and Charbonnier, FH., *J. Appl.Phys.*, **38**, 634-40, 1967.
4.   Rohrbach, F., *Report CERN,* 71-38 (NTIS), 1971.
5.   Hackam, R. and Salman, S.K., *Vacuum*, **23**, 9-20, 1973.
6.   Donaldson, EE. and Rabinowitz, M., *J. Appl. Phys.*, **34**, 319-22, 1963.
7.   Rogowski, W., *Archiv für Elektrotechnik*, **12**, 1-15, 1923.
8.   Bruce, FM., *J.I.E.E.*, **94**, 138-49, 1947.
9.   Harrison, JA., *Brit. J. Appl. Phys.*, **18**, 1617-22, 1967.
10.   Arnal, R., *Ann. Phys. (France)*, **10**, 830-73, 1955.
11.   Mansfield, W.K., *Brit. J. Appl. Phys.*, **11**, 454-61, 1960.
12.   Boersch, H., Hamisch, H. and Wiesner, S., *Z. Agnew. Phys.*, **13**, 450-6, 1961.
13.   Powell, HPS. and Chatterton, PA., *Vacuum*, **20**, 419-29, 1970.
14.   Gerasimenko, VI., *Sov. Phys.-Tech. Phys.*, **13**, 107-12, 1968.
15.   Berland, R., Bert, M., Favrel, J. and Arnal, R., *Proc. IV-ISDEIV*, 56-60, 1970.
16.   McKibben, JL., *Los Alamos Report.* LA-5376MS, 1973.
17.   Hurley, RE., *J. Phys. D: Appl. Phys.*, **13**, 1121-8, 1980.
18.   Jüttner, B. and Wolff, H., *Proc. 7th Int. Conf. Phenom. Ionised Gases (Beograd 1965)*, **1**, 226-9, 1965.

19. Davies, DK. and Biondi, MA., *J. Appl. Phys.*, **37**, 2968-77, 1966.
20. Jüttner, B., Wolff, H. and Pech, P., *Proc. 8th Int. Conf. Phenom. Ionised Gases*, **70**, 1967.
21. Beukema, GP., *J. Phys. D: Appl. Phys.*, **6**, 1455-66, 1973.
22. Hackam, R. and Salman, SK., *Proc. IEE*, **119**, 377-84 and 1747-50, 1972.
23. Williams, DW. and Williams, WT., *J. Phys. D: Appl. Phys.*, **7**, 1173-83, 1974.
24. Allan, RN. and Salim, AJ., *J. Phys. D: Appl. Phys.*, **7**, 1159-69, 1974.
25. Shrednik, VN., Pavlov, VG., Rabinovich, A.A. and Shaikhin, B.N., *Phys. Stat. Sol. (a)*, **23**, 373-81, 1974.
26. Pavlov, VG. and Shrednik, VN., *Proc. VII-ISDEIV*, 209-216, 1976.
27. Cavaille, JV. and Drechsler, M., *Proc. VII-ISDEIV*, 217-221, 1976.
28. Archer, AD and Latham, RV, *Proc. XIV ISDEIV*, **43**, 1990.
29. Looms, J.ST. Meats, RJ. and Swift, DA., *J. Phys. D: Appl. Phys.*, 1, 377-9, 1968.
30. Williams, WT., *Nature: Phys. Sci.*, **231**, 42-4, 1971.
31. Septier, A. and Bergeret, H., *Rev. Gen. Electricite*, **80**, 565-7, 1971.
32. Bergeret, H., Nguyen Tuong Viet and Septier, A., *Proc. VI-ISDEIV*, 112-7, 1974.
33. Gomer, R. and Hulm, JK., *J. Chem. Phys.*, **20**, 1500-5, 1952.
34. Cobourne, MH. and Williams, WT., *Proc. VI-ISDEIV*, 279-84, 1974.
35. Cobourne, MH. and Williams, WT., *Physica*, **104c**, 50-55, 1981.
36. Shkuratov, SI, Ivanov, SN, and Shilimanov, SN, *Surf. Sci.*, **266**, 224-231, 1992.
37. Shkuratov, SI and Ivanov, SN, *Proc. XV-ISDEIV*, 127-131, 1992.
38. Mazurek, B, Xu, NS and Latham, RV, *J. Mat. Sci.*, **28**, 2838-9, 1993.
39. Maitland, A., *Brit. J. Appl. Phys.*, **13**, 122-5, 1962.
40. Beukema, GP., *J. Phys. D: Appl. Phys.*, **7**, 1740-55, 1974.
41. Danloy, L. and Simon, P., *Proc. V-ISDEIV*, 367-72, 1972.
42. Steib, GF. and Moll, E., *J. Phys. D: Appl. Phys.*, **6**, 243-55, 1973.
43. Hackham, R. and Govindra Raju, GP., *J. Appl. Phys.*, **45**, 4784-94, 1974.
44. Alpert, D., Lee, DA., Lyman, EM. and Tomaschke, H.E., *J. Appl. Phys.*, **38**, 880-1, 1967.
45. Ettinger, SY. and Lyman, EM., *Proc. III-ISDEIV*, 128-33, 1968.
46. Beukema, GP., *Physica C.*, **61**, 259-74, 1972.
47. Jüttner, B and Zeitoun Fakiris, A., *J. Phys.D: Appl. Phys*, **24**, 75-756, 1991.
48. Bajic, S, Abbot, AM, and Latham, RV., *IEEE Trans. Elec. Insul.*, **24**, 89-96, 1989.

49. Miller, HC. and Farrall, GA., *J. Appl. Phys.*, **36**, 1338-44 and 2966, 1965.
50. Cox, BM., *J. Phys. D:Appl. Phys.*, **7**, 143-51, 1974.
51. Williams, DW. and Williams, WT., *J. Phys. D: Appl. Phys.*, **5**, 1845 54, 1972.
52. Williams, DW. and Williams, WT., *J. Phys. D: Appl. Phys.,* **6**, 734-43, 1973.
53 Hackam, R., *J. Appl. Phys..*, **46**, 3789-99 1975.
54. Farrall, GA., *IEEE Trans. Parts, Hybrids, Packag.*, **PHP-11**, 134-8, 1975.
55. Jüttner, B., *Beitr. Plasmaphys..*, **19**, 259-65, 1979.
56. Turner, CM., *Phys. Rev.*, **81**, 305, 1951.
57. McKibben, JL and Boyer, K., *Phys. Rev.*, **82**, 315-6, 1951.
58. Linder, EG. and Christian, SM., *J. Appl. Phys.,* **23**, 1213-6, 1952.
59. Murray, JJ., *UCRL Report,* No. 9506, 1960.
60. Prichard, BA., *J. Appl. Phys..*, **44**, 4548-54, 1973.
61. Cooke, C M., *Proc. II-ISDEIV,* 181-93, 1966.
62. Morgan, CG., *Proc. IV-ISDEIV, Addenda,* 1970.
63. Germain, C. and Rohrbach, F., *Proc. 6th Int. Conf. Phenom. Ionised Gases,* 111-7, 1963.
64. Hackham, R., *Int. J. Electron Phys.,* **28**, 79-87, 1970.
65. Govindra Raju, GR. and Hackham, R., *Proc. IEEE,* **120**, 927-36, 1973.
66. Hackham, R. and Altcheh, L., *J. Appl. Phys. ,* **46**, 627-36, 1975.
67. Hackham, R. and Salman, SK., *IEEE Trans. Electr. Insul.,* **EI-10**, 9-13, 1975.
68. Litvinov, EA., Mesyats, GA., Proskurovsky, DI. and Yankelevitch, E.B., *Proc. VII-ISDEIV,* 55-57, 1976.
69. Jüttner, B., Wolff, H. and Altrichter, B., *Phys. Stat. Sol.,* **27a**, 403-12, 1975.
70. Carter, G. and Colligon, JS., *In* "Ion Bombardment of Solids" p.363, Heinemann, London, 1968.
71. Kaminsky, M., *In* "Atomic and Ionic Impact Phenomena on Metal Surfaces", Chapter 1, Springer-Verlag, Berlin, 1965.
72. Chatterton, PA., *Proc. II-ISDEIV,* 195-206, 1966.
73. Powell, HPS. and Chatterton, PA., *Vacuum,* **20**, 419-29, 1971.
74. Hackam, R., *Proc. IEE,* **119**, 377-84, 1972.
75. Cox, BM., *CEGB Res. Report,* R/M/N1O21, 1979.
76. Rabinowitz, M. and Donaldson, EE., *J. Appl. Phys.*, **36**, 1314-19, 1965.
77. Hackam, R. and Salman, S.K., *J. Appl. Phys.*, **45**, 4384-92, 1974.
78. Leader, D., *Proc. IEE*, 100 **2A**, 138-40 and 181-6, 1953.

79. Pivovar, LI., Tubaev, VH. and Gordienko, VI., *Sov. Phys-Tech. Phys.*, **2**, 909-10, 1959.

80. Mason, JH., *Proc. IEEC* , **102**, 254-62, 1955.

81. Thwaites, RH., *Proc. IEEC* , **109**, 600-9, 1962.

82. Loeb, LB., *In* "Electrical Corona, their Basic Physical Mechanism", University of California Press, Berkeley, 1965.

83. Abou-Seada, MA. and Nasser, E., *Proc. IEEE*, **56**, 813-18, 1965.

84. Ryan, H. McL., *Proc. IEEE*, **117**, 283-9, 1970.

85. Cranberg, L., *J. Appl. Phys.*, **23**, 518-22, 1952.

86. Boyle,WS., Kislink P. and Germer, L.H., *J. Apyl. Phys.*, **26** 720-5, 1955.

87. Slivkov, IN., *Sov. Phys-Tech. Phys.*, **2**, 1928-34, 1957.

88. Denholm, AS., *Can. J. Phys.*, **36**, 476-93, 1955.

89. Hawley, R. and Zaky, A.A., *Prog. Dielectri.* **7**, 115-21, 1967.

90. Maitland, A., *J. Appl. Phys.* **32**, 2399-407, 1961.

91. Davies, DK. and Biondi, M.A., *J. Appl. Phys.*, **37**, 2969-77, 1966.

92. Kranjec, P. and Buby, L., *J. Vac. Sci. Technol.*, **4**, 94-6, 1967.

93. Bloomer, RN. and Cox, BM., *Vacuum*, **18**, 379-82, 1968.

94. Fane, R.W., Neal, W.E.J. and Latham, R.V., *J. Appl. Phys.*, **44**, 740-3, 1973.

95. Butcher, EC., Dyer, AJ. and Gilbert, WE., *J. Phys. D*, **1**, 1673-87, 1968.

96. Williams, DW. and Williams, WT., *J. Phys. D: Appl. Phys.*, **5**, 1845-54, 1972.

97. Sudan, RN. and Gonzalez-Perez, F., *J. Appl. Phys.*, **35**, 2269-70, 1964.

98. Jedynak, L., *J. Appl. Phys.*, **35**, 1727-33, 1964.

99. Rohrbach, F., *Proc. I-ISDEIV*, 393-401, 1964.

100. Bolin, PC. and Trump, JG., *Proc. III-ISDEIV*, 50-53, 1968.

101. Rabinowitz, M., *Vacuum*, **15**, 59-66, 1965.

102. Simon, DJ. and Michelier, R., *Proc. III-ISDEIV*, 263-6, 1968.

103. Armstrong, T.V. and Headley, P., *Electron Power*, **20**, 198-203, 1974.

104. Barken, P., Lafferty, JM., Lee, TH. and Talento, JL., *IEEE Trans. Power Appar. Syst.* **PAS-90**, 350-9, 1971.

105. Slade, PG., *IEEE Trans. Parts Hybrids and Packaging*, **PHP-10**, 43-50, 1974.

106. Borgers, SJ., *Proc. IEE* , **99 III**, 307-15, 1952.

107. Venable, D., *Rev. Sci. Instrum.*, **33**, 456-62, 1962.

108. Wijker, WJ., *Appl. Sci. Res.*, **B9**, 1-20, 1961.

109. Kustom, RL., *J. Appl. Phys.*, **41**, 3256-68, 1970.

110. Hill, CE., *Proc. IV-ISDEIV*, 137-41, 1970.

111. Halpern, J., Everhart, E., Rapuano, R.A. and Slater, J.C., *Phys. Rev.*, **67**, 688-95, 1946.

112. Farrall, GA., *J. Appl. Phys.*, **33**, 96-99, 1962.

113. Rosanova, NB. and Granovskii, VL., *Sov. Phys-Tech. Phys.*, **1**, 471-8, 1956.
114. Lloyd, O. and Hackam, R., *Proc. Inst. Electr. Eng.*, **122**, 1275-8, 1975.
115. Lloyd, O. and Fairfield, JH., *CEG.B Res. Report,* R/M/N937, 1977.

3

Diagnostic Studies of Prebreakdown Electron "Pin-Holes"

RV Latham

3. 1   Introduction

As intimated in Chapters 1 and 2, the primary threat to the stability of a high voltage vacuum gap is generally assumed to be associated with the "parasitic" prebreakdown currents that typically precede the spontaneous breakdown of the gap (see Fig 3.3). However, apart from the indirect evidence of the involvement of a "cold" emission process, i.e. as provided by the externally measured current-voltage (I-V) characteristic and the subsequently derived F-N plot illustrated in Fig 3.4, very little else can be deduced from this evidence about the fundamental nature of prebreakdown currents. It follows therefore that technologists, concerned with improving the performance of HV devices, have exerted a sustained pressure for a more detailed understanding of the physical origin of these currents. The past three decades have therefore seen a major effort to develop sophisticated analytical techniques for identifying and studying the fundamental physical processes that occur in a HV gap. Wherever possible, the emphasis has been to interface diagnostic facilities directly with an operational system so that in-situ, dynamic measurements can be made of any particular phenomenon.

The aim of the present chapter is to provide a review of those investigations that have led to important new insights into the physical processes occurring either on electrode surfaces or in the gap between them. In this discussion, it will however only be possible to give outline information about the experimental principles of the techniques cited; the reader is therefore left to source the original publications for further details of the instrumental complexities and ancillary procedures.

3. 2   Spatial Mapping of Electron Emission Sites

Although of practical significance, the total-gap I-V characteristics presented in Figs 3.3 and 3.4 provide no information about the spatial origin of these prebreakdown electron currents on the cathode surface. This question can however be readily answered by employing either some form of imaging facility involving an "active" anode that converts the energy of the impacting electrons into light, or an anode probe technique.

3.2.1   Optical Imaging Techniques

The earliest and most obvious approach was to employ a phosphor coated anode such as is used for the screen of a cathode ray tube. Indeed, it was this simple technique, applied to a planar electrode geometry, that was first employed by Millikan and co-workers [1-3] to reveal the discrete localised nature of the emission process. Subsequently, their approach was refined by employing the combination of a transparent anode grid in conjunction with a following phosphor screen [4,5].

A more recent development of this simple plane-parallel imaging principle, which avoids the undesirable side-effect of having loosely adhering phosphor particles being released into the HV gap, is the "Transparent Anode" imaging technique developed by Latham and co-workers [6-9]. As illustrated in Fig 3.1, the technique offers a simple but very effective means of providing a direct visual display of the emitted electrons. This is achieved by recording the transition radiation that they generate on impact with an optically flat glass anode whose surface has been coated with a transparent layer of tin oxide. By employing a real-time video recording camera, this instrumental facility has proved to be a simple but powerful means of studying both the spatial distribution of emitted electrons from planar cathodes, and also the temporal evolution of a population of such sites [6-9].

As an illustration of the application of this technique, sequence (a) of Fig 3.2 presents three time-lapse images that would typically be recorded at half-hour intervals from a "commercially polished" electrode (see Section 2.5.5) that was emitting a few microamps of prebreakdown current under UHV conditions. From these images, it will be seen that, not only does the emission come from randomly located point emission centres, but that the individual centres, or sites, also exhibit a "switching phenomenon", whereby they are liable to switch on and off randomly with time. In fact, this switching phenomenon is one of the important properties of the emission process that was first identified by this technique [10,11], and will be discussed at length both later in this chapter, and in Chapter 4.

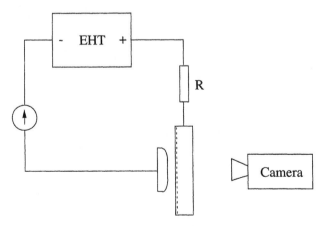

Fig 3.1   Illustrating the operational principle of the "Transparent Anode" imaging technique, as used for displaying the spatial distribution of "electron pin-holes" on broad-area electrodes.

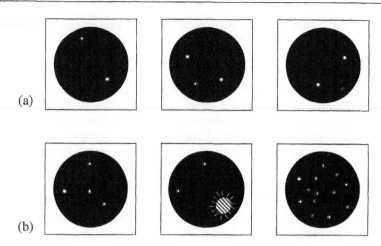

Fig 3.2   Two applications of the "Transparent Anode" imaging technique. Schematic sequences illustrating (a) how "point" emission sites switch on and off randomly with time under constant field conditions [6], and in (b) how breakdown events (i) can generally be spatially correlated with prebreakdown emission sites, and (ii) result in a multiplicity of new emission sites (After [6,8].)

Referring to the images of sequence (b) in Fig 3.2, the technique may also be used to study the breakdown process. Thus frame 1 firstly shows a stable population of prebreakdown sites. If now the gap field is gradually increased, with the camera set for continuous recording, the arc shown in frame 2 will eventually be struck. A careful inspection of these latter two images will reveal that the initiating arc is spatially centred on one of the prebreakdown emission sites; i.e. supporting the long-held belief that a breakdown event is initiated by a prebreakdown electron emission process. Indeed, a more detailed account of the application of this technique will be given in Chapter 5. The final frame of sequence (b) shows the post-breakdown population of emission sites, and vividly illustrates how a breakdown event can frequently result in the creation of a great many additional new emission centres which, in practical terms, would correspond to a catastrophic degrading of the insulating performance of an HV gap.

For the more detailed studies of the prebreakdown emission process to be described below in Section 3.4, it is necessary to incorporate a range of ancillary facilities. Thus, referring to Fig 3.3, a practical experimental system would typically include a heated [7], or cooled [12] specimen stage for studying the effect of electrode temperature, and a combined gas handling and mass spectrometry facility for studying the influence of the gas environment (total pressure and gas species). Finally, it is desirable to incorporate an image processing capability to help with the analysis of the extensive video data that this technique generates.

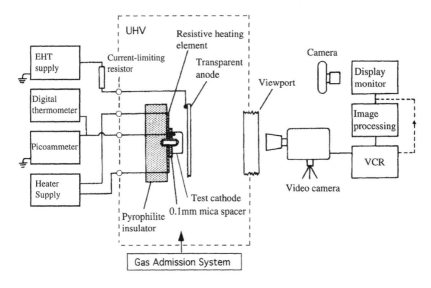

Fig 3.3   Illustrating how a practical "Transparent Anode" imaging system generally incorporates a range of ancillary facilities, e.g. a heated specimen stage, a residual gas analysis capability, video recording and image processing (After [7,9].)

3.2.2   Scanning Probe Techniques

Whilst the optical imaging system described above is very useful for simultaneously observing all members of a population of emission sites, it is not suitable for studying the emission characteristics of the individual sites. This limitation can however be circumvented by employing some form of scanning anode probe technique [8]. An early approach, based on this principle, was pioneered by Cox and Wort [13] with their anode probe-hole technique illustrated in Fig 3.4(a). In this, the cathode C is scanned in a raster pattern in front of an anode A having a central 0.1 mm diameter probe hole behind which there is a Faraday cage collector FC and picoammeter detecting facility to record any current passing through the anode probe hole. Thus, by employing a synchronously scanned x-y recording system, it is possible to generate a comparable spatial display of the active emission sites to that obtained by the optical imaging technique described above. By using this type of scanning system in a "high resolution" mode, it is possible to obtain the type of current contour map of an isolated emission process shown in Fig 3.4(b). As will be described in a later section of this chapter, similar types of map have also been obtained with a scanning micropoint anode technique.

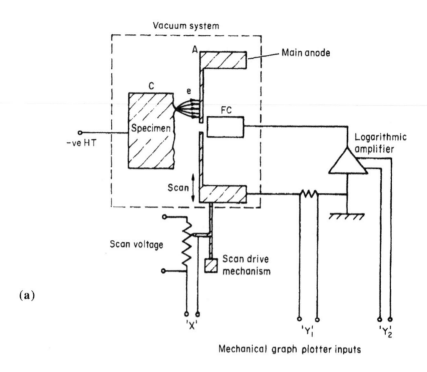

(a)

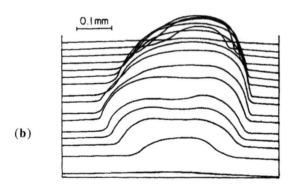

(b)

Fig 3.4   (a) The mechanical and electrical systems of a mechanically-scanned anode probe hole technique for locating the isolated electron emission sites found on broad-area electrodes (b) A typical set of current traces obtained with this facility as a site is scanned across a 0.1 mm diameter probe hole. (From Cox and Wort [13], with permission.)

3. 3   Structural and Material Properties of Emission Sites

The studies described above in Section 3.2 clearly demonstrate that prebreakdown currents originate from some form of microscopically localised "cold" emission process. In this section, we shall present an historical account of how successive generations of specialised analytical techniques have been developed to obtain detailed information about the types of electrode microstructure that are able to promote the "cold" emission of electrons in the anomalously low gap field range of 5-10 MV/m.

3.3.1   TEM-Based Profile Analysis Studies

With the recognition that the emitting sites on broad-area electrodes had sub-micron dimensions, it also became clear that the resolution of optical microscopy was too low for obtaining any detailed information about their microtopography. Accordingly, attention was directed towards the application of electron optical techniques for looking at electrode surfaces. However, in the late 1950s and early 60s when this work was being pioneered, the scanning electron microscope (SEM) had not yet become a commercially available instrument, so it was therefore only possible to use the conventional transmission electron microscope (TEM) for studying the profile of electrode surfaces.

An early report of the application of this technique was by Bogdanovskii [14] in 1959 who was studying the growth of dendritic structures on microelectrodes. However, it was Little and Smith [15] and Jedynak [16] in 1965 who were the first to study the profiles of HV electrodes in a TEM, with follow-up developments of the technique being subsequently exploited by Latham and his co-workers [17,18]. An essential requirement for this type of measurement is to choose a cathode electrode geometry that permits the imaging of the surface structures contained in a single plane that is orthogonal to the electrode surface, i.e. a geometry that offers an "edge" rather than a planar surface to the imaging electron beam. In the system of Latham and Braun [17] illustrated in Fig 3.5, the high voltage electrode assembly, consisting of a filamentary hairpin cathode C separated by a 0.5 mm gap from a planar anode A, was mounted in the specimen chamber of a transmission electron microscope. Thus, the test gap could be studied in-situ in the TEM, with the possibility of interrupting the high voltage cycling at any stage to observe the cathode profile. The most important conclusion to emerge from this type of investigation was that before any voltage is applied between a pair of "test" HV electrodes, the cathode surface appears smooth within the resolution of the imaging system (~ 50 Å) as shown in Fig 3.5(b), and that it is only *after* the occurrence of a microdischarge or breakdown event that the surface becomes "decorated" with the type of microprotrusions shown in the micrograph of Fig 3.5(c).

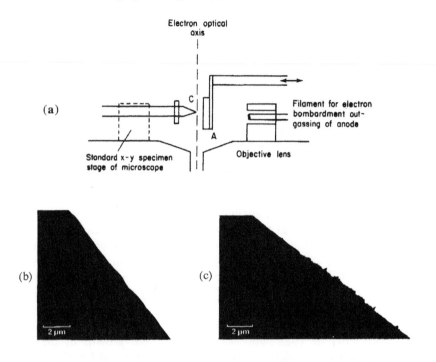

Fig 3.5   (a) The specimen stage of an EM3A electron microscope, as modified for studying the surface profile of a filamentary high voltage electrode. (b) Electron micrographs showing respectively of the smoothly undulating profile of a virgin electrode, and the damaged surface an electrode that has suffered a breakdown or microdischarge event. (From Latham and Braun [17], with permission.)

For the non-uniform point-plane electrode geometry used in this study, it was found that the occurrence of such microprotrusions was accompanied by an increase in the slope of the F-N plot of the gap. Typically, this corresponded to a reduction in the β-factor from an initially high value of ~ 200-300, for which there are apparently no complementary surface microfeatures, to a much lower value of ~ 50-100 which could then be more realistically correlated with the dimensions of the observed microprotrusions. In further support of the proposition that such features can act as "secondary" microemitters, there is the type of evidence shown in the micrograph of Fig 3.6, where it is assumed that the tip of an initially sharp protrusion appears to have become blunted as a result of local melting caused by the resistive heating of an excessive electron emission current.

Fig 3.6   A fusion-induced microfeature commonly found on arced electrodes.

To gain a further insight into the time-scale of the physical mechanism responsible for the formation of these protrusions, Latham and Chapman [18] extended this in-situ profile imaging technique to the use of single-shot HV pulses in the micro-millisecond range. With this development, it was possible to obtain a sequence of micrographs such as shown in Fig 3.7 which illustrate how the microtopography of the cathode surface is progressively degraded between successive pulses from its initially smooth profile of Fig 3.5(b).

Fig 3.7   A sequence of micrographs showing how the profile of an electrode surface is progressively damaged following successive 20 μs, breakdown-inducing high voltage pulses. (From Latham and Chapman [18], with permission.)

This evidence therefore clearly shows that "groups" of these microfeatures form in times that are less than < 20 μs, which was the minimum pulse width available for the experiments.

A complete explanation for these observations had however to wait for the advent of the scanning electron microscope, which provided a complementary high-resolution means of studying the three-dimensional nature of the microtopography of electrode surfaces. With such a facility, it was readily established [19] that the groups of electrode microprotrusions shown in Figs 3.6 and 3.8 are associated with the microcratering of an electrode surface following some form of local fusion process. Thus, referring to Fig 3.8, which is the complementary scanning electron micrograph of the same cathode surface imaged in profile in Fig 3.5(c), it is clearly seen that the microprotrusions are in fact the "frozen" tongues of molten metal projecting from the rim of these "splash" microcraters. Having established that there is strong circumstantial evidence to connect breakdown events with prebreakdown emission processes (see Fig 3.2), it can be concluded from a practical standpoint that the presence of these "whisker-like" features on the surface of an electrode would represent a serious hazard to the subsequent stability of the gap.

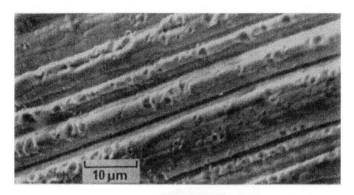

Fig 3.8   A complementary scanning electron micrograph revealing the topographical nature of the electrode damage previously shown in profile in Fig 3.5(c). (From Latham and Braun [19], with permission.)

As to the physical origin of this cathode microcratering, it has firstly to be noted that it is confined to the field-enhancing extrusion ridges of the tungsten filament. The damage was therefore attributed to the "explosive emission" mechanism (see Chapters 5 and 15) that causes the cathode spot to execute a characteristic "hopping" migration during a breakdown event. It was also concluded that the sequence of events was triggered by some sort of "primary" initiating event that, by its explosive nature, was able to ignite the arc and thereby create the first microcrater.

3.3.2   SEM-Based Surface Analysis Studies

With the growing availability of scanning electron microscopy facilities during the early 1970s, attention was directed towards how these new and powerful analytical instruments could be adapted for the identification and study of prebreakdown electron emission sites. The first requirement of any such system was that it should be able to identify which microscopic surface features on an electrode surface are responsible for promoting the emission process. This therefore pointed to the need to incorporate some form of scanning probe system within a SEM that would be capable of identifying in-situ the relevant surface microstructures. The surface imaging capability of this type of instrument could then be exploited to obtain topographical and material information about emission sites.

Referring to Fig 3.9 two alternative systems have been successfully developed: the first employs a scanning anode with a small hole to probe the current density within the gap, whilst the second employs a pointed anode to probe the cathode surface for field-emitting microfeatures. Both of these systems have yielded valuable new information about the microscopic nature of emission sites. Indeed, as will be discussed in Chapter 4, the findings from these probe systems have profoundly changed our physical understanding of the mechanism responsible for prebreakdown electron emission.

(a)   (b)

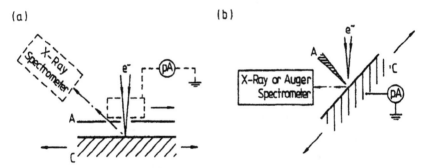

Fig 3.9   Schematic representation of the SEM-based probe technique for the in-situ location of individual emission sites. (a) Using an anode probe-hole, as developed by Cox [20], and (b) a micropoint probe anode, as developed by Athwal and Latham [11]. In both cases, e⁻ is the electron probe of the SEM. (From Latham [8], with permission.)

(i)   The Scanning "Anode Probe-Hole" System

The first direct electron optical information about emission sites was obtained by Cox in 1975 using an anode probe-hole system [20]. As illustrated schematically in Fig 3.10, this employed a plane-parallel "test" gap formed between a specimen cathode C and a stainless steel anode A having 500 μm and 5 μm diameter probe holes that can alternatively be accurately located on the axis of the SEM. Having thoroughly out-gassed both electrodes with an ancillary electron bombardment facility (not shown), the subsequent experimental procedure is to use the micrometer controlled specimen stage to mechanically scan the cathode in a raster pattern in front of the coarse 500 μm probe hole until a reading is obtained on the retractable current-sensor P; i.e.

71

indicating that an emission site has been located approximately on the axis of the SEM. At this stage, the current-voltage characteristic of the site is recorded by varying the voltage across the test gap and noting the corresponding collected current which, with the larger probe hole, represents practically all the emission from the site: subsequently, these data can be used to construct a F-N plot of the emitter. The site may then be more accurately located on the SEM axis by firstly transferring to the 5 μm probe hole and then progressively reducing the d-spacing of the electrode gap to ~ 0.05 mm: at this stage, the main anode and current sensor are removed and the micro-topography of the on-axis region of the electrode is examined by the SEM facility.

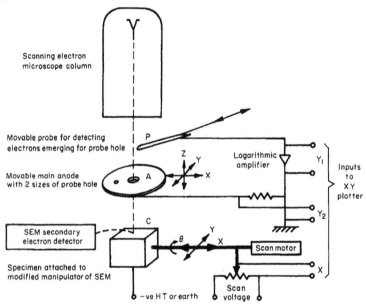

Fig 3.10   An anode probe facility for the dynamic study of field emission sites on broad-area electrodes in-situ in a SEM. (From Cox [20], with permission.)

By using an electron trajectory plotting technique that involves recording how the specimen co-ordinates corresponding to maximum collected current vary as the d-spacing is reduced, it is also possible to locate the position of an emitting site with a spatial resolution that can approach ~ 10 μm. A typical example of the evidence obtained from this technique is illustrated by the micrograph of Fig 3.11(a) in which the circled area represents the localised region of the arced surface of a copper-chrome alloy vacuum switch contact electrode that includes an emitting site. The most important conclusion to be drawn from this and similar observations on other sites, is that the identified region contains no field-enhancing microfeatures corresponding in "sharpness" to the β-factor of ~ 200-300 measured indirectly from the F-N plot of the site.

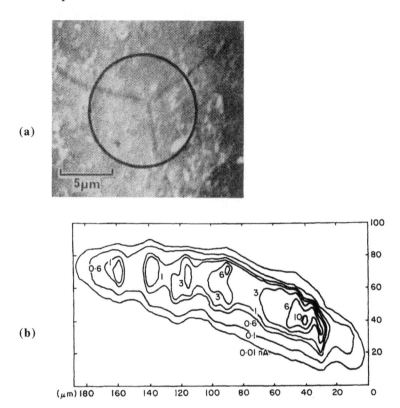

Fig 3.11   (a) A scanning electron micrograph showing the region of an electrode surface (circled) that has been identified as containing an emission site, and (b) the complementary distribution map of the field emission current associated with this site. (From Cox [20], with permission.)

A detailed consideration of the physical implications of this important observation will be delayed until the following chapter, when other more recent complementary data can also be taken into account. However, it can be stated at this stage that the evidence of Fig 3.11 represented an historic watershed in the evolution of the subject, since it strongly suggested that there were other emission mechanisms, apart from the metallic microprotrusion model, for explaining the origin of prebreakdown currents.

In addition to this microtopography information, it is also possible to use the specimen scan facility in conjunction with the fine probe hole current detection regime for obtaining a contour map of the current density from a given site as it appears in the plane of the anode. Thus, Fig 3.11(b) is the complementary distribution of the emission obtained from the site imaged in Fig 3.11(a), and is seen to consist of several sub-emission centres. This is another important characteristic of emission sites that was identified by this technique. As will be discussed below in Section 3.4.2, this finding that emission sites consist of "clusters" of sub-centres has also been confirmed by the use direct projection imaging techniques.

(ii) The Scanning "Micropoint Probe Anode" System

The alternative technique of employing a probe anode, i.e. as shown in Fig 3.9(b), was subsequently pioneered by Athwal and Latham [23,24] with the aim of both improving the resolution of the topographaphical detail of emission sites, and developing a more user-friendly analytical system. The experimental arrangement of the original facility is illustrated in Fig 3.12(a). Here, a standard electrolytically etched tungsten microtip (see Chapter 6), having a typical radius of 20-50 nm, is located on the axis of the SEM by means of a bellows-operated adjustable mounting that has an insulated feed-through to the tip capable of supporting a positive potential of 5-10 kV. The specimen cathode C, which is orientated normally to the probe anode A and at $45°$ to the SEM axis, is mounted on the goniometer stage of the instrument such that its surface can be scanned in a spiral pattern in front of the fixed probe. By arranging for the specimen to be electrically insulated from the earthed mechanical stage, it is possible to use a series picoammeter (electrometer) to monitor any emission current drawn from the specimen. With the probe-to-specimen distance initially set to ~ 0.5 mm and the probe potential to the higher value of ~ 8-10 kV, the first stage of the experimental procedure is to systematically scan the specimen until an emission site is roughly located as indicated by a specimen current of ~ 10^{-8} A. The second stage is to intermittently use the SEM facility for gradually reducing the probe-to-specimen distance to ~ 5-10 µm, ensuring that throughout this sequence of operations the probe potential is progressively reduced to keep the specimen current approximately constant: at this final stage of adjustment, the probe/specimen micro-regime will be as shown in Fig 3.12(b).

The emission site can then be assumed to be mid-way between the points P1 and P2, which can be respectively located in the centre of the field of view by (i) focusing through the tip onto the specimen and (ii) focusing on the tip profile, then, without altering the lens conditions, removing the tip and translating the specimen laterally until the surface is again in focus. With this technique it is possible to improve the resolution of site location to better than ~ 2 µm.

To provide the reader with a perspective of the point-plane electrode regime of this facility, the low-magnification scanning electron micrograph of Fig 3.13(a) shows the microprobe anode (etched from a 0.1 mm diameter tungsten wire) superimposed against the planar cathode. In contrast, the high resolution micrograph of Fig 3.13(b) illustrates the final stage of the technique, where the site being studied has been localised to within the superimposed 2 µm diameter circle. Typically, this is seen to contain a sub-micron insulating inclusion which strongly points to such features being the source of the electron emission.

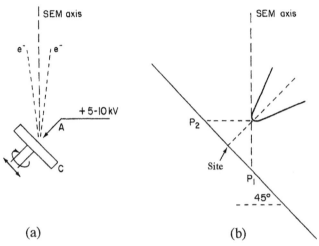

(a)   (b)

Fig 3.12   (a) The experimental regime of an anode microprobe technique for locating electron emission sites on broad-area electrodes in-situ in a SEM. (b) The electrode geometry in the vicinity of an emission site. (From Athwal and Latham [23], with permission.)

Although the resolution of the technique at this stage of development was still not good enough to precisely identify the emission site, this type of result even more strongly confirmed the earlier findings of Cox and co-workers [20-22] that these emission processes are evidently not associated with a field enhancing surface microfeature having a β-value in the range 200-300, such as typically predicted from the corresponding F-N plot of the emitter. From an historical perspective, this conclusion was also further supported by independent optical and electron spectrometry studies of the emission process (see Sections 3.4.3 and 3.5.2) that were being conducted at about the same time, i.e. the late 1970s and early 1980s.

Three further significant observations from these early studies should be briefly mentioned. Firstly, from the routine probe current recordings of the I-V characteristics of these particulate-based sites, and their derivative FN plots, it was typical to obtain β-values of 200-300; i.e. bearing no apparent relation to the geometry of the field emitting particles. Secondly, with the aid of a Kevex X-ray analysis facility, it was frequently possible to identify a "foreign" element within the site region: thus in Fig 3.14(a) the insulating particle was shown to contain silver, probably in the form of an oxide.

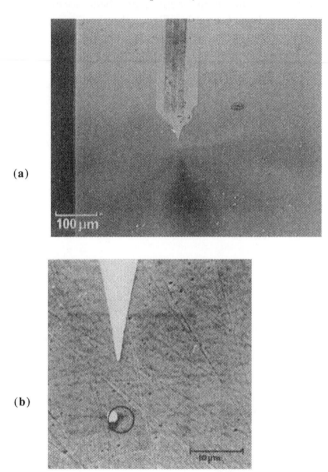

(a)

(b)

Fig 3.13   Scanning electron micrographs showing (a) the anode microprobe electrode regime of Fig 3.9, and (b) and the tip of the probe immediately above the localised microregion of the cathode surface (circled) containing an emission site.

A similar finding had been reported earlier by Allen, Cox and Latham [22] using the anode probe hole technique to identify a site, although in this case the impurity element was aluminium. This result also highlights how the emission site is not associated with one of the many "sharp" features that were present on this particularly rough "as machined" electrode surface.

The second interesting observation involves the identification of a hitherto unidentified property of emission sites: namely, that they require a minimum threshold field to "switch" them on, after which they emit copiously without a further increase in field. Such sites manifest themselves experimentally by the sudden blunting of the probe anode during its systematic scanning of a localised region of the cathode surface under gradually increasing field conditions. The typical outcome of such an event is shown in the micrograph of Fig 3.14(b), which clearly shows how the tip of the probe has been thermally blunted as a result of the excessive heat generated by the sudden collection of a high electron current. As will be discussed in Chapter 6, a similar modification of a tip profile occurs when an excess current is drawn from such an emitter in a point-plane diode configuration: indeed, the profile image of Fig 3.6 shows the same effect.

(a)

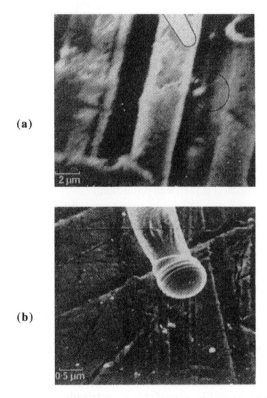

(b)

Fig 3.14   Scanning electron micrographs showing (a) the tip of the probe immediately above the localised micro-region of an "as machined" cathode surface (circled) containing an emission site, and (b) a probe that has suffered thermal instability as a result of collecting a high emission current. (From Athwal and Latham [23], with permission.)

Subsequently, this type of scanning anode probe system was further developed by Fischer and co-workers [25-29]. Thus, referring to the schematic of Fig

3.15, their system had the important practical advantage of being based on a purpose-designed UHV facility, incorporating dedicated SEM and Auger analysis capabilities for the routine structural and material analysis of field emitting microstructures.

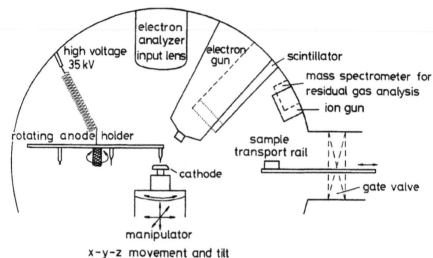

Fig 3.15   The arrangement of a UHV scanning microprobe anode system incorporating in-situ Auger analysis and electron bombardment facilities. (From Niedermann et al [27], with permission.)

The studies conducted with this instrument were principally directed towards characterising the emissive properties of niobium electrodes, where Fig 3.16 is a typical example of the recorded data. In this, Frame (a) was recorded within the analytical facility and confirms both the particulate nature of emission sites, and the fact that they generally appear as "bright" structures in the SEM (see also Figs 3.13 and 3.14); i.e. such particles are apparently electrically insulated from the metallic substrate electrode. In contrast, Frame (b) is a conventional SEM micrograph of the same particle, but recorded in a remote high-resolution instrument. Although of interest in revealing the complex structure of the particle, this latter result does not unfortunately tell us from which particular sub-feature of the particle the field emitted electrons originate. It was not possible therefore to attribute any topographical significance to the previously measured β-value of the site. The final Frame (c) of Fig 3.16 is an Auger line scan of the particle, recorded in the analytical facility, and reveals that it was composed of sulphur.

On the basis of a comprehensive analysis of many such particles on a range of sample niobium and other electrodes, it was concluded that (a) emission sites are always particulate in nature, with a typical size variation of 0.1 to 10 μm, (b) the material composition of particles can be very varied, with carbon being the single most common emission material, accounting for around 20% of all sites, and (c) that this last conclusion is independent of the substrate electrode material [25,27]. More recently, two other groups have established scanning microprobe anode facilities [29,30], and are using them to obtain important new information about how particulate structures promote the cold emission of electrons at anomalously low field. In both cases, the studies are focused on the parasitic fields emission processes that occur with the type of polycrystalline niobium material used in the manufacture of the superconducting RF cavities discussed in Chapter 12.

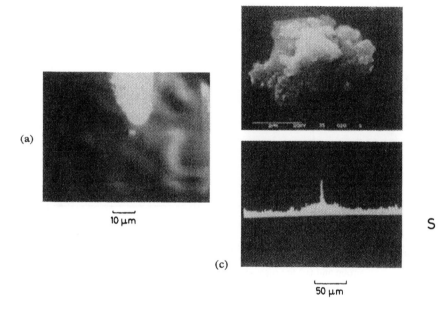

(a)

10 μm

(c)

S

50 μm

Fig 3.16   A sulphur field emission site (a) A low-resolution SEM image, showing the point anode at 10 μm above the site. (b) An SEM image of the same site taken in a high-resolution SEM. (c) An Auger line scan across the particle showing the peak-to-peak height. (From Niedermann [27], with permission.)

3.3.3   STEM-Based Surface Analysis Studies

The scanning tunnelling microscope (STEM) is a powerful new surface analytical tool recently developed by Binnig and Rohrer [31] for obtaining high resolution topographical information. It is similar in concept to an earlier system developed by Young et al [32], and involves a micropoint anode electrode being scanned in close proximity to the surface being studied. Thus, referring to Fig 3.17(a), a positively biased electrolytically etched tungsten tip, i.e. as used in the scanning micropoint anode system of Fig 3.10 and the experiments to be described in Chapter 6, tracks the extended electrode surface at a constant nanometre separation, as determined by the collected tunnelling current. However, on approaching a "site" that has enhanced field emitting properties ($\beta > 1$), the tip has to move away from the surface in order to maintain a constant current under fixed bias conditions. It follows that this "displacement" can be translated into a topographical image of the microscopic field emitting properties of the surface.

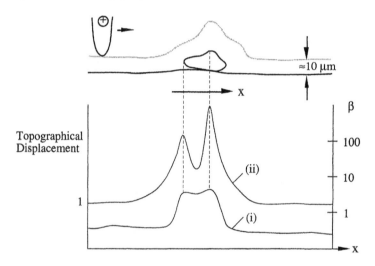

Fig 3.17   A schematic illustration of a typical STEM analysis of a particulate emission site. (a) The displacement path of the scanning probe under constant bias conditions. (b) The variation of (i) the topographical displacement of the probe and (ii) the measured tunnelling current ($\propto 1/\beta$.) (After Niedermann et al.[33].)

It was therefore a natural progression to apply this new technique to a study of the microstructures that give rise to parasitic field emission on broad-area HV electrodes; particularly so, since there were also striking similarities with the existing scanning microprobe system of Fig 3.13. It was therefore a relatively straightforward exercise for Niedermann et al [33] to add a STEM capability to their existing scanning microprobe anode system, previously described above in Section 3.3.2. With the combined availability of scanning electron microscopy (SEM), Auger analysis, and STEM, this group were able to make detailed studies of particulate-based parasitic field emission sites. Thus, as illustrated in Fig 3.17(b), a site is characterised "topographically" in terms of the emission properties of two-dimensional slices through the structure, where local high-β regions of the particle are identified as the source of the site current. By this technique, it has been established that sub-emission centres can correspond to both mid-surface and edge features of a particle.

Because of the detailed and time-intensive instrumental procedures involved in this type of analysis, the technique has only been applied to a limited number of emission sites [33]: equally, it is not capable of responding to the rapid temporal changes that are characteristic of this parasitic field emission phenomenon. It also has to be said that, in many instances, these analyses were "destructive"; i.e. in the sense that the particle forming the emission site was removed during the analysis by the very high local field applied through the probe electrode. Furthermore, this highly non-uniform field distribution is untypical of the local field conditions experienced by emission sites under the normal macro-field operational conditions subjected to typical broad-area HV electrodes. For these sorts of reasons, it is therefore difficult to evaluate the long term potential of this STEM technique as a means of forwarding our understanding of the electronic nature of the basic emission mechanism.

3.3.4   PFEM-Based Surface Analysis Studies

The probe-based studies described earlier in Sections 3.3.1 and 3.3.2 established that the electron "pin-holes" that occur on broad-area HV electrodes are invariably associated with the micron-sized particulate structures that typically "contaminate" electrode surfaces. It was also implied that each of these particles could be thought of as a discrete naturally-occurring, micro-electronic devices. However, these studies did not reveal why typically only 1 in 10^6 or 10^7 of such particles is able to promote the phenomenon; i.e. what critical combination of material and microgeometrical factors is required to create the necessary electronic conditions for the low-field "cold" emission of electrons? To answer these questions, it was necessary to develop a new analytical technique that would provide fresh experimental information about the emission regime, and hence an improved fundamental understanding of the phenomenon.

A key requirement of any new technique is that it should have the ability to obtain high-resolution spatial and material information about where exactly electrons are emitted in relation to the microstructure of the cathode surface. This implies the need to use a dynamic imaging technique in which the specimen surface is under high-field conditions; it therefore excludes such routinely available "passive" techniques as SEM, SIMS, AES, XPS etc. Apart from the STEM approach, discussed in the previous section, there is also the possibility of using photoemission (electron) microscopy PEM. Indeed, this latter approach offers several additional potential advantages, including the ability to obtain dynamic, real-time images of the emission process under properly simulated operational conditions where the whole cathode surface is under macroscopic high-field conditions.

The use of the photoelectric effect to image a surface dates back to the early 1930s, and has formed the basis of several routine analytical microscopes [34]. Recently, however, the technique has received renewed interest in several laboratories [35], including its novel application to the study of cold cathode electron emission processes described in the following paragraphs.

Instrumental Principle

As illustrated in Fig 3.18(a), the microscope developed for the present study employs a three-stage electrostatic imaging column that is mounted in an ultra high vacuum chamber [36,37]. In this design, a planar cathode, in conjunction with the first element of the objective einzel lens, forms a 1 mm "test" HV gap. Furthermore, the elements of this lens are also specially designed to allow the cathode to be directly irradiated by UV photons from a vacuum source. Electrons photo emitted from the surface of the cathode are accelerated through the "anode" aperture, and are formed into a virtual image by the objective immersion lens. This virtual image is then magnified by the following two lens stages and projected onto a 40 mm diameter double channel plate image detector where the local electron flux is enhanced by a factor in excess of 10^4. By using a video camera system to make real-time recordings, image analysis techniques may be used to enhance the recorded data.

Referring to Fig 3.18(b), the microscope can be operated in three distinct modes. In the photoemission mode (PEM), electrons produced by the injection of UV photons are used to form an image that provides local topographical information about the cathode surface: i.e. the instrument is being used as a conventional photoemission microscope [34,35]. For the second mode, the field emission mode (FEM), the UV light source is switched off so that only field- or "dark"-emitted electrons are imaged: in this case, the image generally takes the form of a single or cluster of bright spots. In the third combined mode, both photo- and field-emitted electrons are imaged simultaneously under identical electron optical conditions. It follows therefore that this PFEM mode provides a means of directly relating the spatial location of field-emitted electrons to the local topographical features associated with an emission site.

Instrumental Performance

As an illustration of the type of images obtained with this system, Fig 3.19 presents the four key images associated with the analysis of an isolated carbon particle attached to a mechanically polished OFHC copper electrode. Frame 1 is a calibration image of a reference copper/gold grid, and provides a means of establishing the resolution of the technique, i.e. 0.5-1.0 µm; Frame 2 is a

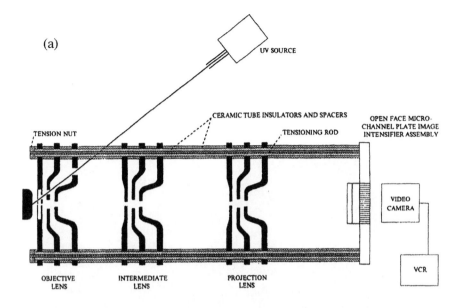

Fig 3.18(a) The instrumental design of the PFEM analysis system.

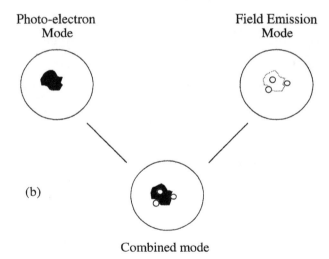

Fig 3.18(b) A schematic illustration of the operational principle, showing the superposition of PEM and FEM images to give a "combined" PFEM image. (From Xu et al. [36], with permission).

discrete locations on micro-particulate structures, and that there are two preferred types of location. Firstly, sub-sites are frequently observed on the upper surface of particles, i.e. as seen in Frame 4 of Fig 3.19; it would therefore appear that these electrons have to be transported through the bulk of the particle from the substrate electrode prior to emission. With the second type of location, electrons are emitted from the edge of a particle with a comet-like appearance, i.e. as illustrated schematically in Fig 3.18(b).

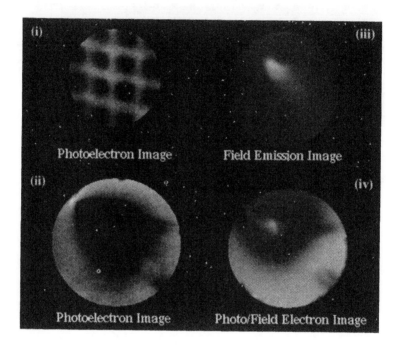

Fig 3.19   A typical PFEM Analysis of a particulate emission site. (i) A PEM image of a reference grid. (ii) A PEM image of the non-emitting particle. (iii) An FEM image of the emitting particle. (iv) A dynamic PFEM image of the field emitting particle. (From Xu et al. [36] with permission).

3.4   Electron Emission Characteristics of Individual Sites

In parallel with the above investigations into the structural and material nature of emission sites, there has also been an on-going enquiry into their electronic characteristics, and particularly the physical basis of the mechanism that can promote the "cold" emission of electrons in the anomalously low field range of 5-10 MV/m. This section is therefore devoted to a review of the three main types of investigation that have contributed to our fundamental understanding of this highly localised microscopic emission phenomenon.

3.4.1 Single-Site I-V Characteristics

Two approaches have been used to measure the current-voltage (I-V) characteristics of individual emission sites. The first employs a standard plane-parallel electrode geometry incorporating a transparent anode imaging facility (see Section 3.2.1), and relies on the fact that the total prebreakdown gap current frequently originates from just one emission site [38,39].

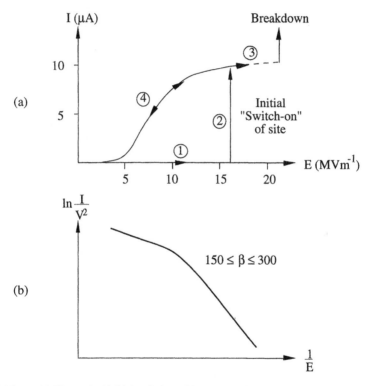

Fig 3.20 (a) The typical initial switch-on I-V characteristic of an isolated emission site and (b) The associated F-N plot of the reversible part of its characteristic (region 3-4).

The second approach employs one of the probe techniques described in the previous section [20,25]. However, this latter approach has two major limitations: firstly, it does not simulate the operational field conditions of HV electrodes, and secondly, the technique does not readily lend itself to studies of the initial turn-on characteristics of individual sites. Accordingly, the following summary account is based on a detailed transparent anode investigation [38,39] that will be more fully discussed in Chapters 4 and 5.

Referring to the schematic of Fig 3.20, it will be seen that during stage 1, when a field is initially applied to a virgin electrode, there is no recorded current (i.e. no emission sites). However, at a certain threshold field, a site suddenly "switches on" and immediately delivers a current of several microamps (stage 2). On subsequently cycling the field (stages 3 and 4), there is no further switch-on process; instead, a smooth reversible I-V characteristic is obtained that yields the well-known type of near-linear Fowler-Nordheim plot shown in Fig 3.20(b). The β-values normally obtained from such plots lie in the range 150-300: however, a possible explanation for the physical significance of these values must await a theoretical analysis of the emission mechanism to be presented in Chapter 4. Typically, the "switch-on field" E_{sw} can be an order of magnitude higher than the subsequent threshold field E_{Th} for emission from this site. It will also be seen from Fig 3.20(a) that the breakdown field E_b for the site can often be very close to the switch-on field E_{sw}. From a technological standpoint, the existence of this switch-on phenomenon is of crucial importance since, without some form of current-limiting provision, an HV gap would be at greatest risk of breakdown during the initial field application.

3.4.2   Projection Imaging of Emitted Electrons

Not surprisingly, several attempts have been made to adapt the projection imaging technique used in the conventional field emission microscope (FEM) (see Chapter 6) by using small-area cathodes to study the parasitic "cold" emission mechanism that routinely occurs at microscopic sites on operational broad-area HV electrodes. An early and particularly ingenious approach was due to Brodie and Weissman [40], who exploited the divergent field properties of a cylindrical electrode configuration to obtain a magnified projection image on an outer phosphor-coated cylindrical glass anode of the emission sites existing on a central tungsten wire cathode. This study revealed that emission sites typically gave images that were composed of several discrete sub-spots; i.e. as was subsequently confirmed by Cox [20]. Also, but not shown, these images were prone to be unstable with time, in that the sub-spots exhibited a "flickering" behaviour.

The more conventional field emission microscope approach was first applied by Tomaschke et al. [41] and later by Jüttner [42]. In the system of Jüttner, illustrated in Fig 3.21(i), an image of the emission sites on a 5 mm diameter rounded molybdenum cathode C (insert a), or on a tungsten microtips of radius ~ 250 nm (insert b), is projected onto a phosphor screen. In this arrangement, the apertured anode A acts as an electron "extracting" electrode such as used in field emission electron guns [43]; i.e. it provides the necessary cathode field but allows a large percentage of the resulting emission to be projected onto the screen.

(i)

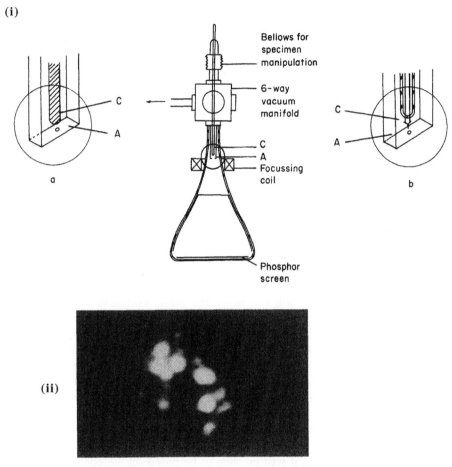

(ii)

Fig 3.21   *(i)* A field emission microscope facility for studying the electron emission from small-area electrodes. Inserts (a) and (b) show respectively the specimen assemblies for "rounded" and "micropoint" cathodes. *(ii)* A typical projection image of the electron emission processes that occur with the "rounded" type of cathode shown (From Jüttner [44], with permission.)

Fig 3.21(ii) is an example of the type of image obtained from emission sites found on these rounded cathodes, and is seen to be characterised by a random distribution of sub-emission centres, each having emission areas with linear dimensions of ~ 5 nm: this result again confirms the same independent finding of Cox [20]. Another important characteristic of these images is that they appear to "flicker", which is caused by individual sub-emission centres

87

randomly switching on and off; not surprisingly, this phenomenon also gives rise to a complementary "pulsing" of the total emission current. It is particularly significant that Tomaschke et al. [41] found that this instability persisted even down to liquid helium temperature since such behaviour is quite contrary to normal experience with metallic micropoint field emitters: in fact, this observation is of considerable significance to the discussion in Chapter 4 of a non-metallic type of electron emission mechanism.

The immediate visual response of this type of technique also makes it a particularly valuable tool for studying the influence of physi- and chemi-sorption processes on the emission properties of the sites found on such semi-broad-area electrodes. For example, Jüttner et al. [44] have conducted a detailed investigation into how the corrosive action of such gases as CO, CO_2 and H_2O in the pressure range 10^{-10}-10^{-7} mbar affects the long-term spatial distribution and stability of the electron emission drawn from these types of cathode. These studies showed that such corrosion processes resulted in an initially bright and well defined emission image from an atomically clean surface becoming progressively "granulated" by the creation of numerous secondary sub-microscopic emission centres thought to be whisker-like structures having emitting area dimensions of ~ 5-10 nm. Accompanying this change, which occurs at a constant gap voltage, it was also found that there is a dramatic decrease in the total emission current level which was attributed to an increase in work function following the initial corrosive action of these gases. There is however a partial recovery of both the emission image and the total current level over a period of 2-3 h, due apparently to the partial removal of these corrosive layers by the sputtering action of positive ions formed by the impact of those electrons that are trimmed by the anode; an explanation that was also supported by Auger spectroscopy studies.

The final projection imaging system to be considered is due to Latham and co-workers [11,45-47]. This differs from the previous arrangements in that it employs a purpose-designed electrostatic lens system to produce a magnified emission image of single sites on a true broad-area planar cathode. Referring to Fig 3.22(a), it will be seen to be based on the anode probe hole system described above in Section 3.3.2. Thus, a site has to be firstly located on the axis of the lens system by scanning the cathode in a raster pattern in front of the anode probe hole until a current is detected by the down-stream electrode assembly. By adjusting the lens voltages, in conjunction with a careful axial centring of the site, a clear emission image is obtained on the phosphor screen, such as shown in Fig 3.22(b). This typically consists of several discrete sub-emission centres which, as had been found by earlier workers, were prone to independently switch on and off, and thus give the impression of a "flickering" image. As will be illustrated towards the end of the following section, there is one type of emission site, namely that produced by particles of graphitic carbon, that gives a characteristic segmentally shaped image [47]. The precise physical reason for this behaviour is not fully understood, although some possible explanations will be discussed in Chapter 4.

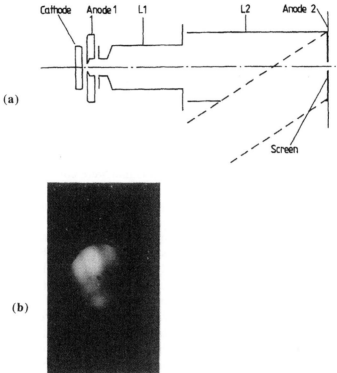

(a)

(b)

Fig 3.22   (a) The electrostatic lens assembly used for obtaining a projection image of the electron emission from a single site [11]. (b) A typical example of a single-site projection image obtained by this technique [11,38-40]. (After Bayliss and Latham [11], with permission.)

3.4.3   Spectral analysis of Emitted Electrons

By the early to mid 1970s, it had become a firm conviction of the present author [17,48] and others [20,21,49,50], that the prevailing metallic micro-protrusion model for explaining the localised "cold" electron emission processes that occur on broad-area electrodes did not provide a satisfactory explanation for much of the experimental data. In particular, it is difficult to justify the unrealistically high β-factor ($\gtrsim 250$) and widely varying values of emitting areas predicted from the F-N plots of the I-V gap characteristic; also, there is an absence of any direct electron optical evidence of microprotrusions of the appropriate dimension on the surface of virgin electrodes [17]. The possibility of there being an alternative emission mechanism began to gain acceptance, particularly following the previously described work of Cox [20], the optical studies to be discussed later in Section 3.5, and the electron spectroscopy studies to be described in the present section.

Early speculations as to what form such a mechanism might take [17,48,49] were centred on the concept of a Malter emission process [51]. This involved the presence of insulating inclusions in or on the cathode surface that could become positively charged by some form of electron emission or positive ion capture mechanism. Such a process would create an internal field sufficient to "heat" electrons injected from the Fermi level of the metal substrate by a further 4-5 eV so that they could escape into the vacuum. However, without any direct experimental evidence to support the existence of such a mechanism, there was a widespread reluctance at that time to make a radical departure from the accepted microprotrusion interpretation of prebreakdown currents. Other mechanisms that were considered as providing possible alternative explanations of "high-β" emission included resonance tunnelling effects [52,53], such as might occur at islands of alkali impurities created by surface diffusion [54], and a lowering of the surface potential barrier by either induced depletion layers [55] or the interaction of charged dislocations at the electrode-vacuum surface, such as had been discussed by Shockley [56] in the case of semiconductors. However, all such suggestions lacked conclusive experimental evidence to confirm that they could indeed play an important role in a vacuum gap.

A particularly powerful experimental technique for resolving this uncertainty was suggested by the extensive literature on the application of electron spectroscopy for studying the various types of electron emission process; i.e. where it had been well established that the electrons emitted by any given mechanism have a characteristic energy spectrum. Accordingly, it was reasoned that if such spectral measurements could be made on the emission from the isolated sites on high voltage electrodes, they would provide a "fingerprint" of the mechanism. From such spectra it would then be possible to positively identify, for example, whether the electrons came from a metallic microemitter or some sort of semiconductor/insulator regime.

Following this reasoning, Latham and his co-workers [8-11,39,45-47,57-68] have, over a fifteen year period, progressively developed the sophisticated UHV ($P \lesssim 10^{-10}$ mbar) analytical facility illustrated in Fig 3.23. This consists of a high-resolution electron spectrometer interfaced with a "test" high voltage gap. The latter has a plane parallel geometry consisting of the specimen cathode and a highly polished stainless steel anode having an axial 1.0 mm diameter probe aperture. A mechanical scan facility allows the specimen to be moved parallel to the anode in the y-z plane, whilst maintaining a known constant gap between the two electrodes. The specimens used in this work are 14 mm diameter discs with a suitably rounded edge profile, and are mounted on a specimen holder that is insulated for applying

90

negative voltages of up to ~ 15 kV. An integral specimen heater is available for both out gassing the test cathode or making experimental measurements at elevated temperatures.

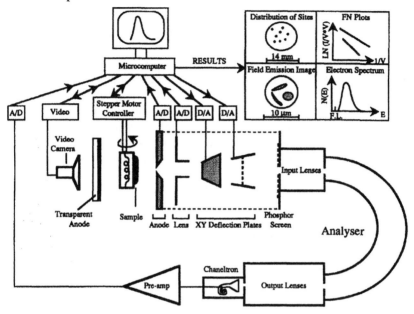

Fig 3.23   A schematic illustration of the advanced and fully automated electron spectrometer facility used for studying the electron spectra of individual prebreakdown emission sites (After [8-11,45-47, 57-68].)

Referring to the more detailed illustration of the electrostatic lens system shown in Fig 3.22(a), the aperture of anode 1 in conjunction with electrode L_1, form an electrostatic lens which has been specially designed as an interfacing element between the test gap and the analyser. It has the two-fold action of (a) focusing the diverging beam of electrons entering the anode aperture into a parallel beam, and (b) decelerating the electrons from a typical energy of ~ 5 keV at anode 1 to the ~ 2 keV requirements at the anode 2 entrance aperture of the analyser. The spectrometer, with a resolution of ~ 30 meV, is based upon a 180° hemispherical electrostatic deflecting element H with associated input and output lenses. The function of the input lens assembly is to decelerate the electrons to the analysing energy of ~ 2 eV and focus them onto the entrance aperture of the hemispherical deflecting element. Those electrons which have the correct energy to be transmitted by the deflecting element are then re-accelerated by the output lenses and focused onto an electron multiplier. Full details of the construction, operation and performance of this instrument can be found elsewhere [8-11,45-47,57-66].

The experimental procedure is initiated by applying a gradually increasing voltage of ~ 5-10 kV across the test gap between the sample cathode C and anode 1 to confirm that there is a finite prebreakdown current, and hence the presence of an emitting site(s) on the specimen cathode. To position a site opposite the probe hole, the specimen is mechanically scanned in a raster pattern L_1 until a current is registered at L_1 (see Fig 3.22(a)). Microadjustments are then made to maximise this current; this ensures that the site is both centralised with respect to the probe hole, and its total emission is being collected by L_1. At this stage, a current-voltage characteristic, and hence a F-N plot of the site, can be recorded. By applying a suitable positive voltage to L_1, the electron beam can then be focused to pass into the analyser input lens, after which, adjustments can be made to the focusing of the analyser in order to obtain the energy spectrum of the electrons emitted from the site.

The first measurements with this system were made on copper specimens [45], where Fig 3.24 compares a typical energy spectrum previously obtained from a reference tungsten emitter at room temperature and a pressure of ~ 10^{-10} mbar using this instrument [66], with that obtained under identical temperature and pressure conditions from a site situated near the centre of one of the copper samples used in the investigation. In both cases, the spectra are stable with time and are reproducible from specimen to specimen. By previously calibrating the instrument against a standard emitter [66], it is also possible to exploit another important feature of this type of spectrometer; namely to reference the position of the Fermi level (FL) of the substrate cathode of each emitting regime on their respective spectra. From a comparison of the two definitive spectra presented in Fig 3.24, three important differences in the characteristics of the two emission regimes can be immediately identified:

(i)   unlike the reference tungsten emitter, all electrons from the site on the copper electrode are emitted from states well below the Fermi level,

(ii)   the half-width of the copper distribution (~ 360 mV) is broader than that of the tungsten emitter (~ 240 mV) and

(iii)   the shape of the copper spectrum is almost symmetrical, i.e. lacking the characteristic sharp high-energy slope found for tungsten.

Indeed, by extending this type of measurement to other electrode materials, it has been confirmed that this type of spectral behaviour is characteristic of all emission sites [57-65].

These early studies revealed several other important spectral properties of emission sites. Thus, from the spectral sequence of Fig 3.25(a), recorded from an "ambient" site under incremental field (and emission current) conditions, and the associated plots of Fig 3.25(b), it will be seen that the spectral half-width (FWHM) together with its shift(s) from the Fermi level, and are both strongly field-dependent [60]. Fig 3.25(a) also illustrates how these field dependences are accentuated by artificially increasing the thickness of the oxide coverage by heating the electrode to ~ 400°C in an enriched oxygen environment.

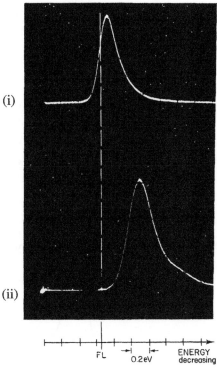

Fig 3.24   Electron energy spectra obtained from (a) a reference tungsten micropoint emitter, and (b) an emission site on a broad-area copper electrode: in both cases, the spectra are referred to the Fermi level (FL) of the substrate cathode. (From Allen and Latham [45], with permission.)

Secondly, Fig 3.25(c) shows how an increase in cathode temperature under constant field conditions leads to (i) an increase in site current, (ii) a spectral shift towards high energies, and (iii) an increase in spectral half-width [67]. Lastly, Fig 3.25(d) illustrates how stimulation by UV photons under constant field conditions leads to an increase in site current, but without any discernible spectral shift [67]. All of these effects are of such importance to the

93

understanding of the basic emission mechanism, that they have been the subject
of more detailed investigations whose findings are discussed in Chapter 4.

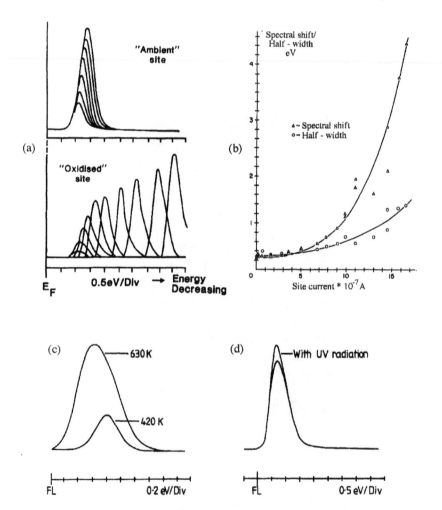

Fig 3.25   (a) A typical sequence of superimposed energy spectra recorded at
increasing gap fields [11]. (b) Plots showing how the half-width (FWHM)
and the shift(s) of the electron spectra with respect to the substrate Fermi
level vary with the site emission current. (c) & (d) Illustrating respectively
the effects of thermal and UV photon stimulation on the spectra of electrons
field-emitted from sites on broad-area electrodes. (From Athwal and
Latham [60], with permission.)

The versatility of the basic instrumental facility shown in Fig 3.23 has been
further exploited to provide a spatially-resolved energy-selective analysis
capability [67,68]. This involved developing the necessary voltage generators

to enable the pairs of x and y electrostatic deflector plates to scan the type of emission image shown in Fig 3.22(b) in a raster pattern over the axial probe hole in the phosphor screen that forms the entrance aperture of the spectrometer.

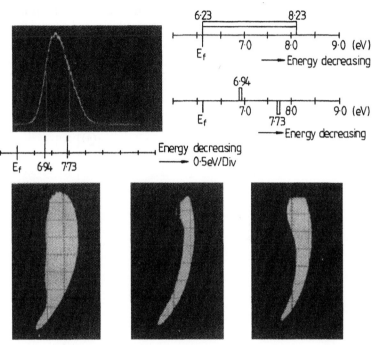

Fig 3.26   Single-segment energy-selective "electronic" images recorded from a carbon emission site on a copper cathode, with (a) an "open" energy window, (b) a high-energy window, and (c) a low-energy window. (From Xu and Latham [67-68] with permission.)

Before concluding this section on electron spectral analysis, mention must be made of a more recent study of Kobayashi and Suzuki [69]. Using the same type of instrumental set-up described above, they measured the energy spectrum of *all* electrons passing through the anode probe hole. Their general finding, which is illustrated schematically in Fig 3.27(a), revealed that a typical spectrum has three components corresponding to the three processes depicted in Fig 3.27(b). Firstly, there is a low-energy peak, corresponding to the secondary electrons generated in the vicinity of the probe hole aperture by the impact of the primary beam. Secondly, there is a broad undulating peak, generated by those electrons that have undergone inelastic mid-gap scattering processes with residual gas atoms. Thirdly, there is the narrow high-energy peak associated with the full-energy field-emitted electrons, that was the subject of the high-resolution studies of Latham and co-workers described above.

Then, by tuning the spectrometer to any desired electron energy, and feeding its output to a synchronously scanned oscilloscope, it is possible to obtain a spatially resolved image of any electron energy. To illustrate the capability of this facility, Fig 3.26 shows how the appearance of a single-segment image obtained from a carbon site [62,68] depends on the analysing energy. This reveals, for example, that the inner concave edge of the image is predominantly formed by low-energy electrons. Here again, the theoretical implications of this and other findings obtained with this technique will be further discussed in Chapter 4.

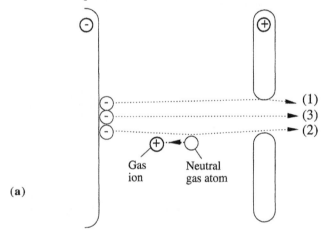

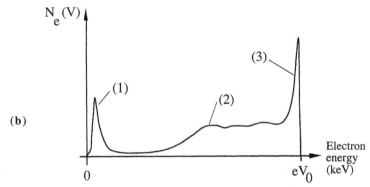

Fig 3.27   (a) Illustrating the three species of electrons that may be transmitted by an HV gap incorporating an anode probe hole. (b) The typical form of the electron energy distribution transmitted by the anode probe hole. (After Kobayashi and Suzuki [69], with permission.)

3. 5   Spectral Studies of Optical Phenomena

It has long been known that light is frequently emitted when an electrically stressed HV electrode gap is viewed under darkened conditions. Generally, this is associated with the flow of prebreakdown currents, although it is not uncommon to observe field-induced electroluminescent effects. The most widely studied process is the "anode spot" that results from the impact of high-energy electrons on the anode surface. However, there have also been studies of "cathode spots" which derive from optical processes that are associated with the underlying prebreakdown electron emission mechanism. This section is devoted to a review of the spectral properties of these two electron-induced prebreakdown optical emission phenomena. (N.B. It is important not to confuse the low-intensity cathode phenomenon referred to here with the very much better known cathode spots associated with the vacuum arc processes discussed in Chapter 15.)

3.5.1   Anode Spots

If a pair of operational (i.e. conditioned) high voltage electrodes are viewed in a darkened environment, it is common to observe the presence of several stable pin-points of light on the surface of the anode [70-74], i.e. such as illustrated in Fig 3.28. These *anode spots*, as they are termed, also occur on unconditioned electrodes, except that here the "spots" are more numerous and unstable. However, as conditioning proceeds, many of the initial population are extinguished, whilst the dominant ones that remain become progressively more stable. Their origin has generally been associated with the anode "hot spots" which are caused by the impact of fine pencils of high energy electrons emitted from microscopically localised cathode sites. However, it has also

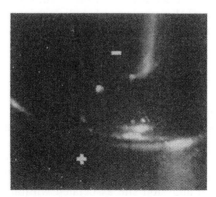

Fig 3.28   Distribution of anode spots on an electrically stressed planar electrode, as viewed with an image intensifier. (From Young [74], with permission.)

been reported by Jüttner [73] that there is another type of anode spot that has a fixed location on the anode surface and is not spatially related to the electron emission sites on the cathode surface; accordingly, these were tentatively attributed to some form of electro luminescent process at thermally stable microinclusions. From studies of the former type of anode spot, it has been confirmed that a breakdown or microdischarge event is frequently associated with the disappearance of a specific spot; i.e. indicating a cathode-initiated process in which a particular emitter has been thermally eliminated [71]. The light from these anode spots was initially thought to be high temperature black body radiation: however, detailed measurements revealed that the short wavelength composition of its spectrum was incompatible with the maximum temperatures likely to be attained at the anode surface due to electron heating [75,76]. Subsequently, its origin was correctly identified as transition radiation resulting from atomic excitation processes in the surface layers of the target following the scattering of the high velocity bombarding electrons [75]. It should also be noted that anode spots have an associated X-ray spectrum; in fact, Zeitoun-Fakiris [76] has used a pin-hole camera technique to obtain an X-ray image of anode spots. The transition radiation phenomena has been analysed theoretically by Boersch et al. [77] who showed that the angular dependence of the spectral power in the emitted radiation $W_\theta(\lambda)$ is given by

$$W_\theta(\lambda) = \left(3.6 \times 10^{-11} P_o / \lambda^2\right) \left\{ \left(\varepsilon' - 1\right)^2 \sin^2\theta \cos^2\theta / \left[\varepsilon' \cos\theta + \left(\varepsilon' - \sin^2\theta\right)^{\frac{1}{2}}\right]^2 \right\}$$

3.1

where P is the power in the incident electron beam, λ is the wavelength being considered and ε' [$= f(\lambda)$] is the complex dielectric constant of the anode surface. Thus, for intensity measurements at a given angle and wavelength, one has

$$W_\theta(\lambda) \propto P_o \qquad \qquad 3.2$$

This latter proportionality was used by Young [74] as the basis of an elegant experimental technique for studying the behaviour of individual broad-area electron emission sites in which the optical intensity of a given anode spot could be directly related to its associated electron current; i.e. it provided a "visual" method of measuring the current-voltage characteristic of the individual electron emission site associated with any of the anode spots shown in Fig 3.28. In the experimental system of Young, illustrated schematically in Fig 3.29, an optical detecting system OD (consisting of a telescopic focusing element, an image intensifier and a photomultiplier, records the intensity of the light emitted at a wavelength λ and at a fixed angle θ from a given anode spot in terms of the photo-multiplier current $i_p(n)$ (dark current corrected). Thus, if the gap voltage is V and I_n is the electron current flowing to the n'th spot, it follows that

$$i_p(n) \propto VI_n = KVI_n \qquad\qquad 3.3$$

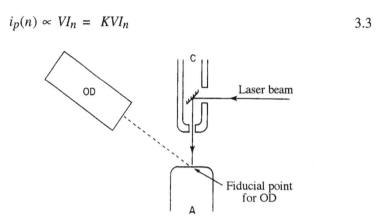

Fig 3.29   The experimental regime used for recording the anode spots shown in Fig 3.28. (After Young [74].)

where K is a constant that can be determined from a calibration measurement in which the photomultiplier current i_{pT}, corresponding to the total intensity from all the anode spots, is compared with the total prebreakdown current flowing across the gap.

Recalling that the F-N representation of the I-V characteristic of an emission process is a plot of $\log(I/V^2)$ against $(1/V)$, it will be seen from equation 3.3 that for the present regime, the F-N characteristic of the n'th site will be given by a plot of $\log(i_p(n)/V^3)$ against $(1/V)$. After correcting for the non-uniformity of the gap field, Young found that the β-values for individual sites typically ranged between 150-250, whilst the "total" β-factor of the gap, as measured from either $\log(I/V^2)$ versus $(1/V)$ or $\log(i_{pT}/V^3)$ versus $(1/V)$ F-N plots, would be ~ 220. From such information, it is possible to identify the sites that are potentially most vulnerable to instability; i.e. those having the highest β-values (see analysis of chapter 5). It is then possible to visually confirm that it is one or more of these that are destroyed with the first breakdown event. The technique also offers a powerful tool for studying how the choice of electrode material and surface preparation determines the incidence of emission sites, and how such sites are affected during electrode conditioning procedures.

3.5.2 Cathode or k-Spots

In the course of a follow-up investigation to the work of Young on anode spots, Hurley and Dooley [78] established the existence of complementary cathode or k-spots, as they were named to distinguish them from the well known phenomenon associated with the vacuum arc. Since this was a surprising finding, not to be anticipated from the metallic microprotrusion model for electron emission, these authors subsequently re-arranged the experimental chamber for a more detailed study of the optical properties of this new cathode phenomenon [79]. Their system, which is illustrated in Fig 3.30, used a plane-parallel electrode geometry inclined at an angle to the axis of the optical detecting system previously used by Young [74], thus providing a line-of-sight to all parts of the cathode surface.

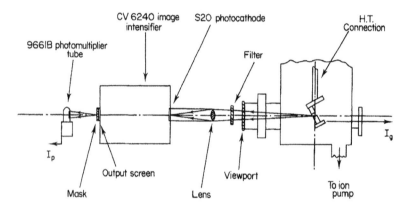

Fig 3.30 Image intensifier and electrode regime used for studying k-spots. (From Hurley and Dooley [78], with permission.)

To gain an insight into the physical mechanism responsible for the phenomenon, it was arranged to record a spectrum of the emission by using a series of high-pass sharp cut-off filters. This technique was then used to study a range of electrode materials (OFHC copper, stainless-steel, dural, gold and molybdenum) and in all cases k-spots were observed that had characteristic emission spectra [79,80]. As an example, that obtained from copper is reproduced in Fig 3.31(a) and is seen to have two dominant lines at 650 and 700 nm, which Hurley tentatively attributed to an accumulation of cuprous oxide, probably at a grain boundary. For comparison, the spectrum of the complementary anode spot is shown in Fig 3.31(b) and is seen to be characteristically shifted to shorter wavelengths. From these studies, it was also noted that breakdown events generally resulted in the disappearance of one or more spots, and that discharges originated from within light emitting regions; i.e. strongly suggesting that k-spots were associated with electron emission processes.

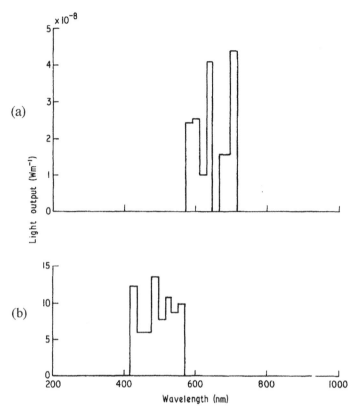

Fig 3.31   The spectra of light emitted from (a) k-spots and (b) anode spots. (From Hurley [80], with permission.)

To confirm this latter possibility, Hurley [80] subsequently replaced the planar electrode geometry of Fig 3.30 with the point-plane arrangement of Fig 3.32, in which a micropoint probe anode is mechanically scanned in front of a planar test cathode until an emission site is accurately located on its surface, as evidenced by a finite specimen current recorded by the electrometer in the low-voltage cathode line. If the same image intensifier viewing system shown in Fig 3.30 is then focused on the cathode surface immediately below the micropoint probe, it is possible to observe luminous k-spots on all occasions that electron emission is detected. It should be added that a complementary anode spot is also invariably observed on the probe anode, although, as already pointed out, its emission spectrum of Fig 3.31(b) is characteristic of transition radiation and therefore readily discriminated from

101

the cathode emission. Another important finding of these probe measurements was that there are two sorts of emission site [81]: type (a) that are characterised by high-β factors ($\gtrsim$ 200) and give rise to breakdown events at a critical current of $\sim 5 \times 10^{-5}$ a, and type (b) that have low β-factors at low fields, but become current-limiting through a negative resistance effect at high fields and consequently do not give rise to breakdown events.

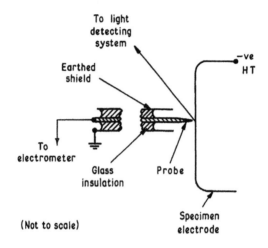

Fig 3.32   An anode probe technique for correlating the electron and optical emission characteristics of individual k-spots. (From Hurley [80], with permission.)

In seeking a physical explanation for the k-spot phenomenon, it was logical for these authors to consider first how the established metallic microprotrusion emission model of that time, discussed briefly in Chapter 2 and more completely in Chapter 4, might give rise to the observed optical effects. In this context, the only apparent option was by thermal radiation from a sub-critically heated emitter: however, such an explanation must be discounted since, in the case of anode spots, it would require a black-body type of emission spectrum which is clearly not supported by the experimental evidence of Fig 3.31(a). The alternative explanation proposed by Hurley and Dooley [79] is that the k-spot emission arises from an electroluminescent effect. This interpretation was strongly supported by measurements of the brightness dependence of a given emission line (as measured by the photomultiplier current i_p) on the applied voltage V which were shown to obey the anticipated Alfrey-Taylor relation [82] by giving a linear plot of log i_p versus $V^{-1/2}$. Since electroluminescence is a "non-metallic" phenomenon, it was clear that this interpretation was incompatible with the simple metallic protrusion model for electron emission. Instead, these authors attributed the emission to a more complex emitting mechanism involving some sort of metal/semiconductor micro-regime at highly localised sites on an electrode surface, possibly associated with impurity inclusions, i.e. similar to the type of regime proposed earlier by Walters et al. [48], Cox and Williams [21] and Farrall et al. [49] from qualitative reasoning based on indirect experimental evidence. It will be recalled from Section 3.4 above that the same conclusions emerged independently from the direct evidence obtained from the early electron spectrometry studies of the emission processes by Latham and his co-workers [45-47].

3.6   Influence of Environmental Factors on Emission Process

In this section, we shall be reviewing a range of technologically-orientated studies in which the various analytical facilities described above have been used to investigate how the basic emission process is influenced by the sort of changing environmental conditions that typical devices have to face during their operational cycle. To take just one example, it is necessary to be able to predict how the performance of a satellite-based travelling wave tube will respond to large temperature fluctuations, mechanical stresses, and ionising radiation. For the interested reader, the practical implications of this topic will be further explored in relation to the specific technologies discussed in Chapters 10 to 15.

3.6.1   Vacuum Conditions

As previously illustrated in Fig 2.4, the most obvious influence of the vacuum conditions is on the I-V characteristic of the gap, where an increase of pressure generally leads to (a) lower switch-on fields for prebreakdown emission sites, (b) noisy prebreakdown currents with frequent step-like discontinuities in their magnitude, (c) the onset of microdischarge phenomena, and (d) lower hold-off fields. At the level of single-site behaviour, a pressure change manifests itself most clearly in the appearance of the projection image of the emission site: thus, referring for example to Figs 3.21 and 3.22, the individual sub-emission processes are seen to become less stable with increasing pressure. The subjective effect of this phenomenon is to cause the image to "flicker", where the mean "flicker frequency" of the emission sub-centres increases significantly with pressure. Typically, it increases from around 0.1 Hz at 10^{-9} mbar to 10 Hz at 10^{-6} mbar, which explains why the noise associated with prebreakdown currents similarly increases with pressure (see Fig 2.4). The effect of pressure on a population of emission sites can be visually demonstrated by using the transparent anode technique previously described in Section 3.2.1. Thus, Fig 3.33(a) and 3.33(b) are schematic representations of pairs of time-lapse images that are typically obtained from electrodes that are operating respectively under low and high pressure conditions [9]. Such evidence clearly demonstrates that the density of these emission "sites", or electron "pin-holes", increases with increasing pressure. It will also be seen that the spatial distribution of these sites is very much more stable at lower pressures.

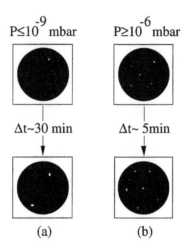

$$P \leq 10^{-9}\ mbar \qquad P \geq 10^{-6}\ mbar$$

$$\Delta t \sim 30\ min \qquad \Delta t \sim 5min$$

(a)   (b)

Fig 3.33   Schematic representations of pairs of time-lapse images obtained from the transparent-anode images technique. These show respectively in (a) and (b) how the spatial stability of populations of emission sites are influenced by pressure.

At higher pressures, sites tend to switch on and off randomly with time, which is one of the processes that accounts for the increase in noise level in the gap current illustrated in Fig 2.4.

It has also been found that this increased instability of sites predisposes them to initiating breakdown events. Coupling this latter observation with the fact that there is a higher density of (unstable) sites at higher pressures, it is possible to understand why the hold-off voltage of a gap generally falls off with increasing pressure. The reader must however be reminded that this argument breaks down when the pressure rises to the semi-vacuum conditions described in Section 2.4, where its insulating properties improves. Returning to the present conditions, it must be added as a postscript that the fall off in hold-off performance is further exacerbated by the fact that a breakdown event can frequently create a profusion of "secondary" emission sites, i.e. as illustrated in Fig 3.2, which are all potential new breakdown sites.

3.6.2 Electrode Temperature Cycling

The effects of temperature cycling have already been touched upon on earlier sections of this book. Thus, in Section 2.2.2 when discussing the general I-V characteristics of an HV vacuum gap, Fig 2.7 illustrated how the prebreakdown emission current depends on electrode temperature, whilst Fig 3.25(b) illustrates how the spectral properties of the emission respond to this parameter. From a more detailed study of the effect of heating an OFHC copper cathode [64,65] under zero-field conditions, Fig 3.34 presents a sample set of data showing how there is a dramatic increase in electrode emissivity for only a relatively modest increase in temperature.

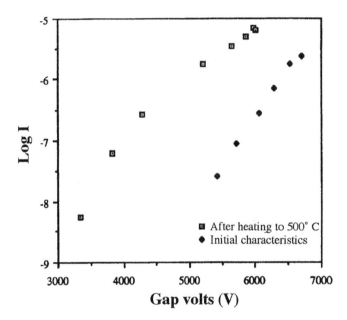

Fig 3.34   The effect of bulk thermal heating under high-field condition on the electrode I-V characteristics. (From Archer [65], with permission.)

By employing a transparent anode technique, it was possible to show that these currents originate from only a small number of ambient emission sites, frequently only one, but that as the temperature rises through 300K, a change occurs in the emission site distribution, with existing sites switching off and new sites switching on. When the electrode is cooled back to ambient conditions, the general pattern of behaviour is for several of the newly switched on sites to remain emitting, and for there to be a significant overall increase in the ambient emissivity. From a practical perspective, it should be further noted that if this temperature cycling was conducted under high-field conditions using a specially designed specimen holder [65], the effects were even more dramatic; in particular, the stability of both the individual sites and the gap itself were significantly lowered.

The effect of temperature cycling under zero-field conditions had also been the subject of an earlier investigation by Niedermann et al. [26,27]. This group employed the anode probe technique illustrated in Fig 3.15 to investigate the effect of electron bombardment heating on populations of emission sites on niobium electrodes. Their important finding is illustrated in Fig 3.35, and shows how the progressive heating of a test cathode up to temperatures of around 750°C results in the stimulation of new emission sites, but that as the temperature climbs to values in excess of 1200°C, all emission sites are generally suppressed.

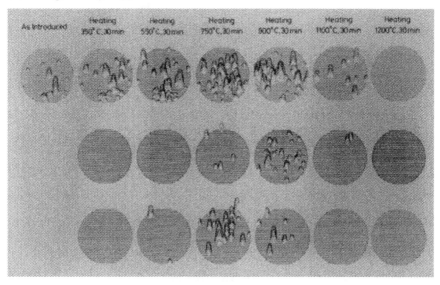

Fig 3.35   Sucessive sequences of emission site "maps", showing the effect of temperature cycling a niobium electrode (From Niedermann et al [28] with permission.)

However, if this same cathode is subsequently re-cycled through a lower temperature excursion of say only 850°C, it is found that a new generation of sites are formed. Furthermore, these new sites are all likely to have a similar material composition (sulphur in the case of Nb electrodes), having been formed by the diffusion of a characteristic impurity material along the grain boundaries to the electrode surface. To remove these "2nd generation" sites, it is again necessary to raise the electrode temperature to in excess of 1200°C. As will be discussed more fully in Chapter 12, this finding has formed the basis of a new technological procedure for pre-processing niobium superconducting RF cavities.

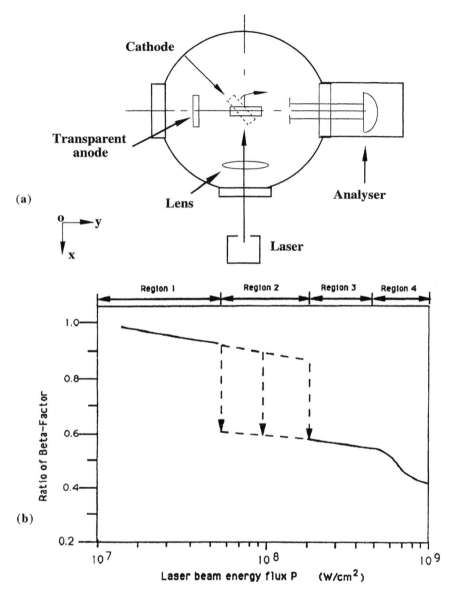

Fig 3.36   (a) A schematic illustration of the scanning laser probe technique and "transparent anode" imaging facility that was used for applying controlled thermal loading to the surface of a broad-area HV electrode. (b) Illustrating how the prebreakdown I-V characteristic of an HV gap is influenced by laser-induced transient surface heating. The "ratio of β-factor is defined as β_α/β_i, where β_i is the initial β-value and β_α the value obtained after dosing treatment. (From Xu and Latham [83], with permission.)

In a more recent study [83], illustrated in Fig 3.36(a), a scanning laser probe has been used to investigate the effect on the emission process of heating only the surface region of an electrode. The practical aim of this investigation was to determine whether the "conditioning" effects achieved by the drastic process of high-temperature bulk heating could be replicated by this transient surface heating technique. Referring to Fig 3.36(b), a test cathode is first characterised in terms of its I-V characteristic. By scanning the laser probe in a raster pattern over the electrode surface, it may then be subjected to increasing levels of uniform thermal loading, or dosing. If the I-V characteristic of the cathode is recorded after each dosing, it is possible to quantify the effect of the treatment. As shown in Fig 3.36(b), there is a minimum threshold dosing level that has to be exceeded if a significant reduction is to be obtained in the β-value of the gap: equally however, there is an upper threshold that has to be exceeded if a permanent and predictable improvement in performance is to be achieved. In its present state of development, the technique is capable of reliably delivering a 40% lowering of the β-value of a gap which, in practical terms, corresponds to a significant potential improvement in the voltage (or field) hold-off performance of a gap.

3.6.3   Gas Desorption Effects

From a consideration of the "active" physical phenomena that can occur in a high voltage gap, it could be anticipated that there will be essentially two processes that can give rise to an increase in the local vapour pressure of a gap. Firstly, during the prebreakdown phase, there will be the thermally activated desorption of occluded gas that will occur principally at the anode "hot-spot" formed by electron bombardment. Secondly, at a breakdown or microdischarge event, there will be a transient increase in the electrode metal vapour pressure indicative of some form of thermal instability at either the cathode or anode (Chapter 5). To verify the existence of the first of these two processes, mass spectrometry techniques have been used to monitor the partial pressure of the residual gas species present in an interelectrode gap as the applied voltage is gradually increased towards breakdown. It should also be noted here that the detection of transient changes in metal ion concentration requires very much more sophisticated techniques, as will be described in Section 7.2(iii).

From measurements made during the prebreakdown phase [84-88], it has been established that, irrespective of the electrode material, there is a characteristic increase in the total pressure during the initial application of a voltage to a gap; i.e. during its "current" conditioning phase. The magnitude of this increase is dependent on the prebreakdown current, but can typically reach half an order of magnitude at an ambient pressure of $\sim 10^{-8}$ mbar with a gap current of $\sim 1\mu A$. However, if the gap voltage, and hence the mean current, is maintained at a constant value, any such pressure increases will very gradually decay away over a period of several hours. It is then found that during any subsequent voltage application to this value, there is no further significant evidence of gas desorption: from this observation it follows that a fully "conditioned" gap is similarly free of desorption processes. From a practical standpoint, it should however be added that the above behaviour pattern can be significantly influenced by the ambient vacuum conditions. Thus, at pressures > 10^{-6} Pa, the decay of the enhanced partial pressure of CO and CO_2 can be very slow, or almost non-existent, during constant electron bombardment conditions [88].

There is also some evidence [89] to suggest that, for a given electrode material, there is a characteristic "voltage threshold" or electron impact energy at which desorption is activated; e.g. 10-12 kV for Al and Bi, 14-17 kV for Cu and 19-20 kV for Ni and Zr. One possible explanation for this effect is that the gas is released from several atomic layers beneath the anode surface, so that differing electron energies are required to penetrate to a given depth among the various electrode materials. Alternatively, the effect may have a simple thermal explanation in terms of a characteristic electron power loading required to raise the anode surface temperature above the value reached by the system during the original UHV bake-out procedure. When the composition of this desorbed gas is analysed by mass spectrometry techniques [84-90], it is found to consist predominately of CO and CO_2 for all electrode materials, with the partial pressure of CO_2 being typically twice that of CO. In addition, there is generally found to be an increase in the partial pressure of H_2O, particularly after electron bombardment [88]. The selective desorption of these particular gases is a well known phenomenon in UHV technology and can be explained both in terms of the high solubility of these species in most metals, and the thermal oxidation of the carbon that is a common residual impurity. This type of gas desorption behaviour, which is characteristic of "commercial vacuum conditions" (i.e. 10^{-7}-10^{-5} Pa), has been further investigated by Suzuki et al. [90] under controlled laboratory conditions using a purpose-designed analytical system.

3.6.4 Electron-Stimulated Ionic Processes

In most practical device applications, the residual gas pressure in the HV gap is sufficiently high for there to be a significant number of mid-gap ionising collisions between electrons and neutral gas atoms; i.e. as illustrated in Fig 3.27. Furthermore, this process will also be greatly enhanced by the additional supply of gas atoms that are continually being desorbed from the anode and other electrode surfaces into the confined space of the electrode gap by the impact of high-energy electrons. It follows therefore that these ions will constitute an additional source of charge carriers, and will thus augment the measured prebreakdown current.

The energy distribution of these ions has been studied by Kobayashi and Suzuki [69] using a spectroscopy system similar in concept to that previously described in Section 3.4.3 for the earlier measurement of electron spectra. However, for ionic spectral measurements, it was necessary to have the probe hole electrode set at a negative potential. In fact, this same system was also used for the complementary broad-range, low-resolution electron spectral measurements briefly mentioned at the end of Section 3.4.3. These studies established that the recorded ionic spectra could be generally characterised as consisting of discrete high and low-energy peaks. The strong high-energy peak, which typically has a FWHM of 7-10 eV, is believed to result from ionisation processes that occur in the vicinity of the cathode presumably associated with the copious supply of gas molecules desorbed from the anode by impacting high-energy electrons. In contrast, the less intense low-energy peak results from ions formed in the vicinity of the cathode, possibly associated with the ion-stimulated desorption of gas molecules from the surface of the probe hole aperture. As noted above in Section 3.6.3, Kobayashi and co-workers [88] have also used a mass spectrometry technique to identify the species of these ions to be predominantly CO, CO_2 and H_2O.

3.7   References

1.   Millikan, RA. and Shackelford, BE., *Phys. Rev.*, **15**, 239-240, 1920.
2.   Millikan, RA. and Eyring, CF., *Phys. Rev.*, **27**, 51-67, 1926.
3.   Millikan, RA. and Lauritsen, C.C., *Proc. Nat. Acad. Sci.* (US), **14**, 45-49, 1928.
4.   Little, RP. and Whitney, WT., *J. Appl. Phys.*, **34**, 2430-2, 1963.
5.   Farrall, GA., *J. Appl. Phys.*, **42**, 2284-93, 1971.
6.   Latham, RV, Bayliss, KH and Cox, BM., *J.Phys. D. Appl.*, **19**, 219-31, 1986.
7.   Bajic, S and Latham, RV, *J. Phy. D. Appl. Phys.*, **21**, 943-50, 1988.
8.   Latham, RV, *IEEE Trans., Elec.Insul.*, **23**, 881-894, 1988.
9.   Latham, RV and, Xu NS, *Vacuum*, **42**, 1173-81, 1991.
10.   Athwal, CS and Latham, RV, *J.Phys. D. Appl. Phys.*, **17**, 1029-34, 1984
11.   Bayliss, KH and Latham, RV, *Proc. Roy. Soc.*, **A403**, 285-311, 1986
12.   Mazurek, B, Xu, NS and Latham, RV, *J. Mat. Sci.*, **28**, 2833-39, 1993
13.   Cox, B.M. and Wort, D.E.J., *Vacuum*, **22**, 453-55, 1972.
14.   Bogdanovskii, G.A., *Sov Phys.-Solid State*, **1**, 1171-6, 1959.
15.   Little, R.P. and Smith, S.T., *J. Appl. Phys.*, **36**, 1502-4, 1965.
16.   Jedynak, L., *J. Appl. Phys.*, **36**, 2587-9, 1965.
17.   Latham, RV. and Braun, E., *J. Phys. D: Appl. Phys.*, **1**, 1731-5, 1968.

18. Latham, RV. and Chapman, CJS., *J. Phys. E: Sci. Instrum.*, **3**, 732-4, 1970.
19. Latham, RV. and Braun, E., *J. Phys. D: Appl. Phys.*, **3**,1663-70, 1970.
20. Cox, BM., *J. Phys. D: Appl. Phys.*, **8**, 2065-73, 1975.
21. Cox, BM. and Williams, WT., *J. Phys. D: Appl. Phys.*, **10**, L5-9, 1977.
22. Allen, NK., Cox, B.M. and Latham, RV., *J. Phys. D: Appl. Phys.*, **12**, 969-78, 1979.
23. Athwal CS and Latham RV, *Physics*, **104C**, 46-9, 1981.
24. Athwal, CS., Ph.D. Thesis, University of Aston, U.K. 1981.
25. Niedermann, Ph, Sankarranen, N, Noer, RJ and Fischer ϕ, *J. App. Phys.*, **59**, 892-901, 1986
26. Nidermann, Ph, Sankarranen, N, Noer RJ and Fischer ϕ, *J. App. Phys.*, **59**, 3851-60, 1988.
27. Niedermann, Ph., Ph.D. Thesis, University of Geneva, 1986
28. Jiminex, JM., Noer, RJ., Jouve, G., Antoine C and Bonin B., *J. Phys. D: Appl. Phys.*, **26**, 1503-10, 1993.
29. Niedermann, Ph. Emch R and Desconts, P., *Rev. Sci. Instrum.*, **59**, 368-9, 1988.
30. Mahner, E, Minatli, N, Piel, H and Pupeter, N, *Appl. Surf. Sci.*, **67**, 23-28, 1993
31. Binnig, G and Rohrer H. *Helv. Phys. Acta.*, **55**, 762-32, 1982.
32. Young, R., Ward, J. and Frederick S., Rev. Sci. Instrum., **B43**, 999-1005, 1992.
33. Neidermann Ph, Renner, CH, Kent AD and Fischer ϕ, *J. Vac. Sci. Technol.*, **A8**., 594-97, 1990.
34. Griffith, OH and Rempfer GF., *SEM*, **8**, 257-337, 1987.
35. Griffith, OH and Engel W., *Lettramicroscopy*, **36**, 1-29, 1991
36. Archer, AD, Xu, NS and Latham, RV., To be published - *Surf. Sci.*, 1994.
37. Xu, NS and Latham, RV., To be published - *Surf. Sci.*, 1994
38. Xu, NS and Latham, RV., *Proc. XVI-ISDEIV*, Moscow, 1994.
39. Latham. RV., *Nucl. Instrum. Methods Phys. Res.*, **A287**, 40-47, 1990.
40. Brodie, I. and Weissman, I., *Vacuum*, **14**, 299-301, 1964.
41. Tomaschke, HE., Alpert, D., Lee, DA. and Lyman, LM., *Proc. II-ISDEIV*, 13-21, 1966.
42. Jüttner, B., *Proc. VI-ISDEIV*, 101-6, 1974.
43. Crewe, AV. Eggenberger E.N. Wall, J. and Welter, L.M., *Rev. Sci. Inst.*, **39**, 576-83, 1968,.
44. Jüttner, B., Wolff, H. and Altrichter, B., *Phys. Stat. Sol.*, **27a**, 403-12, 1975.
45. Allen, NK and Latham, RV, *J. Phys. D. Appl. Phys.*, **11**, 655-57, 1978
46. Bayliss, KH and Latham, RV, *Vacuum*, **35**, 211-217, 1985.
47. Xu, NS and Latham, RV., *J. Phys. D. Appl. Phys.*, **19**, 477-82, 1986.

48.   Walters, CS., Fox, MW. and Latham, RV., *J. Phys. D: Appl. Phys.* **7**, 911-19, 1974.
49.   Farrall, GA., Owens, M. and Hudda, FG., *J. Appl. Phys.*, **46**, 610-17, 1975.
50.   Maskrey, JT. and Dougdale, RA., *Brit. J. Appl. Phys.*, **17**, 1025-34, 1966.
51.   Malter, L., *Phys. Rev.*, **49**, 876-85, 1936.
52.   Duke, CB. and Alferieff, ME., *J. Chem. Phys.*, **46**, 923-37, 1967.
53.   Kellog, GL. and Tsong, TT., *Surface Sci.*, **62**, 343-60, 1977.
54.   Todd, CJ. and Rhodin, TN., *Surface Sci.*, **42**, 109-38, 1974.
55.   Schmidt, LD. and Gomer, R., *J. Chem. Phys.*, **45**, 1605-23, 1966.
56.   Shockley, W., *Phys. Rev.*, **91**, 228, 1953.
57.   Allan, NK, Ph.D Thesis, Aston University, 1979.
58.   Bayliss, KH and Latham, RV., *Proc. Roy. Soc*, **A403**, 285-311, 1986.
59.   Bayliss, KH., Ph.D Thesis, Aston University, 1984.
60.   Athwal, CS and Latham, RV, *Physica*, **104C**, 189-95, 1981
61.   Allen, NK, Athwal, CS. and Latham, RV., Vacuum, **32**, 325-32, 1982..
62.   Xu, NS and Latham, RV, *J. de Physique*, **47**, (C7 95-99), 1986.
63.   Xu, NS., Ph.D Thesis, Aston University, 1986.
64.   Archer, AD and Latham, RV., *Proc XIV-ISDEIV*, 43-46, 1990.
65.   Archer, AD., Ph.D Thesis, Aston University, 1993.
66.   Braun, E., Forbes, R.G., Pearson, J., Pelmore, J.M. and Latham, R.V., *J. Phys. E: Sci. Instrum.*, **11**, 222-8, 1978.
67.   Xu, NS and Latham, RV, *J. de Physique*, **47**, C2 (73-78), 1986.
68.   Xu, NS and Latham, RV, *Surf. Sci.*, **274**, 147-160, 1992.
69.   Kobayashi, S and Suzuki, K., *J. Phys. D: Appl. Phys.*, **19**, 825-34, 1986
70.   DeGeeter, DJ., *J. Appl. Phys.*, **34**, 919-20, 1963.
71.   Alpert, D., Lee, D.A., Lyman, F.M. and Tomaschke, H.E., *J. Vac. Sci. Technol.*, **1**, 35-50, 1964.
72.   Bennette, CJ., Swanson, LW. and Charbonnier, F.M, *J. Appl. Phys.* **38**, 634-40, 1967.
73.   Jüttner, B., *Mber, Deut. Acad. Wiss. Berl.*, **11**(2), 105-7, 1969.
74.   Young, RW., *Vacuum*, **24**, 167-172, 1973.
75.   Chatterton, PA., *Brit. J. Appl. Phys.*, **17**, 1108-10, 1966.
76.   Zeitoun-Fakiris, A., *C.R. Acad. Sci. Paris*, **266**, 828-30, 1968.
77.   Boersch, H., Radeloff, C. and Suerberg, L., *Z. Phys.*, **65**, 464-9, 1961.
78.   Hurley, RE. and Dooley, PJ., *Vacuum*, **28**, 147-149, 1977.
79.   Hurley, RE. and Dooley, PJ., *J. Phys. D: Appl. Phys.*, **10**, L195-201, 1977.
80.   Hurley, RE., *J. Phys. D: Appl. Phys.,* **12**, 2229-45, 1979.
81.   Hurley, RE., *J. Phys. D: Appl. Phys.* 13, 1121-8, 1980.
82.   Alfrey, GF. and Taylor, JB., *Brit. J. Appl. Phys.*, **6**, Suppl. 4, 544-9, 1955.
83.   Xu, NS, and Latham, RV., *Vacuum*, **43**, 99-103, 1992.
84.   Cornish, JCL., Latham, RV. and Braun, E., *Proc. IV-ISDEIV*, **28-32**, 1970.

85. Jüttner, B., *Exper. Tech. Phys.*, **19**, 45-48, 1970.
86. Ghani, A., Ph.D. Thesis, University of Aston, U.K. 1971.
87. Panitz, JA., *J. Appl. Phys.*, **44**, 372-5, 1973.
88. Kobayashi, S, Suzuki, K and Katsube, T, *Appl. Surf. Sci.*, **33/34**, 370-378, 1988
89. Ghani, A., Ph.D Thesis, University of London, 1971.
90. Suzuki, K, Kobayashi, S and Katsube, T., *Appl. Surf. Sci*, **33/34**, 325-334, 1988.

4

The Physical Origin of Prebreakdown Electron "Pin-Holes"

NS Xu

In Chapter 3, the reader was provided with a review of a wide range of experimental investigations that were aimed at characterising the diverse physical phenomena that are associated with the occurrence of parasitic prebreakdown electron emission from a broad-area cathode surface. This analysis also served to give some perspective on the complexity of the physical processes involved. We now have to enquire more closely into the nature of these processes, so that we can acquire a better physical understanding of the experimental findings presented in Chapter 3.

In parallel with these experimental studies, there have also been on-going theoretical programmes aimed at establishing models that describe the physical processes associated with the prebreakdown electron emission phenomenon [1-11]. For example, Latham and co-workers have developed the Metal-Insulator-Vacuum (MIV) and Metal-Insulator-Metal (MIM) models [1-6,8-10], whilst Halbritter has introduced the Dynamic Field Enhanced Electron Emission Model [7]. In all cases, these models contrast strongly with the traditional metallic microprotrusion (MM) model [12,13]. As will be discussed in Section 4.1, the "cold" prebreakdown electron emission mechanisms envisaged by these models are all associated with various types of "contaminating" microstructures that are found on the surface of typical HV electrodes. These field emitting microstructures have been termed as "Electron Pin-holes" [8] to highlight the fact that electrons can flow out of the electrode from these surface features at fields that are two or three orders of magnitude lower than the theoretical threshold value of ~3 × 10^9 V/m required for the MM emission mechanism [12,13].

To give readers a comprehensive account of these theories, the following sections will trace the theoretical developments involved in these models. Accordingly, our discussion will start in Section 4.2 with the historically important MM model, which held centre-stage for over four decades. This will be followed in Sections 4.3 and 4.4 by the MIV and MIM models that have had a profound influence on the technological design and production of modern HV devices and systems.

4.1   Emission Regimes

In order to present a detailed description of these theoretical models, it is firstly necessary to review the experimental evidence presented in Chapter 3, and to identify the physical characteristics of the individual micro-emission regimes that form the basis of these models. Secondly, it is necessary to identify the relative importance of the role that each of these regimes plays in promoting prebreakdown electron emission in an HV gap under a range of operational conditions, e.g. vacuum regime, temperature, external radiation etc.

The key conclusion to emerge from the recent diagnostic studies described in Section 3.3.2 is that, within the field range of ~10-30 MV/m, prebreakdown electron emission originates from certain types of isolated microstructures that adhere to, or are embedded in, the surface layer of an extended cathode surface. In addition, the nature of microstructures that give rise to prebreakdown emission can show considerable variation, depending on the pretreatment of an electrode and its operational history. For a virgin cathode, the findings described in Chapters 2 and 3 indicated that the emission invariably originates from microscopically localised particulate-based sites that have the following typical emission characteristics.

(i)   Individual emission sites have high-β F-N current-voltage characteristics that become non-linear at high fields (Section 2.2.3).

(ii)   Field emitting microstructures exhibit no significant topographic features that would give rise to high-β values (Section 3.3.2).

(iii)   Sub-emission centres (sub-sites) have non-metallic electron energy spectral characteristics with broad FWHMs, and a strongly field-dependent spectral shift from the Fermi level of a cathode material (Section 3.4.3).

(iv)   These electron spectra exhibit a non-metallic like response to both bulk heating and UV photon stimulation (Section 3.4.3).

(v)   Emitting microstructures exhibit associated electroluminescent effects (Section 3.5.2).

The central physical implication of these findings, particularly (iii), (iv) and (v), is that the emission regime itself must include a dielectric or semiconducting medium. On the basis of this reasoning, the two possible surface micro-emission regimes illustrated in Figs 4.1(a) and 4.1(b) have been proposed [1-5].

The first regime of Fig 4.1(a) involves a metal-insulator-vacuum (MIV) microstructure, as initially proposed by Latham and co-workers [1-4], to explain the non-metallic characteristics of field emission electron energy spectral data such as those presented in Section 3.4.3. This regime can relate to an insulating particle either resting on or embedded in a metal substrate [4]. On the other hand, Fig 4.1(b) illustrates a metal-insulator-metal (MIM) regime [5], which was originally proposed by Latham and co-workers [9], and subsequently developed by Xu and Latham [5]. This embodies all of the features of the MIV regime but, in addition, is assumed to have a superficial metallic particle, or flake (e.g. of graphite), that acts like an "antenna" by probing the field above the electrode surface, and thereby producing an enhanced field across the "point-contact" with the insulator [9-10].

On the basis of the diagnostic studies of Chapter 3, three further basic assumptions can be made about the nature of individual sites for the purpose of theoretical modelling.

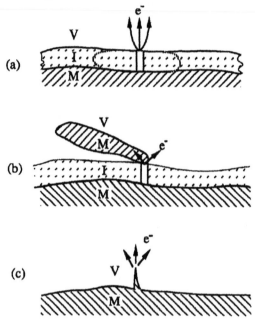

Fig 4.1   Schematic illustration of three possible micro-emissions regimes, or electron "pin-holes".
(a)   the Metal-Insulator-Vacuum (MIV) regime,
(b)   the Metal-Insulator-Metal (MIM) regime, and
(c)   the Metallic-Microprotrusion (MM) regime.

Firstly, it is assumed that, for the MIV regime, the dielectric medium generally takes the form of a "foreign" dielectric surface inclusion [11], or an anomalously thick oxide aggregation [4]. In contrast, for the MIM regime, it is believed that the dielectric medium is generally formed by the ambient oxide layer of the substrate metal and the adsorption layers on the metallic particle. The second assumption, which is common to both regimes, is that these insulating layers typically have a dimension of ~0.1-1 µm in the direction of the applied field; i.e. such that it can support a potential drop of ~ 1-3 V at typical macroscopic gap fields of 10-30 MV/m. Finally, it is assumed that such inclusions are only partly crystalline and almost certainly impure, so that there will be a distribution of both trapping states and donor centres within the inclusion.

For an electrode that suffers a vacuum discharge, or that undergoes a "spark" conditioning procedure, such as described in Section 2.3.4, its surface microstate is more complicated than a virgin electrode. Thus, it is very likely that large numbers of metallic microprotrusions, such as illustrated in Fig 4.1(c), are likely to be created around local arc centres; i.e. in addition to the two types of surface microstructure described above. As shown in Fig 3.8, and later in Chapter 15, these surface microstructures have topographic characteristics that can provide sufficient local field enhancement to give rise to pure metallic field emission at low macroscopic gap fields. The gap phenomena associated with this type of metallic electron emission mechanism have been studied intensively by the Russian group led by Mesyats [12] and the German group led by Jüttner [13], and will be reviewed in both Chapters 6 and 15.

4. 2 The Metallic Microprotrusion (MM) Model

The concept of pure "metallic" Fowler-Nordheim field electron emission, or "cold" electron emission, providing an interpretation of the prebreakdown current-voltage characteristic of HV gaps dates back to the 1920's [14]. According to this model, electrons are assumed to be field emitted from the tip of a microprotrusion at an isolated site on the surface of a broad-area cathode, such as shown in Fig 4.1(c), where the macroscopic gap field is locally enhanced to a value $\geq 10^9$ V/m. In the following section, we shall firstly describe the F-N quantum mechanical tunnelling theory that forms the basis of this type of emission process, and secondly discuss how local field enhancement may be determined from the geometrical parameters of an isolated metallic microprotrusion situated on a planar electrode surface. In fact, an excellent review of both of these two subjects was given by Latham in his earlier book [15].

4.2.1 "Metallic" Field Electron Emission (M-FEE)

The theory of this emission mechanism was originally formulated by Fowler and Nordheim in 1928 [16]. It describes the quantum mechanical tunnelling of electrons through the modified surface potential barrier illustrated in Fig 4.2 that results from the presence of a high external electric field E acting on an atomically-clean metal surface. For tunnelling to occur, the wave function ψ of an electron in the surface of the metal must still be finite beyond the barrier, despite its rapid attenuation. This concept may alternatively be formulated in terms of the Heisenberg Uncertainty Principle [17], i.e. in the form of $\Delta x \, \Delta p_x \approx h/2$. In the present context, Δx is the uncertainty in the position of the electron in the direction of applied field, viz. the barrier width,

119

Δp_X the uncertainty in the electron momentum within the barrier, viz. $\sim(2m\phi)^{1/2}$, where ϕ is the work function of the metal, and h is Planck's constant. For most electrode materials, ϕ may be assumed to be equal to 4.5 eV. Both approaches lead to the requirement that an electron at the Fermi level FL will "see" a barrier width $\Delta x \lesssim 1$ nm, which in practical terms requires an external field $E \gtrsim 3 \times 10^9$ Vm^{-1}. It should also be appreciated that in this emission mechanism the electrons leave the surface with the same initial kinetic energy as they had within the metal.

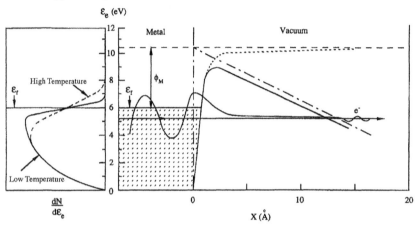

Fig 4.2   An illustration of the quantum mechanical tunnelling mechanism by which a field-emitted electron may escape, or tunnel, from the surface of a metal. It shows the modified high-field potential barrier outside the surface (solid line), where the contributions of the image potential and the applied field ($E \approx 3 \times 10^9$) V/m) are shown by the broken and the broken-solid lines, respectively. In addition, the wave function of a tunnelling electron is depicted. Finally, the inset illustrates the low- and high-temperature electron energy distributions $dN/d\varepsilon_e$ within the metal.

4.2.1 (i) Low-Temperature Theory (T $\lesssim$ 300K)

The derivation of the Fowler-Nordheim (F-N) expression for the field-dependence of the emission current density j_F is based on solid state "free electron" theory, and involves three main stages of mathematical development. Firstly, the electron Supply Function $N(W, \varepsilon_e)$ $dWd\varepsilon_e$ is obtained, which gives the flux of electrons incident on the barrier from within the metal that have total energies in the range ε_e to $\varepsilon_e + d\varepsilon_e$, and normal energy components between W and W + dW. Next, the electron Transmission Coefficient D(W) is derived, which gives the probability that an electron with an energy W will be transmitted by the barrier: the product of these functions $N(W, \varepsilon_e)$ D(W) $dWd\varepsilon_e$ then gives the number of electrons with total energies between ε_e and $d\varepsilon_e$ and normal energies between W and W + dW, that will tunnel through the barrier. Finally, the total number of electrons that tunnel through the barrier is obtained by integrating the above expression over all energies, so that the actual field electron emission current density (Am^{-2}) will be given by

$$j_F = e \int_{all\ energies} dW \int_W^\infty D(W)\ N(W,\varepsilon_e)\, d\varepsilon_e \qquad 4.1$$

where e is the electronic charge. Since the detailed evaluation of $N(W, \varepsilon_e)$, $D(W)$ and this final integral is an extended and largely mathematical exercise that is not directly relevant to the development of the main theme of this text, it will be sufficient to quote the final result; for the interested reader, however, there are comprehensive modern treatments [18,19].

At *low temperatures* $(T \lesssim 300K)$ where only a few electron states above the Fermi level are occupied, as shown by the low temperature electron energy distribution of Fig 4.2, the basic Fowler-Nordheim equation is usually expressed in the form

$$j_{OF} = \frac{1.54 \times 10^{-6}\ E^2}{\phi\ t^2(y)}\ exp\left[\frac{-6.83 \times 10^9\ \phi^{1.5}\ v(y)}{E}\right] \qquad 4.2$$

where j_{OF} is the low-temperature FEE current density, E is the surface electric field (Vm^{-1}) and ϕ the work function of the emitting surface (eV); $t(y)$ and $v(y)$ are tabulated dimensionless elliptic functions [18-21] of the parameter y which is related to E and ϕ by

$$y = 3.79 \times 10^{-5}\ E^{\frac{1}{2}} / \phi \qquad 4.3$$

In practice, $t(y)$ is a very slowly varying function of y and can reliably be taken as unity for the field range likely to be encountered at the tips of microprotrusions in HV gaps, i.e. $3 \times 10^9 \lesssim E \lesssim 10^{10}$ Vm^{-1}; $v(y)$, on the other hand, shows a significant field-dependence which can be closely approximated by the relation [22]

$$v(y) = 0.956 - 1.062y^2 \qquad 4.4$$

for the field range $2 \times 10^9 \lesssim E \lesssim 5 \times 10^9$ Vm^{-1}, which is appropriate for stable field emission regimes. Now, if the field E acts over an emitting area A_e, the emission current I_{OF} from this region will be given by $I_{OF} = j_{OF}A_e$

which, after substituting for j_{OF} from equation 4.2, and re-expressing in natural logarithmic form becomes

$$ln\left[\frac{I_{OF}}{E^2}\right] = ln\left[\frac{1.54 \times 10^{-6}A_e}{\phi t^2(y)}\right] - 2.97 \times 10^9\ \phi^{1.5}\ v(y)\frac{1}{E} \qquad 4.5$$

Substituting lastly for $v(y)$ from equation 4.4, and letting $t(y) \approx 1$, leads to the most useful form of the F-N equation, viz.

$$ln\left[\frac{I_{OF}}{E^2}\right] = ln\left[\frac{1.54 \times 10^{-6}A_e}{\phi} \times 10^{4.52\ \phi^{-\frac{1}{2}}}\right] - 2.84 \times 10^9\phi^{1.5}\frac{1}{E} \qquad 4.6$$

This indicates that if the value of E is known, a plot of ln (I_{OF}/E^2) against $1/E$ would give a straight line with a slope

$$\frac{d\left(ln\left(I_{OF}/E^2\right)\right)}{d\left(1/E\right)} = -2.84 \times 10^9\ \phi^{1.5} \qquad 4.7$$

and an intercept

$$\left[ln\left(I_{OF}/E^2\right)\right]_{E = \infty} = ln\left\{1.54 \times 10^{-6}\ A_e.\frac{10^{4.52\phi^{-\frac{1}{2}}}}{\phi}\right\} \qquad 4.8$$

This observation will be returned to later in Section 4.2.2(ii). From this discussion, it follows that the F-N equation 4.6 may be conveniently expressed in a simplified form similar to that used in Chapter 2, viz.

$$j_{OF} = C_1E^2\ exp\left(-C_2/E\right) \qquad 4.9$$

where C_1 and C_2 are fundamental constants that have the values

$$C_1 = 1.54\ x\ 10^{-6}\xi\frac{10^{4.52\phi^{-\frac{1}{2}}}}{\phi} \qquad 4.10i$$

$$C_2 = 2.84 \times 10^9 \times \phi^{1.5} \qquad 4.10ii$$

It should be noted that many authors ignore the field-dependence of $v(y)$ and, instead, use an approximate form of the F-N equation with $t(y) = v(y) = 1$ in equation 4.2. Thus, whilst this procedure only introduces errors of 5-10% in the slope of such a plot, the intercept can be affected by as much as two orders of magnitude through the factor $10^{(4.52/\sqrt{\phi})}$, so that only a limited physical significance can be attributed to measurements based on this parameter.

4.2.1 (ii) High-Temperature Theory

At higher temperatures (T > 300K), more electrons begin to occupy states above the Fermi level FL, as shown by the high-temperature electron energy distribution of Fig 4.2. Clearly, these electrons will "see" a narrower surface barrier than those at the FL and will therefore have a higher tunnelling probability. In practical terms, this implies that the FEE current density would be expected to show a strong temperature dependence at high temperatures, which is not accounted for in the basic F-N theory. This effect has been considered by several authors in the course of developing unified theories of electron emission, notably Murphy and Good [23], Dyke and Dolan [24] and Christov [25]. For T $\lesssim$ 1500K, the temperature-assisted FEE current density j_{TF} can be accurately described by the analytic expression [24]

$$j_{TF} = j_{OF} \frac{\pi p}{\sin(\pi p)} \qquad 4.11$$

where j_{OF} is given in equation 4.2, and p is a dimensionless temperature and field dependent parameter given by

$$p \simeq 9.3 \times 10^5 \times \phi^{\frac{1}{2}} \times (T/E) \qquad 4.12$$

However, at higher temperatures, where p > 2/3, the approximations leading to the above expression become invalid and the tabulated data of the analyses of Christov [25] have to be used.

4.2.1 (iii) Space Charge Limitation

The validity of the Fowler-Nordheim theory was first investigated experimentally by Haeffer [26], and later in more detail by Dyke and his colleagues [27]. These workers used a point-plane diode geometry resembling a simple field emission microscope, where the dimensions of the laboratory-fabricated micropoint emitter were accurately known so that the microscopic cathode field could be reliably defined. This type of measurement showed that there is a wide range of field for which there is a good agreement between theory and experiment, as evidenced by the linearity of the F-N plot of the diode current-voltage characteristic. However, at fields $\gtrsim 6 \times 10^9$ V m^{-1}, corresponding to $j_F \sim 6 \times 10^{11}$ Am^{-2}, the F-N plot begins to fall away from linearity, indicating the onset of some form of current-limiting mechanism. In a combined theoretical and experimental investigation, Dyke et al. [28] showed that the field-emitting diode characteristic closely followed a "three-halves" power law at these higher emission current densities, which could be satisfactorily interpreted in terms of a space-charge current limitation similar to the well known Child-Langmuir phenomenon that operates in planar and cylindrical diodes. The limiting field E_l at which the low-temperature F-N theory breaks down because of the onset of space charge effects has been estimated by Chatterton [29] as

$$E_l \lesssim 8 \times 10^8 \ \phi^{1.5} \qquad\qquad 4.13$$

which is in good agreement with the experimental findings of Dyke et al. [27,28]. Mention should also be made however of an alternative theoretical explanation of the phenomenon by Nagy and Cutler [30]. Their basic assumption was that the classical image force barrier model cannot be used at small distances from the emitting surface. Accordingly, they re-formulated the F-N emission theory by applying quantum modifications to the electronic potential in the surface region of the metal, and were able to show that a plot of $\ln (I_{OF}/E^2)$ against $1/E$ can be expected to depart from linearity at fields $\gtrsim 5 \times 10^9$ Vm^{-1}.

4.2.2   M-FEE from Broad-Area High-Voltage Electrodes

From the last section, we now know that the theoretical threshold field required for the M-FEE mechanism to provide significant electron emission is $\sim 3 \times 10^9$ Vm^{-1}. On the other hand, the prebreakdown currents that flow between HV electrodes are observed at gap fields that are frequently more than two orders of magnitude lower than this value. Thus, for this electron

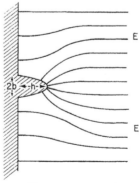

Fig 4.3   Illustrating how a uniform *macroscopic* gap field E is geometrically enhanced to a higher *microscopic* value E_m at the tip of a microprotrusion projecting from a planar electrode surface.

emission mechanism to provide a satisfactory explanation of the pre-breakdown currents, it is necessary to reconcile this apparent paradox. As already discussed, this has been attributed to the assumption that the electrons are emitted from a few isolated microscopic sites having the form of a metallic microprotrusion, i.e. as illustrated in Fig 4.1(c), that can give rise to local field enhancement at their tip surfaces. We now have to consider in more detail how such local field enhancement can be related to the specific geometry of a microprotrusion, and how it may be determined experimentally.

4.2.2 (i) Field Enhancing Properties of Idealised Microprotrusions
In order to analyse the physical properties of this model, it is necessary to assume that the microprotrusion has an idealised symmetrical geometry such as a semi-ellipsoidal microstructure and is orientated normal to a planar electrode, i.e. as shown in Fig 4.3. This figure also illustrates how, by the convergence of the field lines, the uniform *macroscopic* gap field E is geometrically enhanced to a higher *microscopic* value E_m at the tip of the protrusion. The ratio of the microscopic to macroscopic fields is defined as the field-enhancement factor β, i.e.,

$$E_m = \beta E \tag{4.14}$$

where the magnitude of β depends on the geometry and dimensions of a given protrusion. For the semi-ellipse of revolution illustrated in Fig 4.3, the β-factor is given by the detailed expression [31]..

$$\beta = \frac{\left(\lambda^2 - 1\right)^{1.5}}{\lambda \, ln\left[\lambda + \left(\lambda^2 - 1\right)^{\frac{1}{2}}\right] - \left(\lambda^2 - 1\right)^{\frac{1}{2}}} \tag{4.15}$$

where $\lambda = h/b$; i.e. the ratio of the semi-major to semi-minor axes of the ellipse. For needle-like geometries, where $\lambda \gg 10$, the above expression can be approximated by

$$\beta \approx \frac{\lambda^2}{ln\,\lambda - 0.3} \tag{4.16}$$

Alternatively, β may be expressed in the form [32]

$$\beta = \frac{2l^3}{\left(1 - l^2\right) ln\left(\frac{1 + l}{1 - l}\right) - 2l}$$

where $l = [1 - (b/h)^2]^{1/2}$ is the eccentricity of the ellipse. Several other idealised protrusion geometries have been analysed in equal detail, particularly those shown in Fig 4.4, that are based on a spherical emitting surface [31].

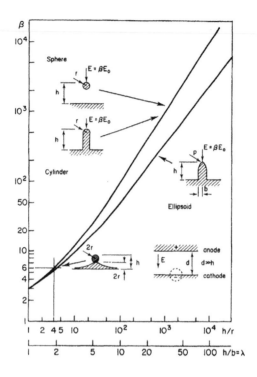

Fig 4.4   A collated representation of the field enhancement factors β associated with various idealised microprotrusion geometries. (From Rohrbach [31], with permission.)

Generally however, it has been shown that if the β-factors are computed as a function of h/r, the ratio of their height h to tip radius of curvature r, they can all be characterised to a good approximation by the unifying expression [33]

$$\beta \simeq 2 + \frac{h}{r} \qquad \qquad 4.17$$

provided $h/r \gtrsim 5$. It has to be added that Miller [34] has shown that β also depends on the electrode separation d through the relation

$$\beta = \beta_\infty \left(1 - \frac{h}{d}\right) \qquad \qquad 4.18$$

126

where β_{∞} is the value of β when d >> h; i.e. the conditions assumed in the above discussion.

4.2.2 (ii) FEE Properties of Idealised Microprotrusions

Consider now a typical practical situation in which the prebreakdown current I flowing between broad-area electrodes is derived from a single microprotrusion at which the electric field is enhanced by a factor β over the macroscopic value E existing at a perfectly smooth cathode surface. From elementary electrostatic field theory, E will be generally defined by a relation of the form E = V/fd where V is the voltage applied between the electrodes, d is their separation and f is a dimensionless function of the geometrical parameters of the electrodes: for a plane-parallel geometry f = 1, for concentric cylinders having cathode and anode radii R_c and R_a respectively one has f = $R_c/(R_a - R_c)$ ln (R_a/R_c), whilst for concentric spheres f = R_c/R_a. Assuming, as in previous discussions, that this single micro-emitter is associated with a plane-parallel electrode geometry, it follows that the microscopic field E_m acting at its tip will be given by

$$E_m = \beta E = \beta \frac{V}{d} \qquad\qquad 4.19$$

If this equation is now used to substitute for E in equation 4.6, we obtain the logarithmic form of the F-N equation expressed in terms of the externally measurable parameters I and V,

$$\ln \frac{I}{V^2} = \ln \left[\frac{1.54 \times 10^{-6} A_e \, \beta^2 \, 10^{4.52\phi^{-0.5}}}{\phi d^2} \right] - \frac{2.84 \times 10^9 d\phi^{1.5}}{\beta} \cdot \frac{1}{V} \qquad 4.20$$

It follows therefore that if the current-voltage data of such a single-emitter gap is presented in the form of an F-N plot, viz. ln I/V^2 versus $1/V$, it will give a straight line with a slope

$$\frac{d\left(\ln \left(I/V^2\right)\right)}{d\left(1/V\right)} = - \frac{2.84 \times 10^9 \, d\phi^{1.5}}{\beta} \qquad\qquad 4.21$$

and an intercept

$$\left[\ln \left(I/V^2\right)\right]_{V=\infty} = \ln \left[\frac{1.54 \times 10^{-6} A_e \, \beta^2 \, 10^{4.52\phi^{-0.5}}}{\phi d^2} \right] \qquad 4.22$$

Thus, if d is known, and the work function ϕ of the emitter is assumed to have the same value as the bulk electrode material, the slope of such an F-N plot provides an indirect measure of the β-factor of the emitter; it follows that this value may then be used to determine the emitting area A_e from the intercept of the plot.

For the more complicated situation where there are n emitters distributed over the cathode surface, each with its own enhancement factor β_i and emitting area $(A_e)_i$, the total prebreakdown current I will be given from equation 4.9 by

$$I = C_1 (V/d)^2 \; \Sigma (A_e)_i \; \beta_i^2 \; exp \left(-C_2 \; d/V\beta_i\right) \qquad 4.23$$

indicating that an F-N plot will no longer be expected to give a straight line. In practice, however, the reverse is found to be true, with a non-linear plot being the exception rather than the rule. An explanation for this apparent paradox was subsequently provided by multi-emitter computer simulation studies, first by Tomaschke and Alpert [35] and later by Farrall [36], which showed that the prebreakdown regime will always be dominated by one of a group of high-β emitters whose β and A_e characteristics will accordingly be reflected in the associated F-N plot [35,36].

4. 3   Metal-Insulator-Vacuum (MIV) Model

Having considered in detail the traditional metallic microprotrusion model, it becomes even clearer than before that this MM model cannot account for the experimental findings described in Chapter 3: in particular, those revealing that the emission characteristics of naturally occurring sites strongly contrast with those predicted by the MM model. Accordingly, we shall now move on to introduce an alternative emission model, namely, the metal-insulator-vacuum (MIV) model. As already pointed out in Section 4.1, this model is based on the surface micro-emission regime illustrated in Fig 4.1(a), which is completely different from the metallic microprotrusion regime. Consequently, it will be necessary to introduce new concepts and theories into our discussion that are not available in the traditional MM model. We shall firstly explain how such a regime can promote electron emission in the field range of 10-30 MV/m. Then we shall derive a current-voltage relationship, similar to the metallic F-N equation. Finally, we shall review some of the new developments of this model that have led to an explanation of the electron energy spectral characteristics, and the effects of thermal and photon stimulation obtained from such an emission regime.

4.3.1   Electronic Properties of the MIV Regime

Before we can describe the physical processes associated with this new theory it is firstly necessary to define the physical representation of such a MIV micro-emission regime. The MIV emission regime can be viewed as a micron-sized electronic device naturally "built" onto the surface of a broad-area cathode-electrode. As will become gradually clear, all parts of this whole system, including the metallic substrate material, the insulating medium, the interface between the metallic substrate and the insulating medium (MI interface), and the interface between the insulating medium and the vacuum (IV interface), are involved in the physical process of electron emission from such a regime. It is therefore necessary to adopt a form of representation that can embrace all parts of the regime. For this purpose, Latham and co-workers [1-6] have used the electron energy band diagram. In the following discussion we shall similarly use this type of representation to describe the electronic properties of a virgin MIV microstructure.

Our discussion will start with a consideration of the insulating medium. Referring to Fig 4.5(a), it can firstly be assumed that the insulating medium contains a large number of traps, since, as mentioned in Section 4.1, such inclusions are likely to be polycrystalline or amorphous materials. Thus, for crystallite sizes of 100 Å, trapping levels as high as 10^{24} m^{-3} are possible due to grain boundary defects alone. Secondly, the medium is assumed to be so impure that it has a high density of donors, typically ~10^{24} m^{-3}. Thirdly, it is also assumed that, for simplicity, these traps occupy a single energy level in the energy band diagram: a similar assumption is made for the donor levels. Finally, it is further assumed that the arrangement of these two energy levels with respect to the Fermi level will be as shown in Fig 4.5(a); i.e. where the trap level lies above the Fermi level, and the donor below it. Bayliss and Latham showed [4] that such an arrangement of energy levels is the only one that has the generality required for explaining the experimentally observed emission characteristics. It is important to mention at this stage the implications of such an arrangement of the energy levels to our problem.

Thus, one should recognise that a trap and a donor are thermodynamically equivalent with regard to their occupancy by electrons. It follows therefore that, with the arrangement illustrated in Fig 4.5(a), the donors do not provide a large number of mobile electrons, i.e. since their states are below the Fermi level, and the traps are essentially empty, with only a few being occupied by electrons excited from the donor centres.

Considering now the metal-insulator interface, it follows from Fig 4.5(a) that a blocking contact will result if it is assumed that $\phi_m > \phi_I$, where ϕ_m and ϕ_I are the work functions of the metallic substrate and the insulator respectively. Thus, on the insulator side of the contact there will be

129

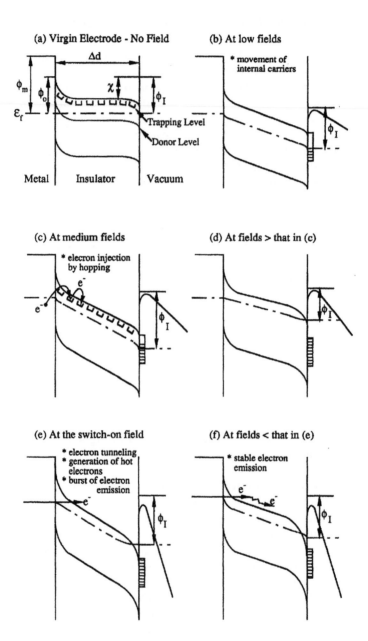

Fig 4.5 A sequence of band diagrams of the MIV emission regime illustrating the switch-on process.

a thin depletion region of width l. In this depletion region both donorstatesand trap centres lie above the Fermi level, and so can be assumed empty. Consequently, there will be a net positive charge with a density equal to $N_d + N_t$, i.e. taking into account the fact that an empty trap typically has the effect of an isolated positive charge. It can be shown that [37], at zero bias, the width of the depletion region may be expressed as

$$\lambda_o = [2\varepsilon_o\varepsilon_r \, (\phi_m - \phi_I)/e_2(N_d + N_t)]^{\frac{1}{2}} \qquad 4.24$$

where ε_r is the relative dielectric constant of the insulator, and ε_o is the permittivity of free space. To show how this width varies with the sum of the densities of traps and donor centres, Table 4.1 gives a few values of λ_o when using $\phi_m - \phi_I = 2$ eV and $\varepsilon_r = 5$.

Table 4.1 Depth of Depletion Region l for Several Values of $(N_d + N_t)$

$N_d + N_t$ (m^{-3})	10^{21}	10^{23}	10^{25}	10^{27}
λ_o (m)	10^{-6}	10^{-7}	10^{-8}	10^{-9}

It follows that the presence of a positive space charge in the depletion region will give rise to an electric field at the metal-insulator interface given by [37]:

$$E_i = -\{2(N_d + N_t)(\phi_m - \phi_I)/\varepsilon_r\varepsilon_o\}^{\frac{1}{2}} \qquad 4.25$$

From the above expression, it can be seen that E_i is proportional to the concentrations of both donor and trap centres. Furthermore, if the externally applied potential drop V across this depletion region is known, the field at the interface can be determined using the following expression

$$E_i = -\{2(N_d + N_t)(\phi_m - \phi_I + V)/\varepsilon_r\varepsilon_o\}^{\frac{1}{2}} \qquad 4.26$$

4.3.2 Switch-On Mechanism

Having presented an energy band diagram interpretation of our current understanding of the electronic properties of the MIV micro-emission regime, we are now able to describe how electrons are emitted from such a micro-regime. Unlike the metallic Fowler-Nordheim regime of Section 4.2, field electron emission cannot occur even when the surface potential barrier at the insulator-vacuum interface is narrow enough for conduction electrons to tunnel through it, since, as mentioned in the last section, very few electrons exist behind the barrier. Thus, it is found that such an emission regime needs to be "switched on" in order to give rise to a steady emission current.

This switch-on process is illustrated in Fig 3.20, where it will be seen that no significant current ($<10^{-8}$A) is measured in a vacuum gap formed by virgin electrodes until a sudden jump occurs in the current, when a site is said to be "switched on". Two conditions are usually required for the occurrence of such a switch-on process: namely, a voltage "surge" superimposed on some threshold gap voltage. It can also be seen from Fig 3.20 that one of the consequences of switch-on is that the current-voltage characteristic becomes reversible with a current of ~10 μA being measured at a gap voltage that is much lower than the threshold switch-on value. This switched-on state normally persists even when a sample is left for a long period without the application of an electric field under UHV conditions at room temperature; i.e. indicating that some permanent change has occurred to the emission regime. In this context, Latham and co-workers [1-6] have proposed that the most important of these changes is the formation of one or more electron conduction channels in the bulk of the insulator. As illustrated in Fig 4.1(a), such a channel grows from a pinpoint on the surface of the substrate to the vacuum surface of the insulator, so that electrons injected into such a channel can be transported to the vacuum surface under the influence of an applied field, and then be emitted into vacuum from the tip surface of the channel. In the following paragraphs, we shall describe how these channels are formed, following the initial application of the field, and how consequently an emission site is switched on.

Starting with the band diagram for the virgin emission regime shown in Fig 4.5(a), where, the metal-insulator (MI) interface is assumed to have a blocking contact (i.e. $\phi_m > \phi_I$), Fig 4.5(b) next illustrates how, with the initial application of a field, any mobile carriers responsible for the finite conductivity of the insulator are moved directionally by the electric force. These include mobile electrons, which will move towards the vacuum interface, and both positive ions and holes which will move towards the MI interface. Some of these electrons will gradually fill up the empty traps in the vicinity of the IV interface, whilst a small portion of them will finally arrive at the vacuum surface of the insulator and fall into the surface states. It is assumed that this low density of accumulated negative charge at the IV interface will not prevent the penetration of the applied field into the bulk of the insulator. On the other hand, the ions and holes will accumulate at the MI interface and, as a result, enhance the internal local field i.e. as predicted by equation 4.25.

If the applied field is gradually increased still further, another process becomes significant: namely, electrons can be injected from the metal substrate into the bulk of the insulator. This injection occurs at relatively low fields, say in the range of 10^7-10^8 V/m, and is by electron hopping [38] over the empty trap centres and donor centres in the depletion layer, i.e. as illustrated in Fig 4.5(c). Subsequently, the injected electrons will fill up the rest of the empty traps. This trap filling process will be slow, since the injection current density is low, and will thus require a number of increments of gap field to complete the process. As the traps are gradually filled up, and space charge is built up in the insulating layer, field penetration becomes difficult, and thus the band diagram becomes rather flat as illustrated in Fig 4.5(d), and both the injection and flow of charges stop.

The gap field now needs to be increased much further beyond this point until the new situation illustrated in Fig 4.5(e) is established. Here, the potential barrier at the MI interface has become thin enough for electrons to tunnel from the substrate to the conduction band of the insulator. Assuming that the barrier height ϕ_0 at the MI interface is 1 eV at room temperature, an electron current with a density of 1 A/m^2 can be injected into the insulator across the MI interface at a field E_i of ~1×10^8 V/m. This corresponds to a gap field of 20 MV/m, if E_i is related to the gap field E by $E_i = (\beta/\varepsilon_r)E$, with $\beta = 25$, and $\varepsilon_r = 5$. At this point, several other conditions have been established that favour hot-electron emission. (a) All the traps are filled up, so that injected electrons can be accelerated in the conduction band without being scattered by the empty trap centres. (b) A much higher field is established in the bulk of the insulator which is able to accelerate conduction electrons so that they become "hot". (c) The vacuum surface barrier has been lowered (easy for hot electrons to surmount) and thinned (easy for low energy conduction electrons to tunnel through). Thus, an increase of field will result in the vacuum injection of a "bunch", or surge, of electrons that will experience a gain in energy from the relatively high field existing in the bulk of the insulator before they reach the vacuum surface.

Under these circumstances, a situation is created where several phenomena reinforce each other to produce a continuous emission current. Thus, as hot electrons are generated near the vacuum surface, the effective surface barrier height at the I-V interface decreases, so that a large proportion of the surface charge, and any charge stored within the conduction band of the insulator, will be emitted in a burst of current. As a result, the external field will penetrate more effectively to the metal-insulator contact and give rise to an increased tunnelling current. At the same time, the surface field will greatly increase relative to the bulk field, i.e. since the electrons at the vacuum surface are preferentially emitted. This will lead to a further lowering of the potential barrier, and an additional increase in the electron temperature; i.e. both effects will tend to push this system towards a highly conducting state This tendency may be further reinforced by the migration of any holes, formed during the scattering processes near the vacuum interface, back to the metal-insulator contact region [39]. Dynamic equilibrium will then be rapidly established by a mechanism similar to that discussed by Mott [40] in his double-injection model; i.e. where the negative charge density decreases as the electrons are accelerated towards the vacuum interface. The dynamic charge distribution will then adjust itself to give the highest density a short distance into the insulator, thus stabilising the vacuum surface field at a high value. This is illustrated by the energy band diagram of Fig 4.5(f), which represents the equilibrium "on" state; i.e. that is responsible for the reversible emission process.

4.3.3   I-V Characteristic

In this section, we shall analyse the emission regime in terms of its electronic parameters, defined above, to obtain an expression for how the externally measured prebreakdown current flowing in a plane-parallel gap varies with the applied voltage.

4.3.3 (i)   Basic Theory
As explained in the last section, electrons are heated to a temperature T_e in the process of being accelerated towards the vacuum surface of the insulator. On the other hand, since the microscopic field E_m at the surface of an emitting channel is high, where typically $E_m = \beta E > 10^8$ V/m^{-1}, the emission of lower energy electrons by tunnelling through the surface potential barrier is also possible. Such an emission regime is analogous to the temperature-assisted field emission situation mentioned in Section 4.2.1(ii) and described by Murphy and Good [23]. Accordingly, the corresponding emission current density can be expressed in the following form (see also equation 4.11):

$$j_e = A\frac{(\pi/N)}{\sin{(\pi/N)}}(kT_e)^2 exp\left\{\frac{B(\beta E)^{0.5}-\phi_I}{kT_e}\right\} \qquad 4.27$$

where the parameter N is defined as $N = CkT_e \, (\beta E)^{-3/4}$, with β being the local field enhancement factor. In equation 4.27, A, B and C are constants defined as:

$$A = \frac{4\pi me}{h^3} \qquad \text{(i)}$$

$$B = \left(\frac{e^3}{4\pi h\varepsilon_o}\right)^{\frac{1}{2}} \qquad \text{(ii)} \qquad\qquad 4.28$$

$$C = \frac{\pi}{\left(4\pi\varepsilon_o em^{-2}\left(\frac{h}{2\pi}\right)^4\right)^{\frac{1}{4}}} \qquad \text{(iii)}$$

For the convenience of developing the present discussion, the symbols A, B and C will be used in all of the following equations, and assumed to have the above significance.

There are a number of considerations that have to be taken into account when calculating the electronic temperature T_e. Thus, Latham [3] related it to the potential drop across the insulator, and proposed the following relation:

$$T_e = \frac{2}{3k} e \, \Delta V$$

4.29

In this expression,

$$\Delta V = \Delta d \, E/\varepsilon_r$$

4.30

where E is the gap field, Δd the thickness of the insulator (where $\Delta d \ll d$), and ε_r is the relative dielectric constant of the insulator. However, this approach did not take into account the loss of electron energy through scattering processes, e.g. with phonons.

Subsequently, Bayliss and Latham [4] proposed (a) that electrons injected from the metallic substrate would be immediately thermalised to the bottom of the conduction band of the insulator through electron-phonon scattering, and (b) that electron heating by the penetrating field occurs only in the surface layer of the insulator in the vicinity of the vacuum. On this basis, they obtained the following expression:

$$(k/e)(T_e - T_0) = \alpha \, \Delta\varepsilon_e$$

4.31

where T_0 is the ambient sample temperature, α is an unknown constant, $\Delta\varepsilon_e = e \, \Delta V_s$, with ΔV_s being the potential drop across the surface layer of the insulator, which may be obtained through detailed measurements of the energy spectra of emitted electrons; i.e. as will be discussed in Section 4.3.4. In this approach, Bayliss and Latham [4] took into account the possible electron energy losses due to scattering with other particles in the surface layer of the insulator. It follows therefore that the unknown constant α should be less than 1.

Recently, Xu [41] considered the most important scattering mechanism responsible for the electron energy losses, namely electron-phonon scattering. This analysis, which only assumed the involvement of acoustic phonons, has led to the following new equation for T_e:

$$T_e = \sqrt{\frac{3\pi T_0^2 \mu^2}{32 v_s}} \, E_s = XE_s$$

4.32

where

$$X = \sqrt{\frac{3\pi T_0^2 \mu^2}{32 v_s}}$$

Here, v_s is the velocity of sound, and μ is the electron mobility. Also, E_s is the averaged electric field in the surface insulating layer which, to a first order approximation, may be obtained from the relation

$$E_s = \beta E / \varepsilon_r \qquad\qquad 4.33$$

where again E is the macroscopic gap field. Substituting for E_s from equation 4.33 into equation 4.32 leads to the following expression relating the electron temperature T_e to the gap field E

$$T_e = X \left(\frac{\beta}{\varepsilon_r}\right) E. \qquad\qquad 4.34$$

Thus, to a first order approximation, the electron temperature T_e is proportional to the macroscopic gap field E.

From the above discussion, one can see the merit of Xu's approach over the earlier two models. Thus, that of Latham [3] obviously overestimates the possible gain of electron energy from the penetrating field, whilst that of Bayliss and Latham [4] has to rely on the measurement of the energy spectra of emitted electrons to determine the unknown constant α. The latter requirement therefore severely limits its practical usage since, very frequently, only a basic I-V characteristic is available.

4.3.3 (ii) Procedure for I-V Data Processing

We shall now explore how the above theoretical model may be used to develop a means of determining the local field enhancement β, the emitting area A_e and the work function of the insulator ϕ_I from externally measured I-V data. Thus, by substituting for T_e from equation 4.32 into equation 4.27, the emission current density can be re-written as follows:

$$j_e = A \frac{\pi \left(Ck\frac{X\beta}{\varepsilon_r}E(\beta E)^{-\frac{3}{4}} \right)}{\sin \left(\pi \left(Ck\frac{X\beta}{\varepsilon_r}E(\beta E)^{-\frac{3}{4}} \right) \right)} \left(k \left[\frac{X\beta}{\varepsilon_r}E \right]^2 \right) exp \left| \frac{B(\beta E)^{\frac{1}{2}} - \phi_I}{\left[\frac{kX\beta}{\varepsilon_r} \right] E} \right| \qquad 4.35a$$

However, it is found that N >> 1 if E > 1MV/m, i.e. in the range of fields of interest. Therefore, the factor $\{(\pi/N)/\sin(\pi/N)\}$ can be approximated by 1.

Thus, equation 4.35a can be re-written in the following simplified form

$$j_e = Ak \left(\frac{X\beta}{\varepsilon_r} E\right)^2 \exp\left|\frac{B(\beta E)^{\frac{1}{2}} - \phi_I}{\left(\frac{kX\beta}{\varepsilon_r}\right) E}\right|$$ 4.35b

Assuming next that the emitting area is A_e, the emission current may then be expressed in the following form:

$$I_e = j_e A_e$$

$$= A_e Ak \left(\frac{X\beta}{\varepsilon_r} E\right)^2 \exp\left|\frac{B(\beta E)^{\frac{1}{2}} - \phi_I}{\left(\frac{kX\beta}{\varepsilon_r}\right) E}\right|$$ 4.36

which may be rearranged to give

4.37

$$\frac{I_e}{E^2} = A_e Ak \left(\frac{X\beta}{\varepsilon_r}\right)^2 \exp\left|\frac{B\varepsilon_r}{kX\beta^{\frac{1}{2}} E^{\frac{1}{2}}} - \frac{\phi_I}{\left(\frac{kX\beta}{\varepsilon_r}\right) E}\right|$$

or

$$ln\left(\frac{I_e}{E^2}\right) = ln\left(A_e Ak \left(\frac{X\beta}{\varepsilon_r}\right) 2\right) + \left|\frac{B\varepsilon_r}{kX\beta^{\frac{1}{2}} E^{\frac{1}{2}}} - \frac{\phi_I}{\left(\frac{kX\beta}{\varepsilon_r}\right) E}\right|$$ 4.38

Finally, by letting $y = ln (I_e/E^2)$, and $x = 1/E$, the above expression may be written in the form

$$y = a_1 + a_2 x^{\frac{1}{2}} + a_3 x$$ 4.39(a)

where

(i)

$$a_1 = ln\left(A_e Ak \left(\frac{X\beta}{\varepsilon_r}\right)^2\right)$$

$$a_2 = \frac{B\varepsilon_r}{kX\beta^{\frac{1}{2}}}$$ (ii) 4.39(b)

and

$$a_3 = -\frac{\phi_I}{\left(\frac{kX\beta}{\varepsilon_r}\right)}$$ (iii)

Thus, one would not expect a perfect straight-line fit to a plot of $\ln(I_e/E^2)$ versus $(1/E)$; i.e. the traditional F-N plot constructed from the I-V data collected from the MIV emission regime.

To derive the field enhancement factor β and the emission area A_e, as one conventionally does with metallic F-N emission I-V data, it would be necessary to follow some form of computer curve fitting procedure to generate a best-fitted curve for the I-V data represented by equation 4.39. Thus, the β-value can be calculated from a known value of a_2 and with assumed values of μ, v_s and ε_r. Once the β-value is obtained, A_e can be calculated with the known value of a_1. One potential bonus resulting from this exercise, in comparison with the traditional metallic F-N curve-fitting procedure, is that it would be possible to avoid assuming a value for ϕ_I, the work function of the insulator. Instead, it may be calculated using the known values of β and a_3.

4.3.3 (iii)   Contact-Limited to Bulk-Limited Transition

In the above discussion, we have considered the field-induced hot-electron emission process at the I-V interface. In this, we assumed that the electrons

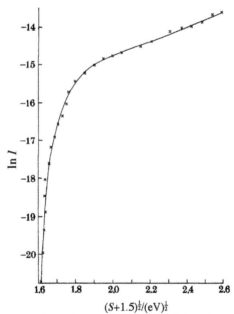

$(S+1.5)^{\frac{1}{2}}/(eV)^{\frac{1}{2}}$

Fig 4.6   The current-voltage characteristic of a typical emission site, showing the transition from contact-to bulk-limited behaviour. NB: The voltage here is the potential drop across the insulator, which may be taken as the shifts of an electron energy spectral peak from the Fermi level of the metal substrate, i.e. as described in the next section. (From Bayliss and Latham [4], with permission.)

emitted into the vacuum are supplied to the I-V interface, and that a dynamic equilibrium exists where the number of the electrons coming into this interface is just equal to those emitted under a given surface field condition.

We shall now consider (a) what controls this supply of electrons, (b) the influence of the electron injection process at the MI interface on the emission current density and, (c) the influence of the bulk conductivity of the insulator. The key experimental evidence revealing the existence of such influences is the contact-limited to bulk-limited transition in the measured I-V characteristic reported by Bayliss and Latham [4], which, for reference is reproduced in Fig 4.6. This demonstrates how, initially, the increase in emission is very rapid with increasing voltage-drop across the insulator, and then gradually becomes much slower. The initial rapid increase corresponds to the electron-injection by field emission at the M-I interface, and so this I-V characteristic is said to be "contact-limited". The latter slower increase of the current is attributed to the domination of the resistive properties of the insulating region; under these circumstances, the I-V charactertic is said to be "bulk limited". We shall now discuss how the transition occurs.

We shall firstly consider the contact-limited I-V characteristic. The current density j_T tunnelling through the metal-insulator contact junction may be expressed in the form derived by Simmons [42-44], namely

$$j_T = \frac{5.56 \times 10^{-14}}{\varepsilon^* \phi_o} N_D \left(V_c + \phi_m - \phi_I\right) \, exp\left[-3.6 \times 10^{13} \left(\frac{\phi_o^3 \varepsilon^*}{N_D \left(V_c + \phi_m - \phi_I\right)}\right)^{\frac{1}{2}}\right]$$

$$4.40$$

where ϕ_m and ϕ_I are respectively the metal and insulator work functions, ϕ_o the contact barrier height, ε^* the high-frequency dielectric constant of the insulator, N_D the donor density and V_c the potential appearing across the contact junction owing to external sources (in this case the applied field). Equation 4.40 predicts a sharply rising current dependence with applied voltage, and hence a rapidly falling contact resistance that is essentially independent of temperature.

When describing the conduction properties of the bulk insulator, consideration must not only be given to donor centres, but also to the effects of electron-trapping states which result from lattice imperfections, etc. This situation has also been thoroughly analysed in the literature; in particular, Simmons [42] has derived the following wide-ranging j-V relation for low-field bulk conduction:

$$j_o = e \, \mu \frac{V_b}{\Delta d} N_c \left(\frac{N_D}{N_T}\right)^{\frac{1}{2}} exp\left[-\left(\frac{\varepsilon_D + \varepsilon_T}{2kT}\right)\right]$$

$$4.41$$

139

where N_C is the effective density of states in the insulator, N_D the donor density, N_T the trap density, ε_D and ε_T the donor and trap energy levels respectively, μ the electron mobility and V_b the potential appearing across the bulk with its length Δd owing to external sources (namely the applied field). This equation applies if $\varepsilon_T > \varepsilon_L > \varepsilon_D$ (ε_L being the position of the Fermi level), where all these energy levels are just beneath the insulator conduction band and nearly all the electron traps are filled. However, for the high-field case, where the Poole-Frenkel effect operates (i.e. field-assisted thermal ionisation of donors and traps), the bulk conductivity becomes field dependent.

Under these circumstances, Simmons showed that the j-V relation may be expressed as:

$$j = j_o \, exp \left[\frac{e}{kT} \left(\frac{eV_b}{\pi \varepsilon \varepsilon_o \, \Delta d} \right)^{\frac{1}{2}} \right]$$

4.42

or

$$j = j_o \, exp \left[\frac{e}{2kT} \left(\frac{eV_b}{\pi \varepsilon \varepsilon_o \, \Delta d} \right)^{\frac{1}{2}} \right]$$

4.43

depending upon the nature and position of the traps [42].

At this point, it should be noted that a contaminated insulator (i.e. one containing donor centres) is required not only to ensure a thin enough depletion region for electron tunnelling at the MI interface, but also to give the bulk insulator a finite conductivity.

Now if the insulator contains a high donor density ($N_D > 10^{24}$ m^{-3}) and a high trap density ($N_T > 10^{25}$ m^{-3}), the depletion region will be very thin and its bulk conductivity will be low in spite of the high donor density (see equation 4.41). Under these conditions, a "contact-limited" to "bulk-limited" transition may be observed in the current-voltage characteristic [42-44]. At low voltages, electrons tunnel into the insulator conduction band and the current rises very sharply with voltage. At this stage most of the voltage appears across the contact, because the contact resistance is much higher than the bulk, and the I-V characteristic will be virtually independent of thickness. Contact-limited conduction will not continue indefinitely however, because the bulk resistance falls much less rapidly with increasing voltage than the contact resistance. Thus, at some voltage V_T, the contact and bulk resistances will be equal and the applied voltage will be shared equally across both regions. A subsequent increase in the voltage will appear mainly across the bulk and, as a result, the current will cease to rise as sharply as for the initial phase; i.e. since it will be controlled by bulk processes that are thickness-dependent. The dominant term in the equations of the bulk-limited case (i.e. equation 4.42 and

4.43) will then arise from the Poole-Frenkel effect, and it is usual to illustrate this by plotting ln I versus $V^{1/2}$, which enables the contact- and bulk-limited regions to be clearly separated. It should be noted that the slope of the bulk-limited region is then inversely proportional to $\Delta d^{1/2}$, where Δd is the insulator thickness. In the above discussions, we have related the emission current to the macroscopic gap field E (see for example equation 4.36), and thus derived a formula for a ln (I_e/E^2) versus $(1/E)$ plot, which is equivalent to the traditional F-N plot (see equation 4.38). The analysis has also been based upon the assumption that, to a first order approximation, the averaged field in the surface layer of the insulator E_s is equal to $(\beta/\varepsilon_r)E$, and this relationship remains unchanged in the range of the fields over which the I-V data are obtained.

4.3.4   Electron Energy Distribution

One of the most decisive pieces of experimental evidence that supports the MIV model is the observation, described briefly in Section 3.4.3, that the electron energy spectra of this type of emission process have non-metallic characteristics.

This finding led to the key conclusion that the associated emission regime must consist of an insulating or semiconducting medium; indeed, it was this observation that prompted the initial development of the MIV model. We shall now consider further how the concepts embodied in the models described in the previous three sections can be used to develop a new theory for explaining the field electron emission spectral characteristics of the MIV emission regime. The impetus for these theoretical developments derived from the comprehensive studies of Bayliss and Latham [4] and more recently Xu [41]. In this context, the following key findings are of particular significance.

The electrons emitted from one of the sub-emission centres of a site, i.e. corresponding to one of the white spots in Fig 4.7(b), have a single-peak spectrum as shown in the spectral sequence of Fig 4.7(c).

Both the spectral shifts S, measured from the Fermi level, and the full-width at half maximum FWHM, f, are strongly field-dependent, as shown by the spectral sequence in Fig 4.7(c) and the plots of Fig 4.8.

The current density, corresponding to the area under a spectral curve, increases with increasing field, as shown by the spectral sequence in Fig 4.7(c).

Thus, the new theory can be anticipated to develop an analytical expression for the single-peaked spectral curve that is capable of describing the field-dependence of the other spectral parameters, such as the FWHM and S. In order to develop a new theory, Bayliss and Latham [4] proposed the detailed energy band diagram, shown in Fig 4.9 to describe the "on" state of a conduction channel.

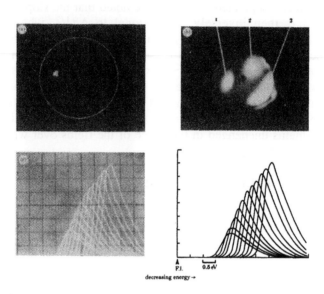

Fig 4.7   (a) A high-resolution map of the emission site distribution on a 15 mm diameter Cu cathode. (b) The emission image of the central site shown in (a). (c) A series of electron spectra taken at 100 V increments of gap voltage from the sub-site 1 in (b). (d) Traces of the electron spectra in (c) FL is the Fermi level of the metal substrate. (From Bayliss and Latham [4], with permission.)

As emphasised in Sections 4.3.2 and 4.3.3, it is assumed that there is internal heating of electrons in the surface layer of the insulator in the vicinity of the vacuum. Therefore, the derivation of the theoretical energy distribution must take this into account. It was assumed in Section 4.3.3 that this emission process can be considered to be closely analogous in concept to the temperature-assisted field emission regime,   considered by Murphy and Good [23]. Unfortunately however, the analysis of these authors was in terms of the normal energy distribution P(W), whereas experimental spectra generally correspond to the total energy distribution, $P(\varepsilon_e)$.

However, the transmission coefficient D(W), derived by Murphy and Good [23] remains valid, and this may be used in conjunction with the total energy supply function $N(W, \varepsilon_e)\, dW\, d\varepsilon_e$, developed by Young [45] to give the required emitted electron energy distribution,

$$P\left(\varepsilon_e\right) d\varepsilon_e = \int_{W\,=\,-\infty}^{W\,=\,\varepsilon_e} D\left(W\right) N\!\left(W,\varepsilon_e\right) dW\, d\varepsilon_e. \qquad 4.44$$

Substituting for D(W) from Murphy and Good [23],

$$D(W) = \cfrac{1}{1 + exp\left\{- \pi \left(\cfrac{E_m}{a_1}\right)^{-\frac{1}{4}}\left[1 + (W/b)\left(\cfrac{E_m}{a_1}\right)^{-\frac{1}{2}}\right]\right\}} \qquad 4.45$$

and for N(W,ε_e) from Young [45],

$$N\left(W,\varepsilon_e\right) d\varepsilon_e \, dW = \frac{4\pi m}{h^3} \frac{d\varepsilon_e \, dW}{1 + exp\left[\frac{\varepsilon_e - \varepsilon_f}{kT_e}\right]} \qquad 4.46$$

and then integrating leads to

$$P(\varepsilon_e) \, d\varepsilon_e = \frac{4mb}{h^3}\left(\frac{E_m}{a_1}\right)^{\frac{3}{4}} \frac{ln\left\{1 + exp\left[\pi\left(\frac{E_m}{a_1}\right)^{-\frac{3}{4}}\left(1 + \frac{\varepsilon_e}{b}\left(\frac{E_m}{a_1}\right)^{-\frac{1}{2}}\right)\right]\right\}}{1 + exp\left[\frac{(\varepsilon_e - \varepsilon_f)}{kT_e}\right]} d\varepsilon_e \qquad 4.47$$

In the above equations, E_m corresponds to the microscopically enhanced surface field at the top of the emitting channel, ε_f is the Fermi energy, T_e is the temperature of the hot-electron population behind the surface barrier, and a_1 and b are constants given by

$$a_1 = \frac{m^2 e^2}{\left(\frac{h}{2\pi}\right)^4 (4\pi\varepsilon_0)^3} = 5.15 \times 10^{11} \text{ Vm}^{-1}$$

and

$$b = \frac{me^4}{\left(\frac{h}{2\pi}\right)^2 (4\pi\varepsilon_0)^2} = 27.2 \text{ eV}$$

Also, the Fermi energy will be equal to $-\phi_I$, i.e. the barrier height at the insulator-vacuum interface.

Equation 4.47 thus provides an analytical expression for the total energy distribution of the field-induced hot-electrons emitted from the MIV regime. Thus, if the microscopic field E_m, the electron temperature T_e and the Fermi level ε_f of the emitting channel are known, it should be possible to compute a spectral curve according to equation 4.47. On the other hand, a simulated curve may be generated by experimenting with suitable values of the parameters E_m, T_e and ε_f, until a fit to a experimentally recorded spectrum is obtained, and then using these values to investigate the general spectral characteristics.

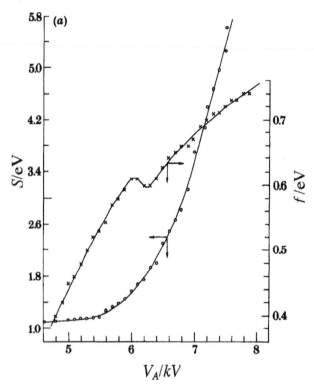

Fig 4.8   The field dependence of the spectral shift S and FWHM of sub-site 1 shown in Fig 4.7(b), as measured from the spectral series of Fig 4.7(c). (From Bayliss and Latham [4], with permission.)

A further theoretical development of this model, as detailed by Bayliss and Latham [4], leads to the following important spectral parameters.

(i)   The shape of a spectrum, which will vary according to the values of T_e and E_m under which it was recorded, can be described by the spectral shape factor

$$N = \frac{\pi k T_e}{b} \left(\frac{E_m}{a_1}\right)^{-\frac{3}{4}}$$ 4.48

(ii)   The FWHM of a spectrum is given by

$$f = \left(\frac{2.328N - 1.303}{N - 1}\right) k T_e$$ 4.49

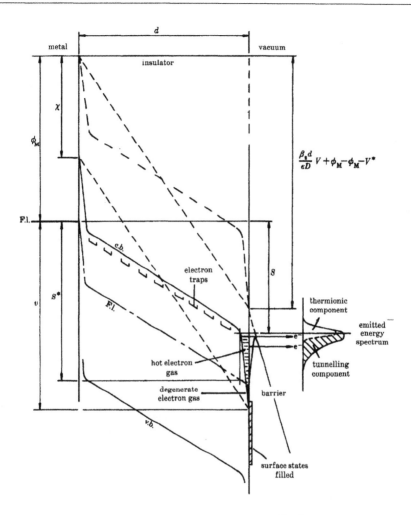

Fig 4.9   A detailed band diagram representation of the "on" state of a conduction channel. FL - Fermi level; vb - valence band; cb - conduction band. (From Bayliss and Latham [4], with permission.)

(iii)   The total spectral shift, defined as the potential difference between the Fermi level of the metallic substrate and a spectral peak, is measured experimentally, and consists of two components: (a) that due to the potential drop across the insulating bulk region and (b) that due to the Schottky reduction term at the surface potential barrier. As illustrated in Fig 4.10, this second component can be expressed as

$$E_p = AkT_e - b\left(\frac{E_m}{a_1}\right)^{\frac{1}{2}}$$ 4.50

145

where A is a variable that can be related to N through the following relation:

$$N = (1 + e^{-NA})\ ln\ (1 + e^{NA}) \qquad\qquad 4.51$$

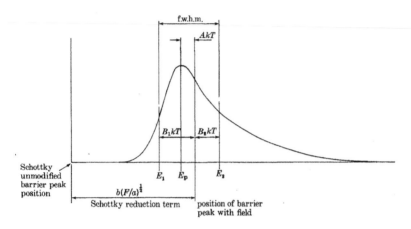

Fig 4.10   A schematic representation of a typical electron energy spectrum, showing the physical meaning of the Schottky reduction term and the other important spectral parameters. (From Bayliss and Latham [4], with permission.)

In fact, it has been noted by Xu [41] that this second term on the right hand side of equation 4.50 may be omitted because it is much smaller than the first term. Finally, with reference to the energy band diagram shown in Fig 4.9, the hot-electron temperature T_e can be related to the channel parameters, the experimentally-determined spectral shift S and the macroscopic gap voltage V_o by the relation:

$$kT_e/e = \alpha[(\Delta d\beta_2/d\varepsilon)\ V_o - S + \chi - \phi_I - V^*] + kT_o/e \qquad\qquad 4.52$$

in which Δd is the thickness of the insulator, β_2 is the enhancement factor of the average field in a conduction channel relative to the field in the region of the insulator outside the channel, d the gap spacing, and χ the electron affinity of the insulator. It is also significant to note that this expression contains three unknown constants, namely α, $(\Delta d\beta_2/\varepsilon)$ and $(\chi - \phi_I - V^*)$.

As already explained in Section 4.3.3(i), the electron temperature T_e can be related directly to the macroscopic field as suggested by Xu [41]. In this way, one of the unknown constants α can then be expressed analytically. However, such a discussion is beyond the scope of the present review, so that the interested reader is referred to the relevant publication [41] for more details.

Having introduced all the important theoretical expressions relating to the field electron emission spectroscopy of the MIV regime, it is now possible to use them to interpret the spectroscopic characteristics summarised at the beginning of this section. Thus, firstly from equation 4.52, one can see that T_e will increase with V_0, despite the corresponding increase of the spectral shift: the net effect of increasing the gap field, or voltage, is to increase the electron temperature T_e. Secondly, since the increase of T_e is dependent on E_m, the spectral factor N can remain essentially unaltered in the range of the applied fields, as one may see from equation 4.48. Thirdly, with N being essentially constant, and T_e increasing with E_m, the FWHM, i.e. f, will increase with increasing E_m, as predicted by the equation 4.49. Finally, since both f and the spectral peak height $P_m(\varepsilon_e)$ increase with increasing E_m, as may be seen from equation 4.47, their product $f \times P_m(\varepsilon_e)$, which is proportional to the area under a spectral curve and corresponds to the current density, will also increase with increasing field. The above discussion therefore demonstrates qualitatively that this new theory can account for the observed experimentally spectral characteristics.

4.3.5 Effects of Thermal and Photon Stimulation

As discussed in Sections 4.3.1 and 4.3.3, the MIV model predicts that the conduction channel of an emission site must contain a high density of electron traps and donors. These theoretical findings have, therefore, prompted further detailed studies to probe the electronic nature of emission sites, and hence the underlying emission mechanisms. In pursuance of this goal, Xu and Latham [46] conducted two important experimental investigations in the early 1980s involving both the thermal and photon-stimulation of the field electron emission process. As already described in Section 3.4.3, these involved using a field emission electron energy spectrometer facility to monitor the changes brought about by such stimulations, and it was demonstrated that these external influences have strong effects on the emission processes [46,47]. To advance the understanding of the detailed physical processes involved, Xu and Latham [48] have recently further developed the early MIV model, and given a theoretical analysis of these findings. This theoretical development will now be presented in the following sections.

4.3.5 (i)   Thermal Stimulation
Before embarking on a theoretical discussion, it is firstly necessary to examine the experimental findings in more detail. Thus, Fig 4.11(b) highlights the changes of the spectroscopic characteristics, shown in Fig 4.11(a), whereby the spectral shift and half-width (FWHM), as directly measured from the spectral sequence referred to above, vary with the sample temperature T_0. From this, it is apparent that the increase in half-width with T_0 is accompanied

by a reduction of the shift of the spectral peak from the Fermi level of the metallic substrate. In addition, as shown in Fig 4.11(c), the area under a spectral curve, which corresponds to the emission current density, increases with increasing T_O.

To explain the change in the spectral shift with T_O, one may assume that, with thermal stimulation, the electrons are still transported through the bulk of the insulator near to the bottom of the conduction band. Thus, the decrease of the spectral shift with increasing T_O may be attributed to the increase in the conductivity of the bulk insulating region, as can be seen from the following expression derived by Simmons [42]

$$\sigma = e\mu(N_CN_D)^{1/2}exp[-(\varepsilon_D + \varepsilon_T)/2kT_o] \qquad 4.53$$

Here, N_C is the effective density of states in the insulator, which can be assumed to be equal to the density of traps N_T, N_D the volume density of impurities per cm^3 (assumed to be fully ionised), ε_T and ε_D the respective average energy levels of the donors and traps, k Boltzmann's constant, and μ the electron mobility which is a function of temperature.

This increase in the conductivity enables a greater current to flow through the bulk of the insulator if the potential drop across it is kept constant. However, the tunnelling process at the MI interface is independent of T_O, as can be seen from equation 4.40 of Section 4.3.3. Thus, an increase in the tunnelling current density will not happen because of an increase in T_O. At the IV interface, an increase in sample temperature has two effects: (a) a lowering of the surface barrier can be expected, and (b) the electronic temperature T_e will rise according to equation 4.52 of Section 4.3.4. These latter two effects push the regime to a higher emissive state. Thus, a re-distribution of the potential drops occurs throughout the whole MIV emission regime. As a result, one would expect a decrease in the potential drop across the bulk of the insulator to enable a greater potential drop at the MI interface for higher current injection from the substrate. The cause of the change in the spectral FWHM may be understood by examining the following expression [4], which has been rewritten from equation 4.49:

$$f = \left(\frac{2.328N(T_o) - 1.303}{N(T_o) - 1}\right) kT_e(T_o) \qquad 4.54.$$

From this, it can be seen that an increase in FWHM may be due to both the spectral shape factor N, and the electron temperature T_e. By carefully examining the spectra shown in Fig 4.11(a), it is found that the spectral shape changes with T_O [46]. Indeed, it has been shown by Xu and Latham [48] that the value of N can be determined for individual spectra.

Finally, according to equation 4.54, the value of T_e can also be determined for each spectrum by measuring f and N. This therefore enables one to construct a plot of T_e versus T, shown in Fig 4.12, from the experimental sequence of spectra shown in Fig 4.11(a). It is significant to note that T_e does not gain the total increment of T_0; an observation that Xu and Latham [48] attribute to the increasing energy loss due to electron-phonon scattering with increasing sample temperature T_0.

(a)

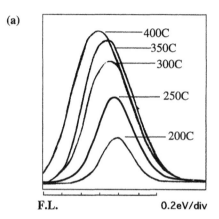

(b) **(c)**

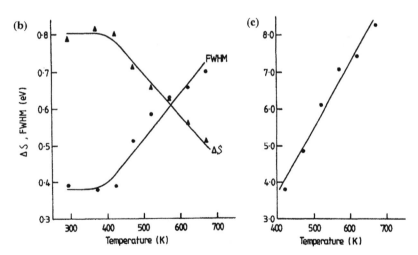

Fig 4.11   (a) A series of spectra recorded at 50°C increments of sample temperature ranging from 150°C to 400°C while the gap field remained constant at 12.4 MV/m. (b) The temperature-dependence of the spectral shift ΔS, and the FWHM, at a constant field of 12.4 MV/m. (c) The temperature-dependence of an emission site current at a constant field of 12.4 MV/m. (From Xu [41], with permission.)

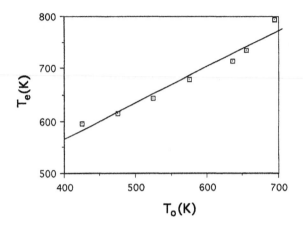

Fig 4.12   Showing how the electron temperature T_e increases at a slower rate than the sample temperature T_o.

4.3.5 (ii)   *Photon Stimulation*

Compared with thermal stimulation, the effect of UV photon-stimulation is more surface specific; i.e. limited by the attenuation of the UV photon beam in the insulating medium. The most important effects of UV photon-stimulation upon the field-induced hot-electron emission processes occurring at a MIV regime is shown in Fig 3.25 of Section 3.4.3. However, studies have also been

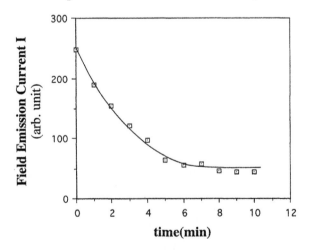

Fig 4.13   A plot showing how the field electron emission current decays with time.

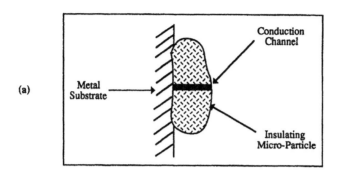

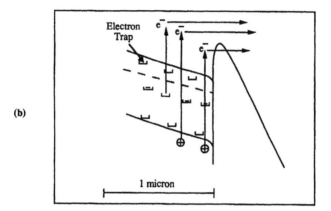

Fig 4.14   Schematic illustrations of (a) a non-metallic electron emission regime, (b) an energy diagram showing possible photo/field emission processes.

made of how the effect of UV photon-stimulation upon the emission process relaxes over a period of time [47]. In order to characterise this latter effect, the height of the spectral peak may be plotted against time after the UV radiation is switched off; i.e. as shown in Fig 4.13. From this, it can be seen that the time for the photon effect to disappear is about 4.5 minutes.

This finding can therefore be interpreted as reinforcing the conclusion that the prebreakdown emission comes from sites that are non-metallic in their electronic nature; i.e. since the decay process demonstrated in Fig. 4.13 could not happen with a metallic emitter.

The physical interpretation of the UV photon-stimulated field electron emission process from a MIV regime is complicated, but can be described with reference to Fig 4.14. Thus without any UV photon-stimulation, the field-induced hot electrons are assumed to be injected into the vacuum from the conduction band. With the excitation of UV photons of 21.2 eV energy,

151

electrons from the conduction and valence bands, as well as from inner gap states, can be lifted up energetically by receiving energy from the photons. Some of these can then go over the surface barrier and be emitted into vacuum, whilst others that have lower energies may be excited to higher energy levels where tunnelling into the vacuum can occur. These processes thus provide a qualitative explanation of the sustained enhancement of the emission by the UV photon beam.

More important, the processes described above result in firstly the generation of holes, through band to band transitions, and secondly, emptied electron traps and acceptor states through other photon impact ionisation processes. As a consequence, a photoconduction process will take place within a surface layer of the conduction channel, the thickness of which will be determined by the penetration depth of the UV photon beam being used. This induced photoconductivity will therefore increase the supply of electrons to the surface where emission is occurring.

Similar to the photon conduction processes observed in semiconductors, the decay of the induced conductivity takes some time to relax after switching off an illumination source [49]. This time constant is mainly determined by the time the electrons spend in electron traps in impure semiconductors [49]. Based on the existing theory of the transition processes involving trapping states [50], one may use the time of decay measured in this experiment to calculate the depth of these trapping levels. Thus, the time that an electron spends in a trap depends on the depth of the trap below the conduction band $(\varepsilon_C - \varepsilon_D)$, where ε_C is the bottom of the conduction band and ε_D the energy level of the electron trapping state, and also on the temperature T. It is generally found that the probability of escape per unit time can be written in the form,

$$P = 1/\tau = Q \ exp - [(\varepsilon_C - \varepsilon_D)/kT] \qquad 4.55$$

where Q is a constant approximately equal to 10^8 s^{-1} [50]. Thus, if the time of decay τ is taken as the average time an electron spends in a trap, and T is assumed to be the ambient temperature, one obtains a value for $(\varepsilon_C - \varepsilon_D)$ of approximately 0.6 eV; i.e. the trap energy level is 0.6 eV below the bottom of the conduction band. Indeed, these are the traps that were involved in the relaxation process observed in this study. This finding therefore provides important information about the electron traps existing in an emission site.

4.3.6 Dynamic Field-Enhanced Electron Emission Model

Having described the most fully developed MIV model, it is also necessary to refer to other possible mechanisms that could be responsible for the field emission of electrons from similar MIV regimes. These include: (i) the filamentary model proposed by Hurley [51], to explain his experimental findings previously described in Section 3.5, (ii) the M-I cathode model of Eckertova [52], and (iii) the dynamic field-enhanced electron emission model of Halbritter [7]. Of these, (iii) is the one that most closely relates to the MIV model of Latham and co-workers described in the previous section, and so will be briefly described in the following paragraphs.

The dynamic field-enhanced electron emission model assumes that there are low-density adsorbates, such as hydrocarbons, on the surface of broad-area electrodes [7]. It is further assumed that electrons will be trapped in such dielectric adsorbates at energy levels $\varepsilon_T < \phi_I$, where ϕ_I is the work function of the dielectric adsorbate. This is equivalent to saying that the energy level of the traps lie below the bottom of the conduction band of the dielectric medium; a similar assumption to that used in the MIV model discussed above. These trapped electrons will be emitted stochastically out of the traps into the conduction band at electric fields E_i in the range of 10^6-10^7 V/m [53], and then be heated to an energy of e Δx $E_i = \Delta\varepsilon >> kT$ in the insulating adsorbate. Such hot electrons can form stationary conducting clouds with $\varepsilon_e \approx 1eV$ in a field range $E_{crit} < E_i < E_{break}$, where according to experimental findings from SiO_2, the possible values of E_{crit} and E_{break} will be ~ 10^8 V/m and ~ 10^9 V/m respectively.

This population of hot electrons can subsequently cause impact ionisation of defects, yielding high conductivity states which can be switched on for $E_i > E_{switch}$ without causing the destruction of the material. The value of E_{switch} is dependent on the ionisation energy ε_{ion} of the defects. Thus, to ionise contaminants with $\varepsilon_{ion} \approx 1$ eV needs a field of $E_{switch} > 10^6$ V/m. It follows therefore that, for $E > 10^6$ V/m, where permanent positive charges enhance the field in the insulating absorbate layer by $\beta^* E = E^*$ (with $\beta^* >> 1$), electrons in the adsorbate will be accelerated to $\varepsilon_e \approx 1$-10 eV and electron attachment states are formed. These excited localised states then decay by the emission of electrons, photons and atoms, with the electron emission giving rise to positive charging, which enhances β^* to β^{**} in the region near to the surface of metallic substrate electrode to cause electron tunnelling from the substrate into the adsorbate. However, only the fastest electrons penetrate through the vacuum surface barrier because of the positive charges in the adsorbate: the other electrons are slowed down and neutralise the positive charges. Thus, the emission is pulsed, where the time constant τ for the firing of the avalanches depends on the details of the switch-on process, the channel dimensions, and the electric field difference $\Delta E = E_i - E_{switch}$. For the typical conditions thought to exist on the surface of HV electrodes, Halbritter [7] estimated that τ would be of the order of nanoseconds.

Although this model can apparently account for some of the experimental observations, it is at present lacking in positive supporting evidence. Thus, the experiments carried out by Niedermann [11] have shown that only in a very few instances is there a suspicion that hydrocarbon particles are responsible for emission processes. Furthermore, its most important conclusion, viz. that there will be a pulsation of the enhanced emission, has not been experimentally observed.

In summarising our review of the mechanisms responsible for parasitic electron emission, it can be said that the semi-quantitative MIV model developed by Latham and co-workers [1-6] provides an explanation for most of the observed phenomena associated with the high-β emission sites. In addition, it offers an alternative approach to the understanding of such well-known properties of high voltage vacuum gaps as the influence of the residual gas species, the physical significance of the "slope" and "intercept" of the traditional F-N plot of the current-voltage characteristic, the various conditioning procedures and, as will be discussed in the following chapter, the actual breakdown process itself.

4. 4   Metal-Insulator-Metal (MIM) Model

This model considers the field-induced electron emission processes occurring at the metal-insulator-metal micro-regime illustrated in Fig 4.1(b). As already mentioned in Section 4.1, this model was initially proposed to explain how a carbon graphite particle artificially deposited on a Cu electrode could promote low-field (< 10 MV/m) "cold" electron emission [9]. Furthermore, it had to explain why the electron energy spectra recorded from such sites have non-metallic characteristics similar to those from a MIV regime, i.e. indicating the involvement of an insulating medium in the emission regime. In an early version of the model, it was proposed that the flake-like structure of the graphite particle would form a metal-insulator-metal (MIM) microstructure with the substrate electrode and its ambient oxide layer [10]. Later, a systematic study of the field electron emission characteristics associated with such an artificial MIM regime was carried out by Xu and Latham [5,10], and confirmed the earlier findings. On the other hand, complementary studies [11] aimed at characterising the chemical composition of individual emission sites also revealed that MIM emission regimes can exist naturally on the surface of a broad-area electrode. More recently, new findings have emerged showing that such MIM microstructures are the predominant electron emission sources on heat-treated broad-area niobium electrodes [54]. As a result of these experimental studies, the original MIM model of Latham and co-workers [9] has been further developed by Xu and Latham [5,10,41]. With reference to Fig 4.15(a), the latest version of the MIM model firstly recognises that a MIM emission regime embodies all of the features of the MIV regime.

154

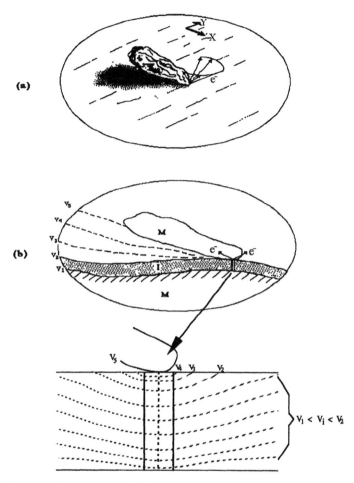

Fig 4.15   (a) Illustrating how coherently scattered electrons are emitted into vacuum from a particulate microstructure on a broad-area planar electrode. (b) A cross-sectional view of a particulate structure, showing the proposed MIM emission regime and the potential distribution (insert) associated with the conduction channel. (From Xu and Latham [5], with permission.)

In addition, it is assumed that the superficial metallic particle, or flake (e.g. of graphite), acts like an "antenna" by probing the field above the electrode surface, and thereby producing an enhanced field across its contact point with the insulator. Thus, in a switch-on process, a conducting channel is preferentially formed in this region. In the steady emission stage, electrons are assumed to be injected from the metallic substrate into the conduction channel, and subsequently accelerated in the channel by the internal field to become "hot electrons" i.e. by the same MIV mechanism described in Section 4.3. In fact, this physical process resembles the concept proposed by

155

Dearnaley et al. [55] to explain the behaviour of purposely-fabricated MIM devices. In relation to the present emission regime (see Fig 4.15) it would at first sight appear that the electrons will be scattered within the top metallic layer and thus lose their energy, i.e. be thermalised. However, as discussed by Xu and Latham [5], there is an additional consideration whereby, if the conditions are appropriate, the electrons may undergo a coherent scattering process, so that they can be emitted into vacuum without losing the kinetic energy gained from the field in passing through the insulating layer. Such emitted electrons will then form an arc-like segmental image as shown in Fig 3.26, and illustrated in Fig 4.16. From this brief description of the MIM model, it becomes clear that the only additional features of the MIM model compared to the previously discussed MIV model are the "antenna effect" and the electron scattering processes within the top metallic layer. Accordingly, in the following sections we shall focus our discussion on the current understanding of these two features. Firstly, we shall consider how the top metallic layer enhances the local field at the metal-insulator contact. Secondly, we shall present further experimental evidence on the coherent scattering of field-induced hot electrons. Finally, we shall discuss the conditions under which these coherently scattered hot electrons can be emitted into vacuum.

4.4.1   Antenna Effect

This effect was initially proposed by Latham and co-workers [9,10], and involves the existence of a protruding metallic flake that is attached at one end to the surface of a planar electrode surface. With reference to Fig 4.15(b), it can be assumed from elementary electrostatic principles that the top metallic flake, which is electrically "floating" due to the blocking contact provided by the insulator, effectively probes the gap field and assumes a potential (V_5) which is approximately equal to the equipotential line at the highest point of the top metallic layer. Through the antenna effect, this potential is "transmitted" to the contact region and gives rise to a local field enhancement across the insulating layer which can be approximated by

$$\beta_{ant} = h/\Delta d \qquad\qquad 4.56$$

where h is the maximum height of the particle above the substrate, and Δd is the thickness of the insulating layer. According to this expression, β_{ant} will be equal to 11 for a flake where h = 1.1 μm and Δd = 100 nm. Thus, if a gap field E of 10 MV/m is applied to the cathode surface, the enhanced field at the MI contact will be $\beta_{ant}E$, i.e. equal to 1.1×10^8 V/m. This field will be high enough to (a) inject an electron current density of 1 A/m² from the metal substrate into the conduction band of the insulating layer, and (b) to subsequently accelerate these electrons so that they become sufficiently hot to cause an initial burst of emission to trigger a switch-on process, i.e. as discussed in Section 4.3.2. This reasoning demonstrates how important the antenna effect is in promoting low-field electron emission.

4.4.2 Coherent Scattering and Field Emission of Hot Electrons

The key evidence in support of the contention that the MIM emission regime promotes the vacuum emission of coherently scattered electrons is the similarities between the arc-like emission images obtained by Xu and Latham [5] from graphite structures and those originals observed by Simmons and Verleber [56], and later by Gould and Hogarth [57], from purpose fabricated metal-insulator-metal (MIM) devices.

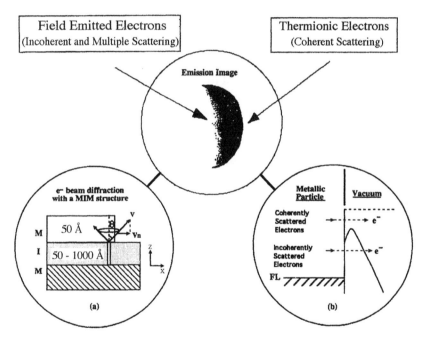

Fig 4.16   (a) Electron diffraction as it occurs with an ideal MIM structure.
(b) Illustrating how electrons are emitted into vacuum over/through the surface barrier, together with the spatial distribution of the field emitted and thermionically electrons within a given segment. (From Xu and Latham [5], with permission.)

To further investigate the physical process associated with the vacuum emission of coherently scattered electrons, the new instrumental technique discussed in Section 3.4.3 was developed. This technique combined the field emission imaging and energy-tuning capabilities of a high resolution electron

157

spectrometer to produce a spatial display of energy-selected electrons emitted from a given site, i.e. as illustrated in Fig 3.26. From this type of measurement, Xu and Latham [5] were able to conclude (a) that the lower energy electrons have a wider spatial distribution, and are predominantly responsible for forming the inner concave part of an arc-shaped image, (b) that, conversely, high energy electrons are preferentially located towards the outer convex edge, and (c) that the profile of the outer edge of the arc-like segment appears to remain unaffected by the energy selection procedure; i.e. indicating that it is imaged by both high and low energy electrons.

In order to explain the formation of this type of image, the coherent scattering model initially proposed by Simmons and Verderber [56,58], and subsequently explored by Gould and Hogarth [57], has been further developed by Xu and Latham [5] to fit the present regime.

4.4.2(i) Vacuum Emission of Coherently Scattered Electrons
The model assumes that an energetic beam of hot electrons, collimated by the conduction channel, impinges on a thin polycrystalline conducting particle at an angle that is approximately normal to its incident surface. As discussed in earlier works by other authors [56-58], the hot electrons referred to above will be coherently scattered in this top metallic particle in a similar way to the electron beam diffraction effects observed in a conventional transmission electron microscope. As a result, the electron beam is diffracted in all directions to form the cone-shaped emerging beam shown in Figs 4.15(a) and 4.16(a).

For this mechanism to operate with the present MIM regime, Xu and Latham [5] proposed that three basic conditions have to be met.

(a)   The electron path length in the top conducting particle must be shorter than its mean free path, i.e. less than 50 Å [59]: this restriction can, however, be relaxed if a channel terminates at a position near to a triple-junction region, such as that shown in Fig 4.16(a).

(b)   Electrons must be forward-scattered, otherwise they will be returned into the insulator. The onset of this condition imposes a requirement on the minimum energy of the hot electrons given by $\varepsilon_{min} = n^2/(2m\lambda^2) \approx n^2/(4mg^2)$. For graphite, where g = 3.34 Å, and assuming that n = 1, this corresponds to 6.7 eV above the ground potential of the graphite, i.e. is equivalent to an electron temperature of 1200°C in the insulator, which is within the range of hot electron temperatures measured in a comparable MIV regime [4].

(c)   Referring to Fig 4.16(b), the energy of an electron approaching the M-V interface must be sufficient to allow it either to escape over the surface barrier or to tunnel through it. The first group of electrons to satisfy this condition are those that are coherently scattered in the conducting particle, and have normal velocities of $v_n \geq \{(\phi + \eta)/2m\}^{1/2}$, where ϕ is the work function of the top conducting particle and η its Fermi energy. These electrons will exhibit a thermionic-type of emission. The conducting particle will therefore appear to be transparent to the coherently scattered hot electrons. Apart from the above group of energetic electrons, there are also a considerable number of other electrons which have suffered energy losses through incoherent or multiple scattering processes, but have not yet been thermalised down to room temperature. Due to the combined effects of the gap field (in a range of 10-30 MV/m), and the existence of surface charges, these electrons will encounter a narrower potential barrier compared to those electrons occupying states around the Fermi level. Accordingly, they will have the possibility of tunnelling through the surface barrier and being field emitted into the vacuum. The electron emission regime will therefore be similar to that proposed by Bayliss and Latham [4] and, as illustrated in Fig 4.16, consist of two components, namely thermionic and field-emitted electrons.

4.4.2(ii) Formation of a Single Segment Image

In order to explain the variation of the energy spectra within the single-segment image, i.e. as shown in Fig 3.26, we have to consider how the electrons behave while they are travelling in the vacuum between the edge of the conducting particle and the phosphor screen. Thus, referring to Fig 4.17(a), an electron escaping from the surface has a total velocity v which can be resolved into components v_z and v_x, respectively normal to and parallel to the plane of the substrate surface. In the absence of a field in the space between the cathode and the phosphor screen, the time for an electron to travel to the screen is determined by the distance between the top metallic particle and screen, and the velocity v_z. Under normal operating conditions, i.e. with the cathode earthed and the anode at a voltage in the range of 5-10 kV, electrons escaping into the vacuum are accelerated in a 0.5mm gap in $\sim 2.4 \times 10^{-11}$ s to a velocity $v_{oz} \sim 1.3 \times 10^7$ m/s, determined by the anode voltage. However, since v_{oz} is very much greater than v_z, it follows that any differences among the normal launch velocities of the electrons becomes insignificant.

For the ideal geometry depicted in Fig 4.17(a), the effect of the transverse velocity component v_x will be for the emitted electrons to form a disc in a Gaussian plane [60], which would then be projected on to the screen by the lens to give a magnified image. However, in the present situation, as illustrated in Fig 4.17(b), where the non-planar cathode topography gives rise to a transverse diverging field at the emitting surface, the angular aperture of

159

the emitted electrons will be further enhanced. In fact, as illustrated in Fig 4.17(b), it will be seen that this diverging field has a dispersing effect on the emitted electrons, whereby those with high lateral energies will be least converged and be imaged on the screen with the greatest lateral displacement. This reasoning therefore provides a qualitative explanation of the experimental evidence of Fig 3.26, which shows how the majority of high-energy "thermionically emitted" electrons are distributed in the sharp convex edge of the segment. Conversely, the low-energy field emitted electrons give rise to the diffuse concave region of the image.

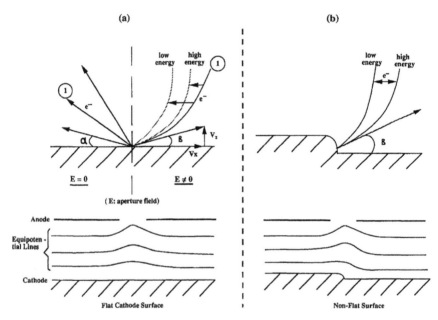

Fig 4.17   Illustration of how the trajectories of electrons of different energies are influenced by an aperture field on (a) a flat cathode surface and (b) a non-flat surface. (From Xu and Latham [5], with permission.)

4.4.2(iii) Another Possible Model

To complete this discussion, reference must also be made to the model of Biederman [61], which offers the basis of an alternative explanation for some of the observed effects. Thus, referring to Fig 4.18, the following sequence of events was proposed. (i) A pin-hole is created in a MIM structure by electron beam heating during the forming processes, by the consequent removal of the material previously forming the top metal film and dielectric layer. (ii) Subsequently, a reversible I-V characteristic is obtained, with electrons travelling along the wall of the channel from the substrate to the top metal film, and then emitted into vacuum from a part of or the whole wall, i.e. as illustrated in Fig 4.18. (iii) With the screen at a positive potential, and the polarity of the applied voltages arranged as shown in Fig 4.18, the orifice of the channel can behave like an electrostatic micro-lens, producing a defocusing effect on the electrons. This set-up thus projects the electrons emitted from the wall on to the screen, whereby the image would be expected to reflect the circular shape of the pin-hole; i.e., they will form arc-like images.

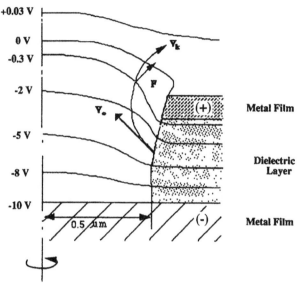

Vo : initial electron velocity

Vk: final velocity

F: electric force acting on electrons

Fig 4.18   A meridian section view of an emitting channel formed in a MIM device, and illustrating how the resulting equipotential lines give rise to electron trajectories that constitute a micro-lens effect [61]. (From Xu and Latham [5], with permission).

Since the electrostatic lens assembly, of Xu and Latham [5] belongs to the projection type, the defocusing effects of the electrostatic micro-lens associated with the pin-hole-channel structure would have been expected to appear in the findings of these latter workers. Equally, the model of Biederman [61] cannot account for the spatial energy distribution shown in Fig 3.26. Accordingly, it has been concluded that the available evidence does not strongly support this micro-lens model.

From the above review of the two models that have been proposed to account for vacuum electron emission from MIM microstructures, it can be seen that the coherently scattering model gives a better account for the experimental findings. However, before concluding this section, a number of remaining problems related to this model should be pointed out. Referring firstly to Fig 4.16(a), it should be noted that only a portion of the electrons forming the cone-shape beam will be emitted into vacuum, with the remainder either returning back to the insulator or decaying in the top metallic flake. It follows that these unemitted electrons will form space charge in both the insulator and top metallic flake, and affect the potential distribution in the MIM micro-regime. This effect will need to be addressed in future studies in order to see how it influences the emission process. Secondly, it is necessary to carry out a theoretical study of the I-V characteristic of the MIM micro-regime, taking into account the following factors: (a) the antenna effect, (b) the space charge effect discussed above, and (c) the field enhancing effect due to the topographical nature of the emitting surface of the top metallic flake. Finally, it is also necessary to conduct further experimental studies into the effect of the material parameters of the top metallic flake upon the electron image, i.e. by deliberately depositing on an electrode particles composed of a range of different materials, including Al, Au and Nb etc. The findings of this type of study would clarify if the emission of the coherently scattered electrons is the only mechanism responsible for prebreakdown electron emission from MIM micro-regimes.

4. 5   References

1.   Allen, NK. and Latham, RV., *J Phys D: Appl Phys,* **11**, L55-57, 1978.
2.   Athwal, CS. and Latham, RV., *Physica,* **104C**, 46-9, 1981.
3.   Latham, RV., *Vacuum,* **32**, 137-140, 1982.
4.   Bayliss, KH. and Latham, RV., *Proc. Roy. Soc.,* **A403**, 285-311, 1986.
5.   Xu, NS. and Latham, RV., *Surface Science,* **274**, 147-160, 1992.
6.   Bajic, S. and Latham, RV., *J . Phys. D: Appl. Phys.,* **21**, 200-204, 1988.
7.   Halbritter, J., *Appl. Phys. A ,* **39**, 49-57, 1986.
8.   Latham, RV. and Xu, NS., *Vacuum,* **42**, 1173-1181, 1991.
9.   Athwal, CS., Bayliss, KH., Calder, R. and Latham, R.V., *IEEE Trans. Plasma Sci.,* **13**, 226-229, 1985.
10. Xu, NS. and Latham, RV., *J. Phys. D: Appl. Phys.,* **19**, 447-482, 1986.
11. Niedermann, Ph., "Experiments on Enhanced Field Emission", PhD Thesis, University of Geneva, 1986.
12. Mesyats, GA. and Proskurovsky, DI., "Pulsed Electrical Discharge in Vacuum", Springer, Berlin, 1989.
13. Jüttner, B., in chapter 15 of this book.

14. Millikan, RA. and Lauritsen, CC., *Proc. Nat. Acad. Sci. (US)*, **14**, 45-49, 1928.
15. Latham, RV., "HV Vacuum Insulation: The Physical Basis", Academic Press, London/New York, 1981.
16. Fowler, RH. and Nordheim, L., *Proc. Roy. Soc.*, **A119**, 173-81, 1928.
17. Gomer, R., "Field Emission and Field Ionisation", Oxford University Press, Oxford, 1961.
18. Good, RH. and Muller, EW., *In* "Handbuck der Physik", Springer-Verlag, Berlin, 21, 176-231, 1956.
19. Van Oostram, AGJ., *Philips Research Reports Supplements*, No.1, 1966.
20. Miller, HC., *J.Franklin Inst.*, **282**, 382-8, 1966.
21. Miller, HC., *J.Franklin Inst.*, **289**, 347-51, 1969.
22. Beukema, GP., *Physica*, **81**, 259-74, 1972.
23. Murphy, EL. and Good, RH., *Phys. Rev.*, **102**, 1464-73, 1956.
24. Dyke, WP. and Dolan, WW., *Adv. Electronics Electron Phys.*, **8**, 90-180, 1956.
25. Christov, SG., *Phys. Stat. Sol.*, **17**, 11-26, 1966.
26. Haeffer, R., *Z. Physik*, **118**, 604-22, 1940.
27. Dyke, WP., Trolan, JK., Dolan, W.W. and Barnes, G., *J. Appl. Phys.*, **24**, 570-6, 1953.
28. Dyke, W.P., Trolan, JK., Martin, EE. and Barbour, JP., *Phys. Rev.*, **91**, 1043-54, 1953.
29. Chatterton, PA., *Proc. Phys. Soc.*, (London), **88**, 231-45, 1966.
30. Nagy, D. and Cutler, P.H., *Phys. Lett.*, **10**, 263-4, 1964.
31. Rohrbach, F., CERN Report, 71-5/TC-L, 1971.
32. Fursey, GN. and Vorontsov-Vel'yaminov, PN., *Sov. Phys: Tech. Phys.* **12**, 1370-6, 1968.
33. Vibrans, GE., Tech. Report 353, Lincoln Lab., MIT (NTIS ESDTDR 64-327), 1964.
34. Miller, HC., *J. Appl. Phys.*, **38**, 4501-4, 1967.
35. Tomaschke, HE. and Alpert, D., *J. Appl. Phys.*, **38**, 881-3, 1967.
36. Farrall, GA., *J. Appl. Phys.*, **41**, 563-71, 1970.
37. Simmons, JG., *J. Phys. D: Appl. Phys.*, **4**, 613-657, 1971.
38. Simmons, J.G. and Verderber, RR., *Proc. Roy. Soc.*, **A301**, 77-102, 1967.
39. Athwal, CS. and Latham, RV., *J. Phys. D: Appl. Phys.*, **17**, 102-1043, 1984.
40. Mott, NF., *Phil. Mag.*, **32**, 159-171, 1975.
41. Xu, NS., "Field-Induced Hot-Electron Emission from Composite Metal-Insulator-Metal Microstructures", Ph D Thesis, Aston University, UK, 1986.
42. Simmons, JG., *Phys. Rev. Lett.*, **22**, 657-662, 1967.
43. Simmons, JG., *Phys. Rev.*, **166**, 912-920, 1968.
44. Simmons, JG., "Tunneling Phenomena in Solids", Eds Burstein, E. and Lundqvist, S., (New York: Plenum Press), Chapter 10, 1969.

45. Young, RW., *Phys. Rev.,* **13**, 110-114, 1959.
46. Xu, NS. and Latham, RV., *J. Phys.(Paris),* **47**, C2, 73-78, 1986.
47. Xu, NS. and Latham, RV., to be published in *Surface Science.*
48. Xu, NS. and Latham, RV., to be published in *Proc. Roy. Soc. (London).*
49. Seeger, K., "Semiconductor Physics" p420, Springer-Verlag, Vienna/New York,1973.
50. Wilson, J. and Hawkes, JFB., "Optoelectronics: An Introduction", p115, Prentice Hall, London/New York, 1989.
51. Hurley, RE., *J. Phys. D: Appl. Phys.,* **12**, 2229-2252, 1979.
52. Eckertova, L., *Int. J. Electronics,* **69**, 65-78, 1990.
53. Halbritter, J., *IEEE Trans. on Electrical Insulation,* **18**, 253-257, 1983.
54. Mahner, E., to be published in the *Proc. of the 6th Workshop on RF Superconductivity,* (CEBAF, Newport News, USA, October, 1993).
55. Dearnaley, G., *Thin Solid Films,* **3**, 161-174, 1969.
56. Simmons, JG. and Verderber, RR., *Appl. Phys. Lett.,* **10**, 197-199, 1967.
57. Gould, RD. and Hogarth, CA., *J. Phys. D: Appl. Phys.,* **8**, L92-L95, 1975.
58. Simmons, JG. and Verderber, RR., Lytollis, J. and Lomax, R., *Phys. Rev. Lett.,* **17**, 675-677, 1966.
59. Savage, ED. and Anderson, DE., *J. Appl. Phys.,* **38**, 3245-3251, 1967.
60. Grivet, P., "Electron Optics", p62, Pergamon Press, London, 1965.
61. Biederman, H., *Phys. Stat. Sol. (a),* **36**, 783-789, 1986.

5

Electron Emission Based Breakdown Mechanisms

NS Xu & RV Latham

5.1 Introduction

Having devoted Chapter 4 to a fundamental discussion of the three important mechanisms that have been proposed to explain the physical origin of parasitic prebreakdown electron emission, it is now necessary to extend the analysis to a consideration of the factors affecting the stability of the associated emission regimes. In particular, we must enquire whether any of the associated physical processes are likely to lead to the creation of a microplasma, since this is the essential "seed" of a breakdown event. The prerequisite of such a phenomenon is the formation of an ionisable medium which, under good vacuum conditions, can only be derived from localised melting or vaporisation processes at the electrode surfaces. Accordingly, it is necessary to consider the localised energy dissipating mechanisms that can lead to thermally activated instabilities on either the cathode or anode. Conventionally, these mechanisms are therefore classified, according to their electrode of origin; thus there are two primary types of breakdown mechanism that need to be considered, namely *cathode-initiated breakdown* and the *anode-initiated breakdown*.

For the metallic microprotrusion MM model illustrated in Fig 4.1(c), it is widely accepted that an associated cathode-initiated breakdown results from the spontaneous vaporisation of the tip of a microprotrusion following its excessive resistive (Joulian) heating by the current I_{OF} flowing towards the emitting surface. The onset of this instability will therefore be associated with a critical microscopic field E acting at the tip (i.e. surface) of the emitter. On the other hand, recent experimental evidence [1-3], to be discussed in Section 5.2, has verified that, for both the metal-insulator-vacuum MIV and the metal-insulator-metal MIM regimes, a *cathode-initiated breakdown* can be attributed to the electrical breakdown of a solid insulating micro-medium.

Irrespective of the detailed nature of the electron emission mechanisms, an anode-initiated breakdown can occur if the local surface temperature of the electron bombarded region (i.e. the anode "hot-spot") approaches the melting point of the electrode material. This condition will therefore be associated with a critical power loading of the electron beam W_c, where the metal vapour pressure can become high enough for the onset of regenerative ionisation processes. (NB: There is also a third type, an indirect mechanism involving the in-flight vaporisation of a microparticle released from the anode "hot-spot", but this will be discussed in Section 7.2.1(iii) as a *microparticle initiating mechanism*).

In this chapter, we shall firstly present some recent experimental evidence that shows how both the MIV and MIM types of DC prebreakdown emission site (DC-PES) can be responsible for initiating a DC or pulsed-field breakdown event. Secondly, we shall review a number of steady-state theoretical analyses of the breakdown mechanisms that can occur with the types of micro-emission regime shown in Fig 4.1. Thirdly, we shall discuss the thermal response of both the cathode and anode under pulsed-field conditions. Fourthly, we shall outline the criteria that may be used in practical situations for predicting the onset of breakdown. Finally, it will be shown that a correlation exists between the prebreakdown electron emission characteristics of a vacuum gap and its subsequent breakdown behaviour.

5.2   Experimental Studies of Breakdown Processes

Experimental observations of the breakdown processes that are initiated at an isolated, negatively biased metallic microprotrusion, or microtip, of a point-plane HV gap have been extensively reported. Thus, excellent accounts of this subject can be found in the reviews of Lafferty [4], and more recently by Mesyats [5]. Interested readers are also referred to Chapters 6 and 15 of this book for a more detailed description of this phenomenon.

In this chapter, our aim is to provide a theoretical explanation of how the non-metallic microstructures shown in Figs 4.1(a) and 4.1(b), are responsible for initiating a DC or pulsed-field breakdown of a broad-area, plane-parallel, vacuum-insulated HV gap [1-3]. To provide a back-cloth for this discussion, it is first necessary to describe in detail the findings of a recent series of experimental investigations that compared the operational performance of an HV gap under DC and pulsed-field conditions.

5.2.1   DC Fields

This type of study was initially carried out by Latham et al. [1], using the transparent anode imaging technique, previously described in Section 3.2.1, in conjunction with an open-shutter photographic camera. With this "low resolution" system, they obtained strong evidence that a spatial correlation exists between a prebreakdown emission site and the subsequent breakdown arc. More recently, Xu and Latham [2] performed a similar type of study, but using the improved UHV experimental system illustrated in Fig 5.1. In this they were able to test the gap that was formed between a transparent anode and a 15 mm diameter Cu cathode, where the transparent anode was used to record the spatial distributions of both prebreakdown emission sites and the locations of breakdown arcs. A video camera with a framing time of 40 ms is used to record the optical images in a real-time sequence, whilst a high-speed oscilloscope (Techtronix 7904) is used to monitor the time-profile of the complementary collapsing gap voltage.

The experimental procedure involves making concurrent recordings of the optical and electrical responses of a test gap. Thus, with the video camera set to record all optical activity in the gap, the system is ready to record the electrical characteristics of the DC prebreakdown processes. This involves

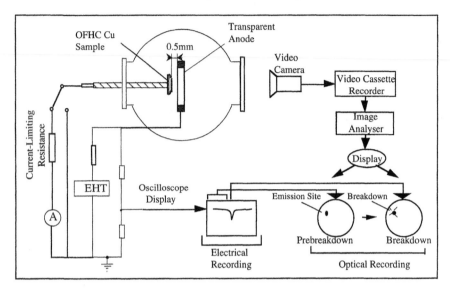

Fig 5.1   The experimental set-up for studying the spatial and electrical correlations that exist between the DC prebreakdown emission characteristic of an HV gap and its subsequent DC breakdown behaviour.

very slowly increasing the field applied to the virgin specimen until its "first generation" of emission sites are switched on to establish a reversible prebreakdown current of ~ 1-10 μA: at this stage an I-V characteristic is recorded in order to derive an effective β-value of the given population of sites. Next, the DC breakdown behaviour of the test gap is investigated by slowly increasing the applied field, until there is a sudden jump in the gap current, accompanied by a collapse of the gap voltage, as recorded on the oscilloscope. The central finding of this study is schematically illustrated in Fig 5.2. This presents in (a) the two sequential video frames associated with a breakdown event and in (b) the associated voltage waveform appearing across the gap. More specifically Frame 1 in Fig 5.2(a) displays the spatial distribution of DC prebreakdown sites just before breakdown, whilst Frame 2 shows the locations of the subsequent breakdown arcs. From a comparison of these images, the important fact emerges that the position of one of these arcs coincides with a prebreakdown site. In fact this study showed that over 80% of all DC breakdown events are initiated at DC prebreakdown emission sites. Frame 2 also shows that a DC breakdown event generally consists of a sequence of arcs that frequently tend to be "clustered" within a local region of the electrode surface.

168

To complement these optical data, Fig 5.2(b) is a representation of a typical oscilloscope trace showing the time profile of the collapsing gap voltage during the 20 ms period of a DC breakdown process. This shows how the initiating arc is followed by a series of "transient arcs", where each voltage collapse is shown as a downward-pointing peak, with the time interval between any two successive being as short as ~4 ms.

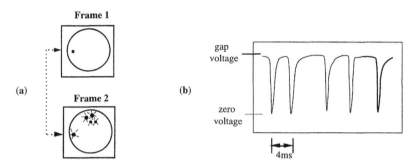

Fig 5.2   Illustrating in (a) how the optical images from two consecutive video frames confirm that a spatial correlation exists between a PES and the primary arc of a breakdown event, and in (b) the form of the associated collapsing gap voltage.

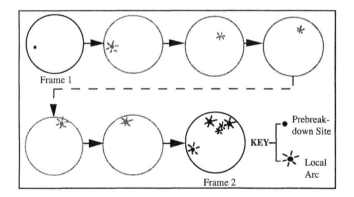

Fig 5.3   A schematic illustration of the time-resolved evolutionary process assumed to be responsible for the types of DC breakdown event shown in Fig 5.2(a).

From a comparison of the gap voltage profile of Fig 5.2(b), and the optical breakdown images of Fig 5.2(a), it will be seen that there is an apparent numerical correspondence between the number of breakdown arcs and collapses of the gap voltage. It is therefore reasonable to assume that a DC breakdown event involves a sequence of spatially random localised arcs as illustrated in Fig 5.3. However, to experimentally verify this hypothesis, it would be necessary to employ a very much higher frame speed recording technique.

On the basis that a spatial correlation exists between a DC-PES and the location of a subsequent "primary" DC breakdown arc, it can be reasonably concluded that a breakdown event is likely to be initiated at the same particulate-like surface microstructures shown in Figs 4.1(a) and (b) that have previously been found to be responsible for DC prebreakdown emission processes [6]. However, this study was not able to identify the causes for the secondary discharges. Very probably, they are breakdown events associated with new electron emission sites formed on the cathode surface as the result of particulate material being injected into the gap during a previous discharge event.

5.2.2   Pulsed Fields

For this type of study, Xu and Latham [3] used the experimental set-up shown in Fig 5.4. In this, a 15 kV 1 μs rise time pulse generator, with its associated electrical circuitry, replaced the DC supply used in the system of Fig 5.1. Otherwise, the optical and electrical recording facilities remain essentially unchanged.

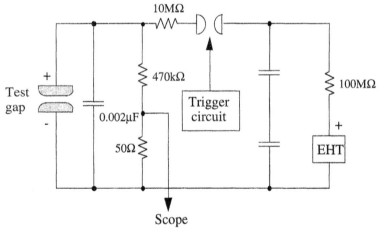

Fig 5.4   The circuitry for pulsed-field measurements.

To investigate the pulsed-field (PF) response of the gap, the following experimental procedure was followed. Firstly, the gap is "characterised" under DC prebreakdown field conditions as described in the previous section. Next, as illustrated in Fig 5.5(a), the amplitude of the pulsed-field is increased incrementally in 500 V steps, starting from a value that is ~75% of the maximum DC field used to record the prebreakdown I-V characteristic of the gap. At each pulsed field level, up to 15 pulses are fired to determine whether or not breakdown events occur. If they don't, the gap is deemed stable, and the field further increased. On the other hand, if breakdown events do occur, then pulses are continuously fired until no more such events are observed at this field level; i.e. the gap has been stabilised, or "conditioned". The magnitude of the pulse is then again incrementally increased and the process repeated.

(a)

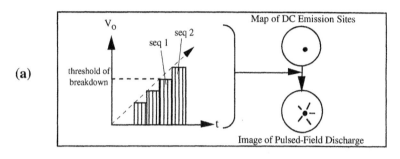

(b)

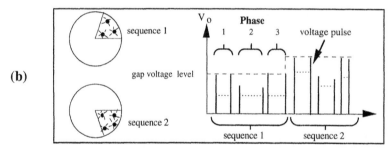

Fig 5.5   (a) Illustrating how successively applied pulse-sequences of incrementally increasing magnitude eventually lead to a PF breakdown event that is initiated at a DC-PES. (b) Illustrating how the breakdown arcs associated with a particular sequence of pulses are clustered within a localised region of the cathode surface.

NB: Each sequence consists of 15 to 25 constant-magnitude pulses, where the individual pulses are shown as single vertical lines; pulses that do not achieve the full gap voltage level correspond to the occurrence of PF breakdown events.

The typical optical images and electrical measurements recorded under these PF conditions are illustrated in Fig 5.5. Here, Fig 5.5(a) illustrates how a PF breakdown arc is again typically initiated at a PES. If this same cathode is now subjected to a further sequence of constant-value pulses, it is found to exhibit three distinct evolutionary phases in its response. Thus, referring to Fig 5.5(b), Phase 1 is characterised by a sequence of ~6 pulses, during which no breakdown events are initiated. This is immediately followed by Phase 2 where typically each of the next 25 or so consecutive pulses gives rise to breakdown events. In the Phase 3, up to a further 20 or more pulses can be

171

applied to the gap without initiating a further breakdown event at this field level; i.e. the gap has become stable or "conditioned" at this field. However, if the magnitude of the pulsed field is now increased, the typical three-phase process described above again repeats itself. From an inspection of Fig 5.5(b), it will also be seen that the PF breakdown arcs associated with any given sequence tend to be "clustered" within a localised region of the cathode surface.

From a technological perspective, this study led to two important findings. Firstly, it established that successively applied pulse-sequences of incrementally increasing magnitude eventually led to a PF breakdown event that is initiated at a DC-PES. Secondly, a sequence of constant-value high-field pulses can de-stabilise prebreakdown emission sites; i.e. modify their electronic properties to the point where they are likely to initiate a breakdown event. In practical terms, a pulsed-field breakdown appears to be typically "delayed" by a sequence of ~6 pulses. This behaviour is likely to result from progressive charge injection processes associated with non-metallic emitting structures [3]: not least, because one would not anticipate a time delay in the occurrence of a breakdown event with metallic emitters.

5. 3   Thermal Response of Emission Regimes to DC Fields

The key conclusion to be drawn from the experimental studies described in Section 5.2 is that "non-metallic" prebreakdown electron emission regimes, including both MIV and MIM microstrucutres, are likely to be predominantly responsible for initiating a vacuum breakdown event in a "passive" HV gap. However, as we shall also see later in Chapter 15, there are other "active" gap regimes where it is assumed that the traditional vacuum breakdown theories based on the MM emission regime apply. In order to understand the physical process of vacuum breakdown, we must consider the thermal dynamics of both the cathode emission regimes, and the local region of the anode that is being bombarded by the emitted electrons, i.e. the anode "hot-spot". Accordingly, we shall start our discussion with the analysis of the steady-state responses of these thermal systems.

5.3.1   Cathode Response

From the discussion of Chapter 4, it is clear that two types of regime have to be considered. Thus, referring to Fig 4.1, it is firstly necessary to analyse the thermal response of the microchannel associated with the MIV and MIM emission models, and secondly, to follow a similar exercise for the microprotrusion of the MM model.

5.3.1 (i)   Microchannel Regime

For both the MIV and MIM emission regimes that were discussed in Section 4.1, it can reasonably be assumed that cathode initiated breakdown will generally originate from the thermal instability of a conduction channel in the insulating medium: i.e. it will be externally manifested as a dielectric breakdown event. The mechanisms responsible for dielectric breakdown have been studied extensively [7-11], in particular for SiO_2 films [7-10]. The following discussion of the thermal response of the present conduction-channel regime will therefore refer extensively to these studies; in particular to the work by Ridley [7], DiStefano et al. [8] and Solomon [9]. In this context, our analysis will firstly derive an expression for calculating the temperature rise resulting from the Joule heating of the electron current flowing through the channel.   Then, we shall discuss two likely mechanisms that could be responsible for the onset of thermal instability.

This Joule heating, and the resulting temperature rise in a conduction channel is a consequence of the energy transfer from the electric field to the electrons at the rate of Ej (Wm^{-3}), where E and j are the field and current density respectively. However, part of this heat will be lost through thermal conduction in the insulating layer, both axially and radially, and another part by radiation into the surrounding medium. To facilitate the analysis of the temperature rise, we shall assume an emission regime as illustrated in Fig 5.6(a). Thus, the insulating particle is considered as a disc of diameter D and thickness l, with all edges rounded. In addition, the conduction channel is assumed to have a diameter of a, and with a length that is the same as the thickness of the disc, namely l. To simplify the analysis, the following four assumptions are also made.

1.   The initial heat production is confined to a cylindrical filament in the insulating layer.
2.   The temperature at the particle outside boundaries remains at ambient.
3.   The radiation losses to both the insulating layer and vacuum can be neglected.
4.   The variation of thermal conductivity with temperature can also be neglected.

Accordingly, the heat flow equation in cylindrical co-ordinates, with no angular dependence, may be expressed as

$$C\rho\frac{dT}{dt} = Ej + K\left[\frac{1}{r^2}\frac{d}{dr}\left(r^2\frac{dT}{dr}\right) + \frac{d^2T}{dz^2}\right]$$

5.1

173

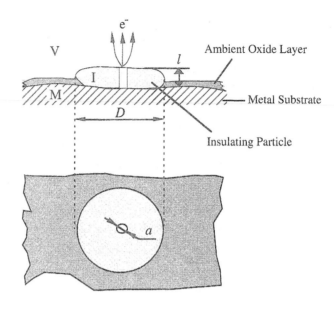

(a)

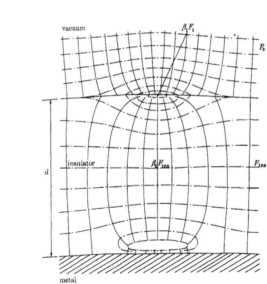

(b)

Fig 5.6   (a) A schematic illustration of a Metal-Insulator-Vacuum emission regime that is assumed for the thermal instability analysis. Here, the insulating particle is assumed as thick disc, and the microchannel as a solid cylindrical rod.
(b) Illustrating the distribution of equal-potential and field lines assumed for an emitting microchannel. (From Bayliss and Latham [12], with permission.)

where C is the specific heat, ρ the density, T the absolute temperature, and K the thermal conductivity. If we now assume that the filament is very thin, radial flow will dominate and, neglecting the temperature variation across the filament, the steady-state solution of equation 5.1 will be

$$T = (Ej/2K)\, a^2\, log\, (D/a) \qquad\qquad 5.2(a)$$

where the ambient temperature, T_0, at the surface of the particle is taken to be zero, i.e. $T_0 = 0$ at $r = D$. T is therefore the temperature rise in the filament, i.e. equivalent to ΔT. When $r \gg a$, heat will preferentially flow to the highly conductive substrate electrode, so that radial flow will be a poor approximation. Thus, it is permissible to replace D by l in equation 5.2(a), so that it may be rewritten as:

$$\Delta T = (Ej/2K)\, a^2\, log\, (l/a) \qquad\qquad 5.2(b)$$

To apply this equation to the present problem, it is necessary to relate the temperature rise to the available and measurable macroscopic parameters. Thus, from Fig 5.6(b), the electric field E can be related to the gap field E_0 through the following equation [12]

$$E = (\beta_2/\varepsilon_r)E_o \qquad\qquad 5.3$$

where β_2 is the field enhancement factor resulting from the geometry of the microchannel in relation to the surface of the substrate electrode, and ε_r is the dielectric constant. Also, the current density j may be related to the total current I flowing in this channel for the gap field E_0 by the expression

$$j = I/[\pi\,(a^2/4\,)] \qquad\qquad 5.4$$

With this assumption, the following expression is obtained for the temperature rise in the channel

$$\Delta T = (2\,\beta_2\, E_o\, I/\pi K\varepsilon_r)\, log(l/a) \qquad\qquad 5.5$$

The implication of equation 5.5 can be easily illustrated by applying it to the practical situation of a field emitting channel having the following characteristics: $l = 1\ \mu m$, $a = 5\ nm$, $K = 1\ W\ m^{-1}\ K^{-1}$, $E_0 = 1 \times 10^7\ Vm^{-1}$, $\beta_2 = 15$, $\varepsilon_r = 5$, and $I = 15\ \mu A$. Substituting these values into equation 5.5 leads to $\Delta T = 660K$; i.e. corresponding to a channel temperature of 960K under ambient condition of 300K.

Since this temperature is far below the melting point of most common insulating media; i.e. typically ~1500K, it follows that there must be some other mechanisms that are responsible for the onset of thermal instability. Two such possible processes are *ion-induced breakdown* and *impact-ionisation-induced breakdown*, each of which will be separately discussed below.

Ion-Induced Breakdown This model was first considered by Riddley for SiO_2 [7], and addresses the effects of the mobile ions produced through dissociation of neutral impurity species due to Joule heating. The sequence of sub-processes involved in this model are:

Field emission → Joule heating → activation of ions → the motion of positive ions to the cathode surface and a resultant field enhancement → enhanced field emission → breakdown

where the basic prerequisite for the overall process to occur is the existence of some form of impurity sites in a conduction channel. This, in fact, is readily satisfied for the regimes we are considering here, since chemical analyses of all emission sites have shown them to be very impure oxides [13]. Although we do not have accurate data on the precise nature of these contaminant microstructures, it can be assumed, for example, that water exists in all naturally occurring sites. In addition, it will be shown later that the ion density required for the onset of the ion-induced breakdown is only 10 ppm, i.e. involving a very low level of concentration.

In order to see how Joule heating activates the ions, we have to enquire into the findings of electrochemical studies of oxide films. In general, contaminants can be divided into two categories; viz. ion and neutral species. The ion species can be either positive or negative, where it is the positive type that are found to be mobile under the influence of an electric field [14], and give rise to a temperature-dependent ionic current density of the form:

$$j = E \, \sigma_i \, exp \, (-H/kT) \qquad\qquad 5.6$$

where E is the electric field, and H is the activation energy.

Having satisfied ourselves that ions are free to move within a channel, we are left with the question of how many ions there are in a typical oxide. Conventional measurements of ion drift give area densities of typically 10^6-10^7 m^2 in clean oxides. This gives rise to a total field enhancement of the order 10^6 V/m, which is three orders magnitude lower than the onset field for Fowler-Nordheim tunnelling to occur. It can therefore be concluded that the influence of this mechanism is negligible.

However, for contaminated oxides, such as those assumed to be involved in the present regimes, a total impurity of concentration of 10^{22} m^3, or an

176

area density of ~ 10^{14} m^2, can be expected to exist in a conduction channel. Thus, the mobile ion area density is expected to be higher than that of the clean oxides.

More importantly, the neutral species of the impurities can become active as a result of their dissociation into ions when the temperature in the channel rises to an appropriate threshold. For example, water contamination is likely to be present in a neutral form, and may be dissociated into ions according to the following reaction:

$$2H_2O \rightarrow H_2O^+ + OH^-$$

Indeed, it has been observed [15] that the total number of mobile ions increases with increasing temperature, with no saturation having been observed up to 400°C. The above reasoning therefore demonstrates that, with rising temperature, the total number of mobile ions can be significantly increased due to the dissociation of neutral contaminants. Furthermore, these experimental findings show that, for an impure oxide, a channel temperature rise to ~300°C, i.e. ~ 600K, will increase the mobile ion area density to a value that is two to three orders of magnitude higher than that at room temperature, reaching a typical value of 10^{13}/cm^2.

Assuming that we can apply the above general findings to our present problem, a temperature rise of 660K will result in the dissociation of the neutral species, so that the ion area density will be at least one order magnitude higher than that at room temperature. Thus, if the area density of ions at room temperature is 2×10^8/m^2, a temperature rise of 660K will increase it to 2×10^9/m^2. Such an ion density will give rise to an extra field of 7.2×10^8 V/m. If this is further enhanced by a factor of β_2 of 15 due to the field-enhancing effect of its geometry on the surface of the substrate, this field will predominate and induce a current density of ~ 10^{12} A/m^2. Putting these new values for the field and current density into equation 5.5, we obtain a temperature of 2×10^4°C.

According to our model, this is the axial channel temperature, which will fall off radially from the filament centre according to equation 5.1. Thus, if the axial temperature is 2×10^4°C, the temperature will remain above 10^3°C out to r = D, and consequently, an almost complete dissociation of ion pairs will occur throughout the cylinder of radius D. Such a heating in the whole of the dielectric particle will cause a liberal flow of positive ions to the cathode, and will induce breakdown filaments at all irregularities in the metal-insulator interface. The breakdown of these filaments will initially result in the local melting of the particle and the electrode, and will finally destroy the whole particle and probably leave the electrode surface with some microcratering damage.

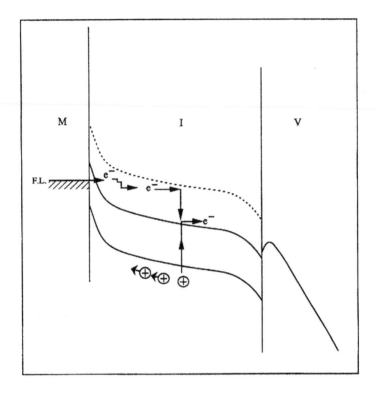

Fig 5.7   Illustrating the physical processes involved in the Impact-Ionization-Induced Breakdown Model.

Impact-Ionisation-Induced Breakdown The description of this process can be based on the impact-ionisation-recombination model for dielectric breakdown proposed by DiStefano and Shatzkes [8]. From this model, a picture can be formulated of the significant processes which occur during breakdown. Thus, referring to Fig 5.7, the following sequence of events is assumed to occur: (i) electrons are first field emitted into the dielectric from the substrate electrode; (ii) the hottest of the electrons ionise the lattice, and leave behind slow-moving holes; (iii) this cloud of positive charge, formed by ionisation and recombination, leads to an increased current of even hotter electrons; (iv) this current then results in similar consequences to the ion-induced breakdown process discussed above; i.e. it leads to a temperature rise through Joule heating, and thus thermal instability. It is also important to note that whilst this model requires strong hole trapping in the dielectric, it only requires low rates of impact ionisation. Thus, it contrasts strongly with the conventional avalanche model where large-scale impact ionisation results directly in the heating and destruction of the insulator.

178

For impact-ionisation to occur, the electrons tunnelling from the substrate into the dielectric medium will need to gain substantial energy from applied field in order to reach a threshold of 2Eg, i.e. twice the band gap energy of the dielectric medium. The corresponding threshold field may be determined using the expression $2E_g = E\lambda$, in which E is the applied field and λ the mean free path of electrons in the dielectric. This type of calculation shows [7] that a typical threshold field for the avalanche breakdown of most oxides is $> 10^9$ V/m.

In summary, the above discussion has demonstrated that the thermal instability of a conduction-channel regime can be caused by an enhanced Joule heating. This can occur at a relatively low gap field of $\sim 10^7$ V/m if the channel consists of neutral impure species. For a purer conduction-channel, the dominant breakdown mechanism will be by impact-ionisation-induced-breakdown, where higher fields may be achieved in the channel. To apply these ideas to our present problem, it can be anticipated that the ion-induced breakdown mechanism is likely to predominate, i.e. because of the impure nature of the prebreakdown emission sites.

5.3.1 (ii)   Metallic Microprotrusion Regime

For a complete analysis of the thermal stability of a metallic microprotrusion it would be necessary to consider four energy exchange mechanisms:

(a)   internal resistive heating resulting from a high current density flowing into the tip of the emitter,

(b)   thermal cooling by heat conduction down the shank of the protrusion into the "cold" reservoir provided by the bulk of the cathode electrode,

(c)   thermal cooling by radiation losses from the surface of the emitter,

(d)   the internal heating or cooling that can result from the Nottingham effect.

This latter process has a similar physical basis to the well known cooling effect associated with thermionic emission, viz. that the energy of an emitted electron will be higher than that of its replacement conduction electron which is supplied at the Fermi level. In the case of low-temperature FEE (T $\lesssim$ 300K), electrons will be emitted from states below the Fermi level so that there will be a heating effect, where the energy dissipated in the lattice $(\Delta\varepsilon)_N$ for each field emitted electron is given by Charbonnier et al. [16] as

$$(\Delta\varepsilon)_N = 1.475 \times 10^{-29} E\phi^{-\frac{1}{2}} \qquad\qquad 5.7$$

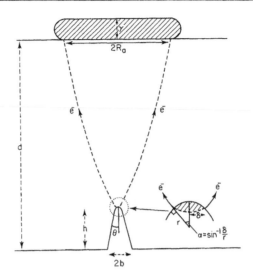

Fig. 5.8 The geometrical relationship between an idealised field-emitting microprotrusion and its associated anode "hot-spot" of radius R_a: b is the radius of the spherical cap from which the electrons are emitted, and γ is the penetration of the electrons in the anode. (From Latham [6], with permission.)

Hence, if it is assumed that all electrons leave the surface with the same energy, the Nottingham power density dissipated in the surface of an emitter by an emission current density j_{OF} will be

$$W_{ON} = 9.2 \times 10^{-11} \, j_{OF} E \phi^{-\frac{1}{2}} \qquad 5.8$$

At high temperatures, electrons will tunnel from populated states above the Fermi level and so give rise to a cooling effect: the transition between heating and cooling is characterised by an inversion temperature T_i (defined later in this section by equation 5.13) which depends on the work function of the emitter and the applied field.

In the original analysis by Dolan et al. [17a] of the thermal properties of the micro-emitter model, only two of the above energy exchange mechanisms were included in their computations, viz. resistive heating and conduction cooling; radiation cooling was also considered, but shown to play an insignificant role. For this restricted model, the temperature difference ΔT between the tip and base of the conical protrusion shown in Fig 5.8 is given by [18]

$$\Delta T = \frac{j_F^2 \, \Omega(T) \, r^4}{4K(1 - \cos\theta)^2} \left(\frac{h - r \cot\theta}{hr \cot\theta} \right) \qquad 5.9$$

180

where j_F is the emission current density, $\Omega(T)$ the electrical resistivity and K the thermal conductivity of the emitter. It soon became apparent however that this was an inadequate approach for obtaining reliable predictions; in particular, this model is inherently unstable at high emission current densities and therefore cannot explain the stable behaviour found in practice for clean emitters under strongly emitting conditions where j_F can approach values of $\sim 10^{12}$ Am^{-2}. Accordingly, it was superseded by a more rigorous treatments that took account of the Nottingham contribution.

The earliest of these was due to Martin and his colleagues [19] in 1960, with important follow-up contributions from Levine [20], Charbonnier et al. [16], Vibrans [21], Brodie [22], Swanson et al. [23], Slivkov [24] and Chatterton [18]. All of these analyses were based on the assumption that any thermal instability will be initiated at the tip of the emitter where the temperature is a maximum; they therefore had as their common central aim the derivation of an expression for the temperature distribution in the vicinity of the apex of the microprotrusion. This therefore involves setting up a differential energy exchange equation and solving it for the boundary conditions appropriate to the chosen protrusion geometry. In practice, this is generally a very complex procedure, particularly if a time-dependent solution is required. However, for slowly varying DC gap voltages, where the sub-microsecond thermal time constant of the microemitter has no significance in controlling the temperature distribution, it is permissible to simplify the problem by using a steady-state solution. Thus, for the case of the truncated-cone type of emitter shown in Fig 5.8, where radiation losses are assumed negligible compared with the heat dissipation by conduction, the solution of Martin et al. [19] for the steady-state temperature difference between the tip and base of the emitter takes the form

$$\Delta T = T_r - T_b = [\Omega(T)/2K] \; (j_F \, r/b)^2 + (\pi k T_r \, r \, j_F/Kb) \cot(\pi p) \qquad 5.10$$

where j_F, K and $\Omega(T)$ have the same significance as in equation 5.9, k is Boltzmann's constant, p is the same field and temperature dependent parameter defined by equation 4.11 and T_r and T_b are respectively the tip and base temperatures of the emitter. Alternatively, this same temperature difference may be given in the following more practical form as a function of the total emission current I_F [23], viz.

$$5.11$$
$$\Delta T = T_r - T_b = [\Omega(T)/2K\pi^2] (I_F/\alpha \, r)^2 + (k T_r \, I_F/K\alpha r) \cot(\pi p)$$

where α is the cone half-angle. It has to be appreciated however that the value of ΔT attained for a given emission current will be very sensitive to the microprotrusion geometry: for example, the thermal coupling of the conical emitter of Fig 5.8 will clearly be more efficient than for a cylindrical emitter of similar h/r ratio, so that the latter will attain a higher tip temperature. The general conclusion from this observation is therefore that long thin microprotrusions will be less stable than those having a broad-base tapering geometry.

The first term in equation 5.11 accounts for the resistive heating and, being proportional to $\Omega(T)(I_F)^2$, increases rapidly with both emission current and temperature. The second term of this equation is due to the Nottingham effect and, because of its direct proportionality to I_F, is clearly less sensitive to increases in the emission current; more important however, it is sensitive to T, falling to zero at the inversion temperature T_i when cotπp = 0 or p = 1/2, thereafter going negative as the effect starts to provide a cooling contribution.

The value of T_i, which is defined by the condition p = 1/2, is based on a theoretical consideration of how the total energy distribution of field-emitted electrons varies with temperature [16,20], and from equation 4.11 leads to

$$T_i = 5.4 \times 10^{-7} E \, \phi^{-\frac{1}{2}} \qquad\qquad 5.12$$

or more exactly

$$T_i = \frac{5.4 \times 10^{-7} E}{\phi^{\frac{1}{2}} \, t(y)} \qquad\qquad 5.13$$

where t(y) is defined for equation 4.2. For most clean metallic emitters that are delivering a high emission current density, E and ϕ can be taken to be typically $\sim 6 \times 10^9$ Vm^{-1} and ~ 4.5 eV respectively, which gives a value for $T_i \sim 1500$K that is almost independent of electrode material. Clearly, if any benefit is to be gained from Nottingham cooling, i.e. by it thermally stabilising the tip so that higher emission current densities can be drawn, T_i must be significantly lower than the melting temperature T_m of the emitter material, as is the case for refractory metals such as tungsten or molybdenum. For less refractory metals such as copper, whose melting point is ~ 1360K, emission cooling can never come into operation, which explains why emitters of this type of material are generally more susceptible to instabilities and hence breakdown events.

The fall-off in the Nottingham heating contribution as $T \to T_i$ is due to an increasing number of electrons tunnelling from states above the Fermi level, so that the low-temperature approximation for the power dissipation W_{ON} given by Equation 5.8 becomes increasingly invalid. Brodie [22] has considered this aspect of the mechanism analytically and gives tabulated data for the high-temperature Nottingham power dissipation W_{TN}. These

182

computations show that for $T/T_i \lesssim 0.9$, the ratio of $W_{TN}/W_{ON} \gtrsim 0.5$ and that it is only when T is very close to T_i that W_{TN} begins to fall off rapidly. It should however be pointed out that this analysis by Brodie [22] is open to criticism [23] since he used the normal instead of the total energy distribution of the field emitted electrons.

The energy exchange associated with the Nottingham effect has been studied experimentally under controlled laboratory conditions by two techniques. In the first, as pioneered by Drechsler [25] and later used by Charbonnier et al. [16], a clean tungsten emitter is welded to a low thermal capacity mounting, usually a thin wire, whose temperature variation can be accurately monitored as heat is either dissipated in or extracted from it. The second technique was developed by Swanson et al. [23,26] and was aimed at investigating the influence of emitter work function ϕ on the effect. To this end, they employed low-ϕ coatings (e.g. zirconium-oxygen on tungsten, giving $\phi \sim 2.7$ eV) to show that, as a result of emission cooling, larger current densities could be drawn from such emitters: they also used other low-ϕ coatings, whose desorption was temperature-dependent (e.g. barium on tungsten, giving $\phi \sim 2.1$ eV), to selectively identify the highly localised region of an emitter where the heating effects are greatest. Whilst both types of experiment confirmed that the predicted energy exchange processes do occur, including the existence of a Nottingham inversion temperature T_i, there was only limited agreement between theory and experiment. In particular, the values of T_i were found to be very much lower than anticipated from the theoretical relation of equation 5.12. Thus, for a clean tungsten emitter with a field of 4.6×10^9 Vm^{-1}, T_i was found to be 550K compared to the theoretical value of 1165K: similarly, the experimental value of T_i for a zirconium-oxygen coated tungsten emitter with a field of 2.4×10^9 Vm^{-1} was ~ 475K as opposed to 775K predicted from theory.

Although no detailed explanation is available for this discrepancy, there are strong indications that the free-electron based theoretical value used for the average energy of a field emitted electron relative to the Fermi level may be in error [26]. These same revised theoretical considerations also suggest that the values of ΔT predicted by equations 5.10 and 5.11 are likely to be too high. In practical terms, the existence of these lower inversion temperatures provides an explanation for the highly stable FEE current densities ($\sim 5 \times 10^{11}$ Am^{-2}) that can be drawn from clean tungsten emitters under controlled conditions (see Chapter 6). Recognising therefore the evidently important role played by the Nottingham effect in determining the limiting conditions for stable electron emission, particularly in the case of the sites found on broad-area electrodes, it should be remarked that the subject warrants further theoretical and experimental investigation aimed at resolving the anomalies outlined above.

5.3.2   Anode Response

The general topic of localised electron bombardment heating by focused beams of impact radius R_a and total beam power W was first treated analytically by Oosterkamp [27]; subsequently, Vine and Einstein [28] developed a more complete theory to cover the case of high energy beams where the effects of electron penetration in the target have to be taken into account. To apply this type of analysis to the specific regime of an anode "hot-spot" associated with the type of field-emitting microstructures found in a high voltage gap, i.e. such as shown in Fig 5.8, it is first necessary to have a knowledge of the electron spot diameter $2R_a$ at the anode. This will be determined firstly by the angular divergence of the electron emission as it leaves the cathode surface and secondly by the magnitudes of the initial transverse velocity components of the electrons, i.e. those parallel to the emitting surface. For the purpose of the present analysis, the following two simplifying assumptions will be made.

(i)   As far as the anode is concerned, the trajectories of the electrons field-emitted from a microchannel embedded in a dielectric medium (see Figs 4.1(a) and (b)) will be made effectively the same as those emitted from an isolated spherically-capped cylindrical microprotusion, i.e. as illustrated in Fig 5.9.

(ii)   Since the field-emitting properties of such microstructures are primarily determined by the radius, r of the emitting tip and the height h of the tip above the planar electrode, it is acceptable to use the analytical treatment developed by Chatterton [18] for an idealised emitter geometry consisting of a truncated cone. Under these circumstances, the emission angle α can be taken as 40°.

Thus, following the analysis of Chatterton [18], and assuming to a first approximation that electrons are emitted normally to the cathode surface (i.e. neglecting their transverse velocity components), that there are no space charge effects, and that the electrons will follow the field lines, one has

$$R_a = \beta \, \delta^{1/2} \qquad\qquad\qquad 5.14$$

or from equation 4.18 where h/r >> 2

$$R_a \simeq (h/r)^{1/2}\delta \qquad\qquad\qquad 5.15$$

Here δ is the radius of the emitting area, as shown in the inset of Fig 5.8.

When account is taken of the initial transverse velocity components, which is essential in most practical situations, it can be shown that [18]

$$R_a = \beta^{1/2}\delta + 2d^{1/2}(\beta\delta\sin\alpha)^{1/2} \qquad\qquad 5.16$$

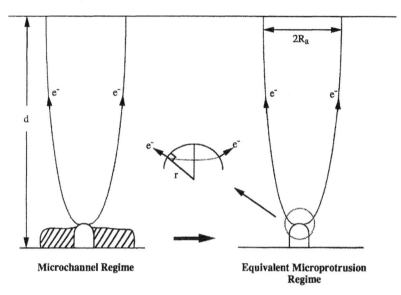

Microchannel Regime **Equivalent Microprotrusion Regime**

Fig 5.9   Illustrating the equivalent microprotrusion regime of a MIV emission structure and the geometrical relationship between this idealised regime and its associated anode "hot-spot".

In fact, R_a is normally dominated by the second term so that

$$R_a \simeq 2d^{1/2}(\beta\delta\sin\alpha)^{1/2} \qquad\qquad 5.17$$

or finally, after substitution for $\sin\alpha = \delta/r$ (see Fig 5.8),

$$R_a \simeq 2\delta(d\beta/r)^{1/2} \qquad\qquad 5.18$$

As an example, electron emission from a truncated-cone-shaped microprotrusion, such as shown in Fig 5.8, with dimensions $h = 0.5$ μm, $r = 5$ nm and hence $\beta = 100$ (from equation 4.18), where it is assumed $\alpha = 40°$ and $\delta \sim r/2 \sim 2.5$ nm, the associated anode spot for a 2 mm gap would have a diameter $2R_a \sim 30$ μm.

The surface volume of the anode electrode in which the beam energy is dissipated (shown shaded in Fig 5.8) will be determined by the electron penetration depth γ which is given by the Thomson-Whiddington Law [29] as

$$\gamma = \frac{V^2}{C} \qquad\qquad 5.19$$

In this equation, V is the accelerating potential of the electron, which in this case corresponds to the gap voltage, and C is a constant defined by

$$C = \frac{g\rho Z}{A}$$

where Z, A and ρ are respectively the atomic number, atomic weight and density of the target material, with the fundamental constant $g = 7.8 \times 10^{10}$ (eVm^2 kg^{-1}). Thus, 10 and 50 kV electrons striking a copper anode would penetrate $\sim$ 0.3 and 8 μm respectively, which illustrates the strong voltage-dependence of γ.

Two analytical approaches have been used to compute the steady state temperature distribution in the anode hot-spot resulting from a constant electron bombardment power loading. The first, as followed by Oosterkamp [27], is really a low-energy approximation applicable to gap voltages V $\lesssim$ 25 kV. Here it is assumed that there is negligible electron penetration with $R_a \gg \gamma$, so that the problem reduces to the classic thermal diffusion problem of a surface disc heat source of uniform power loading on a semi-infinite solid, i.e such as treated by Carslaw and Jaeger [30]. For the more realistic situation of an electron beam having a Gaussian current-density distribution, Vine and Einstein [28] gave the steady-state axial and surface temperature distributions, T(0,x) and T(R,0) respectively as

$$T(0, x) = T_{oo} \, exp \, (x^2/1.44 \, R_g{}^2) \, erfc. \, (x/1.2R_g) \qquad 5.20$$

$$T(R,0) = T_{oo} \, exp \, (-r^2/2.88R_g{}^2) \, I_o \, (r^2/2.88 \, R_g{}^2) \qquad 5.21$$

where T_{oo} is the temperature at the origin of co-ordinates, i.e. on the beam axis in the surface of the target. Clearly, T_{oo} will correspond to the highest temperature reached in the "hot-spot" and is therefore a central parameter in the present discussion of anode initiated breakdown. The value of T_{oo} will be given by

$$[T_{oo}]_1 = 0.414W \, / \, \pi^{\frac{1}{2}} KR_g \qquad\qquad 5.22$$

In equations 5.20 to 5.22, W is the total beam power in watts, K is the mean thermal conductivity of the anode material over the temperature range being considered (Wm^{-1} K^{-1}) and R_g is defined for a Gaussian beam as the radius at which the current density is half the maximum. To account for the power losses through radiation and backscattered electrons, it is necessary to replace W by the actual absorbed power W_a in equations 5.20 to 5.22, where W_a can be conveniently expressed in the form

$$W_a = aW \qquad\qquad 5.23$$

In this relation, a is termed the power-retention factor and is a constant for a given material, ranging from 0.97 for the low atomic number element aluminium to 0.62 for the high atomic number tungsten [28]. Accordingly, the corrected steady-state temperature at the origin of co-ordinates $[T_{oo}]_2$ will be given by

$$[T_{oo}]_2 = 0.414aW/\pi^{\frac{1}{2}}KR_g \qquad\qquad 5.24$$

The second and more general treatment of the temperature distribution in an anode "hot-spot" is particularly relevant to high voltage electron beams $(V \geq 25 \text{ kV})$ where electron penetration becomes significant, i.e. $\gamma \rightarrow R_a$, so that it is no longer permissible to use the surface disc heating model. Thus, from a detailed analysis by Vine and Einstein [28], which considered a volume-distributed heat source, the corrected temperature at the origin of co-ordinates in the surface of the anode is given by

$$[T_{oo}]_3 = \xi\,[T_{oo}]_2$$

$$= (0.25aW/\pi K\gamma)\,\{log(\,1 + 1.4\gamma^2/R_g^2)/[1 - (0.8\,R_g/\gamma)\,tan^{-1}\,(1.2\gamma/R_g)]$$

$$5.25$$

where $[T_{oo}]_2$ has been substituted from equation 5.24. The complicated geometric function ξ will always be <1, i.e. $[T_{oo}]_3 < [T_{oo}]_2$ and has been computed by Vine and Einstein for a wide range of $\gamma/2R_g$; however, Chatterton [18] has shown that it may be closely approximated by the simplified empirical functions:

$$\xi = 0.7\,(2R_g/\gamma)^{0.77} \qquad \text{for } \gamma/\,2R_g > 1$$

$$5.26$$

$$\xi = exp\,(-\,0.3\gamma/R_g) \qquad \text{for } \gamma/2R_g < 2$$

As an illustration of the differing predictions of the "approximate" disc heating model and the more complete latter model, where account is taken of electron penetration and backscatter losses, Fig 5.10 compares their respective normalised isotherms, $[T/T_{oo}]_2$ and $[T/T_{oo}]_3$, for a 2 μm-diameter, 50 kV beam impinging on a copper target for which the penetration depth $\gamma \sim 15.5$ μm. From this figure, it is clear that the largest discrepancies occur within the penetration volume, where the disc model predicts much lower temperature gradients with, consequently, a higher temperature at any given point.

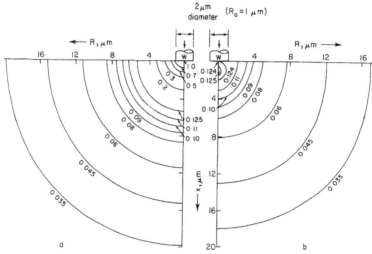

Fig 5.10   Computed normalised isotherms associated with the "hot-spot" produced on an anode (Z = 34) by a 50 kV, 2 μm-diameter beam having a Gaussian current density distribution: (a) according to the disc heating model, and (b) when corrected for penetration and backscatter. (From Vine and Einstein [28], with permission.)

To conclude this discussion, it should be noted that a more complete treatment of this topic would have to take into account the influence of the various electron scattering processes that occur in the target, with particular attention being paid to how the dissipation of the incident electron beam energy varies with depth. An early review of these considerations, as they apply to "massive" targets, was given by Spenser [31]. Subsequently, Bishop [32] used an analytical approach based on Monte Carlo simulation techniques to investigate the energy dissipation in targets of finite thickness such as are encountered in electron microscopy. Later, Vyatskin et al. [33] developed experimental techniques for directly measuring the energy dissipation in massive targets. The important fact revealed by these studies is that, as a result of electron backscattering processes, there is a characteristic depth for a given impact regime at which the rate of energy dissipation is a maximum; its value depending not only on the incident energy of the electrons, but also on the target material, since this determines the range of the electrons.

As an example, we can compare the cases of 20 and 30 keV electrons impacting on a copper target, where the maximum rate of energy dissipation will occur respectively at depths of ~ 300 and 600 nm [33]. The practical implications of this sort of finding become progressively more important at higher energies, i.e. higher gap voltages, where an increasing fraction of the electron beam energy will be dissipated at a greater depth from the surface. For instance, under these circumstances there is the possibility of an explosive "bubble-like" anode process involving the ejection of showers of molten microdroplets from the anode hot-spot such as required for the breakdown initiating mechanism to be described in Section 7.2.1(iii).

5. 4   Thermal Response of Emission Regime to Pulsed Fields

In this section we shall consider the transient thermal response of the cathode microemitting regimes and the associated anode hot-spot under non-equilibrium conditions; in particular, where a gap is subjected to a fast-rise pulsed voltage whose duration t_p is comparable or less than the thermal response times of both the cathode and anode microregimes. By extending the analysis to the use of variable-length, over-voltage pulses, i.e. whose amplitude is sufficient to initiate a breakdown, it is possible to study the temporal evolution of the processes occurring in the gap. The practical importance of such a pulsed-field operational regime is that it offers the possibility of applying higher gap voltages than under DC conditions, and hence drawing higher emission current densities, since neither the cathode nor anode has time to rise to its critical temperature for the initiation of thermal instabilities. As will be described later in Chapter 6, this effect has been impressively demonstrated under controlled laboratory conditions for the case of a point-plane electrode geometry. In practice, it is used to advantage, for example, in high power "flash" X-ray tubes.

5.4.1   Cathode Response

This has been studied extensively by Mesyats and his co-workers [34-45] using both theoretical methods and the type of fast-rise, nanosecond over-voltage pulse technique referred to above. Their findings have led to a three-stage breakdown model involving, firstly, the heating of the cathode emitting site to the point of thermal instability and the onset of an "explosive" electron emission regime; secondly, the formation and expansion across the gap of the microplasma or "cathode flare" created from the vaporised cathode material; and thirdly, the subsequent formation of an "anode flare" that crosses the gap in the reverse direction, and thus helps to bridge it with an arc. A detailed account will be given in Chapter 6 of the experimental investigations into these fundamental processes as they occur under controlled laboratory conditions with a microscopically well defined point-plane electrode geometry. For the present, it is sufficient to indicate that the total breakdown time t_b, defined as the time taken for the gap voltage to collapse, can be expressed as

$$t_b = t_1 + t_2 \qquad\qquad 5.27$$

where t_1 corresponds to the emitter heating time, and t_2 to the time taken for the plasma to cross the gap and strike an arc. The parameter t_2, which is often referred to as the "commutation time", is principally determined by the properties of the cathode-flare, with only a small correction included to take account of the contribution made by the secondary anode flares.

Experimental studies [34] have established that the delay time t_1 to the initiation of pulse breakdown obeys the empirical relation

$$j^2 t_1 = constant \qquad\qquad 5.28$$

where the value of the "constant" depends on the cathode material. It therefore follows that t_1 will be very strongly influenced by the explosive electron emission current density j, and hence the gap voltage: typically, t_1 varies in the range 0.1-100 ns according to the level of over-voltaging. A close approximation to the commutation time t_2 may be obtained experimentally by using high-speed optical techniques [35,36] to measure the time taken for a cathode flare to cross the electrode gap d, where in terms of the cathode flare velocity v_f,

$$t_2 \sim d/v_f \qquad\qquad 5.29$$

Anode flares on the other hand have been shown by these techniques to develop at a time $\sim0.5d/v_f$ ($\simeq t_2/2$) after the initiation of a cathode flare [35-38].

From such optical studies with both broad-area electrodes, and those to be described in Chapter 6 using a point-plane configuration, it has been shown that the magnitude of v_f is principally determined by the space charge density within the flare which, in turn, is closely related to the explosive emission current I(t) that is feeding the flare. Since however in practice v_f shows only a small variation over a wide range of experimental conditions, equation 5.29 indicates that t_2 will be very sensitive to the electrode gap d, typically varying between 20-500 ns for d-values in the range 0.5-10 mm; the corresponding flare velocity is $\sim 2 \times 10^4$ ms^{-1}. For most electrode regimes, I(t) will be space-charge limited and given by the expression [39]

$$I(t) = A [V(t)]^{1.5} v_f t (d - v_f t)^{-1} \qquad\qquad 5.30$$

where $A \sim 3 \times 10^{-5}$ $AV^{-1.5}$ and $V(t)$ is the gap voltage whose time-dependence will be governed by the external circuitry. By establishing an appropriate circuit equation, and solving this in conjunction with equation 5.30, it is possible to obtain an analytical expression for t_2 [35,40]. For the cases considered, which are representative of most conventional experimental regimes, it emerges that

$$t_2 = (d/v_f) f(Z, V_p) \qquad\qquad 5.31$$

where $f(Z, V_p)$ is a slowly varying function of an appropriate input circuit parameter Z and the over-voltage pulse amplitude V_p. However, since $f(Z, V_p)$ is of the order unity, it follows that the approximation of equation 5.29 was justified. An exception to this rule is where a large gap (>3 mm) is being fed from a discharging capacitor whose value is < 30 pf: in this case, the voltage collapse time is less than the commutation time, i.e. the capacitor is discharged before the gap is bridged by the plasma. It should also be added that equations 5.29 and 5.30 are also valid for DC breakdowns when the gap length is $\lesssim$ 6 mm and the applied voltage is $\lesssim$ 120 kV [41,46].

Returning to a further consideration of Equation 5.27, it must be pointed out that it is liable to become invalid with low over-voltage pulses, where it is frequently found that t_b can exceed $t_1 + t_2$ by several orders of magnitude, i.e. as calculated respectively from equations 5.28 and 5.29. Under these circumstances, it is believed that the initial emission current density is too low to cause significant vaporisation of the tip, but is high enough to lower the self-diffusion activation energy of its surface atoms to the point where they can migrate under the influence of the applied field and give rise to a tip sharpening mechanism, and hence the critical conditions necessary for the formation of a microplasma. The characteristic time t_s associated with such a sharpening process is given by [42] as

$$\tau_s = 10 r_o \rho^{\frac{1}{2}} / E \qquad\qquad 5.33$$

where r_o is the initial radius of the emitter (cm), ρ the density of the cathode material (gcm^{-3}) and E the macroscopic gap field (Vcm^{-1}). Thus, taking $r_o = 10^{-5}$cm, $\rho = 10$ gcm^{-3} and $E = 10^6$ Vcm^{-1} as being representative of typical gap conditions, equation 5.33 predicts $\tau_s \sim 10^{-7}$ s, which is of the correct order to account for the high values found for t_b with low over-voltage pulses.

The first complete theoretical analysis of the temperature response of a field emitting micropoint is due to Litvinov et al. [42,43] and was based on a model that assumed the important thermal processes, and hence instabilities, occur in the surface layers of an emitter. Accordingly, their approach was first to establish a differential equation describing the thermal balance in a surface volume following the application of a step-function impulse field, and then to solve this for a set of boundary conditions that closely approximated to those existing in typical experimental situations. To highlight the various physical processes involved, it is convenient to express the basic differential equation in the form

$$\rho C \frac{\partial T}{\partial t} = j_F^2 \Omega(\tau) + \nabla \left\{ (\lambda_e \nabla T_e) + \frac{3 j_F}{5e} . \mu_o \left[1 + \frac{5\pi^2}{12} \left(\frac{kT}{\mu_o} \right)^2 \right] \right\}$$

$$\equiv \quad I \quad + \quad II \quad + \quad III \qquad\qquad 5.34$$

Here, the left hand side corresponds to the nett heat gained per unit volume as a result of electron-phonon interactions, and is expressed in terms of the total thermal capacity C which includes both the electronic and phonon contributions, C_e and C_p respectively: thus

$$\rho C \left(\partial T / \partial t \right) = C_e \left(\partial T_e / \partial t \right) + C_p \left(\partial T_p / \partial t \right) \qquad\qquad 5.35$$

where T_e and T_p are respectively the electron and phonon (or lattice) temperatures, with $T_e = T_p = T$ at $t = 0$ and as $t \rightarrow \infty$. The right hand side of equation 5.34 accounts for the energy dissipating mechanisms in which term I corresponds to the Joulian heating with $\Omega(T)$ being the temperature-dependent resistivity, and terms II + III represent the Nottingham heating effect discussed in Section 5.3.1(ii); i.e. where term II is the mean energy removed by electrons that are field emitted from the cathode surface and term III is the mean energy of the replacement or "supply" electrons (μ_o is the Fermi energy defined at $T = 0$). The detailed form of term II has been given as [42]

$$\lambda_e \nabla T_e = (2kT_i j_F/e) \left[\pi^2 kT_e^2 / 4\mu(T) \, T_i + (\pi T_e/2T_i) \, ctg \, (\pi T_e/2T_i) \right] \qquad 5.36$$

where T_i is the Nottingham inversion temperature defined by equation 5.13 and $\mu(T)$, the temperature-dependence of the chemical potential function, is given by

$$\mu(T) = \mu_o \left[1 - (\pi^2/12) \, (kT/\mu_o)^2 \right] \qquad\qquad 5.37$$

in which μ_o is the Fermi energy at $T = 0$. Equations 5.34 to 5.37 have been solved for the type of idealised microemitting geometries discussed in Section 4.2.2(ii) where the temperature is sufficiently low for j_F to be defined by equation 4.9 or its alternative form

$$j_{TF} = j_{OF} \, (\pi T_e/2T_i) \, sin \, (\pi T_e/2T_i) \qquad\qquad 5.38$$

At emission current densities $\leq 10^{13}$ Am^{-2}, it is found that both the electron and lattice temperatures, T_e and T_p, approach equilibrium values with comparable time constants $\tau_e \sim \tau_p \geq 10^{-10}$ s, i.e. indicating that under these conditions there is an efficient exchange of energy between the electrons and the lattice, so that it is therefore permissible to take $T_e = T_p = T$ and to treat the system as having a single thermal capacitance with $C_p = C$. However at emission current densities $\geq 10^{13}$ Am^{-2}, where T_e can reach several tens of thousands of degrees, the above time constants assume values of $\tau_e \leq 10^{-12}$ s and τ_p 10^{-11}-10^{-10} s, thus indicating that there is a significant lag between the lattice and electron temperatures. However, when these times are compared with the values for the cathode response time t_1 measured experimentally, it is evident that lattice relaxation effects can be neglected in most practical situations. Applying this and a number of other simplifying assumptions, equation 5.34 may be solved [45] to yield the following analytical expression for t_1:

$$j^2 t_1 = a\rho C / H_o \qquad\qquad 5.39$$

Here, $H_o = (\Omega(T)/T)$, ρ is the density and a is a material constant which depends weakly on the work function; in the case of tungsten, for example, $a = 55$. If the values of t_1 predicted by equation 5.39 are compared with those obtained from the empirical expression of Equation 5.28, established from experimental data, agreement is generally very good. Thus, taking again the case of tungsten, experiments give $j^2 t_1 = 4 \times 10^{17}$ A^{2}sm^{-4}, whilst equation 5.39 gives 4.5×10^{17} A^{2}sm^{-4}. It should also be noted that a similar value for the thermal response time of a microemitter was obtained from a computer simulation study by Mitterauer and Till [47] that used a simplified model in which the influence of the Nottingham effect was ignored.

5.4.2   Anode Response

The problem of extending the steady-state analysis of Section 5.3.2 to include the temporal response of the anode hot-spot temperature distribution under pulsed heating conditions has been considered by Charbonnier et al. [48], and in more detail by Dudek [49], who also conducted a complementary experimental investigation. However, to gain an insight into the essential features of the characteristic temperature response of this complex thermal regime, we shall refer here to the findings of the simplified approach of Charbonnier et al. [48] which is appropriate to conditions where the pulse length is less than or comparable to the thermal response time of the anode. In this treatment, the surface disc source diffusion model is solved for the special situation where times are short enough for the flow of heat to remain uni-directional, i.e. normal to the anode surface. The model was further simplified by considering an idealised circular cross-section beam having a uniform power loading w, where space charge effects and backscattered losses are ignored. For such a regime, two functions may be defined that give the temperature increase in the anode.

(i)   For $0 \leq x \leq \gamma$ i.e. for points within the electron range γ of the target surface

$$T_1\ (x,t)\ =\ \frac{w}{2K\gamma}\left[(\gamma + x)^2 f(\alpha_1) + (\gamma - x)^2 f(\alpha_2)\right]$$

5.40

where $f(\alpha_1)$ and $f(\alpha_2)$ are complicated functions defined by

$$f(\alpha) = \left[\exp\left(\frac{-\alpha^2}{2}\right)\right]\left[\alpha(2\pi)^{0.5}\right]^{-1} + \alpha^{-2}\ erf(\alpha) + erf(\alpha) - 0.5$$

5.41

with

$$\alpha_1 = (\gamma + x)\ (2Kt/C\rho)^{-1/2}$$
$$\alpha_2 = (\gamma + x)\ (2Kt/C\rho)^{-1/2}$$

5.42

In these equations, erf signifies the standard "error" function, whilst K, C and ρ are respectively the thermal conductivity, specific heat and density of the target.

(ii)   For $x > \gamma$, i.e. for points inside the bulk of the target and outside the electron penetration range

$$T_2\ (x,t) = \frac{w}{2K\gamma}\left[(\gamma + x)^2\ f(\alpha_1) - (x - \gamma)^2 f(\alpha_2)\right]$$

5.43

The highest anode surface temperature will occur at the centre of the hot-spot (x = 0) which, from Equation 5.40, will be given by

$$T_1\ (0,t)\ =\ \frac{w}{2K\gamma}\left[\gamma^2\ f(\alpha_{01}) + \gamma^2 f(\alpha_{02})\right]$$

$$=\ \frac{w\gamma}{K}f(\alpha_0)$$

5.44

since $\alpha_{01} = \alpha_{02} = \alpha_0 = \gamma(2Kt/C\rho)^{-1/2}$ in this special case. It will be seen from equations 5.40 and 5.43 that the time-dependence of T_1 and T_2 only appears through the quantities α_1 and α_2 in equation 5.41 for $f(\alpha)$, and that this is basically an exponential function in α^2. Accordingly, the quantity

194

$(\gamma \pm x)^2(2K/C\rho)^{-1}$ can be identified as the heat spread parameter which determines how rapidly the thermal diffusion process can dissipate heat energy; alternatively, it is possible to define a local thermal time constant for any point x which gives a measure of the time taken for the temperature at that point to approach its steady-state value. Thus, at $x = \gamma$, i.e. at the inner boundary of the disc heat source, we have from equation 5.42 that $\alpha_2 = 0$ and

$$\tau_\gamma = \frac{2\gamma^2\rho C}{K} \qquad\qquad 5.45$$

Because of the dependence of τ_γ on γ, it follows from equation 5.19 that $\tau_\gamma \propto V_0^4$ and is consequently very sensitive to the energy of the bombarding electrons, where typically, $\tau_\gamma \sim$ nanoseconds for a 10 kV beam and $\sim$ milliseconds for a 1 MV beam. The changing temperature distribution in the surface of the anode, as predicted by this pulsed heat source model, can be conveniently illustrated in terms of the normalised variables (x/γ) and (t/τ_γ). Thus, Fig 5.11 shows a family of normalised temperature distributions corresponding to increasing time intervals up to $t = 16\tau_\gamma$ following the application of a heat pulse [48].

Finally, the model can be used to provide approximate expressions that indicate how the temperature rise in the surface of the anode depends on the duration of the heat pulse t_p.

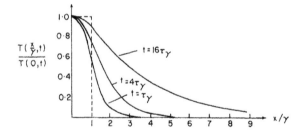

Fig. 5.11   Normalised axial temperature distributions associated with an anode "hot-spot" formed by a pulsed electron beam of varying duration t_p. (From Charbonnier, Bennette and Swanson [48], with permission.)

For very short pulses $(t \ll \tau_\gamma)$, there will be an adiabatic situation with virtually no heat spread, so that at the end of the pulse there will be a uniform temperature rise throughout the disc given by

$$T(0,t_p) \simeq \frac{wt_p}{\gamma C\rho} \qquad\qquad 5.46$$

For intermediate pulses $(t \gtrsim \tau_\gamma)$, there will be an appreciable diffusion of heat into the anode with the surface temperature being given by equation 5.44. However, for situations where t is significantly greater than τ_γ, so that $\alpha_1 \ll 1$, equation 5.44 can be approximated by [48]

$$T_1(0,t_p) \simeq 2w(t_p)^{\frac{1}{2}}/(\pi KC\rho)^{\frac{1}{2}} \qquad\qquad 5.47$$

For very long pulses $(t \gg \tau_\gamma)$; equilibrium conditions will be approached and the anode temperature rise will correspond to the steady-state limit calculated by Oosterkamp [27] viz.

$$T_1(0,t_p) \simeq wR_a/K \qquad\qquad 5.48$$

5. 5   Technical Implications

Having considered in detail the causes of the thermal instability of both the microchannel and metallic microprotrusion emission regime, we shall now turn our attention to how these theoretical findings can be applied to two important technological issues. Namely, whether breakdown events are likely to be cathode or anode-initiated, and whether it is possible to predict the breakdown behaviour of a gap from its prebreakdown characteristics.

5.5.1   Cathode" versus Anode-Initiated Breakdown Criteria

This issue has been extensively studied in relation to the metallic microprotrusion regime, but not to the microchannel regime. In the following section we shall firstly discuss how the practical breakdown criteria are set for the former metallic regime, and then extend our analysis to the latter channel regime.

5.5.1 (i)   MM Microprotrusion Model
As stated in the introductory remarks to this chapter, the basic criterion for the breakdown of a vacuum gap is that the temperature of either a cathode microemitter or anode hot-spot should approach their respective melting points. In the preceding sections, these critical conditions have been independently quantified for the anode and cathode processes. For example, equations 5.10 and 5.11 may be solved to determine the critical (i.e. maximum) emission current density j_{FC} that may be drawn from a conical microprotrusion of known material and geometry before its tip temperature approaches to within $\sim 0.8\text{-}0.9$ of its melting point, i.e. when there is a high risk of breakdown being initiated through thermal or mechanical instabilities.

This value of j_{FC} may then be used to find the corresponding critical field E from equations 4.9 to 4.18. Similarly, equations 5.24 or 5.25 may be used to define the critical power W_c that can be sustained at the associated anode hot-spot; furthermore, there will clearly be a critical emitter current, and hence cathode field, corresponding to W_c.

The initiation of each of these breakdown mechanisms will therefore be associated with a critical gap field, say E_{c1} and E_{c2}, but the question however arises as to whether, for a given microemitter, breakdown will be initiated by an cathode or anode process, i.e. whether $E_{c1} \gtrless E_{c2}$. Qualitative reasoning suggests that long, thin and sharp protrusions will become thermally unstable before they can deliver sufficient emission current for the beam power at the anode to approach W_c, so that such a regime will give rise to cathode-initiated breakdown. On the other hand, short, thick-shanked and rounded protrusions will have good thermal coupling to their base electrode and will therefore be capable of sustaining the necessary emission currents for the beam power at the anode to approach W_c and so precipitate anode-initiated breakdown. This type of reasoning prompted Chatterton [18] and Charbonnier et al. [48], to recognise that there will be a critical emitter geometry, defined by a critical field enhancement factor β_c, which will set the boundary between the initiation of the two mechanisms. Thus, for $\beta > \beta_c$, breakdown will be cathode-initiated, whilst for $\beta < \beta_c$, it will be anode-initiated. Following the analytical approach of Charbonnier et al. [48], which is based on a hemispherically-capped cylindrical protrusion, and uses the approximation of an anode disc heat source model referred to above, the steady-state anode power density is given by

$$w \simeq I_F \, V/\pi R_a^2$$

which, after making the substitutions $I_F \approx \pi r^2 j_F$, $E = \beta(V/d)$ and $R_a = 2(\beta rd)^{1/2}$ becomes

$$w \simeq r j_F E/4\beta^2 \qquad\qquad 5.49$$

Introducing now the limitations imposed by the critical current density j_{FC} and associated critical field E_c (typically $6 \times 10^{11} \mathrm{Am^{-2}}$ and $\gtrsim 5 \times 10^9 \mathrm{\, Vm^{-1}}$ respectively) that can be sustained at a given cathode surface, together with the critical anode power loading w_c, equation 5.42 predicts that for *cathode-initiated breakdown*,

$$\frac{r j_{FC} E_c}{4\beta^2} > w_c \qquad\qquad 5.50$$

197

and for *anode-initiated breakdown*

$$\frac{rj_{FC}E_c}{4\beta^2} < w_c \qquad\qquad 5.51$$

The boundary between the two processes will therefore be defined by $rj_{FC}E_c/4\beta^2 = w_c$, which allows a critical value of the field enhancement factor β_c to be defined for which both processes would be simultaneously initiated, viz.

$$\beta_c = (rj_{FC}E_c/4w_c)^{\frac{1}{2}} \qquad\qquad 5.52$$

Noting that $w_c \simeq KT_m/R_a$ from Equation 5.48 and $R_a \simeq 2(\beta_o rd)^{1/2}$ from equation 5.18 with $\delta = r$, β_c, may be alternatively expressed in the more practical form

$$\beta_c = r\,[j_{FC}E_c d^{1/2}/2KT_m]^{2/3} \qquad\qquad 5.53$$

A similar approach may be used to obtain equivalent expressions for β_c for pulse field operation [48], where it is generally found that these conditions favour cathode-initiated breakdown. Reference should also be made to the experimental investigation of Utsumi [50] which broadly confirmed the conclusions of the above analysis, viz. that high-β emitters ($\beta \gtrsim 500$) invariably lead to cathode-initiated breakdown, whilst low-β emitters ($\beta \lesssim 30$) are found to precipitate anode-initiated breakdown.

When considering the practical implication of the above limiting conditions, it should be noted that calculations of the critical cathode current density and anode power density only vary by a factor of ~ 5 over a wide range of electrode materials from say alumina to tungsten [48]. Accordingly, there will only be a small spread in the values of ~ 5, and hence the macroscopic breakdown field E_b ($=E_c/\beta_c$), among commonly used electrode materials. These considerations alone would suggest however, that high conductivity metals such as copper, that can dissipate heat rapidly, would be more effective as anode materials, whilst refractory metals such as tungsten are to be favoured for cathodes on the grounds of their having lower inversion temperatures [26] with consequently more stable emitting characteristics.

Apart from the inherent approximations of this analysis of breakdown criteria, of which the neglect of space charge effects at high emission current densities is probably the most notable, the approach also suffers from the serious disadvantage of taking no account of emitter geometry; i.e. since there can, in principle, be a wide range of geometries corresponding to a given β-value. This aspect was however considered by Chatterton [18] in a computational analysis that compared the critical behaviour of conical and cylindrical protrusions of similar tip radii r. The main conclusion to emerge from this investigation was that, whereas there is a large reduction in the value of β_c for the thinner cylindrical type of protrusion, with a consequent increase in the likelihood of cathode-initiated breakdown, the critical anode condition is much less sensitive to protrusion geometry. This is not perhaps a surprising finding when it is remembered that a conical protrusion has a better thermal coupling with its base electrode and will therefore be inherently more stable. In assessing the importance of protrusion shape in practical situations it has to be noted that most direct electron optical evidence of protrusions indicates that the conical geometry is very much more typical. For this reason, the analytical expression for β_c of equation 5.53 can still be regarded as a good approximation.

Note should finally be taken of an important observation by Jüttner [46] that although the traditional criterion for anode-initiated breakdown used throughout this chapter is a fundamental requirement, it is not in itself a sufficient condition for there to be a 100% probability of a breakdown occurring. The reason for this is that, in order to explain the origin of the very high currents flowing during a breakdown (10-100 A), it is necessary to assume the occurrence of a secondary process involving the formation of an explosive emission cathode spot through a breakdown between the advancing anode plasma and the cathode surface: an event that could be expected to be extremely rapid, since it is known that cathode spots have nanosecond formation times [51]. However, the probability of this secondary process occurring is less than unity, as can be demonstrated by using a triggering spark to artificially produce an anode plasma in an electrically stressed gap, and noting that this does not always result in the breakdown of the main gap. In contrast, if a similar experiment is performed using a cathode triggering spark, the probability of a breakdown is unity. This observation therefore implies that a breakdown event initiated by an anode microplasma involves a more complicated mechanism than previously assumed, and that statistically, it is less likely to occur than a cathode-initiated event.

5.5.1 (ii)   MIV/MIM Microchannel Model

Despite its technological importance, there has not to date been a detailed theoretical analysis to derive the critical breakdown criteria for the microchannel regime. The main reason for this is that very little quantitative information exists about the physical parameters that determine the properties of microchannels. However, it is possible to approach the question at a quasi quantitative level by a similar line of reasoning to that previously used in Section 5.5.1(i) for analysing the metallic microprotrusion emission regime.

199

Thus, as discussed in Section 5.3.1, the cathode-initiating breakdown mechanism is believed to be predominately controlled by ion-induced current-runaway. Accordingly, the critical condition for the onset of thermal instability in a typical microchannel will be when its axial temperature approaches a sufficiently high value to activate the dissociation processes discussed in Section 5.3.1; typically, this might correspond to $\Delta T_c \sim 400°C$. This condition will be associated with a critical current density j_{FC} flowing in and emitted from the channel, an associated critical internal field and, in turn, a critical external gap field E_c. To obtain an analytical expression for this latter parameter, which is the only externally measurable quantity, equations 5.4 and 5.5 are firstly solved to obtain an expression for ΔT_C of the form

$$\Delta T_C = constant \times j_{FC} E_{ic} \qquad\qquad 5.54$$

where the "constant" includes the material and structural parameters of the channel whose values can be realistically estimated by a "good guess" approach. Substituting for j_{FC} in equation 5.54 from equation 4.27, and noting that $E_c = \varepsilon_r E_{ic}$, one finally obtains an expression that can be solved analytically for E_c. In considering the critical conditions for anode-initiated breakdown, it is reasonable to assume that an emitting microchannel embedded in a dielectric medium behaves like the metallic hemispherically-capped cylindrical microprotrusion regime illustrated in Fig 5.9. Thus, equations 5.24 and 5.25 may be used to define the critical power loading w_c that can be sustained at the associated anode hot-spot. With this known value of w_c, the corresponding critical current density and field may be found. To judge whether, for a given microchannel, breakdown will be initiated by a cathode or anode process, one would need to compare the critical fields for both cases. Thus, if the critical field for cathode process is higher, then breakdown is likely to be initiated by the anode process.

At a more detailed level of analysis, it would be necessary to recognise that the conductivity of a microchannel will not be as high as a metallic microprotrusion of similar geometry. Accordingly, it can be anticipated that the field enhancement induced by the channel will not be as high as an equivalent metallic microprotrusion, and that it will vary with the electronic properties of the channel. It follows therefore that the critical field for a microchannel of given dimensions will be higher than the equivalent metallic microprotrusion. Any future theoretical analysis of the practical breakdown criteria for the microchannel regime should therefore examine more closely how the electronic properties of the channel affect the field enhancement factor β. For example, how is the electron concentration in a channel related to the β-factor. It would also be appropriate to investigate how the geometry, size and tip shape of an emitting channel influence its associated β-factor.

5.5.2 Electrical Correlation Between Prebreakdown and Breakdown Characteristics

Details were given in Section 5.2 of the experimental evidence showing that DC prebreakdown emission from a non-metallic site is responsible for initiating a primary DC arc. Subsequently, in Section 5.3.1, two mechanisms were presented to explain the processes whereby thermal instability is initiated in the non-metallic emission regime. These early studies have led to a more technologically orientated investigation aimed at establishing if an electrical correlation exists between the prebreakdown electron emission characteristics of a gap and the subsequent threshold field at which breakdown occurs.

Table 5.1 Experimental data showing the electrical and spatial correlation between prebreakdown and breakdown characteristics.

Sample no. Parameter	1	2	3	4	5	6	7	8	9	10
Switch-on field (MV/m)	12	12	14	23.3	20.4	16.8	17.2	28.7	14.8	19.2
Breakdown field (MV/m)	10.9	17.8	8.8	15.6	20.7	15.2	14	18.4	10.2	23.7
β-value	399	173	661	242	171	181	201	198	578	158
Correlation	Y	Y	Y	Y	N	Y	Y	Y	N	Y

Thus, referring to the recent investigation conducted by Xu and Latham [2] using the same experimental set-up shown in Fig 5.1, two general conclusions can be drawn from their tabulated data of β-values, switch-on and breakdown fields presented in Table 5.1:

(i) the breakdown field of the first discharge is usually lower than the initial switch-on field of the corresponding prebreakdown emission site, and

(ii) as shown in Fig 5.12, the higher the β-value, the lower the breakdown field.

The first of these findings indicates that it is not a sufficient precaution to use the switch-on field as an upper limit for the operational field range. Rather, to be in a position to predict the achievable upper limit, it is necessary to measure the I-V characteristic, and hence the β-value of the gap. Thus, if such a measurement is carried out on a group of devices under their specific operational conditions, a curve similar to that shown in Fig 5.11 can be constructed. Such a curve can then be used for technical guidance to predict an upper operational limit for each individual device.

As a further precautionary note on the above discussion, it has to be emphasised that the correlation found by Xu and Latham [2] was based on a very limited set of data. Accordingly, similar experiments need to be carried out on a range of device types under their specific device operational conditions, before it will be possible to confirm the reliability of this predictive procedure.

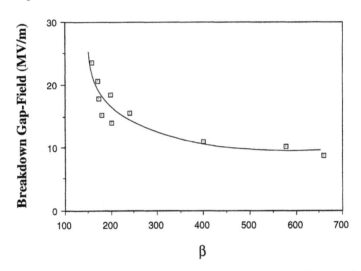

Fig 5.12   Showing how the macroscopic breakdown field (or breakdown gap field E_b) decreases from low-β sites to high-β sites. (From Xu and Latham [2], with permission.)

5.6   References

1.   Latham, RV., Bayliss, KH. and Cox, BM., *J . Phys. D: Appl. Phys.*, **19**, 214-231, 1986.
2.   Xu, NS. and Latham, RV., to be published in *J. Phys. D: Appl. Phys.*, 1994.
3.   Xu, NS. and Latham, RV., *Proc. of XVI - DEIV Conference,* 60-63, 1994.
4.   Lafferty, JM., *Proc. IEEE,* **541**, 23-32, 1966.

5. Mesyates, GA. and Proskurovsky, DI., "Pulsed Electrical Discharge in Vacuum", Springer, Berlin, 1989.
6. Latham, RV., "High Voltage Vacuum Insulation: The Physical Basis", Academic Press, London/New York, 1981.
7. Ridley, BK., *J. of Appl. Phys.*, **46**, 998-1007, 1975.
8. DiStefano, TH. and Shatzkes, M., *J. Vac. Sci. Technol.*, **13**, 50-54, 1976.
9. Solomon, P., *J. Vac. Sci. Technol.*, **14**, 1122-1130, 1977.
10. DiMaria, DJ. and Fischetti, MV., "Excess Electrons in Dielectric Media", edited by Ferradini, C. and Jay-Gerin, J-P., CRC Press, London/Boca Raton, 1990.
11. Marsolais, RM., Cartier, EA., and Pfluger, P.,"Excess Electrons in Dielectric Media", edited by Ferradini, C, and Jay-Gerin, J-P, CRC Press, London/Boca Raton, 1990.
12. Bayliss, KH. and Latham, RV., *Proc. Roy. Soc.*, **A403**, 285-311,1986.
13. Niedermann, Ph., "Experiments on Enhanced Field Emission", PhD Thesis, University of Geneva, 1986.
14. Nicollian, E.H., Berglund, C.N., Schmidt, P.F., and Andrews, J.M., *J. Appl. Phys.*, **42**, 5654-5659, 1971.
15. Kuhn, M. and Silversmith, DJ., *J. Electrochem. Soc.*, **118**, 966-971, 1971.
16. Charbonnier, FM., Strayer, RW., Swanson, LW. and Martin, EE., *Phys. Rev. Lett.*, **13**, 397-401, 1964.
17. Dolan, WW., Dyke, WP. and Trolan, JK., *Phys. Rev.*, **91**, 1054-7, 1953.
18. Chatterton, PA., *Proc. Phys. Soc.*, London, **88**, 231-45, 1966.
19. Martin, EE., Charbonnier, FM. and others, WADD Techn.Rep., 59-20, Linfield Research Inst., 1960.
20. Levine, PH., *J. Appl. Phys.*, **33**, 582-7, 1962.
21. Vibrans, GE., *J. Appl. Phys.*, **35**, 2855-7, 1964.
22. Brodie, I., *Int. J. Electronics*, **18**, 223-33, 1965.
23. Swanson, LW., Crouser, LC. and Charbonnier, FM., *Phys. Rev.*, **151**, 327-340, 1966.
24. Slivkov, IN., *Sov. Phys: Tech. Phys.*, **11**, 249-53, 1966.
25. Drechsler, M., *Z. Naturforsch*, **18a**, 1376-80, 1963.
26. Swanson, LW., Strayer, RW. and Charbonnier, FM., *Surface Sci.*, **2**, 177-82, 1964.
27. Oosterkamp, WJ., *Philips Res. Rep.* 3, pp. 49, **160**, 303, 1948.
28. Vine, J. and Einstein, PA., *Proc. IEEE*, **111**, 921-30, 1964.
29. Whiddington, R., *Proc. Roy. Soc.*, **89**, 554-9, 1914.
30. Carslaw, HS. and Jaeger, JC., "Conduction of Heat in Solids", Oxford University Press, 2nd Edition, Oxford, 1959.
31. Spenser, LV., "Energy dissipation by fast electrons", National Bureau of Standards Monograph, No.1, 1959.
32. Bishop, HE., *Brit.. J. Appl. Phys.*, **18**, 703-15, 1967.
33. Vyatskin, A.Ya, Kabanov, AN., Smirnov, BP. and Khramov, V.Yu, *Rad. Eng. and Electron Phys.*, USA, **22**, 80-4, 1977.

34. Kartsev, GK., Mesyats, GA., Proskurovsky, DI., Rotshteyn, VP. and Fursey, GN., *Sov. Phys.- Dokl.*, **15**, 475-7, 1970.
35. Mesyats, GA. and Proskurovsky, DI., *Sov. Phys. J.*, **11**, 49-51, 1968.
36. Bugaev, SP., Mesyats, GA. and Proskurovsky, DI., *Sov. Phys. - Dokl*, **14**, 605-6, 1969.
37. Bugaev, SP., Koshelev, VI. and Timoleev, MN., *Sov. Phys. J.*, **17**, 193-196, 1974.
38. Baksht, RB., Kokshenov, VA. and Manylov, VI., *Sov. Phys: Tech. Phys.*, **20**, 1069-71, 1975.
39. Bugaev, SP., Litvinov, EA., Lopatin, VV. and Mesyats, GA., "Nanosecond High Power Electron Pulses", Nauka, Novosibirsk, p.52, 1974.
40. Mesyats, GA., *Zhurn, Tekhn. Fiz.*, **44**, 1521-4, 1974.
41. Proskurovsky, DI. and Urike, Ya., *Proc. IV-DEIV*, 78-81, 1970.
42. Litvinov, EA., Mesyats, GA., Shubin, AF., Buskin, LM., and Fursey, GN., *Proc. VI-DEIV*, 107-111, 1974.
43. Litvinov, EA., Mesyats, G.A., Proskurovsky, D.I. and Yankelevitch, EG., *Proc. VII-DEIV*, 55-67, 1976.
44. Litvinov, EA., Mesyats, GA. and Shubin, AF., *Sov. Phys. J*, **13**, 537-40, 1970.
45. Bugaev, SP., Litvinov, EA., Mesyats, GA. and Proskurovsky, DI., *Sov. Phys. Usp.*, **18**, 51-61, 1975.
46. Jüttner, B., Private Communication, 1980.
47. Mitterauer, J. and Till, P., *Proc. VI-DEIV*, 95-100, 1974.
48. Charbonnier, FM., Bennette, CJ. and Swanson, LW., *J. Appl. Phys.*, **38**, 627-33, 1967.
49. Dudek, HJ., *Z. Angew. Phys.*, **31**, 243-50 and 331-7, 1971.
50. Utsumi, T., *J. Appl. Phys.*, **38**, 2989-97, 1967.
51. Jüttner, B., *Beitrags Plasmaphysik*, **19**, 25-8, 1979.

6

The Point-Plane Field Emitting Diode

RV Latham

6. 1   Introduction

As a result of the long-standing importance attached to the microprotusion-based breakdown initiation model, previously discussed in Chapter 5, the past twenty-five years has seen a major effort devoted to investigating the field emitting properties of the filamentary structures that are believed to occur naturally on the surface broad-area HV electrodes [1]. More recently, the importance of these studies has been reinforced by a growing interest in the development of specialised electron tube devices employing cold-cathode structures designed for high-current pulsed-field "explosive emission" applications [2,3]. In this context it is also relevant to include the micro-engineered tip arrays that are being developed for vacuum micro-electronics applications [4]. Indeed, as will be discussed in Chapter 15, this same type of micro-emission regime is also assumed to play a key role in the formation and subsequent behaviour of the cathode spots that "drive" vacuum arcs.

Since it is virtually impossible to perform rigorous measurements on the actual structures that occur randomly (in space and time) on broad-area electrodes, a controlled simulation approach has been followed in which an individual, laboratory-fabricated "test" metallic-micropoint emitter is configured as a point-plane vacuum diode. Apart from providing a well-defined field distribution, this electrode geometry also facilitates the use of complementary electric and optical diagnostic techniques for the dynamic study of gap phenomena. Equally, a micropoint cathode is clearly suited for detailed analytical studies into how its surface profile and structure are influenced by the breakdown process. For the reader who wishes to pursue this topic in greater depth than provided by the present text, Mesyats and Proskurovosky [1] have provided a comprehensive account of the outstanding Russian contribution to this subject.

6. 2   Fabrication of Micropoint "Test" Cathodes

It is fortuitous that the technology for manufacturing such tips had previously been developed for the well known research techniques of field electron emission microscopy, pioneered by Müller [5], and later field ion microscopy. A wide range of metals have been studied using these techniques, and details of the various tip preparation procedures can be found in standard works on these subjects [6,7]. In the case of tungsten for example [8], which is one of the most widely studied materials, tips are electrochemically etched from a fine wire (typically ~ 0.1 mm diameter) in a 2M solution of sodium hydroxide using the arrangement of Fig 6.1.

As illustrated by the inset, the etching process takes place at the meniscus, and as soon as the immersed section is "necked off", the etching current drops and a fast acting electronic switch arrests the etching process before the tip is blunted. Fig 6.2 is a scanning electron micrograph of such a tungsten tip with an emitter radius of ~50 nm, which is comparable with the values found for the fusion induced microprotrusions shown in Fig 3.6(b). Since these tips are extremely fragile, and hence difficult to handle, the short wire specimen is normally spot-welded to the tungsten "cross-bar" of a standard specimen mount before etching.

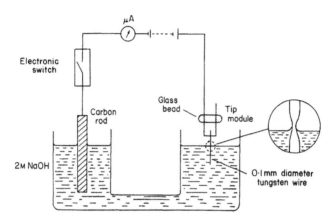

Fig 6.1   The experimental regime used for electrolytically etching micropoint emitters.

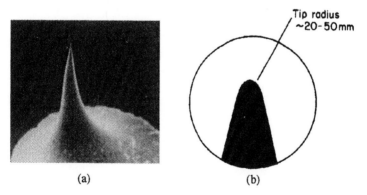

(a)	(b)

Fig 6.2   (a) Scanning and (b) transmission electron micrographs of a micropoint emitter that has been etched from a 0.1 mm diameter tungsten wire.

The basic experimental regime required for studying the emission characteristics of these microtips is illustrated schematically in Fig 6.3. It is seen to consist of a point-to-plane diode configuration, where the microtip cathode C and metallic anode disc A can be heated in situ to high temperatures for their initial out-gassing treatment. This is achieved by passing an electric current through the thermionic "cross-bar" filament TF which serves both to directly heat the cathode to $\gtrsim$ 2500K, and to provide an emission of electrons for the bombardment heating of the anode; the latter being typically insulated via a suitable feed-through to ~ 10 kV. To obtain atomically clean electrode surfaces, the assembly has to be mounted in either a glass or metal UHV chamber capable of reaching pressures $\lesssim$ 10-10 mbar.

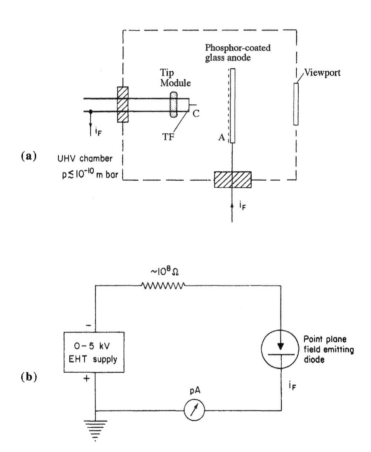

Fig 6.3   (a) A schematic representation of the UHV experimental facility, and (b) the associated electrical circuitry required for studying the field emitting properties of the micropoint emitter C: A is a planar anode and TF is a thermionic filament used for out-gassing both C and A.

In order to give a quantitative interpretation of the data obtained from this experimental regime, it is essential to measure the radius r of the micro-emitter before its assembly in the specimen chamber. This is normally done by mounting the tip module in a suitably modified specimen holder of a transmission electron microscope so that a high resolution profile image of the emitting surface can be obtained. Typically, such a microgap appears as in Fig 6.2(b), from which it is seen that the tip profile closely approximates to a section through a spherically-capped cone. However, the tip radius can also be estimated from empirical relations based on the operating characteristics of the emitter. For example, that of Drechsler and Henkel [9] takes the form

$$r = 0.015 \, \phi^{-2} \, V_{10} \, (\text{nm}) \qquad\qquad 6.1$$

where ϕ is the work function of the emitter measured in electron volts and V_{10} is the voltage that has to be applied across the diode to give an emission current of 10 μA. For this type of emitter, the enhanced electric field E acting at the spherical apex of the tip can then be closely approximated by the relation

$$E = V/fr \qquad\qquad 6.2$$

where V is the voltage applied across the diode and f is a complicated dimensionless geometric function of the tip dimensions and the d-spacing of the diode [10]: however, in cases where d >> r, such as in the field emission microscope, f can be taken as a constant having a value of ~ 5 [6]. It must also be recognised that as one moves over the surface away from the apex of the emitter, the surface field E will fall-off rapidly from the maximum value defined by equation 6.2. As will be described in the following section, the actual surface distribution of E for a given emitter can be obtained from its profile electron micrograph (such as shown in Fig 6.2(b)) by firstly constructing the equipotentials in the immediate vicinity of the emitting surface and then using standard field plotting techniques to determine the local values of E over the tip [10]. Historically, the first quantitative investigation using this type of regime was due to Haefer [11] in 1940 with his classic experimental verification of the Fowler-Nordheim theory of field electron emission. However, with the growing interest in both the potential practical application of micropoint field emission electron sources, and their important role in limiting the insulating capability of high voltage vacuum gaps, Dyke and his co-workers published the first series of important papers describing their detailed fundamental studies on the properties of laboratory-etched micropoint emitters [12,13]. With the advantage of improved vacuum conditions (~ 10^{-12} mbar), and the introduction of sophisticated ancillary experimental techniques for comparing the DC and pulsed-field operational behaviour of such emitters, this work represented a major advance on earlier investigations.

6. 3   DC Emission Characteristics

The external electrical circuitry required for studying the DC current-voltage characteristic of a point-plane field emitting diode regime needs to be no more complicated than that shown in Fig 6.3. In this, a regulated EHT supply provides a variable DC voltage source, a picoammeter measures the total field emission current I_{OF}, and a high series-resistance is included to limit the circuit current should a micro-discharge occur. In practice, it is usual to cautiously "build-up" the characteristic by progressively cycling the applied voltage V through an increasing range until the diode begins to show signs of instability. However, in order to obtain a stable and reversible current-voltage characteristic from the diode assembly shown in Fig 6.3, it is essential to follow an in-situ cleaning procedure *before* applying a high voltage. That used by Dyke and Trolan [13] involved thoroughly out-gassing both electrodes by heating them simultaneously to temperatures in excess of 1300 K until all desorbed gas has been re-absorbed by a tantalum gathering agent, and then to "flash" the emitter by heating it to ~ 2500 K under zero field conditions in order to remove all remaining surface contamination. With a field emission microscopic regime, such as illustrated in Fig 6.3, the effectiveness of this procedure can be directly assessed from the appearance of the emission image on the phosphor screen. For example, an atomically clean, [110] orientated single crystal tungsten emitter would give the characteristic four-lobed image shown in Fig 6.4(i); i.e. reflecting the small variation in work function associated with the specific crystallographic planes that intersect the emitting surface. With clean emitters, it is typically possible to draw stable emission currents of ~ 8 µA and 20 µA from tips having radii of ~ 30 nm and ~ 230 nm respectively. Two alternative approaches have been used for presenting the stable current-voltage data obtained from such emitters. The first, which takes historical precedence, is based on the empirical relation

$$I_{OF} = A \ exp \ (-B/V) \qquad\qquad 6.3$$

where A and B are constants. This formulation was originally due to Millikan and Lauritsen [14] and, because of the excellent linearity of the experimental plots of log I_{OF} against l/V, has subsequently been used by many authors including Dyke and Trolan [13]. It does however have the serious disadvantage that it has not been possible to satisfactorily relate the constants A and B to the physical parameters of the system. In contrast, the alternative approach has the advantage of being based on the well established Fowler-Nordheim electron emission theory; for this reason it has become more favoured by recent authors and will therefore be used in the present treatment.

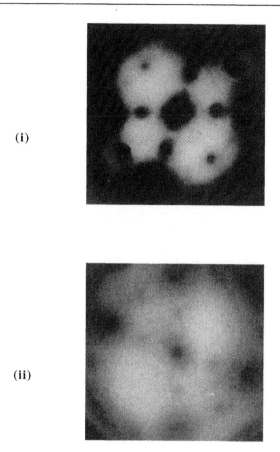

(i)

(ii)

Fig 6.4   (i) The field emission image of an atomically clean tungsten micropoint emitter orientated with the [110] crystal direction parallel to the axis of the wire. (ii) The image of the same emitter, but at higher emission current densities where space-charge blurring is occurring. (From unpublished results of Jüttner.)

Thus for the low-temperature emission regime described by equation 3.9 of Section 3.1(i), the total emission current from a micropoint emitter would be given by

$$I_{OF} = A_e j_{OF} = A_e C_1 E^2 \exp(-C_2/E)$$

in which j_{OF} is the emission current density at the tip of the protrusion where the surface field E has the value defined by equation 6.2, and A_e corresponds to some "effective" emitting area. If however account is taken of the appreciable variation of E, and hence j_{OF}, over the highly curved surface of a micropoint source, i_{OF} should more correctly be expressed in the form

$$I_{OF} = \int_{\substack{emitting \\ surface}} j_{OF}\, dA$$

6.4

To evaluate this integral for the typical microemitter geometry shown in Fig 6.2, and thereby obtain a more meaningful definition of the "emitting area", Dyke et al. [10] applied field plotting techniques to determine the variation of E over the emitter surface and then, assuming a constant work function, used equation 3.2 to compute the corresponding distribution of the low-temperature emission current density j_{OF}. The general findings of these studies can be illustrated schematically with reference to the ideal spherically-capped conical emitter shown in Fig 6.5(i).

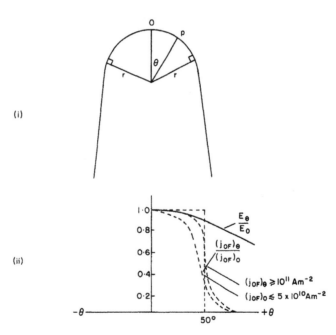

Fig 6.5   (i) The microgeometry of an idealised spherically capped conical emitter.
(ii) The surface variation of (a) the normalised field distribution E_θ/E_O and (b) the low and high emission current density distribution $(j_{OF})_\theta/(J_{OF})_O$. (From Dyke and Trolan [13], with permission.)

Assuming rotational symmetry, all points of constant field on the emitting surface can then be defined in terms of the polar angle θ with an origin of co-ordinates located at the centre of curvature of the spherical cap of the emitter. The normalised field and FEE current density distributions, E_θ and $(j_{OF})_\theta$, then appear as shown in Fig 6.5(ii), where E_O and $(j_{OF})_O$ are respectively the maximum values of E and j_{OF} at the apex of the emitter where $\theta = 0$. From these plots, it can be seen that for all practical purposes, the emitting area A_e can now be defined as the region bounded by $-50° \lesssim \theta \lesssim 50°$ which contributes $\sim 85\%$ of the total emission, and over which the surface field E_θ varies by less than 10%, i.e. we can take

$$A_e \simeq \pi r^2 \qquad\qquad 6.5$$

Fig 6.5(ii) further indicates that with increasing $(j_{OF})_O$, i.e. at higher fields, the total low-temperature emission current I_{OF} can be closely approximated by the product of the peak FEE current density $(j_{OF})_O$ and this newly defined emitting area A_e,viz.

$$I_{OF} = \int_{\substack{emitting \\ surface}} j_{OF}\, dA \cong \pi r^2 \left(j_{OF}\right)_O \qquad\qquad 6.6$$

Substituting for $(j_{OF})_O$ from equation 3.9, and for $E_O \simeq V/kr \simeq V/5r$ from equation 6.2, we obtain finally

$$I_{OF} = 0.04\ \pi C_1\ V^2\ exp\left(-5C_2 r/V\right) \qquad\qquad 6.7$$

indicating that an F-N plot of the I-V characteristics of such an emitter i.e. log (I_{OF}/V^2) versus $(1/V)$, will be linear with a slope of $5C_2 r$ and an intercept of $0.04\pi C_1$ which, significantly, is independent of the emitter radius within the limits of the above approximations.

Experimental plots are found to have the general form shown in Fig 6.6, thus confirming that there is a wide range of fields, viz. $E \lesssim 5 \times 10^9$ Vm^{-1} for which the characteristic is both linear and reversible [8]. Beyond this limiting field E_l however, the emission ceases to obey the Fowler-Nordheim relation but instead begins to follow the well known diode three-halves power law $I_{OF} = PV^{1.5} \propto V^{1.5}$, where P is the perveance of the diode; i.e. indicating that at these high fields the emission is being space charge limited. From their experimental plots, Dyke and Trolan [13] showed that the limiting emission current density for the onset of these space charge effects is

$(j_F)_l \sim 6 \times 10^{10}$ A^{-2}. They also established by a simple calculation that this controlling action occurs within a distance of 2×10^{-8} m from the surface of the emitter where the current density is highest. Under these conditions, the local crystallographic variations of emitter work function, as highlighted by the characteristic emission image of Fig 6.4(i), are smoothed out to give the more diffuse pattern of Fig 6.4(ii).

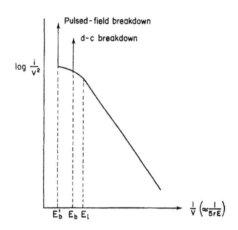

Fig 6.6　　　The typical form of the Fowler-Nordheim plot of the current-voltage characteristics of a micropoint field emitting diode.

As the field is increased beyond E_l, both the emission current and the associated emission image become progressively unstable until, as shown in Fig 6.6, a critical field $E_b \sim 8 \times 10^9$ V m^{-1} and current density $(j_F)_c \sim 5 \times 10^{11}$ Am^{-2} is reached when the diode spontaneously breaks down. The instability associated with this high-field regime, which also manifests itself as a significant "scatter" in the measured values of E_b among emitters, is thought to result from the onset of some sort of time-dependent emitter contamination mechanism that locally modifies the effective work function and hence the emitting regime. Two favoured possibilities are the adsorption of contaminant atoms released from the anode by electron bombardment heating, or the field-assisted surface diffusion of contaminant atoms from the shank of the emitter. This latter type of process can also be activated by the bombardment of ions from the anode and has in fact been considered by Cavaille and Drechsler [15] as a means of promoting crystal growth through the self-diffusion of surface metal atoms to form microprotrusions on broad-area HV electrodes. If the profile of an emitter is re-examined in the electron microscope following a breakdown event, it will be found to have become grossly blunted as shown in Fig 6.7, with clear evidence of tip melting and vaporisation [16], as would follow an "explosive" surface process. As

214

discussed in Chapter 5, the initiation of such an instability will be fundamentally determined by the condition that the emission current density reaches some critical value $(j_F)_c$ where its heating effect (Joulian and Nottingham) is sufficient to cause local vaporisation of the emitter surface and the consequent generation of a microplasma.

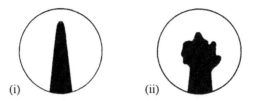

(i)   (ii)

Fig 6.7   A comparison of the profiles of (i) a virgin micropoint emitter and (ii) one that has suffered a breakdown event.

6. 4   Microsecond Pulse Measurements

With the recognition that the electrical breakdown of these micropoint-emitter diodes was a sub-microsecond "explosive" phenomenon involving the thermal response of the emitter, it became evident that high-speed pulse techniques would have to be employed if more detailed information was to be obtained about the evolution of the breakdown mechanism. The early experiments were again pioneered by Dyke and his co-workers in the USA [13,16], who used microsecond pulse techniques to study how the growth of the emission current waveform depends on the magnitude and duration of the applied field. More recently, the Russian groups of Fursey and Mesyats [1,17-22] have used more refined nanosecond pulse techniques to resolve the sequence of events involved in the breakdown mechanism of these diodes. These investigations also incorporated synchronised high-speed optical recording techniques to study the development of the microplasma or "cathode flare" that is a precursor to the actual breakdown of the diode.

6.4.1   Experimental Regime

The practical requirements for this type of impulse measurement are principally dictated by the demands of high speed pulse technology. Thus, if nanosecond rise times are to be achieved, it is essential to minimise all stray capacitances, and to ensure that the external drive and detection circuitry are impedance-matched throughout. In the type of system developed initially by Mesyats et al. [18] and later by Fursey and Zhukov [19], which is illustrated schematically in Fig 6.8, these conditions have been satisfied in part by generally reducing the dimensions of the electrode assembly. More important

215

however, the planar anode of the DC experimental regime of Fig 6.3 has been replaced by a "transparent" earthed grid G, whilst an additional "collector" electrode C has been introduced immediately behind G. This new triode arrangement has the important advantage of reducing the anode-to-cathode capacitance and thereby minimising the interference from the displacement currents that must inevitably flow following the application of a voltage pulse and that would otherwise mask the true electron current.

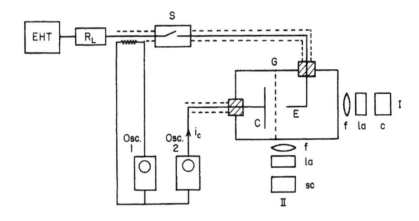

Fig 6.8   The experimental regime used for studying the electronic and optical processes associated with the pulsed-field breakdown of micropoint emitters. (From Fursey and Zhukov [19], with permission).

The omission of an electron bombardment anode heating facility is acceptable for these fast impulse experiments provided the system receives a thorough external bake-out before sealing off the test diode. This is because the thermal response time of the anode is very much longer than the rise time of the emission current so that its surface processes can have no influence on the emitter behaviour. It was also found possible to eliminate the cathode heating arrangement of Fig 6.3, since emitters can be satisfactorily cleaned by a reverse-polarity field-desorption procedure. The orthogonally orientated optical monitoring systems I and II shown in Fig 6.8, incorporating focusing f, light-amplifying la and camera c modules, are used respectively for studying the reference electron emission image displayed on C (i.e. if phosphor-coated), and the photon emission from interelectrode microplasma processes. Referring finally to the associated electrical circuitry, the negative-going high voltage pulses that are applied to the micropoint emitter are derived from an EHT supply that is electronically switched by a fast acting three electrode switch S that is filled with nitrogen at a pressure of 15 atm. Oscilloscopes 1 and 2, that are synchronously triggered from the initial fast

216

pulse resulting from the charging of the transmission line, record respectively the total emission current and the collector current. For minimal pulse distortion, it is vital to impedance-match the transmission lines to the test diode assembly.

6.4.2 Pulsed Emission Characteristics

We shall consider first the earlier experiments pioneered by Dyke and his collaborators [13,16], and later developed by Fursey [23], that used μ-second pulse techniques to investigate the thermal response of micro-emitters. These showed that for emission current densities $j_F \lesssim (j_F)_L$ (see Fig 6.6), the total current transient had the type of waveform illustrated schematically in Fig 6.9(i), where the initial spike is due to the displacement current associated with the charging of the inter-electrode capacitance and the following constant-current plateau region is indicative of stable emission conditions.

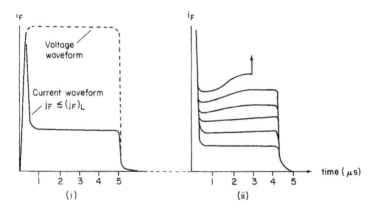

Fig 6.9   (i) The pulsed-field current response of a micropoint emitter when $j_F \lesssim (j_F)_L$. (ii) A family of similar measurements showing the onset of the "tilt" phenomenon as j_F is increased from $(j_F)_L$ to $(j_F)_c$.

If the pulsed-field current-voltage characteristic of the emitter is now measured by this technique for the field range corresponding to $j_F \lesssim (j_F)_L$, the data are found to give an identical F-N plot to the DC behaviour shown in Fig 6.6. When these pulse measurements are extended to higher fields, where the DC emission current density would exceed $(j_F)_L$, the current waveform begins to develop an upward "tilt" as illustrated in the traces of Fig 6.9(ii); i.e. the current exhibits a spontaneous growth at constant field, approaching a saturated equilibrium value with a time constant $\tau \sim 1\text{-}2~\mu\text{s}$. Provided these saturated emission current values are used, the F-N plot of the emitter continues to closely follow the DC behaviour of Fig 6.6 into the non-linear region. However, with pulse experiments, breakdown is more predictable and generally occurs at a somewhat higher field E_b corresponding to a pulsed-field critical emission current density $(j_F)_c \sim 7.8 \times 10^{11}$ Am^{-2}.

(i)

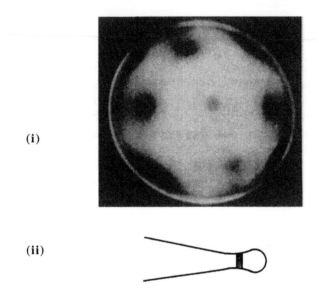

(ii)

Fig 6.10   (i) The space-charge blurred emission image with associated "ring" phenomenon that occurs when $j_F \rightarrow (j_F)_c$, and is coincident with the onset of the "tilt" phenomenon illustrated in Fig 6.9(ii). (From Fursey [23], with permission.) (ii) The characteristic "necked" structure of an emitter that has been overloaded and become thermally unstable. (From Dyke, Trolan, Martin and Barbour [16], with permission.)

The associated emission image, as recorded by the open-shutter camera of optical system I in Fig 6.8, is found to have the same general form as found with DC experiments, including the space charge "blurring" effect of Fig 6.4(ii) that occurs when $j_F \gtrsim (j_F)_L$. There is however an additional feature that develops at the highest pulse fields used, i.e. just prior to breakdown and coincident with the onset of current "tilt". As illustrated in Fig 6.10(i), this takes the form of a bright ring structure, or halo, that skirts the emission image and can reach intensities considerably in excess of the rest of the image [16]. Subsequently, Fursey [23] showed that this halo could in fact be resolved into several well-defined concentric rings.

The evident correlation between these "tilt" and "ring" phenomena led Dyke et al. [16] to attribute them to a thermally enhanced field electron emission mechanism, viz. T-F emission as discussed above, which begins to contribute significantly to the total emission current as local regions of an emitter reach temperatures $\gtrsim$ 2000 K. This explanation was supported by the

218

observations that the profile of emitters that had previously exhibited these effects were characterised by having a "necked" structure as shown in Fig 6.10(ii), and that the electrons forming the ring came from the relatively low-field shaded region situated very near the neck of the emitter where the internal current density, and hence the Joulian heating effect, will be a maximum; i.e. creating a particularly favourable location for the onset of T-F emission. However, more recent scanning electron micrographs of tips that have become thermally unstable, i.e. such as shown in Fig 3.14, have indicated rather that the surface of the shaded neck region of Fig 6.10(ii) acquires a corrugated or "bellows-like" structure whose periodic field enhancing properties could well account for the multiplicity of the rings forming the halo surrounding the emission image shown in Fig 6.10(i). It also follows from this model that the characteristic 1-2 μs rise time τ_t of the "tilt" current must be related to the establishment of a thermal equilibrium in the emitter. An impressive check on this assumption can be made by observing how the maximum stable emission current density depends on the duration t_p of the applied voltage pulse, particularly when the pulse length is reduced to the situation where $t_p \lesssim 0.2\tau_t \lesssim 0.5 \mu$s, so that the temperature increase of an emitter will be limited to well below its maximum equilibrium value. The earliest experiments of this kind using microsecond pulse techniques [16] confirmed that $(j_F)_c$ could be increased by an order of magnitude to $\sim 5 \times 10^{12}$ Am^{-2}: however, the more recent nanosecond experiments to be discussed in the following section showed even more dramatic increases of $(j_F)_c$ approaching $\sim 10^{15}$ Am^{-2}.

6. 5   Nanosecond Pulse Studies of Arc Initiation

Although the above microsecond pulsed field investigations had shown that the vacuum arc is established within a time of $\lesssim 10^{-8}$ s, they were unable to provide information about the final stages of the physical mechanism that leads to the arc. In particular, it was not possible to resolve the time-dependence of the "explosive" growth of the emission current through two to three orders of magnitude, such as occurs from the plateau region on the highest current trace of Fig 6.9(ii) to breakdown. It is not surprising therefore that a fundamental misconception arose in the early understanding of the phenomena; viz. that the arc stemmed from a regenerative T-F emission process that led to a rapid vaporisation of emitter material which, after being ionised, progressively neutralised the electronic space charge that had previously limited the emission current and so created the conditions for a *smooth* but rapid avalanche build-up of the diode current to breakdown. However, with the introduction of pulse techniques having a nanosecond resolution, it was soon shown that the current waveform accompanying arc formation had a quite different form from that predicted by the above model. This finding therefore stimulated the development of new diagnostic techniques for obtaining more information about the physical processes involved and hence a revised explanation of the arc initiation mechanism.

6.5.1   Optical Measurement of Cathode Flare Velocities

The "cathode flare" model, which has now become widely accepted, was based on complementary measurements of the light emission and current growth accompanying the establishment of an arc across an electrode regime such as shown in Fig 6.8. Thus the optical system II, incorporating an open-shutter high speed streak camera SC, records the temporal evolution of all pre-arc optical processes occurring in the gap, whilst oscilloscopes 1 and 2 record respectively the associated emitter and collector currents. The amplitude of the arc-inducing over-voltage pulses used in these experiments is carefully chosen so that "explosive" emission will be initiated just as the pulse reaches its maximum value. From this type of optical observation, Mesyats et al. [24] were able to establish that an arc is initiated from a pin-point of light that appears at the tip of the microemitter and then, as shown in the sequence of Fig 6.11, rapidly develops into a luminous spherically-shaped flare that grows across the gap to establish an arc. The underlying physical mechanism that is now thought to be responsible for this process in the case of an isolated micropoint emitter [17,23] is as illustrated in the more detailed sequence of Fig 6.12.

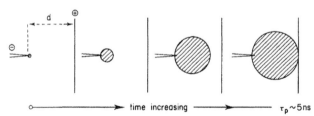

Fig 6.11   A schematic illustration of the initiation and growth of a "cathode flare" until it bridges the electrode gap.

A breakdown event starts with the generation of a localised micro-plasma (electrons and positive ions) that surrounds the micro-emitter following the ionisation of cathode material that has been thermally vaporised as a result of a high T-F electron emission current density. This microplasma then begins to spontaneously expand, but because its constituent electrons have very much higher thermal velocities than the heavy metallic ions, its envelope will rapidly become populated predominantly by electrons. These outer electrons now come under the influence of the gap field and so break away from the plasma sphere to be collected by the anode.

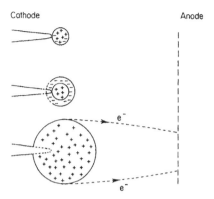

Fig 6.12   An illustration of how it is possible for a charge separation to occur in the expanding microplasma, and hence an enhanced cathode field due to the accumulation of positive space charge.

As a result of this charge separation, the emitter will experience a greatly enhanced field due to the uncompensated positive space charge remaining in the expanding flare, and this, in turn, will lead to an increase in both the emission current density and rate of evaporation of emitter material to levels that are more than adequate for sustaining the growth of the flare until it reaches the anode and establishes an arc across the diode. From a knowledge of the d-spacing of the diode and the photographically measured time for the flare to cross the electrode gap, it was estimated that the expansion velocity of the flare envelope was $\sim 10^4 \, \text{ms}^{-1}$. It should also be noted that the value of this parameter is determined almost entirely by the space charge density; i.e. the externally applied field has very little influence.

6.5.2   Measurement of Explosive Emission Currents

According to the model illustrated in Fig 6.12, an electron current would be recorded at the collector of the experimental regime shown in Fig 6.8 both before and during the time that the flare is crossing the gap: in fact, with a grid that is effectively transparent to electrons, this current could be identified with the total emitter current. However, with the establishment of an arc between the emitter and grid, the gap field would collapse and with it the collector current $i_c (\approx i_F)$; the grid current on the other hand, which would previously have been minimal, will now increase by several orders of magnitude. This behaviour has indeed been confirmed experimentally by complementary electrical studies of the flare phenomenon, where Fig 6.13 illustrates the typical relation between i_c and the applied voltage V.

221

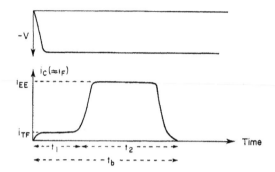

Fig 6.13   The general form of the applied over-voltage waveform and the resulting "explosive emission" current transient, showing the emitter heating time t_1 and the commutation time t_2.

The characteristic initial plateau of the i_c waveform corresponds to a quasi-stable temperature-assisted FEE regime whose duration t_1 is very sensitive to the over-voltage coefficient of the applied pulse α_v, defined as the ratio of the amplitude of the pulse being used to the slow-rise DC voltage that would just maintain stable emission. This regime is terminated at $t = t_1$ by the onset of explosive emission where the current rapidly increased by two or more orders in a time ~1 ns to a new value I_{EE} whose magnitude is also very sensitive to α_v. There then follows a second plateau region of duration t_2, the "commutation time", that can be identified as the time during which the cathode flare is expanding across the gap. Hence, from a knowledge of the emitter-to-grid separation, these measurements were able to confirm that cathode flare velocities are in the range $2\text{-}3 \times 10^4$ ms^{-1} and, as anticipated, are almost independent of the gap field.

As already illustrated in Fig 6.7, this type of event inevitably leads to severe and irreversible damage to the emitter. It follows therefore that if quantifiable information is to be obtained about the dependence of t_1 and I_{EE} on the over-voltage coefficient α_v and the applied field E, it is necessary to conduct single-shot experiments on a series of similar emitters whose initial tip profiles have been electron optically recorded for subsequently calculating E as described in Section 6.2.1. To minimise space charge effects, it is also desirable to use the sharpest possible emitters with tip radii $\lesssim 100$ nm. With this technique, Fursey and Zhukov [19-21] were able to compile families of explosive emission characteristics such as illustrated in Fig 6.14. As will be seen, these transients correspond to only a limited range of α_v from ~ 1.2 to 1.4 and therefore clearly highlight how very sensitive both t_1 and I_{EE} are to E; for example, I_{EE} can readily increase from ~ 0.5 A to ~ 100 A for a 20% increase in E.

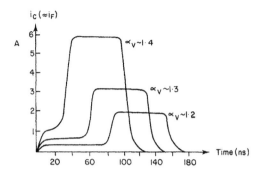

Fig 6.14 A family of idealised "explosive emission" current transients illustrating the strong influence of the over-voltage coefficient α_V of the applied pulse.

It must however be re-emphasised that this sort of dramatic response can only be obtained when fast-rise pulses are used; i.e. indicating that the explosive electron emission mechanism (EEE) must depend on the rate at which energy can be supplied to the system, or more specifically to the flare plasma, since it has proved impossible to explain this finding in terms of the emitter behaviour alone.

This dependence of I_{EE} on E also implies that there should be a corresponding relationship between I_{EE} and j_F; indeed a similar indication follows from a consideration of the important ionising role played by the electron emission in initiating the formation of a microplasma. Accordingly, Fursey and Zhukov [19] were able to show that a plot of log I_{EE} against the emission current density j_F, as calculated from the Fowler-Nordheim theory (equation 3.2) using values of E_m computed from the initial emitter profiles and applied voltage, yielded the good straight line of Fig 6.15 within the measured range $10^{12} \lesssim j_F \lesssim 10^{13}$ Am^{-2}, indicating an empirical relation of the form

$$I_{EE} \propto exp \ (kj_F) \qquad\qquad 6.8$$

where k is a constant. From a knowledge of I_{EE} and the dimensions of the enlarged emitting area, as given by the exploded tip profile of Fig 6.7(ii), it can further be shown that the mean explosive emission current density j_{EE} is typically $\sim 10^{11}$-10^{12} Am^{-2}, which is an order of magnitude lower than the corresponding FEE current densities drawn initially from undamaged emitters. This rather surprising finding can however be plausibly explained on the basis that I_{EE} is not in fact uniform over the surface of an exploded emitter, but rather reaches very high values at a multiplicity of sub-sites where the tip profile locally enhances the surface field.

223

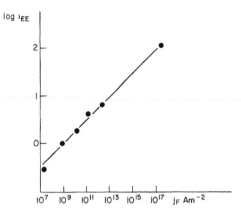

Fig 6.15   Illustrating the exponential dependence of the explosive emission current I_{EE} on the electron emission current density j_F. (From Fursey and Zhukov [19], with permission.)

Although this detailed discussion of the breakdown processes that occur in a point-plane diode was justified on the grounds that it provided a valuable insight into the mechanisms occurring in the more complicated regime of a broad-area gap, there is one important respect in which the model developed in these last two sections has to be qualified. It will be recalled from the discussion in Section 3.4.1 that, with broad-area electrodes, the explosive emission current during the commutation period is space-charge limited and given by equation 5.30. This therefore implies that the degree of charge separation, and hence cathode field enhancement, envisaged by Fig 6.12 cannot occur with this type of regime: indeed, as already pointed out in this section, it is only with the sharpest emitters ($r \lesssim 100$ nm) that space-charge effects can be neglected. The absence of an excessive positive space-charge field in the vicinity of an emitter on a broad-area electrode has also been verified experimentally by Proskurovsky and Puchkrov [25] by showing that the sheath voltage between the cathode and plasma flare is only ~ 50 volts, i.e. of the same order as in DC arcs.

6.5.3   Rate of Vaporisation of Emitter Material

From pairs of micrographs such as shown in Fig 6.7, it is clear that cathode material has been vaporised to "feed" the cathode flare of an explosive emission event. The question however arises as to how the rate of vaporisation of emitter material varies with the evolution of the flare. To obtain this sort of detailed information, and thereby gain further insight into the explosive emission mechanism, a special pulse-arrest technique was developed by Bazhenov et al. [26] and later extended by Proskurovsky et al. [29] and Zhukov and Fursey [20,21]. Starting with a series of freshly prepared emitters having similar initial profiles, each of them was then subjected to progressively longer over-voltage pulses of constant amplitude so that the explosive emission mechanism was arrested at various stages in the I_{EE} current transient of Fig 6.13. These "partially exploded" emitters were then re-examined in an electron microscope and from the change in their profiles it was possible to estimate the mass of material vaporised at each stage. A typical sequence of superimposed profiles might appear as illustrated in Fig 6.16, from which it may be concluded that the highest rate of vaporisation of emitter material occurs during the initial rapid increase of the explosive emission current; thereafter, although the micro-profile may change in detail, with the appearance of numerous sub-micron protrusions extruded from the molten surface of the emitter by the applied field, the macro-profile becomes relatively stabilised with material vaporisation falling off rapidly to a very much lower and nearly constant rate. It should be added that the formation of these sub-micron protrusions and their subsequent role in initiating emitter "side flares" is a very important mechanism in the successful operation of commercial devices such as portable "flash" X-ray tubes and electron accelerators that use explosive emission cathodes. (See also Chapter 10.)

Fig 6.16   A schematic illustration of how the end-profile of a tip is progressively blunted following a succession of over-voltage pulses. (After Fursey and Zhukov [19].)

6.5.4   Oscillatory Instabilities with High Over-Voltage Pulses

If the amplitude of the over-voltage pulse being applied across a diode is increased beyond the level discussed in the previous section, there is a further dramatic rise in the explosive emission current. This therefore leads to an inevitable increase in the physical complexity of the processes occurring in the gap: this is particularly true with broad-area electrode configurations such as exist, for example, in a practical device like a power X-ray tube, where the current levels can temporarily approach the kA or MA range. A characteristic feature of this type of regime is the occurrence of a wide variety of effects associated with the onset of plasma instabilities. Although this topic is of considerable technological importance, an adequate review of its extensive literature is beyond the scope of this book. Accordingly, the interested reader is referred to such standard introductory texts on arc phenomena as those cited in Chapter 15.

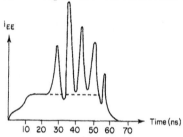

Fig 6.17   A schematic illustration of the oscillations that can appear on the "explosive emission" current transient at high over-voltages. (After Korop and Plyutto [27,28] and Proskurovsky et al. [29].)

It is however relevant to refer briefly to one type of investigation which is a direct extension of the experiments described in the previous sections. Thus, if the explosive emission current transient measurements of Fig 6.14 are extended to even higher over-voltage pulses [27-29], perhaps $\geq$ 100 kV for micronradius tips, a new mechanism comes into operation that gives rise to an oscillatory instability in the current transient. This is illustrated in Fig 6.17, where the amplitude of the superimposed 30-60 nanosecond current spikes can be typically 2-5 times greater than the level of the "plateau" current shown dotted in Fig 6.17. It was also established by mass spectrometry techniques that these current spikes coincide in time with intense bursts of high-energy ionic bombardment of the emitter. More specifically, these ions were also identified as having originated from the vaporisation of emitter material, and that they had subsequently been re-accelerated back towards the cathode to acquire energies comparable with the gap potential. To account for these new experimental observations, it was necessary to revise the simple cathode-flare model of the explosive emission mechanism. Thus, recognising that the expanding flare plasma is electrically neutral and represents an equi-potential volume at virtually cathode potential, it follows that the external voltage V applied across the diode will be dropped across a diminishing "effective" gap bounded by the advancing flare envelope and the anode. This means that by the time the flare has half crossed the "real" gap, a very high field will be acting at its surface; sufficiently high in fact for it to penetrate the plasma and create a flux of "hot" electrons that further depletes the envelope region of negative charge and thereby greatly enhances the charge separation mechanism existing with the simple flare model. This process will therefore leave a highly localised positive space charge density in the vicinity of the plasma envelope and a consequent re-distribution of the gap potential as shown in Fig 6.18(i) and (ii). The resulting positive potential peak, which can approach ~ 0.8 of the voltage applied across the gap, therefore gives rise to a strong electric field that will accelerate the positive ions in its vicinity back to the cathode [27,28], with energies approaching the maximum they could require from the total gap voltage.

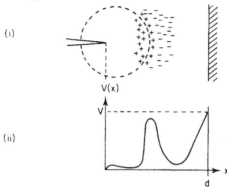

Fig 6.18   To illustrate how a potential "peak" can form at the envelope of the expanding plasma flare where charge separation has occurred. (After Korop and Plyutto [27,28].)

This flow of energetic ions then heats the emitting surface to a very high temperature and so gives rise to the initial current peak observed in Fig 6.17. However, at this stage, the thermally enhanced electron flux is sufficient to neutralise the positive space charge, so that the potential hump of Fig 6.18(ii) collapses and, with it, the heating effect of the positive ions. The emission current therefore falls until it reaches a value where the potential build-up for the next current peak can again be initiated. It follows that these oscillations will continue until the flare has crossed the gap.

6. 6 References

1.   Mesyats, GA and Proskurovsky, DI, *In* "Pulsed Electrical Discharges in Vacuum", Springer-Berlin, 1989.

2.   Posocha, LA and Riepe, KB, *Fusion Technol.*, **11**, 576-610, 1987.

3.   Hsu, TY, Hadizad, P, Liou, RL, Roth, G, Frank, K and Gundersen, MA, *J.Vac, Sci. Technology*, **11**, 1868-1872, 1993.

4.   Brodie, I and Spindt, CA, *Adv. Electronics Electron Phys.* , **83**, 2-106, 1992.

5.   Müller, EW., *Z. Physik*, **106**, 541-50, 1937.

6.   Gomer, R., In "Field Emission and Field Ionisation", Oxford University Press, Oxford, 1961.
7.   Dobretsov, L.N. and Gomoyunova, MV., *In* "Emission Electronics", Kater Press Binding, Israel, 1971.
8.   Van Oostrom, AGS., Philips Res.Rep. Suppl. No.1, 1966.
9.   Drechsler, M and Henkel, E., *Z. Angew. Phys.*, **8**, 341-6, 1954.
10. Dyke, WP., Trolan, JK., Dolan, WW. and Barnes, G., *J. Appl. Phys.*, **24**, 570-6, 1953.
11. Haefer, RH., *Z. Physik*, **118**, 604-22, 1940.
12. Dyke, WP. and Dolan, WW., *Adv. Electronics Electron Phys.*, **8**, 89-185, 1956
13. Dyke, WP. and Trolan, JK., *Phys. Rev.*, **89**, 799-808, 1953.
14. Millikan, RA. and Lauritsen, CC., *Proc. Nat. Acad. Sci. U.S.*, **14**, 45-9, 1928.
15. Cavaille, JV. and Drechsler, M., *Proc. VII-DEIV.* 217-221, 1976.
16. Dyke, WP., Trolan, JK., Martin, EE. and Barbour, JP. *Phys. Rev.* **91**, 1043-54, 1953.
17. Fursey, GN. and Vorontsov-Vel'yeminov, PN., *Sov. Phys.-Tech. Phys..*, **12**, 1377-82, 1968.
18. Mesyats, GA., Rotshteyn, VP., Fursey, GN. and Kartsev, GK., *Sov. Phys.-Tech.Phys.*, **15**, 1202-4, 1971.
19. Fursey, GN. and Zhukov, VM., *Sov.Phys.-Tech. Phys.*, **21**, 176-81, 1976.
20. Zhukov, VM. and Fursey, GN., *Sov. Phys.-Tech. Phys.*, **21**, 182-6, 1976.
21. Zhukov, VM. and Fursey, GN., *Sov. Phys.-Tech. Phys..*, **21**, 1112-17, 1976.
22. Bougaev, SP., Lopatin, VV., Litvinov, EA. and Mesyats, G.A., *Proc. V-DEIV*, 37-41, 1972.
23. Fursey, GN., *Proc. III-DIEV*, 113-118, 1968.
24. Mesyats, GA., Dougaev, S.P., Proskurovsky, DI., Eshkenazy, VI. and Urike, Ya.Ya., *Proc. III-DIEV*, 212-17, 1968.
25. Proskurovsky, DI. and Puchkarov, VF., *Zhurn. Tekhn. Fiz.*, **49**, 2611-18, 1979.
26. Bazhenov, GP., Litvinov, EA., Mesyats, GA., Proskurovsky, D.I, Shubin, AF. and Yankelevich, EP., *Sov. Phys.-Tech. Phys.* **18**, 795-9, 1973.
27. Korop, ED. and Plyutto, AA., *Sov. Phys.-Tech. Phys.*, **15**, 1986-9, 1970.
28. Korop, ED. and Plyutto, AA., *Sov. Phys.-Tech. Phys.*, **18**, 830-1, 1971.
29. Proskurovsky, DI., Rotshteyn, VP., Shubin, AF. and Yankelevich, E.B., *Sov. Phys.-Tech. Phys.* , **20**, 1342-6, 1975.

7

Microparticle Phenomena

RV Latham

7.1   Origin and Characteristics of Microparticles

An inevitable consequence of the standard electrode preparation procedures described in Chapter 2 is that a pre-operational electrode surface will be characterised by having a finite number of microscopic particles loosely adhering to it. These will originate from the various stages of mechanical polishing, and may be in the form of either partially embedded impurity particles of the polishing medium (e.g. alumina or diamond) or filamentary structures that have been torn from the surface: some idea of the complex microscopic structure of a typical "virgin" electrode is given by the scanning electron micrograph of Fig 2.15. Alternatively, they can be merely dust particles that become attached to an electrode surface by Van der Waals forces during the many handling manipulations of its final assembly. Another important source of microparticles that has to be added to this list are those originating from thermal instabilities at either the cathode or anode "hot spots" associated with the electron emission based processes discussed in Chapter 5. However, for "commercially" polished electrodes, it could reasonably be argued that embedded impurities and dust particles probably provide the most prolific source of microparticles: this is particularly true with large gaps where electron emission currents are generally vanishingly small. The electrode geometry can also have an important influence in determining the origin and multiplicity of microparticles. Thus for a symmetric plane-parallel gap formed by identically prepared electrodes having a separation d and supporting a voltage V there will be a uniform gap field $E = V/d$, so that each electrode will, on average, donate a similar number of microparticles per unit area. Consider on the other hand the case of an asymmetric electrode pair that has a non-uniform gap field, say a sphere-to-plane geometry with an axial separation d and supporting a similar voltage V; it is clear that the higher field on the surface of the sphere will result in a selectivity for microparticles to originate from this electrode. It must be added however that this simple argument breaks down if dissimilar materials are used for the electrodes, or their methods of surface preparation differ in any way.

7.1.1   Early Theories

The important role that can subsequently be played by such microparticles in initiating the electrical breakdown of a vacuum gap was first appreciated by Cranberg and quantified in his famous "clump" hypothesis [1], which was the name he coined to describe such microparticles. He recognised that in the presence of a high gap field there would be strong electro-mechanical forces tending to detach such protruding microfeatures from their highly charged parent electrode and that if this occurred they would enter the gap as charged particles to be accelerated to high velocities during their transit to the opposite electrode. At impact, these micro-projectiles would dissipate their kinetic energy U_k as either heat or mechanical shock energy and that if the impact energy density exceeds some critical value, dependent on the colliding materials, there would be sufficient vaporisation of the particle and target electrode to form a localised microplasma that could initiate the breakdown of the gap. Thus, for a particle of given charge Q_p and mass M_p crossing a gap d to which the applied voltage is V

$$U_k = \frac{1}{2} M_p v_i^2 = Q_p V$$

7.1

where v_i is the terminal or impact velocity. It follows that if the cross-sectional area of the particle is A, the impact energy density U_k/A will be given by

$$\frac{U_k}{A} = \frac{Q_p V}{A}$$

where Q_p/A is the charge density on the particle. Now as will be discussed shortly, the charge density acquired by a particle will be proportional to the macroscopic field E, (i.e. $Q_p/A \propto E$), and since in general $E \propto V$, the above can be expressed as

$$\frac{U_k}{A} \propto EV = C'EV \frac{U_k}{A}$$

where C' is a constant. In the particular case of a plane-parallel gap $E = V/d$ so that

$$\frac{U_k}{A} = C' \frac{E^2}{d} = C'V^2 d$$

For a critical impact, where the impact energy density approaches some critical value $(U_k/A)_c$ with $E \rightarrow E_b$ and $V \rightarrow V_b$, E_b and V_b being respectively the breakdown field and potential, Cranberg's hypothesis will be expressed as

$$\frac{U_k}{A} \geq C'E_b V_b \geq C'dE_b^2 \geq C'V_b^2/d$$

or

$$E_b \geq C_c^{\frac{1}{2}} d^{-\frac{1}{2}}$$

7.2(i)

and $\qquad V_b \geq (C_c\, d)^{\frac{1}{2}}$ $\qquad\qquad\qquad$ 7.2(ii)

where $C_c = (1/C'(U_k/A)_c$ is the critical constant as originally defined by Cranberg. The value of C_c clearly involves several numerical constants, including a factor to account for any local field enhancement at the electrode surface where the particle became detached: it will also vary somewhat with the mechanical and thermal properties of the electrode surface. However, a typical value quoted by Cranberg as being appropriate to a wide range of experiments was $C_c \sim 10$ (MV)^{2}m^{-1} for d measured in metres and V_b in megavolts.

In support of his theory, Cranberg [1] collated several sets of independent breakdown data and showed that V_b had the required mean $d^{1/2}$ dependence. He also cited some AC breakdown measurements previously referred to in Section 2.6, where a 2800 MHz, 2 MV source was applied across 5 cm gaps and resulted in very much higher breakdown voltages than could be achieved with a similar amplitude DC voltage: a finding that can be readily explained when it is appreciated that in such an experimental regime, particles would have insufficient time to be accelerated across the gap during any half-cycle of the applied field so that particle-initiated breakdown would tend to be suppressed. Another important feature of this theory noted by Cranberg was that it satisfactorily accounted for the spontaneous breakdown events observed with large gaps where there was no measurable prebreakdown currents and hence no possibility of electron emission based breakdown mechanisms operating.

The basic Cranberg hypothesis just described was further refined by Slivkov [2] who introduced two physical constraints to make the model more physically realistic. Firstly, he assumed that the particle must have sufficient energy to be totally vaporised on impact: this implies that there will be a critical impact velocity and hence a maximum particle radius R_{max} for which this requirement will be satisfied. Secondly, he proposed that the impact must generate sufficient vapour to create the necessary conditions for initiating a localised discharge: this requirement therefore indicates a particle radius R_{min} for breakdown to be initiated. The precise numerical values of R_{max} and R_{min} will inevitably depend upon the particular experimental regime being considered; however their probable range for typical experimental conditions will be discussed in greater detail later in this chapter (see Section 7.3). With the above two constraints incorporated in the basic particle impact model, the modified Slivkov breakdown criterion takes the form

$$V_b = C_s\, d^{0.625} \qquad\qquad\qquad 7.3$$

where C_s is again a constant whose value depends on the experimental conditions. Comparing this relation with equation 7.2, it will be seen that the Cranberg and Slivkov gap dependencies of $V_b \propto d^{1/2}$ and $V_b \propto d$ respectively, are very similar and consequently difficult to discriminate between experimentally. The available experimental evidence does however generally support the above theoretical prediction that Vb should show a power law gap dependence of the form $V_b \propto d^\alpha$, although there is a very considerable scatter in the values found for the power α. From a review of typical experimental data [see Section 2.5.3 it is found that in most cases α lies in the range 0.4-0.7, so that a mean practical formulation of the Cranberg-Slivkov breakdown criterion becomes

$$ V_b = C_{cs}\, d^{0.55} \qquad\qquad 7.4 $$

Although the formulation of these early theories of microparticle initiated breakdown was subject to some criticism, the existence of the physical phenomenon became widely accepted. It was not surprising therefore that a second generation of more refined theories evolved that considered in very much more detail the three main aspects of the mechanism, viz, the origin of microparticles, their in-flight behaviour and their subsequent impact properties.

7.1.2   "Primary" and "Secondary" Microparticles

In the context of the present discussion on the initiation of breakdown by naturally occurring microparticles (i.e. as opposed to those artificially introduced into a gap for experimental purposes), the term "primary" refers to those isolated microparticles that initiate the chain of events that culminates in breakdown: they are in fact the same loosely attached microstructures or embedded impurity granules referred to earlier in this chapter that are inevitably present on a virgin electrode surface. "Secondary" microparticles on the other hand are those that are created as a consequence of a micro-discharge or breakdown event involving the local fusion of electrode material such as occurs in the electron emission-based breakdown-initiating mechanisms discussed in Chapter 3, or from a Cranberg type of explosive impact of a "primary" microparticle in which a shower of molten droplets is produced by the "coronet" effect discussed in the following section (see Fig 7.2(b)) which solidifies to form "secondary" microparticles. It follows that once a population of secondary microparticles has been generated in the gap they will subsequently represent a serious hazard to its stability since each of them is potentially capable of initiating a secondary breakdown. The following sections of this chapter are devoted to an analysis of the physical properties of microparticles.

7.1.3   Microparticle Charge Acquisition and Electromechanical Detachment Criteria

The charge carried by a free microparticle is derived from the field-induced surface charge existing on the original microfeature immediately before its detachment from the parent electrode. Now since the surface charge density σ at any given point is directly proportional to the local microscopic field acting at that point, i.e. $\sigma = \varepsilon_0 E$, it follows that at a protruding microstructure where the surface field is locally enhanced, there will be a correspondingly enhanced surface charge density. This observation implies that the charge acquired by a particle originating from such a feature will be strongly dependent on its shape, with rounded structures having generally lower charge-to-mass ratios than elongated structures. Equally, the electromechanical forces tending to detach such surface structures F_d will induce varying internal elastic stresses that will again depend on their shape, with the most vulnerable structures being those that are only attached to the parent electrode by a constricted stalk so that there is a stress enhancement. These conclusions can be quantitatively illustrated with reference to a detailed theoretical analysis by Rohrbach [3] of the charge acquisition and detachment processes as applied to the idealised geometries of Fig 7.1, i.e. similar to those used for microemitter models in Chapter 3. Thus, for spherically based structures such as Fig 7.1(c) and (d), where $\gamma \; (= h/r) \gtrsim 5$,

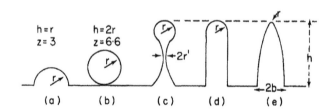

Fig 7.1   A selection of idealised electrode surface microstructures that have been analysed as potential microparticles. (After [3].)

we can use the approximation of equation 4.17, viz. $\beta(\gamma) = 2 + h/r$, in which case the charge acquired by the resulting microparticle would be

$$Q_p\,(\gamma) \;=\; 4\pi\varepsilon_\circ\, h^2 \frac{(\beta - 3)}{(\beta - 2)^2} E \qquad\qquad 7.5$$

where, as usual, E is the uniform macroscopic field in a plane-parallel gap. For geometries where $h \sim r$, such as for the cases illustrated in Fig 7.1(a) and (b) of a spherical particle that is either half-embedded or resting on the electrode surface, the particle charge would be given by

$$Q_p(\gamma) = Z\pi\varepsilon_o r^2 E \tag{7.6}$$

where Z is a numerical factor that has to be computed for individual values of
h/r [4,5], taking the respective values of 3 and 6.6 for the geometries of
Figs 7.1(a) and (b). In the case of the semi-ellipsoidal microfeature of Fig
7.1(e), whose geometry is defined by the ratio $\lambda = h/b$, and for which the field
enhancement $\beta(\lambda)$ is defined by equations 4.15-417, the resulting particle will
acquire a charge of

$$Q_p(\lambda) = \frac{\pi\varepsilon_o h^2 \beta}{\lambda^2} E \tag{7.7}$$

The corresponding expressions for the electromechanical detachment forces
F_d acting on these two types of microgeometry are given by [3]

$$F_d(\gamma) = \frac{4\pi\varepsilon_o h^2 (\gamma - 1)}{\gamma^2} E^2 \tag{7.8}$$

for the spherically-based geometries and

$$F_d(\gamma) = \frac{\pi\varepsilon_o h^2 f(\lambda)}{\lambda^2} E^2 \tag{7.9}$$

for the ellipsoidal geometry, where

$$f(\lambda) = \frac{\beta^2(\lambda)}{2(\lambda^2 - 1)} \left[\frac{\lambda^2}{\lambda^2 - 1} \log \lambda^2 - 1 \right]$$

To establish the critical conditions for microparticle detachment, the
electromechanical stress imparted to the attachment cross-section of the
microstructure is equated with the yield stress of the electrode material. Thus,
for geometries (a) and (d) of Fig 7.1, the required condition will be
$F_d(\gamma)/\pi r^2 \geq \sigma_y$, or

$$4\varepsilon_o(\gamma - 1) E^2 \geq \sigma_y \tag{7.10}$$

235

and for the more vulnerable structure (c), $F(\gamma)/\pi r^2 \geq \sigma_y$, or

$$4\varepsilon_o \left(r/r'\right)\left(\gamma - 1\right) E^2 \geq \sigma_y \qquad\qquad 7.11$$

For the ellipsoidal feature (f) of Fig 7.1, the corresponding condition will be $F_d (\lambda)/\pi b^2 \geq \sigma_y$, which from equation 7.9 becomes

$$\varepsilon_o f(\lambda) E^2 \geq \sigma_y \qquad\qquad 7.12$$

However, in this last case, it has to be noted that, because of its changing cross-sectional area, there is a considerable variation in the stress along the length of the protrusion, reaching a maximum value at its apex where the detachment criterion is

$$\frac{1}{2}\varepsilon_o \ \beta^2(\lambda) E^2 \geq \sigma_y \qquad\qquad 7.13$$

To illustrate the practical implications of these expressions, we can take the example [3] of a 1 cm plane-parallel gap formed by titanium electrodes and supporting a voltage of 600 kV, i.e. $E = 6 \times 10^7$ Vm^{-1}, and assume there to be a fairly blunt microfeature present on one of the electrodes whose β-value is 6. From equations 3.15 to 3.18, this could correspond either to the hemispherically capped cylindrical geometry of Fig 7.1(d) with $h = 0.4$ μm and $r = 0.1$ μm (i.e. $\gamma = 4$), or to the semi-ellipsoidal geometry of Fig 7.1(e) also with $h = 0.4\mu$m and $r = 0.1$ (i.e. $\lambda = 2.0$). It then follows from equations 7.5 and 7.7 that the surface charges concentrated on these alternative structures, i.e. those that would be carried away by them if they became detached, are respectively 2.0×10^{-16} and 4.4×10^{-16} C. From volume of revolution calculations and a knowledge of their density, the respective masses of these potential microparticles would be 1.9×10^{-17} and 1.8×10^{-16} kg, so that their corresponding Q/M values would be 10 and 2.0 C.kg^{-1}; i.e. higher for the narrower cylindrical geometry as would be expected from qualitative reasoning.

However, from equations 7.10 and 7.12, the corresponding mechanical stresses existing in the base cross-section of the protrusions are respectively 3.8×10^5 Nm^{-2} and 1.6×10^5 Nm^{-2} for the cylindrical and ellipsoidal geometries. If these values are then compared with the yield strength of titanium at room temperature, viz. 4.4×10^8 Nm^{-2}, it clearly follows that such microstructures would be mechanically stable under these field conditions. In fact, it is only with more needle-like geometries ($\beta \geq 100$) or the lower-β "necked" type of structure of Fig 7.1(c) where there is an internal stress amplification of r/r' that internal stresses approach the yield strength of typical electrode materials and provide the possibility of microparticles being launched into the gap. It also follows that such an extrusion process would give rise to the filamentary structures illustrated in Fig 7.2, and hence the possibility of a potentially unstable field emitting microprotrusions if the particle originates from the cathode. This reasoning also provides strong grounds for assuming that naturally occurring microparticles are likely to have tail-like structures.

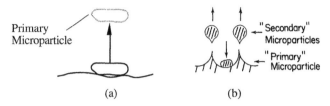

(a)   (b)

Fig 7.2   A schematic representation of the creation of (a) a "primary" microparticle by the field-stripping of a protruding microfeature, and (b) the "coronet" effect in which a shower of "secondary" microparticles result from a high-velocity impact of a primary microparticle.

It will be appreciated that the above analysis only applies to one possible source of microparticles, viz. those that originate from microfeatures that are physical extensions of the bulk parent electrode. An equally important source is that arising from partly embedded impurity granules or dust particles that are attached to an electrode surface through adhesive forces. Boulloud [6] for example has considered the effects of Van der Waals forces and shown that, for large particles, they can be very considerable. For the simple case of a sphere of radius R_p resting on a planar electrode surface, the Van der Waals force F_{vw} is given by

$$F_{vw} = 2A_{vw}R_p \qquad\qquad 7.14$$

where A_{vw} is a constant of value ~0.2 Nm^{-1}. This implies that only particles having dimensions $\lesssim 0.5$ μm will be lifted from the surface of an electrode by the electrostatic force of equation 4.8 for the typical operational field of $E \sim 6 \times 10^7$ Vm^{-1} considered in the previous example. The importance of surface adhesion forces on the microscopic scale is also highlighted by the recent detailed studies of Pollock [7,8] using a micropoint surface probe technique, where it was established that the mean adhesive pressure acting across the contact area can approach ~ 108 Nm^{-2}; i.e. comparable with the compressional yield stress of many common electrode materials. It should be noted however that these adhesive forces can be significantly reduced by the presence of surface contamination, as for example is commonly associated with poor vacuum conditions. Apart therefore from thermally activated microparticles, such as may be released from anode "hot spots", or those that originate from "necked" microstructures, it can be concluded from this discussion of particle detachment that the majority of "primary" microparticles are likely to have sub-micron dimensions. In addition, the contamination of electrode surfaces by adsorbed gas under poor vacuum conditions will tend to lower surface adhesive forces, which suggests that the potential number of microparticles per unit electrode area is likely to be significantly higher with "commercial" vacuum gaps (p $\sim$ 10^{-6} mbar) than with UHV regimes.

7. 2   Dynamic Properties of Microparticles

Having discussed the origin of naturally-occurring microparticles, it is now necessary to enquire whether their subsequent acceleration across the gap by the applied field is a "passive" or "active" event: i.e. whether or not the particle can act as an agent for the initiation of breakdown.

7.2.1   In-Flight Phenomena

Consideration will now be given to possible in-flight mechanisms that would lead to a loss of particle charge and mass; i.e. since all breakdown mechanisms are initiated by some form of charge exchange process (i.e. a discharge) with associated material vaporisation. It must also be appreciated that even if such processes do not actually initiate a discharge, they will still have an important influence on events through their control over the final impact velocity v_i acquired by a particle, and hence the nature of its interaction with the target electrode surface (see Section 7.4).

7.2.1(i)   Particle Charge Modification by Electron Emission
Here, the basic question to ask is whether the initial charge acquired by a particle at detachment Q_p will be conserved during its flight between the electrodes, or whether the local electric field on its surface can reach a sufficiently high value (i.e. $\gtrsim 3 \times 10^9$ Vm^{-1}) for there to be in-flight charge exchange between the particle and electrode by field electron emission (FEE). The possibility that such a mechanism might indeed occur was first suggested from the findings of experiments by Udris [9], Rosanova [10] and Olendzkaya [11] where, by introducing artificial particles into the vacuum gap, it was demonstrated that in-flight microdischarges could occur between a particle and electrode. The mechanism was subsequently formalised as the "trigger discharge" and will be discussed in detail in the following section. However, it has to be stated at this stage that the experiments indicated that the mechanism only operates for relatively large particles (> 100 μm diameter)

238

and then only at very close distances of approach to an electrode. This observation was subsequently explained by the theoretical analysis of Martynov [12], which showed that it was only when the particle was one-diameter from the electrode that the microscopic field associated with a charged particle can approach the threshold of $\sim 3 \times 10^9$ Vm^{-1} necessary for field electron emission to occur. Consideration must also be given to the alternative analysis of Hurley and Parnell [13] whose presentation was contemporary with that cited above. Although the approach of these authors was less rigorous, in that they ignored the positional dependence of the particle in the gap, they introduced the important new concept that the microscopic field between a particle and electrode could be additionally enhanced by a factor β due to the presence of microasperities on the particle surface such as would result from the extrusion process illustrated in Fig 7.2.

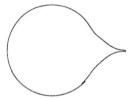

Fig 7.3   An idealised spherical microparticle having a subsidiary conical protrusion that gives rise to a greatly enhanced electric field at its tip and the possibility of in-flight field emission (After [13].)

According to this model, the microscopic field E_m acting at the tip of the approximately conical asperity attached to the spherical particle of radius R_p shown in Fig 7.3 would be given by

$$E_m = \beta \left(3E - Q_p/4\pi\varepsilon_o \, R_p^2 \right) \qquad\qquad 7.15$$

where E is the macroscopic gap field. This expression illustrates how E_m is the resultant of a component 3E arising from the enhancement of E by an uncharged conducting sphere and a second component $Q_p/4\pi\varepsilon_o \, R_p^2$ arising from the actual charge carried by the particle. This expression should however be subsequently compared with equation 7.17 which gives the more exact expression for E_m at close distances of approach. Computations based on equation 7.15 above show that such surface features can have a dramatic influence on the mechanism and could therefore be an important consideration in practical situations where particles are likely to acquire a "necked" structure during their detachment such as in Fig 7.2. Attention was also drawn by this analysis to the fact that a positively charged cathode-bound particle which does not field emit will acquire a higher single-transit impact velocity than a similar particle originating from the cathode that loses charge by in-flight FEE: indeed, computations showed that micron-sized anode bound particles with field enhancing asperities could even have their velocity reversed by this mechanism. The polarity discrimination implied by these observations will be referred to again in Section 7.2.3 when considering the initiation of breakdown by the impact of naturally occurring high velocity microparticles.

However, in the absence of any direct experimental evidence confirming the predictions of this approximate theory it has to be concluded from the more exact calculations of Martynov [12] that in-flight FEE is unlikely to play a significant role with naturally occurring micron and sub-micron particles found in typical high voltage gaps. This therefore implies that the initial charge Q_p can be tentatively assumed to be conserved in the calculations of final impact velocities given in Section 7.2.3.

7.2.1.(ii)   The "Trigger Discharge" Breakdown Mechanism

As indicated in the last section, this breakdown-initiating mechanism was originally proposed by Udris [9] and Olendzkaya [11] following their experiments with mm-sized ion spheres that had been artificially introduced into a test high voltage gap. The mechanism is based on the concept that as a charged particle approaches an electrode, the localised electric field associated with it can reach a sufficiently high value to cause field electron emission and hence the possibility of a "cathode"- or "anode"-initiated discharge (see Chapter 5) being struck between the particle and electrode; i.e. involving the vaporisation of either particle or electrode material. A particularly attractive feature of this mechanism is that it provides an explanation for the type of microparticle-based electrode damage and breakdown events described, for example, by Slivkov [2], where it is positively known that the terminal velocity of particles is too low (< 200 ms^{-1}) to give rise to the impact-initiated breakdown mechanism discussed in Section 7.2.3. In fact, it was concluded by Olendzkaya [11] from her experiments with mm-sized spheres that with such a mechanism it is the gap field rather than the kinetic energy acquired by a particle that determines the breakdown condition.

The trigger-discharge hypothesis was first treated analytically by Martynov [12] for the idealised geometry of a charged sphere approaching a planar anode such as shown in Fig 7.4(a): subsequently, Chatterton et al. [14], following a similar approach to that used earlier by Hurley and Parnell [13], extended the model to the geometry of Fig 7.4(b) where a positively charged sphere is approaching a field-enhancing boss or asperity on an otherwise planar cathode; i.e. an attempt to more closely approximate the real microtopography of an electrode surface. In both analyses, the central aim was to determine the microscopic field E_g existing in the gap between the particle and electrode at close distances of approach. Thus, recalling from the previous section that E_g will be the resultant of a component E_1 arising from the enhancement of E by the particle and a component E_2 arising from the charge Q_p on the particle, Martynov [12]

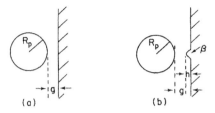

Fig 7.4   The pre-impact particle-electrode regimes used by (a) Martynov [12] and (b) Chatterton et al. [14] in their analyses of the trigger discharge mechanism.

obtained the following relation for the variation of E_g with gap separation g

$$E_g = E_1 + E_2$$
$$= Ef_1\left(g/R_p\right) + E_s f_2\left(g/R_p\right) \qquad 7.16$$

Here, E is again the uniform macroscopic gap field and E_s is the field on the surface of the charged particle in the absence of E and the conducting electrode, i.e. $E_s = Q_p/4\pi\varepsilon_\circ R_p^2$ whilst f_1 and f_2 are geometric functions that determine how the proximity of a conducting electrode influences E_1 and E_2. At close distances of approach (g ~ R_p) they increase rapidly and, for $g/R_p < 0.3$, can can be approximated by $f_1 \simeq (\pi^2/6)f_2 \simeq 1.29(R_p/g)^{0.8}$ so that equation 7.16 becomes

$$E_g = 1.29\left(R_p/g\right)^{0.8}\left(E + 3Q_p/2\pi^3\varepsilon_\circ R_p^2\right) \qquad 7.17$$

In this treatment by Martynov, no account was taken of the curvature of the particle and it was assumed that when g ~ 0.3R a uniform field existed in the particle-electrode interspace. The corresponding expression for E_g in the regime of Fig 7.4(b), where the boss is characterised by a field enhancement factor β and g << R, is given by Chatterton et al. [14] as

$$E_g \ \beta/g\left[\left(R_p + g\right)E + Q_p/4\pi\varepsilon_\circ R_p\right] \qquad 7.18$$

where Q_p was taken as the charge acquired by a spherical particle in contact with a plane electrode (Fig 7.1), which from equation 7.6 is $Q_p = (2\pi^3/3)\varepsilon_\circ R_p^2 E = 6.58\pi\varepsilon_\circ R_p E$. Also, the field enhancement factor β will clearly vary with g for the geometry being considered and therefore has to be represented by the Miller expression [15] $\beta = \beta_\infty (1-h/g)$, where β_∞ is the value of β for $g \gg h$.

From numerical evaluations of equations 7.17 and 7.18, it becomes evident that the implicit supposition of the trigger hypothesis, viz. that a discharge is struck between the particle and electrode, can only occur at close distances of approach when E_g exceeds a value of $\sim 3 \times 10^9 \, \text{Vm}^{-1}$ when significant field emission can occur. It also follows that the critical gap separation at which this occurs will increase with increasing particle radius through the dependence of E_g on both the ratio R_p/g and Q_p ($\propto R_p^2$). In the case of the Martynov model [12], computations confirm the associated experimental observations [16,17] that particles have to be $\gtrsim 100\mu m$ in diameter for the phenomena to occur: however, with the modification introduced by the theoretical model of Chatterton et al. [14], it would appear that the mechanism can operate with $\sim 20 \, \mu m$ diameter particles if the microasperity is assumed to have a β-value of $\sim 5-10$.

This more recent analysis of Chatterton et al. [14] contained two further important developments in the theory of this discharge model. Firstly, by taking account of the diminishing value of Q_p in equation 7.18 that follows the onset of field emission, it has been shown by computation for a given particle radius that there is a particular gap separation g at which the microscopic gap field EG passes through a maximum: for example, in the case of 40 and 80 μm radius particles moving in a macroscopic field $E = 2 \times 10^7 \, \text{Vm}^{-1}$ and approaching a boss with a β-factor of 7, E reaches maximum values of ~ 6.5 and $7.0 \times 10^9 \, \text{Vm}^{-1}$ at $g \sim 1$ and 2 μm respectively. Secondly, the "anode" versus "cathode" breakdown criteria discussed in Chapter 5 were applied to the microregime of Fig 7.4(b) in an attempt to establish more precisely how the discharge is initiated. It was generally concluded from these computations that whereas "cathode"-initiated breakdown, i.e. the vaporisation of the field emitting microasperity, was unlikely except for large particles of radius >500 μm, the necessary condition for "anode"-initiated breakdown, i.e. the vaporisation by electron bombardment of material (either gas or metal) from the surface of the incident particle, appears to be satisfied for particle diameters > 20 μm. This analysis, in common with that of Martynov [12], highlights how sensitive this breakdown-initiating mechanism is to the particle radius R_p: thus, for say aluminium particles in a 1 cm gap, the breakdown voltage is only ~ 40 kV for $R_p = 80 \, \mu m$ but increases to ~ 200 kV for $R_p = 20 \, \mu m$. It should be added that this prediction provides one explanation of why arced electrodes, with their profusion of large microparticles formed from droplets of molten metal, have generally lower breakdown voltages. Although the above analyses of the in-flight electron bombardment process indicate that the local temperature rise of small microparticles with radii $\lesssim 10 \, \mu m$ is too low for the initiation of breakdown by the trigger discharge mechanism, it was nevertheless recognised by Beukema [18] that there is an alternative mechanism whereby even a sub-critical heating of a particle could create the necessary conditions for a subsequent breakdown event. This assumed that the local temperature of the electron-bombarded surface region of a positively-charged, cathode-bound microparticle could be raised sufficiently to cause melting and and the consequent welding of the particle to the electrode on impact. Under these circumstances, it follows that such a feature would give rise to a local enhancement of the surface field with the associated possibility of it becoming a new emission site that could subsequently initiate a breakdown if the microscopic field exceeded the critical value of $\sim 7 \times 10^9$ Vm^{-1} discussed in Section 5.3.1. It was also pointed out that the creation of such a site would be even more likely if the impacting microparticle had a surface asperity that further enhanced the field. From a practical stand-point, such a mechanism would clearly play an important role in determining the voltage hold-off properties of real gaps, since it applies to particles having micron-size dimensions that are thought to be typical of those occuring naturally.

The validity of this concept was investigated by Beukema from both a theoretical and experimental point of view [18]. In the first place, he set up a theoretical model based on a sphere-plane impact regime and showed that the predicted local temperature rise for the range of gap fields and particle geometries appropriate to practical systems were sufficient to promote particle welding. On the experimental level he developed an ingenious technique involving the use of various combinations of conditioned and unconditioned electrodes that provided convincing corroborative evidence for the existence of this microparticle process. Thus, it was demonstrated for example that the emitting properties of a fully conditioned cathode, as measured by its β-factor defined in Section 4.2.2(i), could change dramatically when it was used in conjunction with an unconditional anode; typically, the application of a relatively low gap field could result in an increase of β from say ~ 40 to 100. From other supportive evidence, such as obtained from SEM studies of electrode surfaces and the observation that there is frequently a correlation between the occurrence of current pulses and the changes in β, it was also possible to infer that material in the form of microparticles is transferred from an unconditioned to a conditioned electrode on the first application of a gap field; a conclusion that was subsequently to be confirmed by the direct experimental techniques described later in Section 7.3.2.

7.2.1(iii)   The "Particle Vaporisation" Breakdown Mechanism
From a theoretical consideration of the anode processes resulting from localised electron bombardment [19] and the subsequent experimental observation described in Section 7.3.1 that neutral electrode vapour originating from the anode is liberated in a gap a few microseconds before its

breakdown [20], Davies and Biondi developed their "heated-anode-particle" model [20,21] illustrated schematically in Fig 7.5. In this a field-emitting microprotrusion on the cathode, characterised by an enhancement factor β and emitting radius r, gives rise to an anode hot-spot as described it is then assumed that thermal instability occurs first at the anode, such that a semi-molten droplet is extruded from its surface by the applied field and in passing through the "necked" structure shown in Figs 7.1 and 7.2 will acquire a high initial positive charge Q_p that will cause it to be accelerated towards the cathode.

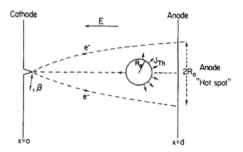

Fig 7.5   A schematic illustration of the "heated-anode-particle" breakdown initiating model; the idealised spherical microparticle is created at the anode hot-spot formed by the electron emission from a cathode microprotrusion having a geometrical field enhancement factor β. (From Davies and Biondi [21,22], with permission.)

Since however this microparticle will be moving in the path of the on-coming electron beam it will simultaneously experience a progressive charge neutralisation (which will have the effect of prolonging its transit time) and a rapid increase in temperature as a result of electron bombardment heating. In the limit, this latter process could lead to the mid-gap vaporisation of the particle with an associated increase in the local vapour density that is sufficiently high for the onset of regenerative electron-atom ionisation processes and the consequent breakdown of the gap. It follows therefore that the mechanism is really a development of the "anode"-initiated breakdown process described in Section 5.3.2, where the microparticle is behaving as an anode.

    A complete theoretical analysis of this model [22] involves the solution of the following set of six inter-related time dependent differential equations. The first accounts for the thermal energy exchange processes at the particle, in which the heat gained by electron bombardment is equated with that lost by surface evaporation, radiation and the well known thermionic emission cooling U_{Th} [23],

$$\frac{dT}{dT} = \frac{1}{M_p C}\left[F_1\, F_2\, V_p\, i_F - 4\pi R_p^2 G\left(L - \frac{6kT}{\rho R_p}\right) - 4\pi R_p^2 \sigma_s\, T^4 - U_{Th}\right] \quad 7.19$$

Here T is the temperature of the particle; t is its flight time after leaving the anode; F_1 and F_2 are respectively the fraction of the total cathode field emission current i_F intercepted by the particle and the fraction of the incident bombardment energy retained by the particle after back-scattered and secondary emission losses (largely determined by the electron range in the particle as given by the Widdington Law [24]; G is the rate of loss of mass per unit area; V_p is the potential of the particle relative to the cathode; k and σ_s are the Boltzmann and Stephan constants; C, L and p are respectively the specific heat, latent heat of vaporisation and density of the particle material, and M_p (= $4\pi\rho R_p^3/3$) is the particle mass: it follows therefore from the definition of G that

$$\frac{dR_p}{dt} = -\frac{G}{\rho} \quad 7.20$$

Next, if i_{Th} is the current emitted thermionically from the particle determined by the Richardson-Schottky relation [25], and if i_c is defined as the fraction of the intercepted field emission current that is actually collected after secondary electron emission, the net rate of accumulation of negative charge will be given by

$$\frac{dQ_-}{dt} - \frac{dQ_+}{dt} = i_c - i_{TH} \quad 7.21$$

Finally, the velocity and acceleration of the particle will be respectively

$$\frac{dx}{dt} = v_p \quad 7.22$$

and

$$\frac{dv_p}{dt} = E\,\frac{Q_p(t)}{M_p} \quad 7.23$$

where $Q_p(t) = Q_p - Q_-(t) + Q_+(t)$, being the particle charge at any given instant after detachment. After making appropriate substitutions for V_p, F_1, F_2, i_F, i_c, i_{TH} and U_{Th} in equations 7.19 to 7.23, they have been solved numerically for the contrasting cases of a non-refractory copper particle of radius 0.5 µm and a similar sized refractory molybdenum particle [22]. The findings are illustrated in Fig 7.6 where it has been assumed that at t = 0, $T = T_0 = T$, the boiling point of the particle material.

245

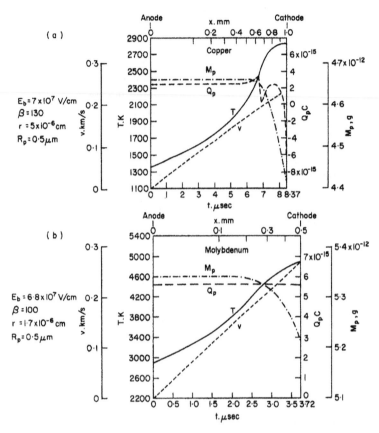

Fig 7.6 Collated data showing how the mass M_p, charge Q_p, velocity v and temperature T of the microparticle involved in the "heated-anode particle" model vary as it crosses the gap from anode to cathode: (a) corresponds to the case of a copper particle, and (b) to a molybdenum particle [22]. From Davies and Biondi [22], with permission.)

These computations were also specially chosen to fit some previous experimental breakdown measurements [19,26] where, for Cu, the microscopic breakdown field was calculated from a measured β value of 130 to be $E_b = 7 \times 10^9$ Vm^{-1}, whilst for Mo, where β = 100, $E_b = 6.8 \times 10^9$ Vm^{-1}. When interpreting the data presented in this figure it should be noted that we are dealing with a cathode-bound particle where the conventional positions of the cathode and anode shown in Fig 7.5 have been transposed in order to have a left-to-right reading.

Thus, for both materials, these plots clearly show that after a few microseconds, as the particle approaches the cathode, M_p begins to fall off steeply as T → T_{max}, indicating a rapid vaporisation process and a consequent pressure rise in the gap that would create the necessary condition for an arc to be initiated. The more rapid increase of T with the Mo particle can be attributed to its lower thermal capacity.

From an extensive analysis of this model covering a wide range of values of the variable parameters, it has been concluded that breakdown can be initiated by this mechanism for particle sizes of $R_p \lesssim 3$ µm. With larger particles io always exceeds i_{Th} so that they become negatively charged and return to the anode before their temperature reaches a high enough value for significant vaporisation to occur.

7.2.2 Single-Transit Impact Phenomena

The physical nature of the interaction between a microparticle and an electrode surface is principally determined by the impact or terminal velocity v_i of the particle which, from equation 7.1, is given by $v_i = (\sqrt{2Q_pV/M_p})$ where Q_p and M_p are respectively its initial charge (assumed conserved) and mass of the particle, with V being the gap voltage. To gain some insight into the range of impact velocities likely to be encountered in practice, the above equation may be evaluated for the idealised particle geometries analysed in Section 7.1.3 Thus, considering for example the case of a semi-ellipsoidal microstructure, the particle charge will be given by equation 7.7, its mass by $M_p = 2\pi h^3\rho/3\lambda^2$ (ρ being the density of the particle material), so that

$$v_i = \left(\frac{3\varepsilon_o\beta(\lambda)}{h\rho d}\right)^{\frac{1}{2}} V = \left(\frac{3\varepsilon_o\beta(\lambda)}{r\lambda^2\rho d}\right)^{\frac{1}{2}} V \qquad 7.24$$

since the radius of curvature of the tip $r = b/\lambda$ or $h = r\lambda^2$. The practical implication of this equation can be illustrated [3] by considering the three typical high voltage gap regimes of d = 1, 10 and 100 mm with V = 20, 100 and 1000 kV respectively, where in all cases the electrode material is titanium (ρ = 4.5 × 10^3 kg m^{-1}) and the vulnerable surface microstructure is characterised by $\beta(\lambda) = 6$ and r = 0.1 µm. We find that v_i increases from 180 ms^{-1} for the 30 kV low voltage gap (E = 2 × 10^7 Vm^{-1}) through 280 ms^{-1} for the intermediate voltage 100 kV gap (E = 10^7 Vm^{-1}) to v_i = 900 ms^{-1} with the 1 MV high voltage gap (E = 10^7 Vm^{-1}). A similar range of impact velocities are obtained from estimates based on a spherical microparticle geometry [27]. It must however be pointed out that this type of estimate is very sensitive to the β and r values chosen for the initial microfeature: sharper structures will be more highly charged and hence acquire very much higher impact velocities. This therefore implies that there is likely to be a large variety in the types of particle/electrode interaction in most practical situations.

7.2.2(i)   Low Impact Velocities ($v_i \lesssim v_c$)

These are defined as the range for which permanent plastic deformation of either projectile or target is absent, so that there will be a semi-elastic non-destructive bouncing impact. The limiting value of v_i for this type of event, defined as the critical impact velocity v_c, will depend on the material constants of the colliding surfaces; for the case of similar materials, which is likely to be appropriate to most high voltage gap situations, its value has been given by Cook [28] from a hydrodynamically-based theory as

$$v_c \cong \left(\frac{8\sigma_y}{\rho}\right)^{\frac{1}{2}}$$

7.25

where σ_y is the yield strength of the colliding materials and ρ their density. Introducing appropriate material constants into this equation [29] predicts values of v_c ranging from ~ 200 ms^{-1} for a "soft" material such as copper ($\sigma_y = 5.5 \times 10^7$ Nm^{-2}) through ~ 500 ms^{-1} for stainless steel ($\sigma_y = 2.8 \times 10^8$ Nm^{-2}) to as high as ~ 1 kms^{-1} for titanium ($\sigma_y = 8 \times 10^8$ Nm^{-2}) a material commonly used for high voltage electrodes. Whilst these predictions appear to be in reasonably good general agreement with the findings of the controlled microparticle impact studies described in Section 7.4, it should be pointed out that they differ by over three orders of magnitude both from the values of v_c predicted by classical elastic impact theory [30] and those found from the corresponding macroscopic sphere-plane impact experiments [31] where typically v_c ~ 0.05-0.1 ms^{-1}. This discrepancy, which is discussed in greater detail later in this chapter, can however be explained on the basis that impact events on the microscopic scale have to be thought of as single crystal collisions, whereas on the macroscopic scale collisions will be between poly-crystalline systems whose elastic modulii have been found by Gane [32] to be significantly lower.

It has also been pointed out by Menon and Srivastava [29] that for any given operating gap where V is constant, particles will have to be smaller than some critical dimensions if their impact velocities are to exceed v_c . Thus, for the ideal case of a spherical particle resting on a plane electrode, we have from equation 7.6, $Q_p = 6.6\pi e_o ER_p^2$, which, when substituted into equation 7.1, gives $R_p \cong 10\varepsilon_o \div v_i^2 \rho d \, V^2$. This therefore implies that there will be a critical particle radius $(R_\rho)_c \propto V^2/d$ below which $v_i > v_c$. For a 1 cm gap supporting 200 kV one finds $(R_p)_c$ varies from 0.7 µm for copper electrodes through 0.14 µm for stainless steel to 0.05 µm for titanium.

In the absence therefore of any direct impact damage, and excluding the possibility of the conventional trigger discharge mechanism described in Section 7.2.2(ii) operating for the anticipated particle dimensions of < 5 µm, there remains only one other mechanism by which a low-velocity microparticle ($v_i < v_c$) could directly initiate a breakdown. This is by the "hot-welding" model of Beukema [18], also described in Section 7.2.2(ii) in which a positively charged particle is heated in-flight by electron bombardment and, as a result, sticks to the cathode to form a potentially unstable field emitting site. Apart from this, there is however another indirect mechanism to be discussed in Section 7.2.4, whereby a particle undergoing a semi-elastic bouncing impact could initiate a breakdown following a subsequent transit of the gap.

7.2.2 (ii) Intermediate Impact Velocities ($v_c \lesssim v_i \lesssim 5v_i$)

In this velocity range, it is assumed that most of the incident kinetic energy of the particle will be dissipated irreversibly in the permanent mechanical deformation of the colliding surfaces. This type of particle impact event will therefore be highly inelastic, giving rise either to identation microcraters in the surface of the target electrode, or a protruding microfeature if a particle becomes impacted or welded to the surface. A direct experimental demonstration of these processes may be obtained from the controlled laboratory studies of microparticle impact phenomena to be described later in this chapter. In both cases, if such a particle impacts on the cathode electrode, there is the possibility of creating an "instantaneous" electron emission site, as would be evidenced by an abrupt change in the I-V characteristic of the gap [18]. It follows that if this were to occur it would bring with it the subsequent risk of breakdown being initiated by any of the mechanisms discussed in Chapter 5. In the case of target indentation, such emission sites will tend to be located around the rim of the microcrater where the electrode surface is most distorted; for the embedded microparticle situation, any electron emission will occur at the sharpest point of the newly formed protruding microfeature.

Whilst corresponding impact events on the anode electrode will cause similar surface damage, they cannot of course give rise to electron emission processes: accordingly anode-bound microparticles must be regarded as less dangerous from a breakdown point of view. This discrimination is also reinforced by the in-flight field emission considerations of Section 7.2.2(i), where it was pointed out that anode-bound particles are more susceptible to charge loss and will hence have lower impact velocities for a given gap voltage. For these reasons, it could be anticipated that with gaps having non-uniform geometries, there would be a polarity effect in their voltage hold-off capability, with the more stable regime being that where the higher surface field is arranged to act on the cathode, since this would favour the detachment of anode-bound particles (see Section 7.2.1). It must also be pointed out however, that as with low-velocity impacts, there remains the possibility of particles undergoing inelastic "reflections" and multiple transits of the gap (see

249

Section 7.2.3), thus masking the polarity effect just outlined.

7.2.2(iii)   High Impact Velocities (v ≳ 5v_c)

At these high velocities, corresponding to the conditions covered by the Cranberg hypothesis, an ever increasing fraction of the incident kinetic energy of the particle will be dissipated irreversibly in thermal processes, particularly the vaporisation of particle or electrode material. In this connection, Cook [28] has shown that the extreme situation of complete vaporisation of the particle will occur when

$$v_i > 4 \left(L_v\right)^{\frac{1}{2}} \qquad\qquad 7.26$$

where L_v is the latent heat of vaporisation of the colliding materials. For titanium, as an example, where $L_v \simeq 7MJ\ kg^{-1}$, this would correspond to an impact velocity of ~ 10 kms^{-1}. It must be emphasized however, that experimental evidence from simulation studies (see Section 7.4) strongly suggests that the necessary local rise in vapour pressure for initiating breakdown can occur for particle impact velocities significantly lower than 10 kms^{-1}. For refractory metals such as molybdenum the critical velocity is probably ~ 5 kms^{-1}, but could be as low as 1.5 kms^{-1} for non-refractory metals such as copper. Sub-critical impact events are assumed to be observed as microdischarges as opposed to complete breakdown. Direct confirmation of this vaporisation of particle or electrode material has been provided by mass spectrometry analyses of the species of metallic ions (both charged and neutral) that are present in the gas burst that accompanies a simulated high-velocity microparticle impact (see Section 7.4).

On this picture, the localised cloud of metal vapour resulting from a high velocity impact event will be rapidly ionised by a combination of thermal and field-based mechanisms so that a microplasma is formed in the vicinity of the impact zone. At this stage regenerative processes come into play that allow the microplasma to expand under the applied field to fill the gap and so initiate breakdown by the formation of an arc between the electrodes. In fact, the latter stages of this mechanism closely resemble the sequence of events operating in the electron emission induced breakdown mechanism discussed in Chapter 5. As with "intermediate velocity" impact events, it is important to note that cathode-bound particles will again have a greater probability of initiating breakdown by this mechanism since the regenerative ionisation processes that sustain the growth of the microplasma are likely to be considerably enhanced by electron emission from cathode sites created by the impact. An impressive experimental demonstration of this selectivity was obtained by Slattery et al. using artificially generated charged microspheres [33]. This and other simulation studies of the impact ionisation and electrode microcratering processes will be fully described later in this chapter. Before concluding this discussion, the reader should be reminded that the critical condition of this type of microparticle-initiated breakdown is determined by the voltage across the gap rather than the field, i.e. as required by the Cranberg hypothesis of equation 7.2. Experimental details of this type of behaviour are given in Section 2.5.3.

7.2.3   Multi-Transit Impact Phenomena

7.2.3(i)   Particle Bouncing

The phenomenon of particle bouncing between the high voltage electrodes of a vacuum gap was first demonstrated experimentally by Olendzkaya [11] using mm-sized steel spheres that had been artificially introduced into the gap. She showed that below the breakdown voltage of such a gap these particles would oscillate between the electrodes and give rise to a measurable current in the external circuit, although no details were given of how its magnitude was related to the particle charge, velocity, or its oscillation frequency. However, the potential importance of the contribution from such a charge transfer mechanism to the total prebreakdown current was subsequently considered by Boulloud [6] and shown to be far too small to account for the magnitude of typical prebreakdown currents. The experiments of Olendzkaya [11] also showed that whilst the breakdown of the gap is prematurely initiated by the presence of these particles, it was by a trigger discharge mechanism rather than by a "total voltage" effect as would be required by the Cranberg impact hypothesis.

7.2.3(ii)   Particle Kinetic Energy Enhancement

The concept that particle bouncing could also provide the basis of a mechanism for enhancing the kinetic energy of a microparticle once released in a vacuum gap was first formulated by Latham [34] as a model for explaining how a Cranberg type of impact-initiated breakdown mechanism could operate in low voltage ($\lesssim$ 15 kV) gaps where single-transit impact velocities are well below the threshold for generating adequate impact ionisation (see Section 7.2.2). An important feature of such a mechanism is that it would be particularly relevant for micron and submicron sized particles where breakdown could not be prematurely initiated by the trigger discharge mechanism of Section 7.2.1(ii): in fact, this observation is strongly supported by microcratering evidence discussed in Section 7.3.

The two basic requirements of this energy enhancing model are that during a bouncing impact there is an efficient reversal of both the particle momentum and particle charge. Thus ensures firstly, that a particle is launched into a second transit of the gap with a high initial velocity and secondly, that such a "reflected" particle will be re-accelerated by the applied field. Thus, referring to Fig 7.7, which illustrates the principle of the model, it will be seen that a particle of initial charge Q_p and mass M_p acquires an initial single-transit impact velocity $v_1 = (2Q_{p1}V/M_p)^{1/2}$, a rebound velocity of ev_1 (where e is the coefficient of restitution between the particle and electrode) and, assuming conservation of particle mass, a second-transit impact velocity of $v_2 = [(ev_1) + 2VQ_{p2}/M_p]^{1/2}$ where Q_{p2} is the magnitude of the "reversed" charge acquired by the particle during its contact with the oppositely charged electrode. For energy enhancement at this second impact we evidently require $|v_2| > |v_1|$ which reduces to the condition

$$e^2 + \left|\frac{Q_{p2}}{Q_{p1}}\right| > 1 \qquad\qquad 7.27$$

It follows that if v_2 is still less than the critical impact velocity v_c, defined by equation 7.25, there can be further bouncing impacts until the particle velocity builds up to a value in excess of v_c, when one of the previously discussed breakdown mechanisms can be initiated.

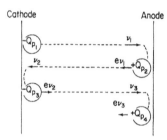

Fig 7.7   An illustration of the progressive kinetic energy enhancement model of Latham [34], in which a microparticle undergoes successive bouncing impacts with associated momentum and charge reversal.

The criterion defined by equation 7.27, which is a general relation applying to any of the subsequent bouncing events illustrated in Fig 7.7, highlights how this energy-enhancing mechanism depends firstly on the mechanical properties of the electrodes through e, and secondly, on the electrical properties of their surfaces through the ratio Q_{p2}/Q_{p1} which is a useful measure of their control over the charge reversal process. An attractive feature therefore of this model is that it provides a physical explanation of why certain materials provide more stable electrode pairs than others. For example, the well established technological practice of using stainless steel rather than say copper electrodes, could be anticipated on the basis of the present model since these materials are characterised by having surface oxide films that are respectively insulating and semiconducting [34];

i.e. the contact charge exchange process and hence the energy enhancing mechanism would tend to be more inhibited with stainless steel than copper. In Section 7.4, laboratory simulation experiments with artificially generated charged microspheres will be described that provide an impressive confirmation not only of the energy enhancing mechanism itself, but also of the importance of the atomic state of an electrode surface in controlling the charge exchange process.

7.2.3 (iii)   Theoretical Basis of Charge Exchange

To gain further insight into the fundamental physical processes that occur during a semi-elastic particle impact event, we shall now analyse the

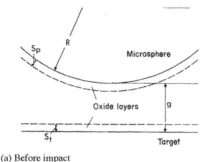

(a) Before impact

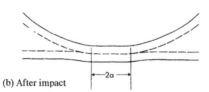

(b) After impact

Fig 7.8   The idealised model used for analysing the mechanical and electrical interaction between a metallic microsphere and a planar target having respectively surface oxide layers of thickness s_p and s_T: (a) before impact ($t < 0$), (b) during impact ($t \sim t_c /2$). (From Latham and Brah [37], with permission.)

mechanical and electrical impact response of the idealised sphere-plane geometry shown in Fig 7.8(a): in fact, as will be seen from Fig 7.30, this is a good approximation of the microregime associated with the simulation studies of this phenomenon (see Section 7.4.5). As a further simplification of the model, it will also be assumed that the particle and electrode are the same material and hence covered by identical semiconducting or insulating surface oxide films of thickness s, which is typically 30-50 Å (i.e. s << R_p).

Mechanical Considerations:

At the instant of maximum depression during a semi-elastic impact, i.e. when the particle is momentarily at rest, the contact surfaces will be reversibly deformed as shown schematically in Fig 7.8(b), where $\Delta z \ll R_p$. This regime is described by the well established equations of classical impact theory, as originally derived for macrosystems by Hertz (see Andrews [35]). Thus, if Y is Young's modulus of the impacting material and $\overline{F}$ is the mean force transmitted to the target during its elastic deceleration, the radius of the circle of contact will be

$$a \cong \left(\overline{F} R_p / Y \right)^{\frac{1}{3}}$$

7.28

Now for an initial impact velocity v_i and total contact time t, the mean force $\overline{F}$ acting during a uniform deceleration period ($t_c/2$) will be approximated by

$$\overline{F} \cong M_p v_i / \left(t_c / 2 \right)$$

7.29

where $v_i = (2Q_p V/M_p)^{\frac{1}{2}}$ and, from the analysis of Hertz

$$t_c \cong 4.5 \left(5M_p / \pi Y R_p^{\overline{2}} \right)^{0.4} v_i^{0.2}$$

7.30

From discussions in Section 7.4, and by Latham et al. [36] elsewhere, it emerges that when these equations are applied to microsystems where impact events are likely to be between single grains, it is more appropriate to use "single crystal" values for the elastic constants which can exceed the corresponding "bulk" (i.e. macroscopic) values by factors ~ 25. Hence for the example of a 1 μm copper microsphere impacting on a copper electrode with a velocity of 50 ms^{-1}, one finds $t_c \sim 10^{-9}$ s and a ~ 3 × 10^{-8} m. It should however be added that the recent micro-contact experiments of Pollock et al. [7] indicate that the above estimate of a is probably too low, since no account was taken of possible plastic deformation arising from the additional compression of the surfaces caused by contact adhesive forces: for similar reasons, the estimate of t_c is also likely to be too low.

Electrical Considerations:

For the ideal case of Fig 7.8, where both the spherical particle and planar electrodes are atomically clean, charge exchange during contact would only be limited by the relaxation time τ of the metal which is typically ~ 10^{-14} s. Since therefore, $\tau \ll t$, it follows that $Q_{p_1} = Q_{p_2} = Q_{p_3}$ etc. in Fig 7.7 and will be equal to the equilibrium charge $Q_\circ$ acquired by a spherical particle resting on a planar electrode; i.e. $Q_\circ = \frac{1}{3}\left(2\pi^2 R_p^2\varepsilon_\circ\right)E = 6.6\ R_p^2\varepsilon_\circ$ as given by equation 7.6. However, referring to Fig 7.8(b), it will be seen that, in practice, transitory electrical contact is established between the particle and electrode via an oxide junction whose surface area will vary continuously between zero and a maximum of πa^2 during the compression and recovery stages of the impact. To a first order of approximation therefore the mean area of the junction during the time of contact t_c can be taken as $\pi a^2/2$ and, if the oxide films are assumed to be incompressible, the junction thickness can be taken as $2s$. For a regime such as this, charge may be exchanged between the particle and electrode by either direct ohmic conduction through the composite film or quantum mechanical electron tunnelling across the junction. However, from a detailed consideration of the electrical properties of surface oxide films and the relative efficiencies of these alternative mechanisms, it was shown by Latham and Brah [37] that the process is dominated by the tunnelling mechanism: it follows therefore that we are dealing with a metal-oxide-metal (MOM) tunnel junction situation whose DC properties have been thoroughly analysed for example by Simmons [38-40]. The electrical behaviour of this type of tunnel junction is conventionally described in terms of its DC tunnel resistivity $\rho_t(\Omega m^2)$ defined as

$$\rho_t = Vp/jt \qquad\qquad 7.31$$

where V_p is the potential of the particle relative to the planar electrode, i.e. the bias voltage existing across the junction, and j_t is the tunnel current density: it follows therefore that the instantaneous current flowing across this junction will be

$$i_t = \pi a^2 j_t = \pi a^2 V_p/\rho_t \qquad\qquad 7.32$$

Although ρ_t is a complicated function of the thickness and electrical properties of the oxide [38-40], it may be approximated for the present purposes by the exponential dependence [37]

$$\rho_t = \rho_{ot}\,exp\,(-KV_p) \qquad\qquad 7.33$$

where K is a constant ($\sim 2\ \Omega m^2 V^{-1}$) that can be taken as independent of the oxide material, and ρ_{ot} is another constant that depends on the thickness, band-gap and dielectric properties of the oxide film [38-40]. In applying this steady-state model to estimate the total charge Q_T transferred by the transient tunnelling current flowing during a bouncing impact, it is necessary to integrate equation 7.32 over the interval t = 0 to t = t_c, i.e.

$$QT = \int_0^{t_c} i_t dt = \pi \int_0^{t_c} \left(aV_p\left(t\right)/\rho_t\right) dt$$

$$\equiv \left|Q_{p_1}\right| + \left|Q_{p_2}\right| \qquad 7.34$$

at the first bouncing impact of Fig 7.7. If now $V_p(t)$ is assumed to decay according to an exponential relaxation law, we shall have

$$V_p\left(t\right) = V_{op}\, exp\left(-t/\tau_t\right) \qquad 7.35$$

where V_{op} is the initial potential of the particle at the instant of contact and τ_t is the time constant associated with the tunnelling charge exchange process: the value of τ_t can be assumed to be $\lesssim t_c$ since it has been shown from controlled laboratory studies of this phenomenon (Section 7.4) that the charge reversal process approaches completion during such a bouncing impact. Hence, assuming a mean contact area $\pi a^2/2$ and substituting for ρ_t and $V_p(t)$ in equation 7.34 from equation 7.33 and 7.35, and integrating we find

$$QT \cong \left[\pi a^2\, \tau_t\, exp\left(KV_{op}\right)\right]/2\rho_{ot}K \qquad 7.36$$

Finally, the field calculations described earlier in Section 7.2.2(ii) in connection with the impact trigger discharge mechanism may be used to evaluate V_{op}. Thus, from equation 7.17, which gives the axial field E_g in a vacuum gap between a particle and electrode, we have for the present oxide gap of dielectric constant ε.

$$V_{op} = 2s.\, E_g = 2.58s\left(R_p/g\right)^{0.8}\left(E + 3Q_p/2\pi^3\varepsilon_o\,\varepsilon R_p^2\right) \qquad 7.37$$

From an inspection of these last two equations, it will be apparent that the efficiency of this charge reversal process, as measured by Q_T, will be determined by the properties of the oxide film through the inverse dependence of Q_T on ρ_{ot} and ε. Thus an insulating film (large band gap and hence a large ρ_{ot}) having a high dielectric constant ε can be anticipated to promote the process much less efficiently than say a low band gap semiconducting film. It follows therefore that the likelihood of breakdown being initiated as a result of successive energy-enhancing transits of a gap will be much higher for electrode pairs having semiconducting oxide films (e.g. copper) than for those having insulating films (e.g. stainless steel).

As a specific example of the application of this theory, consider a 1 μm diameter microsphere moving in a 1 mm electrode gap across which there is a potential of 10 kV (or $E = 10^7$ Vm^{-1}) and assume that its initial detachment charge $Q_{p_1} \equiv Q_o = 2.5 \times 10^{-16}$ C as calculated from equation 7.6. If the thickness s of the oxide films on the particle and the electrode surfaces are taken as 30 Å, their dielectric constant as 3, their band gap as ~ 1 eV (i.e. semiconducting), and the tunnelling resistivity ρ_t is approximated from the published data of Simmons [38-40] as ~5 × 10^{-2} Ω m^2 computations using equation 7.37 then show [37] that V_{op} ~ 7 V. Substituting finally these values into equation 7.36 with $\tau_t \simeq t_c \simeq 3 \times 10^{-9}$ s (from equation 7.30) we find $Q_T \simeq 3.5 \times 10^{-16}$ C. If this value is now compared with that associated with an "ideal" impact between atomically clean surfaces where $Q_T \simeq 2Q_o = 5 \times 10^{-16}$C it follows that the charge reversal process will be ~ 70% completed during a "real" impact with such an oxidised surface: under these circumstances, it can also be inferred from equation 7.27 that there will be a high probability of the kinetic energy of the particle being enhanced during its "return" transit of the gap. As will be described in Section 7.4, these predictions are in good agreement with the direct findings of laboratory simulation studies [37] with charged microspheres impacting on copper electrodes whose ambient oxide is known to be semiconducting: on the other hand, comparable studies with stainless steel electrodes whose ambient oxide film is insulating, have confirmed that both the charge exchange process and energy enhancement mechanism are significantly inhibited. This type of evidence is therefore taken as lending strong support to this tunnelling explanation of how charge is exchanged during the bouncing impact of a microparticle and how the electrode surface properties play a vital role in controlling the process.

7.2.4 Microparticle Behaviour with Impulse Voltages

Before concluding this review of microparticle-initiated breakdown mechanisms, it is necessary to consider what additional factors have to be taken into account when, instead of DC voltages, fast and linearly rising impulse over-voltages, such as illustrated in Fig 7.9, are applied across large gaps ($\gtrsim$ 2 mm) where microparticles are thought to play a dominant role.

7.2.4(i) Impact-Initiated Breakdown

Following the analysis of Farrall [41], a particle is assumed to become detached from an electrode at some time t_d on the rising edge of the pulse, where t_d is small compared with the rise time t_r of the pulse: the voltage V_d corresponding to $t = t_d$ is also identified with the Cranberg DC breakdown voltage given by equation 7.2. Because of the finite transit time t_t of the particle, the impulse breakdown event will be delayed by this time interval and be experimentally observed at a time t_b, where $t_b = t_d + t_t$. This reasoning therefore explains how the impulse breakdown voltage V_b' of a given gap can be higher than its corresponding DC value. It also follows from this model that if the duration of the pulse is less than the transit time of the particle, breakdown cannot occur by this mechanism since the impact event will occur under zero-field conditions where any resulting microplasma will expand "passively" without any possibility of an arc being struck between the electrodes.

Since the Cranberg relation has been shown to provide a useful guide for predicting the breakdown behaviour of DC gaps, it is important to enquire how its form has to be modified when using impulsed over-voltages. Accordingly, Farrall [41] has analysed the cases of pulses having either a *constant rise rate* dV/dt or *constant rise time* t_r, where it is assumed that breakdown is initiated by single transit impacts. Under these circumstances therefore, Fig 7.9 indicates that the impulse breakdown voltage V_b' will be given by

$$V_b' = V_d + t_t \frac{dV}{dt} \qquad\qquad 7.39$$

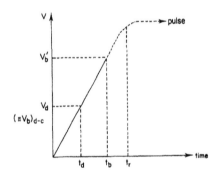

Fig 7.9   Impulse voltage waveform, showing the typical relation between the DC and impulse breakdown voltages, V_b and $\mathbf{V_b}$ respectively, for a given gap regime.

In this equation, dV/dt is a known experimental variable and the transit time t_t is determined from the equation of motion of the particle across the gap, viz.

$$M_p \frac{d^2x}{dt^2} = \frac{Q_p}{d} \left(V_c + t \frac{dV}{dt} \right) \qquad\qquad 7.40$$

where for a spherical particle initially in contact with a planar electrode

$Q_p = \frac{2}{3}\pi^3 R_p^2 \varepsilon_o E$ (see equation 4.6) and $M_p = \frac{4}{3}\pi\rho R_p^3$. To obtain the theoretical dependence of V_b on d and other experimental parameters, these last two equations may now be independently solved for the two impulse regimes referred to above. For "typical" experimental conditions, such an analysis [41] shows that to a good approximation, *constant rise rate* pulses result in

$$V_b' \propto R_p^{\frac{1}{3}} d^{\,5/6}$$
                                                                        7.41

and for *constant rise time* pulses

$$V_b' \propto R_p^{\frac{1}{3}} d^{\,5/2}$$
                                                                        7.42

It will be seen that the gap dependence in both of these expressions differs significantly from Cranberg's DC relation $V_b \propto d^{1/2}$. they also indicate a size-dependence which was not accounted for in the original Cranberg hypothesis. Since the examples of complementary experimental impulse data are very limited, it is difficult to give a reliable appraisal of the validity of this theory. However, from a review of impulse breakdown data covering a range of ill-defined operating conditions, Farrall [41] concluded that the gap-dependence of V_b' does in fact show a variation from $d^{0.4}$ to $d^{1.9}$ which is in reasonably good agreement with the maximum predicted theoretical range of $d^{0.5}$ to $d^{2.5}$.

7.2.4 (ii)   The "Detachment" Trigger Discharge

Another important phenomenon that can only occur with impulse voltages is the initiation of breakdown by a trigger-like discharge that is struck between a negatively charged particle and its parent cathode-electrode almost immediately after detachment. The phenomenon was demonstrated experimentally by Poshekhonov and Solov'yev [42] by introducing artificial microparticles into a gap and observing the delay in the incidences of breakdown associated with both cathode and anode-launched particles. Apart from typical delay times of 0.5-2 m found for both polarity particles, which can be satisfactorily explained in terms of particle transit times, there was also a very high incidence of breakdown events associated with cathode-launched particles that had delay times of only 1-3 μs after the onset of the voltage pulse; such events were therefore identified as being initiated by this alternative type of trigger discharge.

These same authors [42] were able to provide a physical explanation for this detachment trigger mechanism in terms of the local microscopic field E_g existing in the gap between the particle and its parent electrode immediately following separation. The microgeometry of this ideal sphere-plane regime

259

will be identical to that shown in Fig 7.4(a), so that the resultant field E_g will again be given by the sum of the two components E_1 and E_2 of equation 7.16, defined for the analysis of the "impact" trigger discharge given in Section 7.2.1(ii). However, in the present case, where the particle is leaving the electrode, these two components will be oppositely directed so that

$$
\begin{aligned}
E_g &= E_1 - E_2 \\
&= 1.29 \left(R_p/g \right) 0.8 \left(E - 3Q_p/2\pi^3 \varepsilon_o R_p 2 \right)
\end{aligned}
\qquad 7.43
$$

where Q_p is the particle charge and E is the macroscopic gap field. In the absence of any in-flight charge exchange, Q will be the charge acquired by the spherical particles in contact with the planar electrode, viz. $Q_p = \frac{2}{3}\pi^3 \varepsilon_o R_p^2$ so that in a DC situation where E is constant, the above equation indicates that E = 0, i.e. no discharge can occur. However, for fast-rise impulse voltages, E will no longer be constant with time, being now defined by $E(t) = (l/d)(dV/dt)$, and consequently $E_g \neq 0$. It follows therefore that if $E_g \rightarrow 10^9$ Vm^{-1}, electron emission will occur between the cathode and particle, thus providing the possibility of the same breakdown-initiating vaporisation processes discussed in Section 7.2.1(ii) for the impact trigger mechanism.

An obvious consequence of this electron emission mechanism is that the negative charge on the particle will increase during its flight from the cathode, thereby promoting an associated increase of E_2. It therefore follows from equation 7.43 that this effect will exert a controlling influence on E_g, limiting it to some maximum value $(E_g)_m \sim 10^9$ Vm^{-1} set by the field emission threshold. The actual charge carried by the particle will in fact increase from its initial detachment value $Q_p = \frac{2}{3}\pi^3 \varepsilon R_p^2 E$ at $t = t_c$ to a maximum value $(Q_p)_m$ at a time to after detachment when the charging current (i.e. the electron emission) ceases to flow. The value too can be determined from the condition $dQ/dt = 0$ whilst $(Q_p)_m$ will be given to a good approximation by

$$
(Q_p)_m = \frac{2}{3}\pi^3 \varepsilon_o \left[\frac{1}{d}\left(\frac{dV}{dt}\right)\left(t_d + t_o\right) - \frac{E_m}{1.29\left(R_p g\right)0.8} \right]
\qquad 7.44
$$

As would be expected from the time-dependence of E, viz. $E(t) = 1/d \frac{(dv/dt)}{dt}$, experiments confirm [42] that the probability of breakdown being initiated by this mechanism increases rapidly with the rate of rise of the impulse voltage (dV/dt). In addition it was found that the present "detachment" trigger discharge mechanism, like the "impact" mechanism, only plays a significant role in the initiation of breakdown where particle diameters are in excess of 100μm. An explanation for this observation has been provided by computations of the duration t of the charging current. These show that this parameter falls off rapidly with decreasing particle diameter so that below some limiting value of $t_o \sim 3$ µs the trigger discharge cannot be sustained for a sufficient time to initiate an arc across the main gap. For the typical pulse rise rates of $(dV/dt) \sim 10^{10}$ Vs^{-1} used in the experiments cited [42], calculations have confirmed the experimental finding that this limit can only be exceeded for particle diameters $\gtrsim 100$ µm.

It should be finally noted that in a "real" gap situation, microparticles are likely to have been torn from an electrode surface with the associated formation of the field-enhancing filamentary structures illustrated in Fig 7.2. Therefore, since such a microgeometry represents a more favourable field emitting regime, and since also both the emitter and effective anode have very much lower thermal capacities and consequently more liable to vaporisation, it follows that the threshold conditions for this type of trigger mechanism to operate are likely to be significantly lower than for the artificial sphere-plane regime.

7. 3   Experimental Studies of Microparticle Phenomena in HV Gaps

Since the early 1950s, when Cranberg [1] formulated the first theory of microparticle-initiated breakdown (see Section 7.1), there has been a sustained effort to devise experimental techniques for detecting the presence of the hypothesized high-velocity charged particles. In fact, this has proved to be a very challenging problem and it is only very recently that direct evidence of their existence has become available through the development of in-flight detecting systems. Prior to this, it was only possible to refer to indirect evidence which mostly took the form of observing changes in the microtopography of electrode surfaces that were attributable to the impact of such particles.

7.3.1   Particle Detection by Indirect Techniques

The earliest experiments of this type were those aimed at obtaining positive evidence of material transfer between electrodes following breakdown events. Two techniques have been employed, although both share the common approach of using an electrode gap formed by dissimilar electrode materials. In the first, such as was used by Anderson [43], positive identification of the phenomenon has been obtained from an optical spectroscopy analysis of the elemental composition of electrode surfaces which revealed traces of "foreign" elements corresponding to the complementary electrode material. In the second approach, radioactive tracer techniques [44] were used to detect material transfer. Neutron activation analysis is another technique that has been used to obtain evidence of material transfer across a gap [16,46], whereby one of the electrodes is made artificially radioactive by the bombardment of fast neutrons so that any traces of it can subsequently be detected on the unradiated electrode. In the case of copper electrodes, for example [16], it is possible to use the β^+ activity ($T^{1/2}$ = 9.8 min) of ^{62}Cu produced by the reaction ^{63}Cu(n,2n)^{62}Cu. It must be added however, that the technique lacks spatial resolution and has a minimum mass detection limit of ~ 10^{-13} kg which probably corresponds to upwards of 50 microparticles. Nowadays, with the ready availability of the scanning X-ray microprobe surface analysis technique, it is a relatively routine experiment to confirm these earlier findings; also, with the high spatial resolution of these instruments ($\lesssim$ 1 μm^2), it is possible to establish that "foreign" species are highly localised on an electrode surface, as would be expected from the microparticle impact model.

The other major source of indirect evidence for microparticle phenomenon also stems from the advent of scanning electron microscopy as a readily available analytical technique. With this facility it is possible to study the microtopography of electrode surfaces both before and after breakdown events and thereby respectively identify the characteristic structural changes resulting from microparticle impact processes.

Fig 7.10   A scanning electron micrograph of a 0.25 μm diamond polished copper anode showing the type of microcratering damage that typically follows a breakdown event. (From Latham and Braun [45], with permission.)

There have been several such reports in the literature [16,47] whose general findings can be typified by the micrograph shown in Fig 7.10 which shows the localised "splash" microcratering damage on a planar anode used in conjunction with the hairpin cathode of the experimental system previously described in Section 3.3.1 Fig 7.10 is of particular significance to the present discussion since it is possible to identify embedded microparticles at the centre of these craters, i.e. strongly suggesting that they originated from either the high velocity impact or trigger discharge process discussed in Section 7.2.1 (ii). Both of these explanations are however open to some doubt, since on the one hand the estimated single-transit impact velocities of micron sized particles in the present regimes are too low ($\lesssim 300$ ms^{-1}) to cause the observed fusion cratering [45], and on the other hand, theory indicates that a trigger discharge is unlikely to occur with such small particles. Despite these doubts, it is nevertheless difficult to discount a microparticle-based explanation for at least some of this microcratering since, as discussed above, considerable evidence has been accrued from investigations using dissimilar electrodes suggesting that relatively large numbers of particles are in fact exchanged between electrodes. Another important feature of the micrographs of Figs 3.8 and 7.10 is the distribution of loosely adhering "secondary" microparticles that have evidently been formed from molten droplets generated during the fusion process. On the basis therefore of the various microparticle mechanisms discussed elsewhere in this chapter, it is reasonable to anticipate that such secondary microparticles could play an active role in the subsequent operation of the gap.

In another novel approach for obtaining indirect evidence of microparticle processes, Theophilus et al. [48] devised the experimental regime illustrated in Fig 7.11 in which any particles generated between the high voltage electrode assembly L_1 and L_2 during a prebreakdown voltage cycling can escape from the gap through small perforations in L_2 to become attached to the collecting plate P.

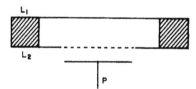

Fig 7.11   An electrode regime used for collecting the naturally occurring microparticles that are active in a gap during the initial application of a high voltage. (After Theophilus et al. [48].)

By examining this plate before and after an experiment in an SEM, these investigators were able to show that a large number of microparticles having dimensions that were typically ~ 3 μm had been active in the gap during the prebreakdown phase. However, since the experiment was conducted in poor vacuum conditions (~ 10^{-5} mbar) with a consequently greater likelihood of the occurrence of microdischarge events, some doubt must exist as to whether the collected sample of particles can be reliably assumed to be passively-launched "primary" microparticles. The interpretation of the findings was somewhat complicated by the presence of insulating ceramic particles among the collected sample; these could either have been derived from the alumina polishing medium, or have indicated that the insulating mounting of the electrodes was playing an active role in the experiment.

7.3.2   Particle Detection by Direct Techniques

The earliest attempts to record the actual transit of naturally occurring "primary" microparticles across a high voltage gap relied on the fact that they would be electrically charged and that it should therefore be possible to use fast electronic techniques to detect them as a charge transient superimposed upon any continuous prebreakdown current. Such a transient could then be used to directly characterise the microparticles in terms of their charge, mass and velocity. However, whilst this approach was conceptually very attractive, it had to be abandoned since no conclusive evidence was obtained, presumably due to the limitations of the electronic detecting systems which demanded the incompatible specifications of the highest charge sensitivity and the fastest response time.

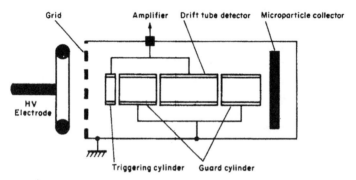

Fig 7.12   An experimental regime for dynamically recording the charge and velocity of naturally occurring microparticles generated in a high voltage gap. (From Texier [49], with permission.)

Recently however, Texier [49] has developed an alternative electrical technique for the dynamic detection of charged microparticles. In this system, illustrated schematically in Fig 7.12, one of the electrodes of a plane-parallel gap is replaced by a stainless steel grid and is followed by a capacitive drift tube detector system, similar to that later described in Section 7.4, for recording the charge and velocity of any microparticles that escape from the HV gap. In addition, there is a plate for collecting these particles so that their size and shape can be checked in a complementary SEM analysis. It was found that when a slowly increasing voltage is initially applied across the gap between the grid (cathode) and a freshly polished anode electrode A, large numbers of loosely adhering particles, typically ranging in size from 1-8μm, are extracted from the gap well before breakdown is approached, and that a further cycling of the gap voltage over this sub-breakdown range yielded no further particles; i.e. indicating that the initial voltage application had "electrostatically cleaned" the anode electrode of all loosely adhering "primary" microparticles.

A complementary elemental analysis of this sample of particles yielded the important but somewhat unexpected result that they were principally composed of the material used for the anode electrode A; unlike the findings of Theophilus et al. [48] described above, particles of the alumina polishing medium appeared to be of only secondary importance. Another interesting finding was that the size of the collected particles decreased with the applied voltage, which is in general agreement with Boulloud's theoretical reasoning as to the role played by Van der Waals adhesion forces [50]. As an extension of the technique, it was found, not surprisingly, that if the gap voltage is subsequently increased to breakdown, fresh bursts of "secondary" microparticles are detected whose characteristics appear to depend somewhat upon the target material. Whilst this technique undoubtedly represents an important new tool for the in-situ study of microparticle phenomena, more reliance could be placed on the findings of any future application if it were (i) operated under UHV rather than the present "commercial" ($\sim 10^{-5}$ mbar) vacuum conditions, (ii) incorporated some form of prebreakdown current/microdischarge monitoring facility to ensure that the apparent "primary" microparticles were not associated with electrical instabilities and (iii) reversed the polarity of the electrodes so that the field-enhancing fine grid wires [51] could not play an active role as a source of electron emission.

An alternative approach to the dynamic detection of in-flight microparticles has been to use a laser-based optical scattering technique to detect the light scattered by microparticles as they cross an electrode gap. Here again however, the findings from early investigation [52,53] were inconclusive and it is only comparatively recently that Jenkins and Chatterton [54] have succeeded in obtaining reliable results from an improved experimental regime employing higher laser power and an improved detection efficiency. In the system used by these authors, illustrated schematically in Fig 7.13, a 4 W laser beam, with 1.6 W in the 514.4 nm line, is directed between the electrode gap to impinge on a suitable light dump so that only scattered radiation is collected by the lens and recorded by the photomultipler PM. To obtain quantitative information about the particle size, the facility was calibrated by measuring the magnitude of the signal produced by the light scattered from particles of known diameter, artificially introduced into the system: particle velocities were estimated from the duration of the light pulse, which corresponded to the time taken for a particle to cross the known diameter of the laser beam in the gap. The initial findings of this study showed that the first AC voltage application to virgin electrodes is characterised by the passage of large numbers of "slow" particles having velocities in the range 15-30 ms^{-1}, thus confirming the findings of Texier [49]. By using an impulse voltage technique, it was also demonstrated that a breakdown event is typically precursed by the detection of a microparticle crossing the gap ~ 5 μs before the voltage collapse; however, unlike the "slow" particles referred to above, this latter type have very much higher velocities (80-800 m.s^{-1}) and smaller diameters (0.1-0.2 μm).

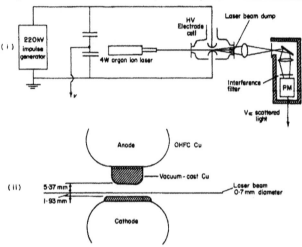

Fig 7.13   (i) The general lay-out of a laser facility for detecting the transit of microparticles across a planar high voltage gap, and (ii) the detailed electrode geometry. (From Jenkins and Chatterton [54], with permission.)

Although not conclusive, this correlation between particle and breakdown events must be regarded as direct evidence in support of the general premise that breakdown can be microparticle initiated. This investigation also confirmed that as spark conditioning proceeded, there was the expected rapid fall off in the number of particles crossing the gap. An unfortunate limitation of the present form of the technique is that it cannot discriminate between anode and cathode launched particles, or verify that they are electrically charged.

    Finally, in this discussion of in-flight microparticle detection, it will be recalled from Section 7.2.2(iii) that Davies and Biondi proposed a mechanism whereby microparticles can be vaporised by electron bombardment during their transit between electrodes and thereby induce breakdown. To obtain some experimental verification of this model, these same authors developed a spectroscopic technique for recording transient increases in the density of metal ions in a gap immediately prior to a breakdown [55]. Copper resonance radiation from an external source is focussed into a narrow beam to pass through the 1 mm gap formed between a pair of plane-parallel copper electrodes.

266

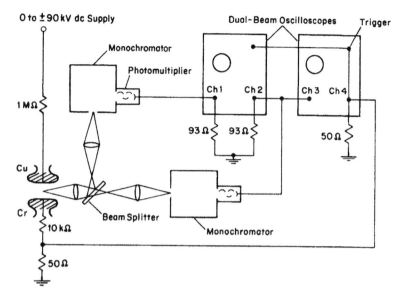

Fig 7.14   An optical spectrometry facility for detecting and discriminating between the generation of anodic or cathodic metal vapour in a high voltage gap during a breakdown event. (From Davies and Biondi [56,57], with permission.)

The transmitted light then enters a monochromator coupled to a photomultiplier so that it is possible to record rapid intensity variations in a given line of the copper resonance radiation. Such an "open shutter" system therefore provides a sensitive means of detecting transient increases in the vapour density of copper atoms in the inter-electrode gap, since the latter will constitute a strongly absorbing medium for the copper resonance radiation.

To investigate the temporal correlation between the detection of metal vapour and the incidence of electrical discharge events as the gap voltage is gradually increased, the output signal from the photomultiplier, and that derived from a potentiometric resistor chain of 10 kΩ and 50 Ω carrying the total gap current in the low-voltage side of the circuit, are simultaneously monitored with a twin-beam oscilloscope. With this system it was established that a burst of neutral metal vapour is commonly liberated into the electrode gap ~ 2-3 microseconds before a breakdown or microdischarge event.

Since this relatively long time period is inconsistent with the anticipated delay time of several nanoseconds for the avalanche growth of ionisation processes that would follow an "explosive" breakdown initiating event, Davies and Biondi were able to interpret these findings as strongly supporting their "heated-anode-particle" breakdown mechanism discussed in Section 7.2.2(iii); i.e. in which a microparticle extruded from an anode "hot-spot" is vaporised by electron bombardment during its subsequent flight between the electrodes.

Further evidence for the anodic origin of the initial vapour release was obtained from a development of this technique [56,57] illustrated in Fig 7.14, in which dissimilar electrode materials (copper and chromium) were used in conjunction with twin monochromator detecting systems tuned respectively to the resonance lines of Cu and Cr. With this system these authors were able to establish that, although both anode and cathode vapour are liberated at a breakdown, the anode species consistently appears in the gap ~ 30-150 nanoseconds before cathode vapour; i.e. consistent with the model of breakdown being initiated in the vapour medium generated from the anode particle, followed by erosion of the cathode by the ensuing discharge.

7. 4   Stimulation Studies

From the discussion of Section 7.2, it is clear that microparticle initiated breakdown is a well established concept. However, for most of the theoretical mechanisms considered, it is very difficult to obtain more than indirect supportive evidence for their occurrence, since the degree of control that can be built into direct measurements made on the complex microscopic system of an extended-area high voltage gap is very restricted. A more profitable approach for verifying the predictions of these theories has been to devise experimental techniques in which microparticle phenomena are studied in isolation under controlled laboratory conditions that simulate those existing in operational high voltage gaps. As will be discussed in Chapter 14, these same techniques have been exploited in space-orientated technologies to simulate the effects of micrometeorite impact damage.

The simplest type of experiment merely involves placing micron-sized particles either directly on the surface of pre-assembled electrodes, or in some form of specially prepared reservoir in the surface of an electrode [58-60], and showing that the voltage hold-off capability of such a gap is significantly lower than an identically prepared control gap that is free of artificial microparticles. In a very much more sophisticated version of this approach that was developed for studying the trigger discharge mechanism discussed in Section 7.2.1(ii), Martynov and Ivanov [16] used single 100 μm diameter iron spheres that were electro-magnetically held to an electrode surface until the gap was suitably stressed. Upon release, a charged sphere is accelerated across the gap and, on approaching the opposite electrode, is discharged through a visible microspark or "trigger discharge" that is temporarily correlated with the breakdown of the gap. By reversing its charge during the ensuing impact, the sphere subsequently oscillates between the plane-parallel electrodes, giving rise to a succession of trigger discharge and associated breakdown events. Although this technique provided an impressive demonstration of this breakdown mechanism, its findings are only of limited relevance to most practical gap regimes since, as discussed in Section 7.2.1(ii), the mechanism can only operate with large particles (>100 μm diameter) which must be considered to be extremely rare occurrences following modern electrode polishing and cleaning procedures.

An alternative approach, that has subsequently proved to be very much more valuable for studying the impact behaviour of the micron and submicron diameter particles thought to be responsible for initiating breakdown events in operational gaps, employs the concept of using "externally" generated microparticles of known size, charge and velocity, that can be either injected into a high voltage gap or used in other experimental regimes for studying such phenomena as impact ionisation and impact damage. The findings from these and other studies based on this approach will now be reviewed in the following sections of this chapter.

7.4.1   The Dust-Source Microparticle Gun

Since the basic requirement for this type of investigation must clearly be some form of microparticle "gun", it was fortuitous that at about the same time that these microparticle-based breakdown theories were being established, an almost ideal particle source had been independently developed for micrometeorite simulation studies in connection with the American space research programme.

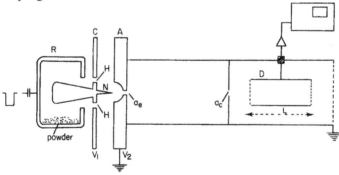

Fig 7.15   Schematic representation of the dust-source microparticle gun and drift-tube capacitive detecting system. (After Shelton et al. [61] and Friichtenicht [62].)

The original design of this device, which is so named because the particles are derived from a reservoir of fine powder, was due to Shelton et al. [61], with subsequent developments by Friichtenicht [62]. Its operational principle can be understood with reference to Fig 7.15, which shows the electrode configuration of the device. The powder, which normally consists of readily available carbonyl-iron microspheres of 0.1-5 ,μm diameter, with corresponding masses of 10^{-16} - 10^{-11} kg, is contained in a re-entrant cup-shaped reservoir electrode R into which projects a bull-nosed snout from the charging electrode C maintained at a fixed positive potential V_1. To "fire" the gun, a negative-going pulse of ~ 10 kV amplitude and 10-20 ms duration is applied through an external capacitor to R, thereby establishing a temporary electric field within the powder reservoir. As a result, the particles become positively charged by contact with the inner surface of R and, because of the mutually repulsive forces between them, "explode" to fill the reservoir, with a few escaping through the extraction holes H into the high-field gap between C and the accelerating electrode A held at potentials V_1 and V_2 respectively: typically, the "gun voltage" V_g ($\equiv V_2 - V_1$) is ~ 10-15 kV with V_2 set at earth potential. The "extracted" particles execute bouncing excursions between C and A, and as a result, a few acquire a high positive charge by coming into a glancing contact with the tip of the charging needle N located at the geometrical centre of the hemispherical aperture in A of radius 1-2 mm. These positively charged particles therefore experience a radial accelerating field so that those with approximately paraxial trajectories emerge from the exit aperture a of the gun into a field-free space; a supplementary collimating aperture a is also included to further select only those particles with axial trajectories. The needle itself is typically etched from a tungsten rod into a conical end profile with an approximately hemispherical tip of radius R_t ~ 25 μm. If, therefore, the charging geometry is taken as a small sphere of radius R_p in contact with a larger sphere of radius R_t, an order of magnitude estimate of the equilibrium charge Q acquired by the microsphere can be made using the relation [61]

$$Q_p = \frac{2\pi^3}{3} \varepsilon_o \left(V_2 - V_1\right) \frac{R_p^2 R_t}{\left(R_t + R_p\right)^2} \qquad 7.45$$

where for a 1 μm diameter microsphere accelerated through 15 kV we find Q_p ~ 10^{-15} C; in practice, however, Q will be significantly less than this since, with a glancing contact, the charging process will be incomplete. It should finally be noted that although the device could, in principle, be used for generating negatively charged particles, by simply reversing the polarity of the electrode voltages, this mode of operation is not to be recommended since a negative needle voltage gives rise to high field electron emission currents that rapidly destroy the sharp tip of the needle and cause electrical breakdown of the gun.

The charge and velocity of the axial particles generated by this device are measured by a capacitive detector module [61] consisting of a coaxial drift-tube D faced with transparent metal grids that is connected to an external voltage-sensitive preamplifier of gain g followed by a storage oscilloscope. As indicated above, the overall charge sensitivity of this detecting system should be $\lesssim 10^{-16}$ C, which nowadays can readily be achieved with a FET input operational amplifier used in the non-inverting mode as illustrated in Fig 7.16. When a particle of charge Q_p enters the detector an equal charge will be induced on the drift tube, so that if the capacitance of the tube to earth is C_D, the voltage ΔV appearing at the output of the preamplifier is $g(Q_p/C_D)$. Thus knowing g and CD, which are constant parameters of the system, and measuring ΔV on the oscilloscope, Q_p may be determined.

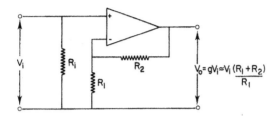

Fig 7.16   The FET input operational pre-amplifier used for observing the voltage produced on the drift-tube detector by a charged microparticle.

A typical trapezoidal-shaped signal obtained with this arrangement is shown in Fig 7.17. (Had the facing grids been omitted from the drift tube, the pulse would have been more rounded with less steep leading and trailing edges.) The length of the pulse t_t represents the transit time of the particle through the drift tube, so that knowing its length l, the particle velocity v is given by l/t_t.

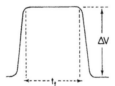

Fig 7.17   The trapezoidal-shaped signal resulting from the passage of a single positively charged microparticle through the drift tube detecting system shown in Fig 7.15.

Finally, if the particle charge is assumed to remain constant throughout its passage from the charging needle N, energy conservation requires that

$$\frac{1}{2} M_p v^2 = Q_p V_1 \qquad\qquad 7.46$$

from which the particle mass M_p can be calculated. Alternatively, the particle radius RP can be determined if its density ρ is known, since $M = \frac{4}{3}\pi R_p^3 \rho$ for these spherical particles. From an analysis of a large sample of the type of single particle traces shown in Fig 7.17, it is found that this type of gun delivers a spectrum of microspheres ranging from those with sub-micron dimensions that have typical charges and velocities of ~ 10^{-16} and 1.5 kms^{-1} to the largest 5 μm diameter spheres that can have charges approaching ~ 5×10^{-13} C and velocities as low as ~ 100 ms^{-1}.

(i)   (ii)

Fig 7.18   Histograms illustrating (i) the distribution of particle charges N(Q$_p$) and (ii) velocities N(v) delivered by a dust-source gun using a carbon fibre charging arrangement. (From Brah, A.S. and Latham, R.V. [63], with permission.)

In a more recent UHV version of this dust source gun, which was specially developed within the author's research group for obtaining a spectrum of low-velocity particles [63,64], the needle electrode was replaced by a slightly splayed bunch of 7 μm diameter carbon fibres. With this charging system, the operating voltage of the gun can be reduced from the 10-15 kV necessary for the needle charging system to only 0.5-1 kV for a similar spectrum of particle charges; i.e. representing over an order of magnitude improvement in the charging efficiency of the gun. This therefore has the important consequence that the particle velocity spectrum is reduced to a range of ~ 20-150 ms^{-1}. Another significant advantage of this fibre charging system is that it provides a much more efficient device in terms of the probability of obtaining a paraxial particle per activating pulse. To illustrate the type of performance characteristics obtained from dust source guns, Figs 7.18(i) and (ii) give respectively the charge and velocity spectra for a typical sample of 200 consecutive particles produced by this recent low-velocity carbon fibre gun: the spectra obtained for the needle charging guns have a similar form, although the velocity spectrum is naturally shifted to a higher range. It is also possible to further expand the range of particle velocities obtained from any basic gun module by using either post acceleration or post declaration facilities.

Thus, with a megavolt Van der Graaff powered accelerating and focussing module interposed between a needle-type gun and the drift tube detecting system, as shown schematically in Fig 7.19(i), hypervelocity particles may be obtained; e.g., 2 MV acceleration can give particle velocities up to ~ 40 km^{-1} [61,62,65-67].

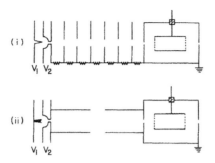

Fig 7.19   Dust-source microparticle gun systems incorporating (i) post acceleration and (ii) post deceleration.

Alternatively, with a two-tube decelerating lens used in conjunction with a carbon fibre based gun as illustrated in Fig 7.19(ii), it is possible to obtain particle velocities ~ 1 ms^{-1}. To this list of ancillary facilities, there should also be added the post-gun particle mass and velocity selecting system developed by Rudolph [68].

7.4.2   Particle Injection Experiments

The most direct and convincing laboratory demonstration that the breakdown of a stable high voltage gap can be initiated by the impact of a high velocity microparticle was provided by a technique originally developed by Slattery et al. [33]. In this, positively charged microspheres are accelerated to the hypervelocity range of ~ 2-6 kms^{-1} by the type of 2 MV system shown in Fig 7.19(i) and then injected into the 10-30 kV plane-parallel high voltage gap regime illustrated in Fig 7.20. The stainless steel electrodes of this 1-2 mm gap were inclined at 45° to the incident particle direction in order that their impact zone is shifted from the surface region where the electric field lines are most distorted due to the presence of the 1 mm diameter aperture in the low-voltage electrode. Particle-initiated discharges were detected with an oscilloscope by monitoring the current transients flowing in the 0.1 Ω resistor connected between the low-voltage electrode and earth. The findings of this investigation indicated that for a given gap voltage V there was a minimum particle radius $(R_p)_{min}$ and impact energy U below which breakdown initiation was very rare. For example, when V = + 15 kV results showed that $(R)_{min}$ ~ 0.9 μm and U_{min} ~ 0.8 × 10^{-7} J; if, however, V was increased, $(R_p)_{min}$ and U_{min} were significantly lowered, indicating that smaller particles were producing discharge events. It was also found that these thresholds were dramatically lowered if the polarity of the target electrode was reversed: thus, in the above example, if V = -15 kV, both $(R_p)_{min}$ and U_{min} are halved. The authors provided a qualitative interpretation of their data in terms of the

Cranberg-Slivkov particle impact model discussed in Section 7.1, which imposes the constraint that, before a particle can initiate breakdown, it must have a minimum energy sufficient to produce a gas cloud that is large enough to permit charge multiplication while simultaneously satisfying the conditions at the Paschen minimum. No explanation was proposed at the time for the

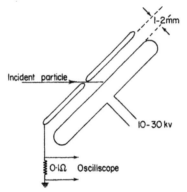

Fig 7.20   Inclined electrode regime used for studying microparticle initiated breakdown in a plane-parallel high voltage gap. (After Slattery et al. [33].)

polarity effect; however, in the light of more recent experimental findings discussed later in this section, and the concepts discussed earlier in Chapters 5 and Section 7.2, three possible mechanisms might be considered. Firstly, the gap field will give rise to a final acceleration or deceleration of the particle, so that the impact velocities for a negative high voltage electrode are likely to be significantly higher than when it has a positive polarity. Secondly, the 45° angle between the incident particle direction and gap field means that a particle trajectory will depend on the polarity of the target electrodes which could similarly affect the final impact angle could be similarly and hence the amount of impact ionisation generated [67]. Thirdly, the impacting microparticle could give rise to a new field- emitting microprotrusion at which the local microscopic field is enhanced beyond the critical value associated with "cathode" initiated breakdown. In a more recent experiment of this type, Chakrabarti and Chatterton [69] investigated in more detail how particle-induced discharges and the associated electrode damage are related to the size and velocity of the incident particle, and to the target electrode material. They used the cylindrical electrode geometry shown in Fig 7.21, in conjunction with a 50-60 kV microparticle gun (see Fig 7.19) that delivered positively-charged iron microspheres with diameters and velocities ranging between 0.5-150 μm and 600-10 ms^{-1} respectively. The 0.05 mm diameter central cathode wire, which was the target electrode, was either gold or tungsten and maintained at 0-60 kV with respect to the 3 cm diameter concentric stainless steel anode: the cathode wire could also be externally retracted to provide a changing impact region for any given sequence of experiments. A photomultiplier was also included for determining whether specific discharge events gave rise to an emission of light from the gap. For the range of cathode fields used in these experiments, viz. up to around 4×10^7 Vm^{-1}, it was found that microdischarge and breakdown events with an associated light output were only observed for low-velocity microparticles having diameters ~ 10-25 μm: a complementary SEM examination of the relevant impact region of the cathode wire also revealed that such events were linked with the occurrence of fusion induced microcratering.

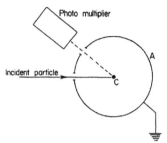

Fig 7.21   Electrode regime used for studying microparticle initiated breakdown in a cylindrical high voltage gap. (After Chakrabarti and Chatterton [69].)

Since however this damage could not be realistically explained on the basis of a Cranberg-Slivkov impact process with such slow particles, the authors concluded that this type of breakdown event was most likely initiated by the trigger discharge mechanism discussed in Section 7.2.1(ii). Although there was considerable evidence of particle melting and/or electrode indentation damage for the smaller particles with higher velocities, no incidence of discharges or breakdowns were recorded. This was a somewhat surprising finding, particularly in the case of the softer gold target wire, where the damage was more pronounced and appeared to include many microfeatures that could give rise to the type of unstable FEE processes discussed in Chapter 5.

7.4.3   Impact Ionisation Measurements

Although the basic assumptions of the Cranberg-Slivkov particle impact breakdown initiating mechanism discussed in Section 7.1 are widely accepted, it is nevertheless clearly desirable to have direct corroborative evidence of the associated phenomena, particularly the ionisation processes. In fact, there have been several major laboratory investigations directed at measuring the transient ionisation that accompanies the impact of high velocity microparticles [62,65-67,70,71]. In all cases, these were conceived as simulation studies of micrometeorite impact phenomena, and used the type of hypervelocity, positively-charged microsphere facility illustrated in Fig 7.19(i). Most of these measurements were however made with a low-field

275

impact regime, and so it must be recognised at the outset of this discussion that their findings may not be strictly relevant to the high voltage gap situation, where the high electric field and electrode surface charges could well influence both the ionisation mechanisms and the development of the plasma. It is possible, for example, that as a result of charge separation, there could be a build-up of positive space charge in the vicinity of the impact zone which might give rise to a burst of "explosive" electron emission and associated ions, such as occurs during the development of the cathode flare described in Section 5.3: i.e. the level of ionisation associated with a "passive" very low field impact event is generally likely to be significantly lower than that associated with an equivalent event in a high-field impact regime.

Our discussion of the main features of this type of investigation will be based on the work of Smith and Adams [71] as providing a recent and representative study of the phenomenon. Their experimental facility is shown schematically in Fig 7.22, in which, for convenience, the two alternative operational regimes have been combined on the same figure. An incident particle with a velocity in the range 0.05 to 10 kms^{-1} is first characterised during its transit through the electrostatic drift tube detector D_t; it then impacts on a slatted molybdenum target T which allows any impact ionisation products to escape in the forward direction through the electrically insulated grid G_2.

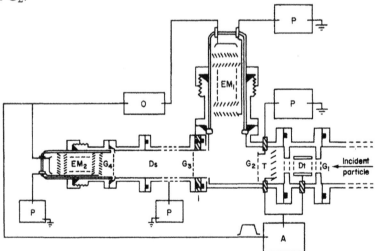

Fig 7.22   Schematic diagram of a combined apparatus for: (a) measurements of the ionisation resulting from the impact of hypervelocity microparticles, and (b) time-of-flight mass spectrometry studies of the ionisation products: G_1 are the screening grids, D_t is the capacitive drift-tube detector, G_2 the charge extraction grid, G_3 the deceleration grid, D_S the drift space, EM_1 and EM_2 the electron multipliers, I an insulating ring, A the amplifier, O the two-channel storage oscilloscope and P are the power supplies. (From Smith and Adams [71], with permission.)

However, by biasing G_2 either positively or negatively with respect to the target, it is possible to extract either negative or positive ions respectively from the impact-generated plasma. The quantity of charge extracted from the target region is measured in terms of the resulting change in potential of the target as recorded by the amplifier A, which is also used to monitor the potential of the drift tube detector D. With this arrangement, the storage oscilloscope O records the sequential signals from D_t and T, which appear as in Fig 7.23(i) and (ii) for negative and positive ion collection respectively: in both signals, the particle of charge Q_p enters and leaves the drift tube at t_1 and t_2, and finally impacts on the target at t_3. The liberated charge q_i is then obtained from these oscilloscope traces by subtracting the charge deposited on the target by the particle (viz. Q_p) from the total target signal. electron multiplier EM_1 provides an alternative facility for monitoring q_i which, because of its greater sensitivity ($\sim 10^{-17}$ C), is particularly valuable for extending the range of measurements to the low levels of ionisation found for low-velocity impacts that are beyond the charge detecting threshold of the target amplifier ($\sim 10^{-15}$ C). By reducing the viewing resistor of EM_2 so that its response time was $\lesssim 10^{-7}$s, the facility could be put to the further valuable use of measuring the rate at which charge was collected from the impact region.

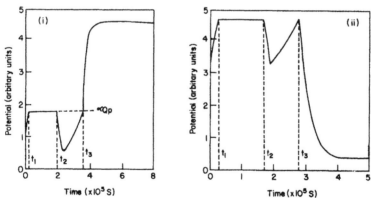

Fig 7.23   Variation in the potential of the capacitive detector tube and target as a function of time when the potential of grid G_2 (see Fig 7.22) is (i) positive and (ii) negative with respect to the target. t_1, t_2 and t_3 indicate the times at which the hyper-velocity particles enter and leave the capacitive detector and impact on the target respectively. (a) M = 6 $\times$ 10^{-13} kg, v = 2.8 kms^{-1}; (b) M = 1.6×10^{-13} kg, v = 3.4 km s^{-1}. (From Smith and Adams [71], with permission.)

It was concluded from this [66,71] and similar types of investigation [67,70] where refractory metal targets were used, that both the positive and negative charges released by impact ionisation processes follow an empirical relation of the form

$$q_i \propto M_p^{\alpha} \, v^{\beta} \qquad\qquad 7.47$$

where, to a first approximation, α and β can be taken as constants; several investigators [66,70,71] have in fact detected a second-order velocity dependence of α. This relation was based on data derived from particles impacting normally on the target; however, if q_i is measured as a function of the angle of particle impact θ, Dietzel et al. [67] found that there is a favoured value of θ (30°-50°) at which q_i can approach four times its value at normal incidence for a given impact velocity. Accordingly, a more correct representation of equation 7.46 would be

$$q_i \propto f(\theta) \, M_p^{\alpha} \, v^{\beta} \qquad\qquad 7.48$$

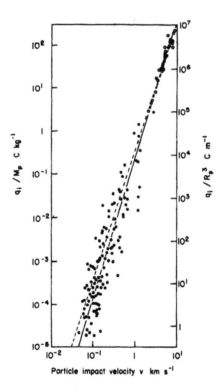

Fig 7.24 q_i/M_p and q_i/R_p^3 plotted against particle velocity for the impact of spherical iron particles on molybdenum (closed circles). Also included are the data of Auer et al. [70] (dashed line and open circles) and of Friichtenicht [62] (solid line) for the impact of iron particles on W. The solid line also represents a straight line least squares fit to the present data. (From Adams and Smith [66], with permission.)

The "maximum" associated with this angular effect has been explained qualitatively by these authors in terms of the compromise between the enhanced ionisation to be expected from the increase in contact area that occurs with increasing θ, and the reduction in the normal component of the particle momentum relative to its transverse (i.e. non-destructive) component that also occurs with increasing θ. A quantitative form of equation 7.46, for the case of normal incidence ($f(\theta) = 1$), was determined by Smith and Adams [71] from the type of log-log plot of q_i/M_p or q_i/R_p^3 against v shown in Fig 7.24 as

$$q_i \propto M_p^{1.33} \, v^{3.2} \text{ for } \quad 0.5 \lesssim v \lesssim 1 \text{ kms}^{-1}$$

$$q_i \propto M_p^{0.85} \, v^{3.2} \text{ for } \quad v > 1 \text{ kms}^{-1}$$

7.49

where 0.5 kms^{-1} is the velocity threshold below which impact ionisation becomes vanishingly small.

The increased scatter of data points on Fig 7.24 at lower velocities is a common feature of microsphere impact experiments and will be highlighted again in the following section. In the present context, its origin is discussed in terms of the partition of the incident kinetic energy, typically 10^{-11} to 10^{-7}J (or 10^8-10^{12} eV), among the various dissipating mechanisms available at impact, e.g. heat conduction, elastic compression, fusion and vaporisation processes (the latter giving rise to the measured ionisation products). Thus, there are two explanations for this scatter that appear particularly plausible: firstly, at lower velocities, contact times become longer, so that a relatively larger fraction of the incident energy is dissipated irreversibly by thermal conduction processes, and secondly, at lower velocities, certainly below 0.5 kms^{-1}, semi-elastic bouncing impacts begin to occur (see Section 7.4.5) where a large fraction (possibly 30-50%) of the elastic strain energy is recoverable [72] and so is no longer available to "feed" the other dissipating mechanisms that can give rise to ionisation. It is also significant that both of these low-velocity phenomena would tend to produce an increase in the mass power index α, as is in fact found experimentally. The ionisation collection time measurements made by Smith and Adams [71] show that there is a variable build-up time for the ionisation in the range 5-20 μs as the potential of the extracting grid G_2 is changed from 300 to 60 volts. Although it is tempting to speculate that the low-field value of 20 μs might reflect the duration of the particle impact process, the authors tend to discount this possibility but rather attribute it to the influence of V_{G_2} in controlling the growth and extraction of the plasma.

Referring again to Fig 7.22, the second type of measurement made with this apparatus [71] used a 3-5 amu resolution time-of-flight mass spectrometry facility to identify the ion species present in the impact generated plasma. In this regime, the grid G_2 has to be maintained at a potential ~ -300 v to obtain an efficient extraction of ions. Since, however, this potential gives the ions too high energies for operating the mass spectrometer facility at maximum

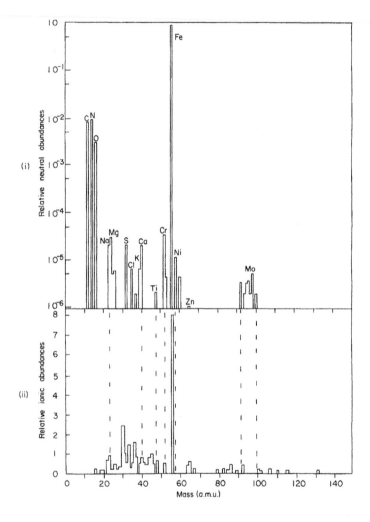

Fig 7.25   (i) Relative abundances of the elements in the iron projectile material. (ii) A histogram of the relative abundances of ions produced at the impact of eight iron projectiles in the velocity range 2 to 4 kms^{-1} on a molybdenum target. The broken lines connect the two mass scales at some of the important masses. Note that the ordinate in (a) is logarithmic whilst that in (b) is linear. (From Smith and Adams [71], with permission.)

resolution, a decelerating field is provided between G_2 and G_3 so that they enter the 50 cm long field-free drift space between G_3 and G_4 with energies of several tens of electron volts. The distributed ion current, resulting from the separation of the ion species in the drift space, is recorded on the electron multiplier EM_2.

A typical spectrum of the relative abundancies of the recorded ion species for a particle velocity range 2-4 kms^{-1} is shown in the histogram of Fig 7.25(ii) for comparison, the chemical analysis of the particle material obtained by a spark-source mass spectrometer is shown in Fig 7.25(i). The important conclusion to be drawn from this result is that the predominant ion species present in the impact plasma is the Fe^+ peak at 56 amu derived from the particle, with a barely significant contribution from the molybdenum target material at 92-100 amu.

This type of finding is in agreement with the earlier investigations of Hansen [65] and of Auer and Sitte [73], using similar incident particles, but with tantalum and tungsten targets respectively, who also concluded that the target plays a relatively unimportant role in the impact ionisation processes for velocities $\lesssim$ 10 kms^{-1}. It is only when the incident particle velocity approaches $\sim$ 20 kms^{-1} that the recorded spectra begin to show evidence of target material being ionised: in fact, it is this property that has been exploited in satellite-borne detection systems for analysing the elemental composition of micrometeoroids [74,75].

To explain this somewhat unexpected selectivity, it is necessary to refer to hydrodynamic projectile impact theories such as have been discussed by Dietzel et al. [67] and Smith and Adams [71] in connection with hyper-velocity microsphere impacts. This topic will be considered in more detail in the following section, but it can be stated at this stage that for the combination of materials and impact velocities used in all the cited investigations, theory predicts that the impacting particles will be preferentially "shocked" and thus lead to the selective ionisation processes found experimentally. To observe the opposite effect, where the target material is ionised, it would be necessary to use, say, a copper target with the same iron microspheres.

The identification of the subsidiary peaks appearing in the spectra of Fig 7.25(ii) that don't directly correlate with those in Fig 7.25(i) has been the subject of considerable qualitative discussion [65,70,71]. However, there is a consensus among these investigators that the two most likely explanations are firstly, the existence of multiply charged ions, and secondly, the general lowering of ionisation potentials due to the action of the strong space-charge based electric field that could exist in the localised impact microplasma (cf. the conditions prevailing in the cathode flare regime discussed in Section 5.3). In connection with the quantitative application of the spectroscopic impact ionisation data to the development of a space-launched micro meteoroid detector, there has also been extensive discussion as to whether the ion spectra of Fig 7.25(ii) are representative of their relative concentration in the initial impact microplasma [65,67,70,71]. This hinges upon the temperature and density of the initial microplasma, since the magnitude of these parameters will determine whether recombination processes are likely to play an important role in degrading the plasma before its composition is measured.

In fact, the evidence is somewhat conflicting on this point. On the one hand, estimates of the initial plasma temperature [65], based on measurements of the relative concentrations of the ions ejected from the plasma, are in the range $3\text{-}20 \times 10^3$ K for impact velocities of 1-40 kms^{-1} which, if the initial microplasma volume is assumed to be comparable with the indentation volume [75], strongly suggests that recombination effects would be important, particularly at higher impact velocities. On the other hand, there is no evidence to suggest that the collected charge q_i departs significantly from the empirical relation of equation 7.47 at high velocities, as would be expected if recombination processes began to operate. To resolve this contradiction, it has to be assumed that the initial plasma volume is considerably larger than the indentation volume, so that the initial plasma density will be sufficiently low to prohibit significant recombination. Support for this conclusion is provided by the work of Dietzel et al. [67], who developed a scanning X-ray microanalysis technique for a detailed study of how the particle debris is topographically distributed about the impact zone on the target surface. Their results show, for example, that for impact velocities ≥ 5 kms^{-1}, particle material is deposited over a surface area that is more than twice the cross-sectional area of the particle.

7.4.4   Impact Damage

From the earlier discussion of Chapter 5 and Section 7.2 it is clear that any physical process that changes the microtopography of an electrode surface in such a way as to create field enhancing microfeatures must be regarded as a potential agency for initiating one of the main breakdown mechanisms. Thus, microprotrusions formed on the cathode can become unstable field electron emitters and lead to either cathode- or anode- initiated breakdown (see Chapter 5), whilst anode protrusions, for example, can greatly increase the probability of the trigger discharge mechanism occurring if microparticles are present in the gap (see Section 7.2). In both cases, such microfeatures are frequently loosely bound to their parent electrode and therefore offer the possibility of being torn off by the electro-mechanical forces of the applied field to become potentially energetic "secondary" microparticles. The detailed nature of this impact damage, and how it depends on the physical parameters of the colliding surfaces has also been the subject of extensive microsphere simulation studies. In fact, this formed an important aspect of several of the investigations cited in the previous section.

The least drastic type of permanent electrode damage is the smooth-profiled indentation microcrater shown schematically in Fig 7.26(i) that is formed when particle impact velocities only marginally exceed the critical value v_c beyond which elastic recovery of the interacting surfaces is incomplete; i.e. plastic flow takes place. For the experimental regimes studied to date, which mainly consist of iron microspheres impacting on a wide range of mechanically polished electrode materials [67,72,75] it has been generally concluded that $150 \lesssim v_c \lesssim 350$ ms^{-1}; significantly, a similar order of magnitude is predicted from the theoretical approach of Cook [76] referred to in Section 7.2.2.

Apart from the possibility that the microparticle involved in such an impact event might "stick" to an electrode surface by some form of cold-weld process and so create a field enhancing feature, it can otherwise reasonably be assumed that the type of damage associated with this velocity range poses little direct threat to the stability of a high voltage gap. However, at velocities in excess of ~ 500 ms^{-1}, the work of Rudolph [75] and later Dietzel et al. [67] showed that, in addition to severe plastic deformation, with the particle becoming partly or completely embedded in the target, localised melting of the impacting surfaces progressively contributes to the damaging process, so that the crater profile now begins to take the form illustrated in Fig 7.26(ii).

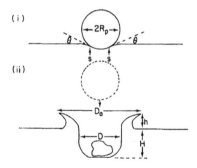

Fig 7.26   Microcrater regimes produced by microparticles whose impact velocities (i) only marginally exceed the critical value v ($\lesssim 300$ ms^{-1}), and (ii) are well in excess of v_c, such that there is extensive plastic deformation and progressive melting of the impacting materials. (After Rudolph [75].)

Viewed from above with a scanning electron microscope, these craters appear as shown in the micrograph of Fig 7.27; i.e. very similar to the familiar "splash" micro craters found on electrode pairs that have broken down (see Figs 7.10 and 7.27), and whose subsequent high voltage hold-off capability is drastically reduced. From a detailed analysis of how the dimensions of the Fig

283

7.26(ii) type of crater depend on the incident particle radius R_p, mass M_p and impact velocity v_i for a given target material, Rudolph [75] established the following empirical relations.

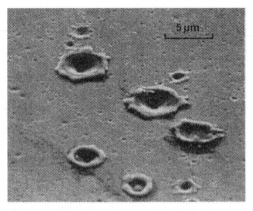

Fig 7.27 A scanning electron micrograph of the type of high impact velocity microcrater illustrated in Fig 7.26(ii).

For $v_i \lesssim 1 \text{ kms}^{-1}$

$$D \propto 2R_p \qquad \text{(i)}$$

whilst for $1 \text{ kms}^{-1} \lesssim v_i \lesssim 10 \text{ km. s}^{-1}$

$$D \propto R_p v_i^{2/3} \qquad \text{(ii)} \qquad\qquad 7.49$$

$$\text{or} \propto M_p^{1/3} v_i^{2/3}$$

$$(H + h) \propto D \qquad \text{(iii)}$$

$$D_a \propto 1.5D \qquad \text{(iv)}$$

where the constants of proportionality in equations 7.49(ii) and 7.49(iii) are functions of the target material, and are tabulated for Pb, Cd, Ag, Cu and Al [75]. The change in the form of these relations at $v_i \sim 1 \text{ kms}^{-1}$ was identified with the onset of visible deformation of the embedded particle: a conclusion that was also supported by X-ray microanalysis studies of the distribution of projectile material over the target surface [67] and the velocity threshold for the detection of impact ionisation [66].

An understanding of the physical mechanisms that determine the character of microparticle impact damage may be had by considering the well established hydrodynamically-based theories of projectile impact. These broadly deal with how the material constants and velocity of the colliding media determine the partition of the incident kinetic energy between the projectile and the target, and hence their respective shock responses. From a discussion by Dietzel et al. [67], dealing with the impact of iron microspheres on a range of target materials (Fe $\rightarrow$ Au, W, Cu, Al and basalt), data were presented showing how the immediate post-impact energies of the target and projectile are divided between "kinetic" and "internal" (e.g. elastic strain energy) components. Thus, for Fe $\rightarrow$ Au and Fe $\rightarrow$ W, the Fe projectile retains ~ 55% of its initial kinetic energy with ~ 10% appearing as kinetic and ~ 45% as internal: for basalt and Al the reverse situation applies, whilst Cu has intermediate properties. In all cases at higher impact velocities, an increasing fraction of the trapped internal energy of the projectile is converted into irreversible heating, where the heating of the projectile is shown to be approximately proportional to its initial kinetic energy.

In a complementary discussion, Smith and Adams [71] considered the behaviour of the shock waves that are generated around the perimeter of the contact disc, i.e. at s in Fig 7.26(i). These will remain attached to this contact circle whilst its velocity along the interface during the compression process exceeds the velocity of the shock waves in either of the colliding materials. This will generally be true in the initial stages at low contact angles θ, but, for a given impact velocity, a point is reached where this inequality reverses for one of the materials. When this happens, its shock wave will become detached and result in highly shocked, high-energy material being exposed to the surface with the consequent ejection of vaporised and ionised material with accompanying crater formation. Theory shows for example, that for Fe $\rightarrow$ W or Mo, this detachment first occurs for the Fe-projectile, whilst for a Fe $\rightarrow$ Cu or Al combination the target shock wave is the first to become detached.

7.4.5   Particle Bouncing

As discussed in Section 7.2.3(ii), Latham proposed a further development of the original Cranberg-Slivkov single-transit, impact-initiated breakdown mechanism whereby the kinetic energy of a microparticle can be enhanced to a critical value during a multi-transit bouncing sequence between the electrodes of a high voltage gap in which there is a reversal of particle charge and momentum at each impact (see Fig 7.7).

An associated theoretical analysis of the proposed model further showed that the effectiveness of this energy enhancing mechanism would be strongly dependent on the microscopic electrical and mechanical properties of the electrode surfaces. To test the validity of this concept, a laboratory investigation was established to study the bouncing impact behaviour of charged microspheres under controlled experimental conditions that closely

(i)

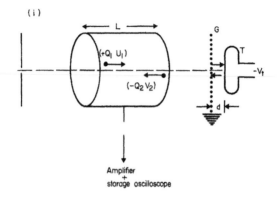

Amplifier
+
storage oscilloscope

(ii)

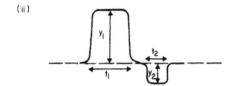

Fig 7.28   (i) A schematic representation of the experimental regime used for studying the particle bouncing phenomenon: it shows the drift-tube detector, the high-field test gap and the incident and reflected particles (Q_1, u_1) and (Q_2 v_2) respectively.
(ii) The sequential trapezoidal signals obtained from such a bouncing even (From Latham et al. [36], with permission).

simulate those existing in a high-voltage gap.

The basic principle of the experimental technique used for these measurements is shown in Fig 7.28. A positively charged microsphere with an initial charge Q_{p1} ($\sim 10^{-15}$ C) - or Q_1 to simplify the notation in this section - and velocity u_1 ($\sim 1\text{-}10$ ms^{-1}), derived from the low-velocity source illustrated in Fig 7.19(ii), is recorded on its first transit through the drift tube detector as a trapezoidal pulse of height y_1 and length t_1. It then enters a high field gap d (~ 1.5 mm) formed between a transparent stainless steel earthed grid G and the negatively-charged target electrode T maintained at a negative potential -V_T, typically ~ 5-10 kV. If, after being accelerated across the gap to an impact velocity v_1, the particle then bounces off the target with a reversed negative charge -Q_2 and rebound velocity u_2, it will be re-accelerated by the gap field to acquire a final velocity v and be recorded as a sequential negative-going trapezoidal pulse of height -v_2 and length t_2. Since the magnitude of v_2 will depend upon both $|Q_2|$ and u_2, which in turn are determined by the mechanical and electrical interactions of the impacting surfaces, it is convenient to study the bouncing phenomenon in terms of the two dimensionless parameters: (i) $|Q_2|/|Q_1|$ - the ratio of reversed to incident charge, which is a measure of the charge exchange during impact, and (ii) $|v_1|/|u_2|$ - the ratio of the rebound to incident velocities (i.e. the coefficient of restitution e), which is a measure of the momentum exchange during impact. In these experiments, the $|Q_2|/|Q_1|$ value is obtained directly from the detector signal as (y_2/y_1), whilst the e-value has to be computed indirectly. Thus, noting from Fig 7.28 that $u_1 = L/t_1$ and $v_2 = L/t_2$, one has

$$v_1^2 = u_2^2 + 2Q_2 V_T/M_p \qquad\qquad 7.50$$

which after substituting for M_p from equation 7.45 gives

$$v_1 = u_1 \left(1 + V_T V_1\right)^{\frac{1}{2}} \qquad\qquad 7.51$$

where V_1 is the effective accelerating voltage of the gun (see Figs 7.15 and 7.19). Similarly

$$v_2^2 = u_2^2 + 2Q_2 V_T/M_p \qquad\qquad 7.52$$

which after substituting for M_p from equation 7.46 becomes

$$v_2^2 = u_2^{\ 2} + u_1^{\ 2} \left(Q_1/Q_2\right)\left(V_T/V_1\right) \qquad\qquad 7.53$$

so that finally from equations 7.51 and 7.52

$$e = |u_2|/|v_1|$$

$$= \left\{\left[(L/t_2)^2 - (y_2/y_1)(V_T/V_1)(L/t_1)^2\right] / \left[(L/t_1)^2\ 1 + (V_T/V_1)\right]\right\}^{\frac{1}{2}}$$
$$\qquad\qquad 7.54$$

The UHV experimental facility that was developed by Brah and Latham [63,64,77,78] for this investigation is shown schematically in Fig 7.29. It is based on the low-velocity microparticle module described in Section 7.4.1, incorporating a carbon fibre dust-source gun M_g, decelerating electrostatic lens E1, gravity-correcting deflector plates P_g and drift tube detector D_t with its associated amplifier A. To give the reader some perspective of the microregime involved in these experiments, Fig 7.30 shows a 4.2 μm diameter microsphere resting on a 0.25 μm diamond-polished electrode surface.

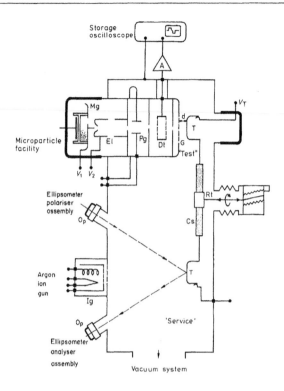

Fig 7.29   The UHV experimental facility for studying the bouncing impact of positively charged microspheres in a high voltage gap. An argon ion gun I_g provides in-situ cleaning of target surfaces, whilst optical ports O_p permit the ellipsometric monitoring of their atomic state. (From Latham and Brah [78], with permission.)

Trace A of Fig 7.31 shows a typical example of a trapezoidal pulse obtained from this system for the passage of a single positively charged microsphere. After leaving this module, the particle enters a high voltage gap d formed between the earthed grid G and target electrode T which is maintained at a potential $-V_T$ via a spring contact and an insulated lead-through to an external EHT supply; typically $d \sim 2$ mm, $-V_T \sim 2\text{-}10$ kV so that the gap fields are in the range $(1\text{-}5) \times 10^6$ Vm^{-1}. There is also an in-situ choice of the pre-selected targets mounted on ceramic insulating supports C_s that form the arms of an eight-spoke externally rotatable turret assembly. With this arrangement, a given target can be rotated from a "service" position in the lower part of the vacuum chamber, where an argon ion gun provides an in situ etching facility for cleaning electrode surfaces, to the "test" position in the upper part of the chamber, where it is on the axis of the microparticle source.

288

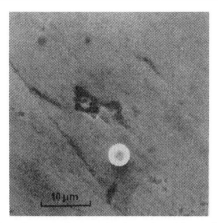

Fig 7.30 A scanning electron micrograph of a carbonyl-iron microsphere resting on the 0.25 μm diamond-polished surface of a titanium electrode.

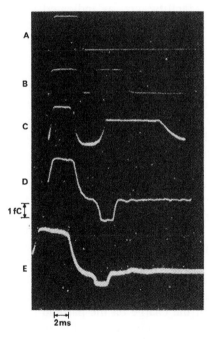

Fig 7.31 Oscilloscope traces illustrating the important micro-particle events. In each case, positive particles have been used with an oscilloscope time base setting of 2 ms/division and an equivalent charge sensitivity of 1 fC/division. A, a single paraxial particle of velocity 10 ms^{-1} and charge 2.5 fC; B, in-flight momentum reversal $|V_T| > |V_1|$; C, bouncing impact with charge modification $|V_T| > |V_1|$; D and E bouncing impacts with charge reversal and velocity enhancement $|V_T| < 0$ (From Brah and Latham [63], with permission.)

289

The experimental chamber also includes optical ports O_p for an external ellipsometry assembly [79] that is used to monitor the state of oxidation of a target surface whilst it is in the "service" position. Full details of this technique are beyond the scope of this text, and can be found in the literature [80,81]. Very briefly, however, it depends upon the fact that if a beam of plane polarised light from the polariser assembly is incident on a conducting surface, in this case the target T, it will be reflected as elliptically polarised light, where its ellipticity, as measured by the analyser assembly, is determined by the wavelength of the light and the optical constants of the surface, viz. its complex refractive index n. From a knowledge of the n-values for the metal substrate and its surface oxide, it is then possible to determine the thickness of the oxide film: in its most sophisticated form, the technique can be made sensitive to the detection of less than a monolayer coverage. However, for the experiments described here, the facility is only used to measure the thickness of the ambient oxide film present on untreated electrode surfaces, and to determine when it has been removed by the argon ion electrode cleaning facility; viz. as indicated by the ellipticity of the reflected light approaching a constant value with time of etching. Depending on the potential of the target V_T relative to that of the charging electrode V_1, three types of particle "reflection" can occur.

(i)   For $V_T > V_1$, a particle will be decelerated by the gap field and brought momentarily to rest in mid-gap before being re-accelerated on its return transit. Consequently, there will be complete in-flight momentum reversal with conservation of charge, so that the detector will see two sequential pulses of identical shape as shown in trace B of Fig 7.31.

(ii)   For $0 < V_T < V_1$, the particle will again be decelerated by the gap field but will now impact on the target to loose some of its incident momentum and charge. Thus, although the "return" particle will be re-accelerated across the gap after impact, its final charge and velocity will both be reduced, as indicated by the sequential pulses of trace C of Fig 7.31.

(iii)   For $V_T < 0$, a positively charged particle will be accelerated by the gap field and impact on a negatively charged target, so that there is now the additional possibility of the charge on the particle being wholly or partially reversed during a bouncing impact. If this occurs, it will then again experience an accelerating field for its return transit of the gap with the possibility of its kinetic energy being enhanced as previously hypothesized in Section 7.2.4(ii). Two examples of this type of event, first recorded by Brah and Latham [63,77], are shown in traces D and E of Fig 7.31, and clearly constitute an experimental demonstration of the proposed model. These traces also illustrate the wide variation in charge reversal efficiency and velocity enhancement that can occur; in trace D, whilst $Q_2/Q_1 \sim 0.5$ the velocity has only been enhanced by a factor of ~ 1.6, whereas in trace E, with $Q_2/Q_1 \sim 0.25$, there has been a three-fold velocity enhancement. It is clearly this latter type of event which has a major significance in the discussion of microparticle-initiated breakdown, and has therefore been the subject of a detailed study.

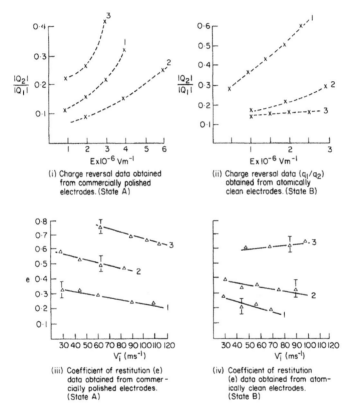

(i) Charge reversal data obtained from commercially polished electrodes. (State A)

(ii) Charge reversal data (q_1/q_2) obtained from atomically clean electrodes. (State B)

(iii) Coefficient of restitution (e) data obtained from commercially polished electrodes. (State A)

(iv) Coefficient of restitution (e) data obtained from atomically clean electrodes. (State B)

Fig 7.32 The electrical and mechanical impact data obtained from a range of target materials, showing the influence of the electrode surface preparation: in each case, curves, 1, 2 and 3 correspond respectively to copper, stainless steel and titanium targets. (After Brah, Mohindra and Latham [82].)

To illustrate the important conclusions to emerge from these measurements, we shall compare the bouncing interaction of carbonyl iron microspheres on 0.25 μm "commercially" polished targets of titanium, stainless steel and copper; the first two of these materials being known to provide stable high voltage electrodes, with the third being acknowledged to be unsuitable for this use. In addition, we shall highlight how the surface state of a target has an important influence on the bouncing mechanism by contrasting the above characteristics with those obtained from the same targets when their ambient oxide coverage has been removed; i.e. where the microspheres are impacting on atomically clean surfaces.

Since, as previously explained, the response of a surface is most profitably characterised in terms of the $|Q_2|/|Q_1|$ and e-values of the particle interactions, it is then appropriate in the context of the high voltage vacuum gap to investigate how these parameters depend on the macroscopic electric field acting on each of the target surfaces, which in this case corresponds to the gap field. Accordingly, Fig 7.32 is a unified presentation of the mean trends of the $|Q_2|/|Q_1|$ and e-values for the "commercially" polished (state A) and atomically clean (state B) surfaces of each target material, as computed from extensive experimental data [64,78,82]. It is however important to emphasise that these measurements are characterised by a very large scatter among the individual bouncing events recorded at a given field [72]. This is not perhaps a surprising finding when it is appreciated how complicated and varied are the local microscopic properties of the type of polycrystalline electrode surface used for targets in the present experiments. For example, there will be a non-uniform oxide coverage that will give rise to a wide variation in the microscopic electrical properties of the surface, an irregular microtopography that will result in complicated particle contact regimes, and varying degrees of work-hardening that will have a large effect on the local mechanical properties of the surface [17]: in addition, there is the spread in particle diameters (0.1-5 µm) which is likely to compound with these other factors.

The most obvious conclusion to be drawn from the plots of Fig 7.32(i) and (ii) is that the $|Q_2|/|Q_1|$ values of all surfaces and hence the charge reversal mechanism, are very sensitive to the macroscopic gap field E.

For *low fields* $E \lesssim 5 \times 10^5$ Vm^{-1}, which is a regime where breakdown is not normally encountered and therefore of little significance to the present text, the incidence of charge reversal is generally very rare, although it has been observed for titanium [64] and a limited number of other target materials [28].

For *intermediate fields* $5 \times 10^5 \lesssim E \lesssim 5 \times 10^6$ Vm^{-1}, charge exchange becomes more significant through a complex field-dependent mechanism that is also strongly influenced by the choice of target electrode material and its surface state.

At *high fields* $E \gtrsim 6 \times 10^6$ Vm^{-1}, all experimental regimes tend to approach a common charge reversal characteristic that shows only a slow variation with applied field and no longer depends significantly on the physical properties of the target electrode. Under these high field conditions, it has been suggested [78] that the charge acquired by a rebounding particle will approach the equilibrium value it would acquire if it remained in contact with the electrode; i.e. as defined by equation 7.6. It should also be noted that the restricted range of macroscopic E-fields for which measurements are possible with state B surfaces is a consequence of the well known gap instability that results from the ion etching surface cleaning treatment of electrodes. For example, whilst 10 kV can typically be supported across a 1 mm gap with a state A target, only ~ 5 kV can be used for a state B target if breakdown is to be avoided. A quasi quantitative physical explanation of the main features of such a family of charge reversal characteristics shown in Fig 7.32(i) and (ii) is provided by the theoretical model of the process developed in Section 7.2.3(ii), where it was shown how the electron tunnelling between target and particle is determined by the local microscopic field E_g and the properties of the intermediate oxide film. In a situation where the particle and target have different work functions, ϕ_p and ϕ_T, as in the present experiments, there will be an additional "internal" contribution to E_g that can either enhance or oppose the "external" contribution to E_g derived from the gap field and charge on the particle. This effect will therefore give rise to either a "forward" or "reverse" biasing of the junction and be most evident in the lower field ranges when the external E_e and E_i contribution to E_g are comparable. On this picture therefore, three types of tunnelling regime are possible:

(i) $E_g = E_e$, corresponding to $\phi_p = \phi_T$ where $E_i = 0$

(ii) $E_g = E_e + E_i$, corresponding to $\phi_p > \phi_T$ so that the junction is "forward-biased"

(iii) $E_g = E_e - E_i$ corresponding to $\phi_p > \phi_T$ so that the junction is "reverse-biased".

In considering the interpretation of the data of Fig 7.32(i) and (ii) it has first to be noted that the mean values of the quoted work function for the three materials under ambient conditions (i.e. equivalent to state A) are

$$\phi_{Cu} \sim 4.45 \text{ eV}, \ \phi_{Fe} \sim 4.47 \text{ eV}, \ \phi_{St.St.} \sim 4.5 \text{ eV}, \ \phi_{Ti} \sim 3.95 \text{ eV}$$

It therefore follows from the above discussion that since $\phi_{Fe} \sim \phi_{Cu} \sim \phi_{St.St}$, the impact data obtained from the copper and stainless steel targets correspond to tunnelling regime (i) and that the enhanced charge reversal observed for State B targets has to be interpreted in terms of the reduced thickness of the intermediate oxide film as discussed by Latham and Brah [78]. In the case of the State A Ti-target, the tunnelling junction will be "forward biased" since $\phi_{Fe} > \phi_{Ti}$, and this will give rise to an enhanced charge reversal response that is clearly evident in the experimental characteristics of Fig 7.32(i). The increase of the $|Q_2|/|Q_1|$ ratio for the State B Ti-target, which is in sharp contrast to the behaviour of the other materials can only be explained if it is assumed that the ion etching process has resulted in a significant increase in the T_i work function so that the tunnelling regime is altered from being "forward" to "reverse" biased: this effect is then assumed to predominate over the enhancement expected from the reduction of the oxide thickness. In support of this explanation, reference can be made to measurements which confirm that the necessary changes in the work function of T_i can occur following Ag-ion etching, the effect being attributed to ion impregnation [84,85]. As a final observation on the plots of Fig 7.32(i) and (ii), it should be noted that the absence of charge reversal events at low fields can largely be explained by the fact that under such conditions particles invariably "stick" to the target surface [86]; a phenomenon that is thought to result from some form of particle-discharge welding process such as that proposed by Beukema [87] and described in Section 7.2.1(ii), or that associated with electron scattering processes in the oxide layer of the MIM junction contact regime.

Referring now to Fig 7.32(ii) and (iii), the significant lowering of the e-values for all three materials between their state A and state B surfaces can readily be explained in terms of the influences of the ambient surface oxide film which is generally known to be mechanically harder than its substrate metal. In addition, the reduction of the e-values with increasing impact velocity follows the behaviour associated with macrosystems and has been discussed elsewhere by Latham et al. [72].

In considering the implication of these findings to the practical problem of the breakdown of commercially polished (state A) high voltage electrodes by microparticle impact following multiple bouncing discussed in Section 7.2.4, it has to be assumed that particles will generally be of the same materials as their parent electrodes, so that $\phi_p = \phi_T$ and there will be no "forward" or "reverse" biasing effects: the charge reversal mechanism will therefore be controlled by the properties of the ambient oxide film [78]. If however the operational gap field exceeds $\sim 6 \times 10^6$ Vm^{-1}, which is frequently the case in practice, Fig 7.32(i) indicates that all state A surfaces approach a common charge reversal response. Under these conditions any differences in the velocity enhancement from the bouncing mechanism must therefore arise from differences in the e-values. It would therefore appear that mechanically harder surfaces, such as would be provided by the refractory metals, would be more likely to promote the mechanism than the softer non-refractory metals.

Before concluding this discussion on microparticle bouncing phenomena, reference must be made to the investigations of Texier [83,88,89] who, using a similar simulation technique, but with higher impact velocities (50-200 ms^{-1}), has also studied the mechanical and electrical impact behaviour of micronyl iron microspheres on a comparable range of mechanically polished target materials (Al, Cu, Ni, Au). These measurements were however made under "commercial" vacuum conditions (~ 10^{-5} mbar) without any in-situ monitoring or treatment of the target surface state: also, they were restricted to an impact regime having a relatively low target field of $-2 \times 10^5 < E < +2 \times 10^5$ Vm^{-1}. The mechanical interaction was studied in terms of the coefficient of restitution and the dimensions of the associated indentation craters on the target surface, and shown to follow a semi-elastic type of impact behaviour as originally described by Tabor [90] for macro-systems. In common with the approach of Latham and Brah [78], the complementary electrical interaction between a particle and target was investigated by comparing the charge carried by an incident and reflected particle: however, the resulting data were interpreted using a simplified model based on the contact potential difference existing between the colliding surfaces.

7.5 References

1. Cranberg, L., *J. Appil. Phys.* , **23**, 518-22, 1952.
2. Slivkov, IN., *Sov. Phys.-Tech. Phys.*, **2**, 1928-34, 1957.
3. Rohrbach, F., *Rpt. CERN*, **71-38** *(NTIS)*, 1971.
4. Jeans, JH., *In* "The Mathematical Theory of Electricity and Magnetism", 5th Edn., Cambridge University Press, 1958.
5. Lobedov, NN. and Skal'skaya, IP., *Sov. Phys.-Tech. Phys.*, **7**, 268-71, 1962.
6. Boulloud, A., *Le Vide*, **99**, 240-3, 1962.
7. Pollock, HM., Shufflebottom, P. and Skinner, J., *J. Phys. D: Appl. Phys,* **10**. 127-38, 1977.
8. Pollock, HM., *J. Phys. D: Appl. Phys.*, **11**, 39-54, 1978.
9. Udris, Ya. Ya., *Radio Engng. Electron Phys.*, **5**, 226-37, 1960.
10. Rosanova, NB., *Bull. Acad. Sci. USSR*, **11**, 1262-4, 1962.
11. Olendzkaya, NF., *Radio Engng. Electron Phys.*, **8**, 423-9, 1963.
12. Martynov, Ye.P., *Elektronnaya Tekhnika*, Ser.**10**, 3-10, 1968.
13. Hurley, RE. and Parnell, TM., *J. Phys. D: Appl. Phys.*, **2**, 881-88, 1969.
14. Chatterton, PA., Menon, MM. and Srivastava, KD., *J. Appl. Phys.*, **43**, 4536-42, 1972.
15. Miller, HC., *J. Appl. Phys.*, **38**, 4501-4, 1967.
16. Martynov, Ye.P. and Ivanov, VA., *Radio Eng. Electron Phys.*, **14**, 1732-7, 1969.
17. Martynov, Ye.P., *Sov. Phys-Tech. Phys.*, **18**, 1364-8, 1972.
18. Beukema, GP., *J. Phys. D: Appl. Phys.*, **7**, 1740-55, 1974.
19. Davies, DK. and Biondi, MA., *J. Appl. Phys.*, **39**, 2979-90, 1968.
20. Davies, DK. and Biondi, MA., *J. Appl. Phys.*, **41**, 88-93, 1970.
21. Davies, DK. and Biondi, MA., *J. Appl. Phys.*, **42**, 3089-3107, 1971

22. Davies, DK. and Biondi, MA., *Proc. VII ISDEIV (Novosibirsk)*, 121-9 1976.
23. Nottingham, WB., *Phys. Rev.*, **59**, 907-10, 1941.
24. Widdington, R., *Proc. Roy. Soc.* **A89**, 554-60, 1914.
25. Schottky, W., *Z. Phys.*, **14**, 63-106, 1923.
26. Williams, DW. and Williams, WT., *J. Phys. D: Appl. Phys.*, **5**, 1845-54, 1972.
27. Latham, RV. and Braun, E., *J. Phys. D: Appl. Phys.*, **3**, 1663-70, 1970.
28. Cook, MA., *In* "The Science of High Explosives", Reinhold, New York, 1958.
29. Menon, MM. and Srivastava, KD., *Proc. VI DEI (Swansea)*, 3-10, 1974.
30. Davies, RM., *Proc. Roy. Soc.,* **A317**, 367, 1970.
31. Tabor, D., *Proc. Roy. Soc.*, **A192**, 247-74, 1948.
32. Gane, N., *Proc. Roy. Soc.* **A317**, 367-91, 1970.
33. Slattery, JC., Friichtenicht, JF. and Hanson, DO., *Appl. Phys. Letters*, **7**, 23-8, 1965.
34. Latham, RV., *J. Phys. D: Appl. Phys.* **5**, 2044-54, 1972.
35. Andrews, JP., *Phil. Mag.*, **9**, 593-610, 1930.
36. Latham, RV., Brah, AS., Fok, K. and Woods, MO., *J. Phys. D: Appl. Phys.*, **10**, 139-151, 1977.
37. Latham, RV. and Brah, A.S., *J. Phys., D:Appl. Phys.*, **10**, 151-167, 1977.
38. Simmons, JG., *J.Appl. Phys.* **34**, 2581-90, 1963.
39. Simmons, JG., *Trans. Met. Soc. AIME*, **233**, 485-496, 1965.
40. Simmons, JG., *J. Phys. D: Appl. Phys.*, **4**, 613-57, 1971.
41. Farrall, GA., *J. Appl. Phys.*, **33**, 96-99, 1962.
42. Poshekhonov, P.V. and Solov'yev, V.I., *Radio Enging. Electron Phys.*, **16**, 1545-49, 1972.
43. Anderson, HW., *Elec. Eng.* **54**, 1315-1320, 1935.
44. Latham, RV and Braun, E., *J. Phys. D: Appl. Phys.*, **3**, 1663-70, 1970
45. Browne, PF., *Proc. Phys. Soc.*, **B68**, 564-6, 1955.
46. Tarasova, LV. and Razin, AA., *Sov. Phys.-Tech. Phys.*, **4**, 879-85 1959.
47. Chatterton, PA. and Biradar, PI., *Z. Ang. Phys.*, **30**, 163-70, 1970.
48. Theophilus, GD., Srivastava, KD. and Van Heeswijk, R.G., *J. Appl. Phys.*, **47**, 897-9, 1976.
49. Texier, C., *J. Phys. D: Appl. Phys.*, **10**, 1693-1702, 1977.
50. Boulloud, A., *Le Vide*, **99**, 240-3, 1962.
51. Feynman, RP., Leighton, RB. and Sands, M., *In*, "Feynman Lectures on Physics", Addison Wesley: Massachusetts, **2**, 710-714, 1970.
52. Piuz, F., *CERN Report TC-L/Int.*, **72-8**, 1972.
53. Smalley, J., *J. Phys. D: Appl. Phys.*, **9**, 2397-401, 1976.
54. Jenkins, JE. and Chatterton, PA., *J. Phys. D: Appl. Phys.*, **10**, L17-23, 1977.
55. Davies, DK. and Biondi, MA., *J. Appl. Phys.*, **41**, 88-93, 1970.

56. Davies, DK. and Biondi, MA., *Proc. VI-ISDEIV*, **45-50**, 1974.
57. Davies, DK. and Biondi, MA., *J. Appl. Phys.*, **48**, 4229-33, 1977.
58. Heard, HG. and Laner, EJ., *UCLRL Report*, **2051**, 1953.
59. Olendzkaya, NF., *Rad. Engng. Electron Phys.*, **8**, 423-9, 1963.
60. Little, RP. and Smith, TS., *Proc.II-ISDEIV*, **41-49**, 1966.
61. Shelton, H., Hendricks, Jr. CD. and Wuerker, RF., *J. Appl. Phys.*, **31**, 1243-7, 1960.
62. Friichtenicht, JF., *Nucl. Instrum. Meth.*, **28**, 70-8, 1964.
63. Brah, AS. and Latham, RV., *J. Phys.E: Sci. Instrum.*, **9**, 119-23, 1976.
64. Brah, AS., Ph.D. Thesis, University of Aston in Birmingham, 1978.
65. Hansen, DO., *Appl. Phys. Lett.*, **13**, 89-91, 1968.
66. Adams, NG. and Smith, D., *Planet Space Sci.*, **19**, 195-204, 1971.
67. Dietzel, H., Neakum, G. and Rauser, P., *J. Geophus. Res.*, **77**, 1375-95, 1972.
68. Rudolph, V., *Z. Naturforsch.*, **21a**, 1993-98, 1966.
69. Chakrabarti, AK. and Chatterton, PA., *J. Appl. Phys.* , **47**, 5320-28, 1976.
70. Auer, S., Grun, E., Rauser, P., Rudolph, V. and Sitte, K., *In.*"Space Research VIII, North Holland", Amsterdam, **606-16**, 1968.
71. Smith, D. and Adams, NG., *J. Phys. D: Appl. Phys.*, **6**, 700-719, 1973.
72. Latham, RV., Brah, AS., Fok, K. and Woods, MO., *J. Phys. D: Appl. Phys.*, **10**, 139-150, 1977.
73. Auer, S. and Sitte, K., *Earth Planet Sci. Lett.*, **4**, 178-83, 1968.
74. Bandermann, LW. and Singer, SF., *Rev. Geophys.*, **7**, 759-97, 1969.
75. Rudolph, V., *Z. Naturforsch.*, **24a**, 326-331, 1969.
76. Cook, MA., *In* "The Science of High Explosives", Reinhold, New York, 1958.
77. Brah, AS. and Latham, RV., *J. Phys. D: Appl. Phys.*, **8**, L109-11, 1975.
78. Latham, RV. and Brah, AS., *J. Phys. D: Appl. Phys.*, **10**, 151-67, 1977.
79. Fane, RW., Neal, WEJ. and Latham, RV., *J. Appl. Phys*, **44**, 740-3, 1973.
80. Winterbottom, AB., *J. Iron Steel Inst.* Lond. **156**, 9-21, 1950.
81. Archer, RJ., *J. Opt. Soc. Am.*, **52**, 970-81, 1962.
82. Brah, AS., Mohindra, S. and Latham, RV., *Proc. VIII-ISDEIV*, **C6.**, 1-9, 1978.
83. Texier, C., *Proc. VII-ISDEIV*, 92-96, 1976.
84. Smith, T., *Surface Sci.*, **27**, 45-9, 1971.
85. Chang, YW., Lo, WJ. and Samorjai, GA., *Surface Sci.*, **64**, 588-94, 1977.
86. Latham, RV. Cook, S. and *Surf. Sci.*, **179**, 503-26, 1987.
87. Beukema, GP., *J. Phys. D: Appl. Phys..*, **7**, 1740-55, 1974.
88. Texier, C., *Rev. Phys. Appl. (France)*, **13**, 13-22, 1978.
89. Texier, C., *Rev. Phys. Appl. (France)*, **13**, 165-70, 1978.
90. Tabor, D., *Proc. Roy. Soc.* , **A192**, 247-74, 1948.

8

High Voltage Performance Characteristics of Solid Insulators in Vacuum

H Craig Miller

8. 1   Introduction

Insulators are used in many applications which require them to support significant voltage differences while exposed to vacuum. The voltage hold-off capability of a solid insulator in vacuum is usually less than that of a vacuum gap of similar dimensions and depends upon many parameters. These include (a) the properties of the insulator itself - material, geometry, surface finish, attachment to electrodes; (b) the applied voltage waveform - duration, single or repetitive pulse and (c) the history of the insulator-processing - operating environment, nature of previous voltage applications.

In practice, the vulnerable area is the surface of the insulator, since its bulk voltage hold-off capability is normally better than that of a similar-sized vacuum gap. Accordingly, this review will concentrate on experimental results relevant to voltage breakdown along the surface of an insulator in vacuum (surface flashover). For the interested reader, there have been several previous reviews of the electrical behaviour of insulators in vacuum [1-4].

The content of this chapter is broadly divided into three sections covering the theoretical, experimental and discussional aspects of the subject. Accordingly, the opening Theory section reviews the established models that have been developed to explain the phenomena of surface flashover, with particular emphasis being placed on the physical mechanisms responsible for the initiation, development, and final growth of the discharge. A detailed discussion of more recent theories of surface flashover that are based on insulator charging effects may be found in Chapter 9. The Experimental section presents and discusses some experimental results pertinent to surface flashover. Subsequently, the Discussion section presents some implications of theory and experiment for practical applications of insulators where surface flashover is a potential threat. Several suggestions are made regarding how to choose the material, geometry, and processing when designing an insulator for a particular application. Also, some specific techniques are recommended for improving the hold-off voltage of insulators.

8. 2   Theory-Mechanisms of Surface Flashover

From a traditional perspective, the initiation of a surface flashover is widely assumed to begin with the emission of electrons (generally by field emission or thermally assisted field emission) from the cathode triple junction (i.e. the interface where the insulator, cathode, and vacuum are in close proximity). While there is general agreement among this genre of theory on this initiating mechanism of surface flashover, there is considerable disagreement concerning the details of the intermediate or development stage of the discharge. The final stages of surface flashover are predominantly thought to occur in desorbed surface gas or in vaporised insulator material. (An excellent recent review by Anderson concisely presents the main details of various theories concerning surface flashover of insulators in vacuum [5]).

Probably the most generally accepted mechanism for the intermediate stage is an electron cascade along the surface of the insulator [1,2,6], as illustrated in Fig 8.1(a). According to this, electrons field-emitted from the triple junction strike the surface of the insulator producing additional electrons by secondary emission. Some of these secondary electrons will again collide with the insulator surface producing tertiary electrons. Continuation of this process results in a cascade along the surface of the insulator that develops into a secondary electron emission avalanche (SEEA). This secondary electron emission avalanche, in turn, leads to a complete breakdown. The term (SEEA) is frequently used to refer to this electron cascade mechanism.

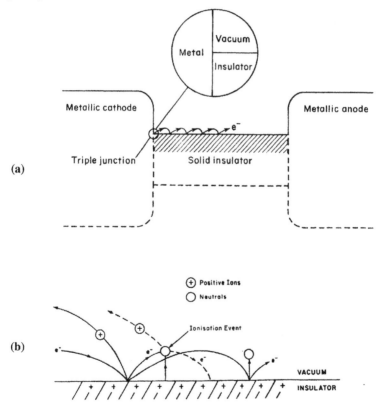

Fig 8.1   An illustration of (a) The electron hopping mechanism thought to be responsible for initiating SEEAs. (b) The mechanism by which gas molecules are desorbed from the surface of the insulator by electron-stimulation and subsequently ionised by electron impact. (After Anderson and Brainard [6].)

Most SEEA theories predict that significant surface charging of the insulator is associated with the SEEA. The initial electron cascade produces surface charging which can strongly affect the further development of the SEEA.

As illustrated in Fig 8.1(b), Anderson and Brainard postulated a development of this model whereby a SEEA could lead to a complete breakdown [6]. The surface of a ceramic in vacuum is normally covered with a layer of adsorbed gas, amounting to a monolayer or greater. The SEEA electrons bombarding the surface of the insulator desorb some of this adsorbed gas, forming a gas cloud which is then partially ionised by the electrons in the SEEA. Some of the resulting positive ions drift towards the cathode to enhance the electric field at the triple junction. This then results in an increased electron emission from the triple junction and thus increases the current along the surface of the insulator. Consequently, a regenerative process occurs that leads quickly to surface flashover of the insulator. In this model, desorption of adsorbed surface gas (or vaporised insulator material for organic insulators such as PMMA-polymethylmethacrylte) is a key process leading to the surface flashover of an insulator.

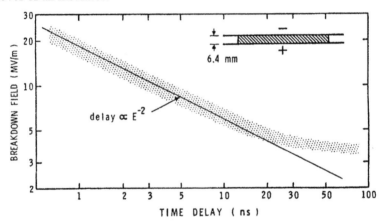

Fig 8.2   Relationship between the prebreakdown time delay and the applied electric field for PMMA insulated, d = 6.4 mm [6]. (From Anderson and Brainard, [6], with permission.)

Anderson and Brainard also used their model to predict the behaviour of the time delay between the application of a voltage pulse and the flashover of the surface: they found that, for nanosecond voltage pulses, this time delay should be proportional to the inverse square of the applied electric field. Experimentally, they observed the predicted behaviour for short (nanosec) voltage pulses, but as the voltage pulses became longer, a point was reached where, as illustrated in Fig 8.2, the breakdown field became much less dependent on the pulse length. They attributed this break to an effect of the masses of the major positive ions involved in the flashover mechanism. In the short-pulse range an increase in pulse duration means that more positive ions are produced, with the overall electric field acting to sweep positive ions towards the cathode. However, if the pulse length is less than the transit time of ions in the gap, then more positive ions can accumulate at or near the insulator, which accumulation lower in turn results in the required breakdown field. When the pulse duration has increased to a length where ions could be swept from the gap as rapidly as new ions are produced, then further accumulation of ions in the gap is greatly decreased; thus, once the pulse duration has reached this critical value a further increase in pulse duration have much less effect on the breakdown field. With bipolar waveforms (alternating square waves) the inverse-square dependence of flashover delay on applied electric field continued for pulse durations of hundreds of nanosec [7]. This is a consequence of the alternating voltage waveform allowing some of the ions to remain for times long after they would have been swept away by a unidirectional voltage, thus allowing the accumulation of charge to continue.

A similar, but not identical, mechanism for surface flashover was postulated by Cross [8]. Emission from the triple junction initiates a SEEA that charges the insulator surface, and produces an additional electric field that acts to increase the electron emission from the cathode. However, in this model the surface flashover is triggered by the emission of a large burst of electrons from the triple junction which charge successive areas on the insulator negatively, and thus causes the next electrons in the burst to strike the insulator closer to the anode. This process continues until electrons from the burst strike the anode. At this point, a channel of excited gas exists from the anode to the cathode, and the final flashover occurs along this channel. In this context Nevrovskii suggested that the electron cascade in a SEEA should be divided into two regions [9]. A region near the surface where low energy electrons create secondary electrons, heat the surface, and cause the thermodesorption of gases, and a region further away from the surface where electrons gain sufficient energy to ionise the desorbed gas.

Several investigators have presented mechanisms other than the SEEA for explaining the surface discharge development process. Bugaev and colleagues [10,11] believed that breakdown occurred within the adsorbed gas layer. Electrons emitted from the triple junction induce a gas discharge avalanche in this layer of desorbed gas, producing a plasma and thus breakdown. This breakdown mechanism is similar to that of a streamer in a gas discharge. In this theory the electron avalanche occurs in the gas layer, not at the surface of the insulator.

Avdienko and Malev's explanation for breakdown of high resistivity insulators ($\rho \geq 10^{13}$ ohm-cm), again took the initiating event as electron emission from the triple junction, with the final surface flashover occurring in desorbed surface gas [12,13]. However, they postulated that, in uniform fields, the original electrons emitted from the triple junction are enough to desorb sufficient gas to promote the prebreakdown process. Thus, SEEAs are not necessary, and surface charging has only a slight effect on surface flashover. In non-uniform fields, SEEAs may be necessary, but only to produce a positive charge on the surface that is sufficient to keep the travelling electrons striking the surface and desorbing gas.

In contrast, Avdienko and Malev [12] explained flashover across insulators of moderate resistivity ($\rho \geq 10^{12}$ ohm-cm) with long (> 1 millisec) voltage pulses in terms of thermal breakdown in a thin surface layer of insulator; thus, for this type of surface flashover, desorbed gas was not necessary.

Another postulated mechanism for the intermediate phase of surface flashover is the propagation of electrons in the conduction band of an insulator that is under the influence of the applied field [14,15]. According to this model, the electrons gain kinetic energy and, as soon as this exceeds the band gap of the insulator, they begin to undergo inelastic collisions, and thus create an electron cascade along the surface (but just inside the insulator). A fraction of these cascade electrons will be emitted into the vacuum, whereupon the external electric field derives them towards the anode. From an external point of view, one sees a steadily increasing flow of electrons along the insulator. This is similar to a SEEA, but its cause is quite different. This model also predicts the formulation of a positive surface charge on the insulator. The sub-surface cascade of electrons will produce a number of holes in the valence band, some of which are trapped to form a positive surface charge. The ionisation cascade also stimulates desorption of surface gases, which are assumed to be adsorbed in traps at or near the surface of the insulator. Unlike the SEEA, where the surface gases are desorbed by external electron impact, in this model the electrons cascading just inside the insulator's surface make inelastic collisions with the traps, transferring enough energy to liberate the adsorbed gas molecules. Final flashover then occurs in the desorbed surface gas, as in other models.

A similar mechanism, based on collisional ionisation occurring in a surface layer of the insulator, was presented by Jaitly and Sudarshan [16]. Because collisional ionisation in a material normally requires electric fields much higher than those at which surface flashover is observed, they made the reasonable assumption that, since the insulator-vacuum interface represents a discontinuity in the structure of the insulator, the conduction mechanisms in the surface layer of the insulator are different from those in the body of the insulator. Conduction in a thin surface layer is facilitated by the presence of defects, unsaturated bonds, traps, etc. which allows collisional ionisation to begin in this layer at a much lower electric field than would be required in deeper layers of the insulator. Their theory postulates that at sufficiently high voltages, carriers are produced at the cathode triple junction. Then, because of the relatively low bulk conductivity, a negative space-charge region is produced near the cathode. However, since the thin subsurface layer has a higher conductivity, the majority of the injected current flows in this layer. Collisonal ionisation then produces positive surface charges in the subsurface layer. As the applied voltage increases, the collision-ionisation process becomes more efficient, which enhances the field at the triple junction and thus increases the current injected into the surface layer. At a critical value of the applied electric field, this electron emission/collision-ionisation feedback process becomes self-sustaining, resulting in a high conductivity plasma in the surface layer. This is the final stage of surface flashover. Note that, in this model, final flashover does not require desorbed surface gas, rather it occurs in a regeneratively established high conduction plasma in the surface layer.

All subsurface electron cascade models presumably depend upon the effects a dielectric-vacuum interface can have in modifying the properties of a surface layer of a dielectric relative to the properties of the bulk material; i.e. as discussed above and in more detail elsewhere [16].

As discussed in detail in the following chapter, Blaise and Le Gressus also assume the presence of a subsurface electron cascade (avalanche); however, they consider such an avalanche, while essential to the initiation of surface flashover, as playing a minor role in its development [17,18]. The primary mechanism whereby the flashover develops is attributed to rapid charge detrapping from localised sites, and the corresponding relaxation of energy of polarisation, via a collective many-body process. This theory also predicts the observed slight time delay (nanosec) observed experimentally [19] between the initial rise in luminosity and the initial rise in current.

Most work concerning surface flashover in vacuum has considered the cathode-initiated flashover mechanism just discussed. However, under appropriate conditions, a surface flashover may be initiated at the anode, and will then develop differently from a cathode-initiated flashover. Anode-initiated surface flashover is thought to involve bulk breakdown in a way related to treeing failure of insulators [5]. Anode-initiated surface flashover has been mainly considered by Anderson [20], who describes his model as follows: "Flashover initiation is assumed to require the generation of small area plasma adjacent to the insulator surface. The plasma is maintained near the anode potential, either from being connected to the anode electrode or through electron emission. As a result, the edge of the plasma contributes a strong electric field component parallel to the insulator surface. Filamentary branches develop because the electric field at their tips presumably exceeds the dielectric strength of the material. Localised breakdown at the branch tips, the cause of the surface damage, generates the new plasma necessary to carry the field enhancement forward. The growth of filaments into the bulk of the insulator is assumed to be arrested when the surface plasma advances past their point of entry, and the electric field is reduced. Photoemission, photodesorption or photoconductivity may play a role in this flashover mechanism since ultraviolet light is undoubtedly produced".

It seems quite probable that no single theory is capable of explaining all cases of surface flashover in vacuum (as pointed out by Anderson [5]), but rather that depending upon specific experimental conditions (geometry, dielectric material, voltage waveform) a particular mechanism could dominate the surface flashover.

8. 3   Experimental Insulator Performance

This section will review a range of experimental investigations aimed at understanding how an insulator responds to an applied electric field. It will therefore be of direct interest to both the research community and those involved in the design of practical high voltage insulators.

8.3.1   Surface Charging

Surface charges on insulators have been observed by many investigators [21-28] and have been measured for both DC and pulsed voltages. It is generally proportional to the applied voltage and positive, though negative charging has been seen in experiments which could resolve the charge distribution over small areas. Surface charging seems to occur fairly rapidly, though different charging times have been seen [22-24]. These variations in charging time may reflect differing experimental conditions. It seems quite likely that while single avalanches can charge surfaces in nanoseconds, they may only charge a narrow strip and thus many avalanches could be required to achieve a final surface charge state [29]. Indeed it should be noted that a voltage pulse which does not cause breakdown can still produce a residual charge on the surface of an insulator.

8.3.2   Applied Voltage Waveform

Waveforms which have been used for investigating the surface flashover characteristics of insulators include rectangular (trapezoidal) pulses, lightning impulses, DC, AC (with frequencies from power to rf), and bipolar (alternating pulse polarity). Generally, with pulsed voltages, the surface flashover voltage decreases with increasing pulse duration in the nanosec region, as shown in Fig 8.3. This decrease of hold-off voltage with pulse duration can continue out to DC voltage, or go through a minimum at a particular pulse duration and then increase for DC voltage. Experimental results displaying both behaviours have been obtained [30-34], where often the lowest breakdown voltages occur for power frequency AC (50/60 Hz).

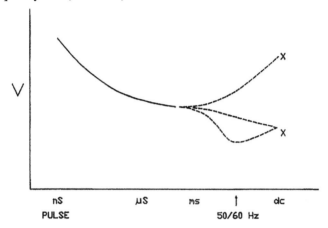

Fig 8.3   Typical behaviour of the flashover voltage versus voltage duration.

8.3.3   Pre-stress

The application of a DC voltage to an alumina insulator before applying an impulse voltage was found to have a strong effect upon the flashover voltage [35]. If the DC pre-stress was applied with the same polarity as the subsequent impulse voltage (1/50 μs), then the flashover voltage was increased, being nearly doubled for a pre-stress just below the flashover voltage. Conversely, if the DC pre-stress had the opposite polarity, the flashover voltage was decreased, being quite low for an opposing pre-stress just below the flashover voltage. Cross and Srivastava [35] found that the effect of the pre-stress decayed rapidly, disappearing within 50 μs after the pre-stress voltage was removed. This explained why an AC (60 Hz) pre-stress produced results depending only on the location on the AC waveform where the impulse voltage was applied. The effect of the pre-stress disappeared if the junction of the insulator with the electrodes was recessed, which led these authors to attribute their results to charge injection at the electrodes. Similar results for the effect of DC pre-stress on impulse (1.2/50 μs) flashover were found using sapphire and glass-ceramic insulators [36].

Pillai and Hackam investigated the effect of DC pre-stress on subsequent DC or AC flashover [37]. There was a 60 to 90 s interval between the removal of the DC pre-stress and the application of the AC or DC breakdown voltage. They found an aiding DC pre-stress had no effect upon the DC flashover voltage, but an opposing pre-stress reduced it significantly. They attributed these results to the surface charges left on the surface by the DC pre-stress.

Such charges would be expected to decay very slowly on high resistivity materials such as the PTFE, quartz, and glass-ceramic insulators that they used. Such DC pre-stresses had no effect on a subsequent AC flashover. However, if the DC pre-stress was maintained when the AC voltage was applied (producing a DC + AC waveform), then the flashover voltage was increased significantly, to about the value for DC voltage alone.

8.3.4   Conditioning

It is frequently found that after a surface flashover has occurred, subsequent flashovers occur at higher voltages. Most investigators in surface flashover have recognised the existence of such a "conditioning" in effecting their work. Usually they have reported the fully conditioned voltage values, but some have given the first and conditioned voltage values, or sometimes even a whole sequence of conditioning data [34,38-43] .

First flashover data are needed for applications where the level of the first flashover voltage is important. It has been found that if a conditioned insulator is left unstressed by voltage for a long enough time, its flashover voltage will be less than the fully conditioned value. Thus, for applications where an unstressed insulator has to suddenly withstand a voltage, one cannot usually depend on the conditioned value of voltage, but must adopt a more conservative rating. Conversely, applications where the insulator is under continuous voltage stress, or where it is allowed to recondition itself, the conditioned value can be utilised. In practical terms, conditioning has been attributed to the removal of breakdown sites, surface gas, or surface contaminants.

Conditioning can also occur without actual flashover, but under such circumstances it is usually slower than when flashovers are allowed to occur. However, such non-breakdown conditioning offers the significant practical advantage that it is much less likely to damage the surface of the insulator [39]. As will be discussed in Chapter 9, Le Gressus et al. [44] explained conditioning with stepped voltage increases, but without flashover, as polarisation of the insulator by the electric field, thus decreasing its instantaneous response to a subsequent increase in electric field. Conditioning with power frequency AC has been found to be quite effective in increasing the subsequent flashover voltages for impulse, DC, or AC [40], in fact, it is more effective than conditioning by impulse or DC flashovers themselves [34] .

8.3.5   Discharge Delay

As discussed previously, and shown in Fig 8.2, the relationship between the minimum applied electric field which will cause surface flashover, and the time delay until surface flashover occurs, is of the form $E_{min} \sim t^{1/2}$ for short times, with E_{min} becoming fairly independent of t for longer times. The point at which the E vs t plot departs from a straight line is a function of the length of the insulator, with longer insulators the break occurs at later times [6]. Another way of expressing the relationship between E and t is: if a voltage pulse with a fixed duration is used, then as the applied voltage is increased above the minimum value necessary to cause flashover, the associated time delay decreases as the overvoltage increases [10,11] . This delay can reach seconds for very long pulses (DC).

8.3.6   Discharge Speed

It has generally been observed that surface flashover begins with a luminous front travelling from the cathode to the anode. When this luminosity reaches the anode, then a much brighter luminosity starts from the anode and proceeds to the cathode, usually at somewhat higher speed than the initial cathode $\rightarrow$ anode luminosity.

One of the earliest measurements of surface flashover velocity was made by Bugaev, Iskol'dskii, and Mesyats [10]. They used fast-rise(< 1 ns) pulses with ceramic insulators. The surface flashover began with a luminous front travelling from the cathode to the anode at a speed $\geq 2.7 \times 10^5$ m/s. Upon the arrival of this front at the anode the flashover was completed in less than a nanosec. In this context, Anderson [45] measured initial cathode $\rightarrow$ anode propagation velocities of 2×10^7 m/s along a PMMA insulator. These were probably the velocities of the fastest electrons, i.e. electrons leaking off the leading edge of the propagation avalanche [29].

Cross investigated DC flashover along alumina insulators [8], and measured the velocity of the initial cathode-anode front to be 4×10^7 m/s; the return anode-cathode front was slightly faster, 4.8×10^7 m/s. This work was later extended to include impulse (1/50 μs) flashover [46], where the initial cathode-anode velocity was 1.5×10^6 m/s, i.e somewhat slower than for DC flashover; the anode-cathode return front velocity was ~10^7 m/s.

Thompson et al. [47] observed flashover velocities for pulsed (10 nanosec) flashover along KDP (potassium-dihydrogen-phosphate), measuring an initial cathode-anode speed ~10^7 m/s, with the anode-cathode return speed $\geq 2.5 \times 10^7$ m/s. With a +45° insulator they observed a different behaviour. The field distortion started at the anode and travelled to the cathode at a slower speed of 8.3×10^6 m/s. Significantly, this slower speed for nonuniform fields was also observed by Bugaev and Mesyats [11].

The roughness of the surface also has an effect upon the discharge speed. Anderson [48] found that roughening the surface of a PMMA insulator with sandpaper reduced the propagation velocity of the "leading edge electrons" (i.e. those at the front of the developing electron avalanche) from 2.1×10^7 m/s to 1.6×10^7 m/s. As will be discussed in Section 8.3.10, surface roughness also influences the response of an insulator to a magnetic field.

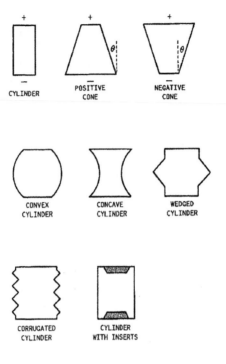

Fig 8.4.   A selection of common insulator geometries

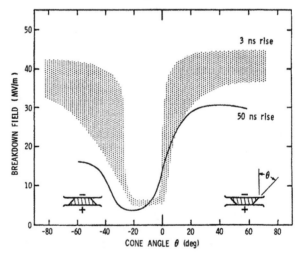

Fig 8.5   Surface flashover voltage versus insulator angle for 50 nanosec rise [50] and 3 nanosec rise [54] voltage pulses applied to PMMA conical insulators. (After [50] and [54].)

310

8.3.7 Geometry

8.3.7(i)   Shape

The exact shape of an insulator can have a strong effect upon its surface flashover behaviour [36,39,43,49-53]. Some typical insulator geometries are shown in Fig 8.4, where the simplest shape, a cylinder, generally has lower hold-off voltages than do more complex shapes. The poorest insulator performance seems to be for cones with slight negative angles. To illustrate this, the effect of cone angle on flashover voltage is shown in Fig 8.5 for 50 and 10 nanosec pulses [50,54].

Fig 8.6 shows the effect of adding a needle in contact with the anode or cathode end of the insulator to serve as a copious source of electrons. The presence of such a needle reduces the breakdown field to a flat minimum from 0° to -40°; at more negative angles it then tends upwards as $(\cos \theta)^{-1}$, i.e. in proportion to the component of the applied field parallel to the surface of the insulator. The extra source of electrons lowers the flashover field for almost all angles, though it has less effect for large angles (> 70°). At positive angles > 65° the flashover voltage is less for an anode needle than for a cathode needle. These results strongly suggest that (at least for nanosec pulses) changing the cone angle affects the flashover field (or voltage) by modifying the local field at the triple junction, as well as by changing the conditions for electron propagation along the insulator.

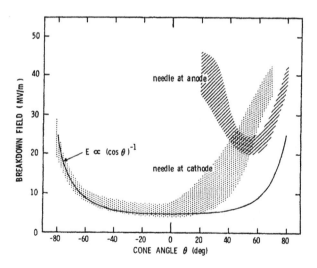

Fig 8.6.   Surface flashover field vs insulator angle for insulators with a needle touching the insulator. PMMA conical insulators, d = 6 mm, needle touches insulator 0.5 mm from electrode. Voltage pulse: 3 nanosec rise/10 nanosec delay to flashover. (After [54].)

311

Some representative results for the surface flashover voltages of cylinders, -45° cones, and +45° cones as a function of voltage waveform are given in Table 8.1. These results indicate that the effect of cone angle is greatest for short voltage pulses, but the more modest increases seen for DC and power frequency AC voltages could be significant in practical applications.

One method of changing the fields near the insulator, especially at the triple junctions, is to put metal inserts into the ends of an insulator [28,39,55,56]. This procedure can significantly increase the surface flashover voltage of the insulator compared to a plain cylinder of the same material [55]. A related method of improving insulator performance is by shaping the electrodes, so that the ends of the insulators are recessed into the electrodes [27,53,55,57-59]. This significantly reduces the fields at the triple junctions. Similar field reductions can be obtained by the use of shields, which often serve not only as field modifiers, but to prevent particles, ultraviolet, soft X-rays, etc. from striking the surface of the insulator [60]. Chamfering the cathode end of the insulator can be effective in increasing its voltage hold-off performance [52]. This probably occurs because electrons accumulating on the surface of the chamfer reduce the electric field at the cathode triple junction [52,61]. As mentioned previously, the flashover voltage varies with the length of the insulator, usually according to a power law $V \sim L^{\alpha}$, where α is typically near to 0.5 [33,40,43,50] .

8.3.7(ii)   Area
The flashover voltage of insulators decreases with increasing insulator diameter. This was observed for 1.2/50 μs impulse, DC, and 60 Hz AC voltages [33], and probably occurs because the increasing length of the triple junction provides more relatively weak points for the initiation of surface flashover. A similar explanation would apply to the observed reduction in 50 Hz AC flashover voltage of a coaxial gap as the number of insulating spacers across the gap is increased [62]. This behaviour is analogous to the variation of vacuum breakdown voltage with electrode surface area. However, both groups of experimenters were careful to keep their electrode areas and gap lengths constant, thus ensuring that their observed phenomena reflected insulator effects only. Cross and Mazurek [63] attributed this effect to an increase in available flashover energy (i.e. increased local capacitance), rather than to an increased number of potential flashover sites.

8.3.8   Material

Considerable work has been done on the effect of the insulator material on surface flashover [34,40,41,50,55,58,64-70], showing that the material can have a strong effect on surface flashover for pulsed, DC, and AC voltage waveforms. Different investigators have found ratios of 2:1, or better, when comparing flashover voltages among various materials. Unfortunately, they haven't always agreed on the relative positions of different materials. For both organic and inorganic materials, the degree of homogeneity of the material seems to be important, with the more homogeneous materials tending to make better insulators. It has also been demonstrated that there is an inverse relationship between the relative permittivity (dielectric constant) κ of a material and its flashover voltage [41,58,70]. (The relative permittivity κ is defined as the ratio of the permittivity of the material ε to the permittivity of vacuum ε_0, i.e. $\kappa = \varepsilon/\varepsilon_0$.)

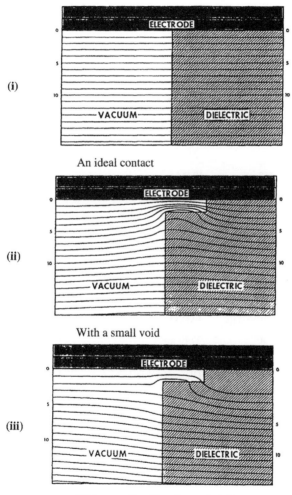

(i)

An ideal contact

(ii)

With a small void

(iii)

With a small void and a graded dielectric constant

Fig 8.7   Equipotential plots illustrating the importance of voids at the triple junction.

313

Table 8.1. Surface flashover voltages of ±45° cone insulators for various voltage waveforms, relative to the values for cylindrical insulators.

Reference	Voltage waveform	Material	Relative	flashover	voltage
			Cylinder	-45°cone	+45°cone
Anderson[54]	10 ns	PMMA	1	$3\frac{1}{2}\times$	4.0×
Watson[49]	30 ns	7740 glass	1	+ 3%	2.2×
		7070 glass	1	+12%	2.2×
Smith[92]	30 ns	Epoxy resin	1	3×	5×
		PMMA	1	-	3×
		Glass	1	-	1.6×
Milton[50]	50 ns	PMMA	1	+15%	$2\frac{1}{4}\times$
Watson [49]	75 ns	Epoxy resin	1	$2\frac{1}{2}\times$	3.0×
Golden Kapentanakos[51]	0.75 μs	Oiled PMMA	1	1.7×	2.4×
Smith	0.8 μs	PMMA	1	2×	3×
		Glass	1	+15%	2×
Pillai/Hackam[55]	1.2/50 μs	Glass-ceramic	1	+10%	+27%
Yamamoto et al.[43]	2/50 μs	PMMA	1	$2\frac{1}{2}\times$	$3\frac{1}{2}\times$
Milton [50]	5 μs	7740 glass	1	3×	6×
		PMMA	1	5×	6×
		Polycarbonate	1	4×	6
		Polystyrene	1	3×	$3\frac{1}{2}\times$
Avdienko[64]	~10/50 μs	PMMA	1	$2\frac{1}{2}\times$	3×
Shannon et al[39]	DC	7740 glass		+10%	-
		7070 glass	-	1	+10%
Pillai/Hackam	DC	Glass-ceramic	1	Little effect	+27%
Jaitly/Sudarshan[52]	DC	PMMA	1	+60%	+30%
Khamidov et al.[27]	DC	Ceramic	1	+24%	+38%
Lee et al.[53]	60 Hz	PMMA	1		+ ≥67%
Mazurek/Tyman[103]	60 Hz	porcelain	1	~+25%	(Corrugated cylinder)
Pillai/Hackam[55]	60 Hz	Glass-ceramic	1	Little effect	
Hamidov et al.[104]	50 Hz	-	1	+55%	(Convex cylinders)
Khamidov et al.[27]	50 Hz	Ceramic	1	+65%	

Thus Fig 8.7(i) shows the equipotential lines for an ideal triple junction, and highlights how they run parallel to the electrode and are uniformly spaced. In contrast, Fig 8.7(ii) shows a triple junction for a practical case, where very small voids can exist at the junction between the insulator and conductor. Note that the presence of such a small void strongly concentrates the equipotential lines, i.e. it greatly increases the local electric field. The degree of concentration depends upon the relative permittivity (dielectric constant) of the insulator; that shown in Fig 8.7 is alumina with the relative permittivity taken as $\kappa = 10$. Fig 8.7(ii) graphically illustrates why the triple junction is often a weak point in real insulating systems, whilst Fig 8.7(iii) shows the effect of a graded relative permittivity at the triple junction.

It is apparent that such grading significantly reduces the field concentration effect of a small void at the triple junction. This effect explains why the insertion of a layer of an insulating material (with a relative permittivity greater than that of the main insulator) between the conductor and the main insulator can improve the performance of an insulating system [71]. It also explains why it is desirable to use a penetrating metallise at the face of an alumina insulator which will be joined to an electrode to form a triple junction [72]. Fig 8.7 suggests that the mechanism for the observed strong effect of κ upon the hold-off voltage of an insulator is that higher values of κ intensify the adverse field magnification effects of small voids at the triple junction.

Blaise and Le Gressus [17] attribute the better performance of materials with lower values of κ to an increased trap energy, and therefore an increase of the field required to initiate dielectric relaxation and subsequent flashover. These two explanations are not contradictory, but complementary. The κ-effect, in conjunction with the material homogeneity effect, may explain the tendency for many organic materials and glasses to have higher flashover voltages than alumina ceramics.

8.3.8(ii)   Silicon
Silicon is discussed separately because surface flashover for silicon appears to occur in a manner unlike other dielectrics. Nam and Sudarshan [73,74] observed three distinct phases in flashover of silicon (using 0.39/10 µs double exponential pulses). Their first phase was conduction (leakage current) in the bulk insulator. The second phase was a partial breakdown occurring at many sites on the cathode. Direct injection of electrons at these sites produced extremely high localised fields. This bulk injection mechanism initiated the final surface flashover. Peterkin et al. [75] working with shorter pulses (20 nanosec rise, 250 nanosec duration), concluded that the surface flashover current in silicon flowed primarily inside the semiconductor's surface, with the accompanying luminosity coming from a plasma located just outside the surface. The surface flashover required intense localised heating of the silicon,

which led to a filamentary distribution of the flashover current. Thus, unlike most materials where the final stage of surface flashover occurs in a region of desorbed surface gas or vaporised insulator material just outside the surface, in silicon this final stage occurs in a thin layer of solid silicon at the surface.

8.3.9   Illumination Effects

Surface flashover of insulators is usually accompanied by bright luminosity, but such luminosity has typically been considered to be a consequence, rather than a cause of surface flashover. Some recent work suggests that in specific cases, shortwave (λ < visible) radiation can have a strong influence on surface flashover. The exposure of angled polymer insulators to an ultraviolet laser light pulse was found to trigger surface flashover for sufficiently intense light (fluence ~10-100 mJ/cm^2) [76, 77], with the critical fluence being a function of material, geometry, and electric field. In this context, the presence of X-rays in plain and insulator-bridged vacuum gaps was observed to contribute an appreciable photoelectric component to the prebreakdown current [42].

8.3.10   Magnetic Field Effects

Magnetic fields can strongly influence surface flashover. In the presence of a magnetic field normal to the insulator surface (and thus also normal to the applied electric field), the breakdown paths are inclined to the electric field as shown in Fig 8.8; a behaviour analogous to the Hall effect, as pointed out by Anderson [48].

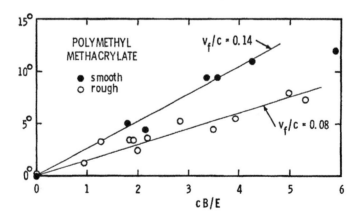

Fig 8.8.   Deflection of the flashover path by a magnetic field, showing the effect of surface roughness on the electron propagation velocity. d = 19 mm. Nanosecond voltage pulses. (After [29].)

With B oriented such that the E × B force would drive the electrons into the surface of the insulator, the surface flashover voltage was reduced [64]. Korzekwa et al. [78,79] investigated the effects of magnetic fields on surface flashover for glass and several organic insulators using 0.6/50 μs voltage pulses. The flashover voltage increased by 25-120% with E × B driving electrons away from the insulator, while with E × B electron drift into the insulator there was an initial decrease (0 to 40%) followed by an increase in flashover voltage as B increased [79]. This effect seemed to be concentrated at the cathode end of the insulator, since a permanent magnet, which produced an appreciable magnetic field only near the cathode, still increased the flashover voltage. Magnetic fields are also helpful with angled insulators, although the effect may be greater for negative angles than for positive angle insulators [80]. It must also be noted from Fig 8.8 that surface roughness also has a major influence of the effects of magnetic fields.

8.3.11   Surface Treatment

If an insulator is made of a material with an inherently smooth surface (such as glass or quartz), then roughening the surface of the insulator, especially the part of the surface near the cathode, can raise the breakdown voltage significantly for DC [65], 50 Hz [59], and microsecond pulse [32] voltages. Polycrystalline ceramics are often ground to shape or size after firing, and so the possible effects of such grinding on surface flashover should be considered. A 99.9% alumina insulator, with an as-fired surface roughness of 0.8 μm, had its flashover voltage lowered considerably when the surface was ground to a 0.25 μm finish. Further polishing to a 0.05 μm finish increased the flashover voltage, but it didn't reach the original value [81]. The initial decrease in flashover voltage was attributed to the creation of additional surface defects by the grinding operation, and the partial recovery to the removal of some of the severely damaged layers by the final polishing operation.

The application of a more conductive (or semiconductive) coating to the insulator surface has also been found helpful [30,31,59,82-89]. Coatings which penetrate into the outer layers of the insulator are often preferable to those which remain on the surface, since insulators with such doped surface layers are generally more resistant to surface damage. Most surface coating treatments have involved inorganic materials, usually ceramics. A surface treatment which increased the surface flashover voltages of organic insulators was presented by Hatfield et al. [69,78].

Jaitly et al. [90] found that firing certain alumina insulators in wet hydrogen significantly reduced their flashover voltages, while subsequent firing in dry hydrogen raised the flashover voltage again, in some cases to higher than the initial values. These effects presumably reflect changes in surface chemistry of the alumina insulators.

The effectiveness of surface treatments is dependent on the coating materials and technique, the material of the insulator, and the condition of the surface to which it is applied. Several possible mechanisms by which surface treatments could increase the surface flashover voltage have been postulated: (i) decrease the secondary electron emission coefficient of the surface; (ii) in making the surface dielectrically more uniform; (iii) decreasing the surface resistivity [1-3,83], (iv) decreasing the amount of gas adsorbed onto the surface, or increasing the adsorbing energy [6]; (v) allowing controlled detrapping of space charge from the surface layers [17].

It is generally desirable to clean the surface of an insulator to remove possible contaminants. Such cleaning may include chemical etching, solvent washing, plasma cleaning, glow discharge cleaning, vacuum or hydrogen firing, etc. Bake-outs of the assembled system are generally desirable if the insulator can tolerate the necessary temperatures.

A surface flashover can damage the surface of an insulator. In some cases this damage is so severe that the insulator can no longer perform its function. Such damage can occur in a single flashover, but more commonly is the cumulative effect of several (up to hundreds) of surface flashovers. Insulator surface damage has plagued most workers in this field, and a few papers which specifically address it are cited in the following references [34,39-41,50,66,69,91]. It should also be noted that surface treatments can also decrease the likelihood of the insulator suffering permanent damage from flashover [50,92].

8.3.12   Surface Gases

Most theories of surface flashover predict that the final stage of flashover involves gas desorbed from the surface of the insulator. Therefore, one would expect to see a burst of gas accompanying flashover of an insulator, and indeed such gas bursts have frequently been observed. Similar quantities of gas are produced by DC and μs pulse flashovers [94]; thus, referring to Table 8.2, $2.5 - 6.5 \times 10^{13}$ molecules are desorbed from DC surface flashover on alumina [93] and $2.5-5 \times 10^{13}$ molecules from 20 μs pulses on alumina or Pyrex [TM] 7740 glass.

Table 8.2. Total quantities of gases released by normal surface flashovers, and by similar voltage pulses which did not cause flashover; results are given for unbaked and baked insulators (glass and alumina), and for 20 μs voltage pulses [94].

	QUANTITIES OF GAS DESORBED $(10^{12}$ molecules)	
	Flashover	**No flashover**
Baked	25-32	~0.07
Unbaked	48	1.8-8

Note in Table 8.2 that, while there is a large difference between baked and unbaked insulators in the quantities of gases evolved by 20 μs voltage pulses which did not cause breakdown, the quantities evolved by surface flashovers on baked or unbaked insulators were within a factor of two. This suggests that the quantity of adsorbed surface gas which must be desorbed in surface flashover is relatively independent of the amount of gas present. That is, assume that for a particular insulating system a quantity of gas Q (assumed to be desorbed surface gas, but could include vaporised surface material) had to be released from the surface for a surface flashover to occur. Now consider an identical insulator, but with less gas (or more tightly bound gas) on its surface.

Then, to release the same amount of gas Q, a more energetic voltage pulse (higher voltage) would have to be applied. Thus the insulator with more easily released surface gas would have the lower surface flashover voltage. This would explain why baking an insulator, cleaning it with UV, glow discharge etc., or just pumping on it for longer periods before applying voltage [95], can significantly improve its hold-off voltage. Such treatments would reduce the amount of adsorbed surface gas. Therefore, if originally a particular voltage would release enough gas to induce surface flashover, after treatment an increased voltage would be required to release the same quantity of gas.

The compositions of the gases produced by DC prebreakdown currents and surface flashovers have also been investigated and are discussed in detail elsewhere [3,94]. Some investigators attribute the source of these gases to adsorbed background air and oil vapour [67,96]. However, in clean systems, the surface gas was not attributed to adsorbed background gas, but rather, gas which was attached to the insulator surface during processing and handling [94]. This conclusion was based upon the differences in composition between the system background gas and the gas produced by surface flashovers, i.e. as shown in Table 8.3. In special cases, as when the insulator surface was a crystal freshly cleaved in vacuum, gas produced by the surface flashover would be adsorbed background gas [97].

Table 8.3. The maximum range of gas fractions observed in the system background and in the gas bursts produced by surface flashovers resulting from 20 μs voltage pulses applied to. alumina or glass samples, where the - insulator surface was cleaned with UV-Ozone or Plasma [94].

Gas	GAS FRACTIONS (%)	
	Background	**Flashover**
CO_2	3-7	32-50
CO	4-7	20-30
H_2	16-21	23-30
CH_4	3/4-1	3-8
N_2	52-64	0-5
H_2O	8-13	1-4
Ar	2.5-3	0
He	1/3-2/3	0
Pressure	0.7-1.2	3.4-4.4
$(\times 10^{-7}$ torr)		

8.3.13   Temperature

Decreasing the temperature of an insulator has been shown to increase its flashover voltage [96,98]. Experiments with cooling one end of the insulator suggest that the temperature effect is mainly at the cathode end [46]. The increase in flashover voltage is attributed to the effect of the lowered temperature on the adsorbed surface gas [46,96,98]. Heating a ceramic from 273 K to 873 K can lower its flashover voltage by a factor of five [99], which was attributed to the change in surface resistivity of the ceramic with temperature.

8. 4   Discussion-Technological Implications

While theories of surface flashover differ considerably in their mechanisms and assumptions, their predictions, concerning the important physical factors in flashover, agree on several critical points. These predicted critical points, and the experimental confirmation thereof, are listed in Table 8.4, and can be quite helpful in choosing or designing insulators for use in vacuum.

Table 8.4 Theoretical predictions and experimental confirmation of methods to reduce surface flashover of insulators.

Theoretical predictions	Experimental results
Importance of cathode triple junction	
A. Decrease macroscopic field	Shape rlectrodes Recess insulator Install metal inserts in insulator Use external metal shields
B. Decrease microscopic field	Minimise coids at junction Pressure on soft insulators Smooth interfaces Bonded interfaces Graded permittivities
Importance of desorbed gas for final stage of surface flashover	
A. Minimise quantities of adsorbed Gas	Fire insulators before assembly Bakeout dystem Condition insulator
B. Minimise sites available for adsorption	Original formation of insulator Minimise surface damage during fabrication
C. Maximise binding rnergy of adsorbed gas	Chemical treatment of insulator quasimetallizing

In the opinion of this author, the experimental results confirm the importance of field emission from the cathode triple junction as the primary initiating mechanism of surface flashover of insulators in vacuum; however, in particular cases, bombardment of the insulator surface by electrons or ions originating elsewhere in the system could be more important. Indeed, with suitably non-uniform fields, the flashover could originate at the anode triple junction. Thus, conditions at both triple junctions are very important in controlling the surface flashover of insulators. It follows therefore that geometrical modifications which weaken the electric field at the triple junctions can sometimes be helpful in reducing surface flashover.

The final flashover is believed to occur in gas desorbed from the surface of the insulator. Possible contributions of vaporised insulator material, specially for organic insulators, have also been observed. Thus, minimising the quantity of gas adsorbed on the surface of an insulator, or causing the gas to be bound more tightly to the surface, can contribute to improving the voltage hold-off performance of insulators in vacuum.

In designing insulators for a specific application, perhaps the first question to ask is "has the system to be baked out to high temperature (>300°C)?", i.e. since such bake-outs usually require the use of inorganic insulators. If an inorganic material is required, then the mechanical and thermal requirements of the application will suggest the use of a glass or a ceramic. If organic materials can be used, then it is preferable to choose a homogeneous material, and one with a relatively low vapour pressure. PMMA is often chosen, probably because of its relatively good performance, and the practical advantage in some applications of using a transparent insulator. In applications where surface flashovers may occur during operation, an insulator material strongly resistant to damage may be preferred to one with a higher surface flashover voltage. Whether the final choice of material is organic or inorganic, a fabricated insulator with low densities of defects (traps, voids, etc.) would be desirable. With all materials, care should be taken to keep them as clean and free from contaminants as possible. Baking out of the system (including the insulator), even if only to 100°C is usually desirable, if possible.

Next, the geometry of the insulator should be chosen. Sometimes this is specifically determined by the application, but if not, geometries other than a plain cylinder offer advantages. A cone with a +45° angle gives the highest flashover voltage, several times greater than a cylinder for nanosec pulses. There is considerably less advantage for longer pulses, DC, and AC voltages, but such cones can still sustain higher voltages than cylinders. However, cones take up more space than cylinders, unless the small end of the cone is made much smaller than the equivalent cylinder diameter, which is a disadvantage for many applications. Thus, one must balance the gains from choosing a more complex geometry against the additional effort and expense involved.

In any case considerable care should be taken in attaching the insulator to the electrodes, so as to minimise the electric fields at the triple junction. With glass, the glass-metal seal can be buried in the glass. With other materials, the use of recessed electrode-insulator interfaces, shields, or conducting inserts in the insulator ends have proven helpful. Another possible method of lowering the fields at the triple junction is by grading the relative permittivity (dielectric constant), which can be done by, (a) inserting a thin layer of material (with a permittivity higher than that of the insulator) between the insulator and the electrode (done mainly with organics) or, (b) doping the region of the insulator at the triple junction so as to grade its permittivity (done mainly for ceramics, specially aluminas, where a penetrating metallise may be used).

Appropriate surface treatments can increase the surface flashover voltage or the damage resistance of a material. Properly oriented, magnetic fields can significantly increase the surface flashover voltage. Again, in particular cases, the possible improved performance must be balanced against the cost and complication of the treatments or magnetic fields.

The conditioning of the insulator can be useful in many cases. If the system can be conditioned (or reconditioned), before use, this is quite helpful. Even if the application is one where the insulator must be idle for long periods before sustaining voltage, initial conditioning can be helpful, perhaps by removing surface contaminants or destroying initiating sites at the triple junction.

Though, in particular cases, operating an insulator at an elevated temperature may help in keeping contaminants from accumulating on the surface, the surface flashover voltage of insulators tends to decrease as the temperature is increased.

Insulators are not generally used in isolation, but as part of a system. Thus, surface flashover, while important, is only one factor to be considered in choosing an insulator. Economic factors, such as size and cost, and other considerations such as vacuum breakdown, bulk insulation, etc. are often critical. Examples of where insulators are considered as parts of systems e.g. for use in klystrons and travelling wave tubes, are found in [89,100-102].

8.5   Summary and Conclusions

This chapter has reviewed the nature and consequences of surface flashover of insulators. It has presented a brief sketch of theoretical mechanisms of surface flashover, mainly emphasising pertinent experimental results and implications of theory and experiment for practical applications of insulators where surface flashover is involved. Field emission from the cathode triple junction is believed to be the primary initiating mechanism of surface flashover of insulators in vacuum although, with suitably non-uniform fields, the flashover can originate at the anode triple junction. The flashover then develops either by a secondary electron emission avalanche (more probable for short pulses) or by a subsurface electron cascade (more probable for DC flashover). The final stage of flashover occurs in desorbed surface gas (or possibly in vaporised insulator material), although for materials such as silicon, the final stage may occur in a highly conducting surface layer. It was shown that the hold-off voltage of insulators depends upon many parameters, including (a) the properties of the insulator itself - material, geometry, surface finish, attachments to electrodes, (b) the applied voltage - waveform, duration, single pulse or repetitive, and (c) the history of the insulator processing, operating environment, and previous applications of voltage. Several suggestions have been made regarding how to choose the material, geometry, and processing procedures when selecting an insulator for a particular application. Also, some specific techniques for improving the hold-off voltage of insulators have been recommended.

Acknowledgements

Discussions with, and comments by, Drs R.A. Anderson, C. Le Gressus, and T.S. Sudarshan were appreciated and helpful. Dr. Anderson also supplied several figures.

8. 6 References

1. Hawley, R., *Vacuum* , **18**, 383-390, 1968.
2. Latham, RV., High Voltage Vacuum Engineering, Academic Press, London/New York, 1981.
3. Miller, HC., *IEEE Trans. Electr. Insul.* **24**, 765-786, 1989.
4. Miller, HC., *Le Vide* , **Suppl. #260**, 250-257, 1992.
5. Anderson, RA., *Proc., XIV ISDEIV* , Santa Fe (U.S), p 311, 1990.
6. Anderson, RA. and Brainard, J.P., *J. Appl. Phys.*, **51**, 1414-1421, 1980.
7. Anderson, RA. and Tucker, W.K., *J. Appl. Phys.*, **58**, 3346-3349, 1985.
8. Cross, JD., *IEEE Trans. Electr. Insul.* **EI-13**, 145-148, 1978.
9. Nevrovsky, V.A., *Proc. XIV , ISDEIV,* Sante Fe (U.S), p. 365, Santa Fe, 1990.
10. Bugaev, SP., Iskol'dskii, A.M., and Mesyats, G.A., *Sov. Phys. Tech. Phys..*, **12**, 1358-1362, 1967.
11. Bugaev, S.P. and Mesyats, G.A., *Sov. Phys. Tech. Phys.* **12**, 1363-1369, 1967.
12. Avdienko, AA. and Malev, MD, *Vacuum* , **12**, 643-651, 1977.
13. Avdienko, AA. and Malev, MD., *Sov. Phys. Tech. Phys.,* **22**, 986-991, 1977.

14. Vigouroux, JP., Lee-Deacon, O., Le Gressus, C., Juret, C., and Boiziau, C., *IEEE Trans. Electr. Insul.*, **18**, 287-291, 1983.
15. Vigouroux, JP., Le Gressus, C., and Durand, J.P., *Scan. Electron. Mic.*, **II**, p. 513, 1985.
16. Jaitly, NC. and Sudarshan, TS., *J. Appl. Phys.*, **64**, 3411-3418, 1988.
17. Blaise, G. and Le Gressus, C., *J. Appl. Phys.*, **69**, 6334-6339, 1991.
18. Blaise, G., *Le Vide,* **Suppl. 260**, 1-27, 1992.
19. Bommakanti, RG. and Sudarshan, TS., *J. Appl. Phys.*, **66**, 2091-2099, 1989.
20. Anderson, R.A., *Conf. Electr. Insul. Dielec. Phen.*, p. 173.,1979.
21. Boersch, H., Hamisch, H., and Ehrlich, W., *Z.. Angew. Physik*, **15**, 518-525, 1963.
22. Brainard, JP. and Jensen, D., *J. Appl. Phys.*, **45**, 3260-3265, 1974.
23. de Tourreil, C.H., Srivastava, KD., and Woelke, UJ., *IEEE Trans. Electr. Insul.*, **EI7**, 76-179, 1972.
24. Arnold, PA., Thompson, JE., Sudarshan, TS., and Dougal, RA., *IEEE Trans. Electr. Insul.*, **23**, 17-25, 1988.
25. Sudarshan, TS., Cross, JD., and Srivastava, KD., *IEEE Trans. Electr. Insul.*, **EI12**, 200-208, 1977.
26. Jaitly, NC. and Sudarshan, TS., *IEEE Trans. Electr. Insul.*, **23**, 261-273, 1988.
27. Khamidov, N., Agzamov, A., Kasimov, S., and Ishanova, S.S., *Zes. Nauk. Poli. Poznan. Elektryka*, **38**, 221-229, 1989.
28. Chalmers, ID., Lei, JH., and Bethel, JW., *Proc. XIV Int. Sym. Disch. Electr. Insul. Vac.* p. 324, Santa Fe, 1990.
29. Anderson, R.A., *Appl. Phys. Lett.*, **24** , 54-56. 1990
30. Cross, JD. and Sudarshan, TS., *IEEE Trans. Electr. Insul.*, **EI-9**, 146-150, 1974.
31. Sudarshan, TS. and Cross, JD., *IEEE Trans. Electr. Insul.*, **EI-11**, 32-35, 1976.
32. Kalyatskii, II. and Kassirov, GM., *Sov. Phys. Tech. Phys.,* **9**, 1137-1140, 1964.
33. Pillai, AS. and Hackam, R., *J. Appl. Phys.* , **58**,146-153, 1985
34. Kuffel, E., Grzybowski, S., and Ugarte, R.B., *J. Phys. D: Appl. Phys.*, **5**, 575-579, 1972
35. Cross, JD. and Srivastava, KD., *IEEE Trans. Electr. Insul.*, **EI-9**, 97-102, 1974.
36. Pillai, AS. and Hackam, R., *J. Appl. Phys.*, **56**, 1374-1381, 1984
37. Pillai, AS. and Hackam, R., *IEEE Trans. Electr. Insul.*, **EI-18**, 29-300, 1983.
38. Gleichauf, PH., *J. Appl. Phys.*, **22**, 535-541, 1951
39. Shannon, JP., Philp, SF., and Trump, JG., *J. Vac. Sci. Techn.*, **2**, 234-239, 1965

40. Grzybowski, S., Thompson, JE., and Kuffel, E., *IEEE Trans. Electr. Insul.*, **EI18**, 301-309, 1983.
41. Suzuki, T., *Japan J. Appl. Phys.*, **13**, 1541-1546, 1974
42. Jaitly, NC. and Sudarshan, TS., *IEEE Trans. Electr. Insul.* **23**, 231-242, 1988.
43. Yamamoto, O., Hara, T., Nakae, T., and Hayashi, M., *IEEE Trans. Electr. Insul.*, **24**, 991-994, 1989.
44. Le Gressus, C., Valin, F., Henriot, M., Gautier, M., Durand, J.P., Sudarshan, TS., and Bommakanti, RG., *J. Appl. Phys.*, **69**, 6325-6333, 1991.
45. Anderson, RA., *J. Appl. Phys.*, **48**, 4210-4214, 1977
46. Cross, JD., Srivastava, K.D., Mazurek, B., and Tyman, A., *Can. Elec. Eng. J.* **7**, 19-22, 1982
47. Thompson, JE., Lin, J., Mikkelson, K., and Kristiansen, M., *IEEE Trans. Plasma Sci.*, **PS8**, 191-197, 1980
48. Anderson, RA., *Appl. Phys. Lett.*, **24**, 54-56, 1974
49. Watson, A., *J. Appl. Phys.*, **38**, 2019-2023, 1967
50. Milton, O., *IEEE Trans. Electr. Insul.*, **EI7**, 9-15, 1972
51. Golden, J. and A. Kapentanakos, C., *J. Appl. Phys.*, **48**, 1756-1758, 1977
52. Jaitly, NC. and Sudarshan, TS., *IEEE Trans. Electr. Insul.*, **EI22**, 801-810, 1987.
53. Lee, R., Sudarshan, TS., Thompson, JE., Nagabhushana, GR. and Boxman, RL., *IEEE Trans. Electr. Insul.*, **EI18**, 280-286, 1983
54. Anderson, RA., in: *Proc.VII Int. Sym. Disch. Electr. Insul. Vac.* p. 252, Novosibirsk, 1976.
55. Pillai, AS. and Hackam, R., *J. Appl. Phys.*, **61**, 4992-4999, 1987
56. Vogtlin, GE., in: *Proc.XIV Int. Sym. Disch. Electr. Insul.Vac.* p. 307., Santa Fe, 1990.
57. Sudarshan, TS. and Cross, JD., *IEEE Trans. Electr. Insul.*, **EI-8**, 122-128, 1973.
58. Kofoid, MJ., *AIEE Trans. Pwr App. Sys.*, **Part 3**, 999-1004, 1960.
59. Moscicka-Grzesiak, H., Gorczewski, W., Stroinski, M., and Fekecz, J., in: *Fifth Int. Sym. High Volt. Eng.* (92.06), Braunschweig, 1987.
60. Kotov, YA., Rodionov, NE., Sergienko, VP., Sokivnin, SY., and Filatov, AL., *Instrum. Exp. Tech.*, **29**, 415-418, 1986.
61. Pillai, AS. and Hackam, R., *IEEE Trans. Electr. Insul.*, **EI19**, 321-331, 1984.
62. Juchniewicz, J., Mazurek, B., and Tyman, A., *IEEE Trans. Electr. Insul.*, **EI14**, 107-110, 1979.
63. Cross, JD. and Mazurek, B., *IEEE Trans. Electr. Insul.*, **23**, 43-45, 1988.
64. Avdienko, AA., *Sov. Phys. Tech. Phys.* , **22**, 982-985, 1977.
65. Gleichauf, PH., *J. Appl. Phys.*, **22**, 766-771, 1951.

66. Borovik, ES. and Batrakov, BP., *Sov. Phys. Tech. Phys.*, **3**, 1811-1818, 1958.
67. Akahane, M., Ohki, Y., I to D., and Yahagi, K., *Electr. Engng Jpn*, **94**, 1-6, 1974.
68. Takahashi, H., Shioiri, T., and Matsumoto, K., *IEEE Trans. Electr. Insul.*, **EI20**, 769-774, 1985.
69. Hatfield, LL., Leiker, GR., Kristiansen, M., Colmenares, C., Hofer, W.W., and DiCapua, MS., *IEEE Trans. Electr. Insul.*, **23**, 57-61, 1988
70. Akahane, M., Kanda, K., and Yahagi, K., *J. Appl. Phys.*, **44**, 2927, 1973
71. Lee, R., Sudarshan, TS., Thompson, JE., and Nagabhushna, GR., in: *IEEE Int. Sym. Electr. Insul .*, p. 103, 1982.
72. Miller, HC. and Furro EJ., Proc. VIII-ISDEIV, Albuquerque (U.S), D6, 1978.
73. Nam, SH. and Sudarshan, .S., *IEEE Trans. Electr. Insul.*, **24**, 979-983, 1989.
74. Nam, SH. and Sudarshan, TS., *IEEE Trans. Electron Dev.*, **37**, 2466-2471, 1990.
75. Peterkin, FE., Ridolfi, T., Buresh, LL., Hankla, B.J., Scott, DK., Williams, PF., Nunnally, WC., and Thomas, BL., *IEEE Trans. Electron Dev.*, **37**, 2459-2465, 1990.
76. Enloe, CL. and Gilgenbach, RM., *IEEE Trans. Plasma Sci.*, **17**, 550-554, 1989.
77. Enloe, CL. and Gilgenbach, RM., *IEEE Trans. Plasma Sci.*, **16**, 379-389, 1988.
78. Korzekwa, R., Lehr, F.M., Krompholz, HG., and Kristiansen, M., *IEEE Trans. Plasma Sci.*, **17**, 612-615, 1989
79. Korzekwa, R., Lehr, FM., Krompholz, HG., and Kristiansen, M., *IEEE Trans. Electron Dev.*, **38**, 745-749, 1991
80. VanDevender, JP., McDaniel, DH., Neau, EL., Mattis, RE., and Bergeron, KD., *J. Appl. Phys.*, **53**, 4441-4447, 1982
81. Bommakanti, RG. and Sudarshan, TS., *J. Appl. Phys.*, **67**, 6991-6997, 1990.
82. Fryszman, A., Strzyz, T., and Wasinski, M., *Bull. Acad. Polon. Sci. Ser. Sci. Tech..*, **8**, 379-383, 1960.
83. Miller, HC. and Furno, EJ., *J. Appl. Phys.*, **49**, 5416-5420, 1978.
84. Miller, HC., *IEEE Trans. Electr. Insul.*, **EI15**, 419-428, 1980.
85. Brettschneider, H., *NTG. Fachber.*, **85**, 187-191, 1983.
86. Brettschneider, H., *IEEE Trans. Electr. Insul.*, **23**, 33-36, 1988.
87. Hatfield, LL., Boerwinkle, ER., Leiker, GR., Krompholz, H., Korzekwa, R., Lehr, M., and Kristiansen, M., *IEEE Trans. Electr. Insul.*, **24**, 985-990, 1989.
88. Banaszak, Z. and Moscicka-Grzesiak, H., in: *Proc. XIV Int. Sym. Disch. Electr. Insul. Vac.* p. 357, Santa Fe, 1990.

89.   Vlieks, AE., Allen, MA., Callin, RS., Fowkes, W.R., Hoyt, E.W., Lebacqz, JV., and Lee, TG., *IEEE Trans. Electr. Insul.*, **24**, 1023-1028, 1989.

90.   Jaitly, NC., Sudarshan, TS., Dougal, RA., and Miller, H.C., *IEEE Trans. Electr. Insul.*, **EI22**, 447-452, 1987.

91.   Miller, HC., *IEEE Trans. Electr. Insul.*, **EI20**, 505-509, 1985

92.   Smith, ID., *Proc. Int. Sym. Insul. High Volt. Vac.*, p. 261, Cambridge, USA, 1964.

93.   Pillai, AS. and Hackam, R., *J. Appl. Phys.*, **53**, 2983-2987, 1982.

94.   Miller, HC. and Ney, RJ., *J. Appl. Phys.*, **63**, 668-673, 1988.

95.   Gorczewski, W., Moscicka-Grzesiak, H., and Stroinski, M., in: *Proc.V Int. Sym. Disch. Electr. Insul. Vac.* p. 399, Poznan, 1972

96.   Ohki, Y and Yahagi, K., *J. Appl. Phys.*, **46**, 3695-3696, 1975

97.   Kassirova, OS. and Kuzova, NK., in: *Proc.VI Int. Sym. Disch. Electr. Insul. Vac.* , p. 195, Swansea, 1974.

98.   Wankowicz, J., *Cryogenics*, **23**, 482-486, 1983.

99.   Grishutin, GS., Zhurtov, VM., Pokrovskaja-Soboleva, A.S., Shapiro, AL. and Shebsuhov, AA., *Proc. ISDEIV Int. Sym. Disch. Electr. Insul. Vac.*, p. 347, Berlin, 1984.

100   Bommakanti, RG. and Sudarshan, TS., *IEEE Trans. Electr. Insul.*, **24**, 1053-1062, 1989.

101   Wetzer, JM., Danikas, MG., and van der Laan, PCT., *IEEE Trans. Electr. Insul.* **24**, 963-967, 1989.

102   Wetzer, J M., Danikas, MG., and van der Laan, PCT., *IEEE Trans. Electr. Insul.* **25**, 1117-1124, 1990.

103   Mazurek, B. and Tyman, A.,*Proc ISDEIV Int. Sym. Disch. Electr. Insul. Vac.* p. 265., Novosibirsk, 1976.

104   Hamidov, N., Arifov, UA., Isamuhamedov, .D., Kelman, MC., and Agsamov, A., *Proc. VII Int. Sym. Disch. Electr. Insul. Vac.*, p. 282. Novosibirsk, 1976

9

Charge Trapping-Detrapping Processes and Related Breakdown Phenomena

G Blaise & C Le Gressus

9.1   Introduction

Traditionally, the origin of the surface flashover and bulk breakdown of a dielectric has been considered to be different because the consequences of these phenomena appeared different: inter-electrode discharge and tracks observed on insulators in vacuum are associated with flashover whereas, partial discharges, treeings and punchthrough are associated with bulk breakdown. Thus, it is not surprising that different models have been proposed to explain flashover and bulk breakdown. However, as the formation of a space charge has been recognised as having a determining influence on the breakdown and ageing of insulators [1], a number of questions have emerged relating to the capability of theories and of characterisation techniques to provide pertinent information on charging. For example:

-    What is a trap and what is the charging mechanism?
-    Do the current theories of trapping allow the charging properties of well defined materials to be predicted?
-    Is a low charge concentration in the range of 10^{-6} to 10^{-9} charge/atom as usually measured after breakdown or after poling, representative of the charging properties of a material, knowing that the dipolar rupture field is attained at concentrations in the range of 10^{-2}-10^{-3} charge/atom?
-    What is the nature and density of the defects required to interpret the size effect of the breakdown field; i.e., is the dependence of the breakdown field on sample size?
-    How can the charge density be linked to factors that change the breakdown probability, such as the shape of electrodes and the wave form of the applied field?
-    How does charging modify the internal energy of the material?
-    Is the breakdown process pre-determined by atom and ion diffusion, chemical reaction, phase nucleation and phase transition? If it is, what is the relationship between the characteristic times of these various processes and the time of failure?
-    Do long delay times to failure ($\geq 10^{-8}$ s), compared to typical electronic collective processes (10^{-15} s), support the contention that intrinsic breakdown is electronic in nature? A distinction is made in reference [1] between intrinsic breakdown and defect dominated breakdown.

All of these questions converge on one fundamental question, which is to know *whether the space charge is the cause or the consequence of breakdown.*

A unified explanation of breakdown has been recently proposed, and supported by:

-   the development of a space charge physics [2] in which the description of the trapping phenomenon allows the nature of the breakdown motive force to be elucidated.
-   the development of a space charge characterisation technique giving the relevant breakdown parameters [3]; significantly, the technique does not require the application of an external electric field, which proves that the motive force of breakdown cannot be found in the effect of an applied field on electrons.

Furthermore, the proposed model confirms that surface treatments for improving the breakdown strength [4] have significant effects on the relevant breakdown parameters [5].

In formulating this radically new approach it has been necessary:

-   to recognise the limitations of the current breakdown theories (section 9.2.
-   to investigate the dielectric polarisation/relaxation mechanisms,(section 9.3.
-   to investigate the secondary electron emission mechanisms in order to confront the new approach with the well accepted idea that the secondary electron yield δ and the breakdown strength V_B are correlated: i.e. the higher δ, the lower V_B (section 9.4).
-   to develop a technique (i) for verifying the space charge physics principles, (ii) to measure the secondary electron yield, (iii) for controlling the various stages of the breakdown process, (iv) for measuring the pertinent high voltage parameters, (v) for showing the effect of various surface treatments on the flashover strength (Section 9.5).

9. 2   Review of Breakdown Models

The oldest flashover model [6], called the "secondary electron cascade model (SEC)", appears to be unsatisfactory from a theoretical point of view, because, (i) it does not reflect all the physical processes involved in electron-insulator interactions as, for example, trapping, charge spreading and related energetic phenomena [7], (ii) there are other desorption processes than the electron stimulated desorption (ESD) mechanism described by the model for example, thermal desorption and electric field stimulated desorption [8].

Having been thoroughly investigated for adsorbed layers on metals [9], ESD has been accepted as a cause of insulator damage via inter-atomic Auger transition processes [10]. In fact it has been demonstrated that many-body effects due to charge detrapping is the major mechanism of damage [26].

The SEC model is also unsatisfactory from a several practical points of view because:

(i)   The surface charging of the insulator does not prove that it has been irradiated by electrons: Poling, ion neutralisation and metastable de-excitation are also causes of charging [12]. Furthermore, there is a lack of experiment data relating the secondary electron yield to the surface potential

(ii)   Many practical observations are not explained: e.g. anode-initiated discharges, the time lag between the application of the field and the occurrence of breakdown, the probability of breakdown when the field is switched off, the desorption of gas embedded in deep layers [4].

To counter the controversy associated with the "secondary electron cascade" model, another model derived from the pioneering work of Fröelich [13], namely the "ionisation-collision cascade" model (ICC) is often cited [14]. However, a remarkable review showing the impossibility for the theory to fit practical observations has been made in reference [1]. As no new experimental result has been found to refute the conclusion of this review, there is no reason to persist in interpreting flashover with the ICC model. However, two results obtained in the context of the ICC development deserve to be discussed.

1.   Since it has been demonstrated that hot electrons cannot gain enough energy to ionise the valence band of wide band gap insulators, it has been proposed that they could generate defects [15]. The question then arises as to how a hot electron behaves in the vicinity of an intrinsic defect, i.e., does it become trapped or does it ionise the defect?

2.   Photon emission around 300-500 nm which occurs in an electrically stressed insulator, has been attributed to hot electron-defect interactions and to charge carrier recombinations involving defects [16]. But, photon emission cannot be used as unequivocal evidence of ionisation-collision effects because photons are also emitted when insulators are subjected to any other type of stress. The luminescence spectrum comprises two types of lines: broad lines at 300-500 nm attributed to charge carrier recombination and sharp lines which are characteristic of impurities (e.g. chromium in alumina). Moreover, recent results [17] have shown that broad lines which are sensitive to the crystal field, are also modified in shape and energy by a space charge field. This illustrates

the complexity of luminescence phenomena: probably, the luminescence combines charge carrier recombination and structural modification of highly stressed materials since it has been predicted by theorists that dielectric relaxation could also produce photon emission [11].

9.3   The Charging of Insulators

9.3.1   Ohmic Charging

The ohmic charging of a dielectric is often discussed in terms of the time constant $\tau = \rho\varepsilon$ (ρ being the resistivity and ε the dielectric constant) which characterises the charge decay of a short-circuited parallel-plate capacitor filled with the dielectric. In this case, the charge decay is controlled by the flux of free charge carriers available in the material. As the intrinsic density of free charge carriers is extremely small, the resistivity is large. The wide range of resistivities, varying from some 10^5 Ω cm to 10^{18} Ω cm means that τ can vary from some 10^{-7} s to several days. As a consequence, ohmic charging is only expected to occur in insulators having very high resistivities.

In fact, the problem of ohmic charging is not related to the flowing of intrinsic charge carriers, as is the case in a shorted capacitor, but to the flowing of injected charge carriers. So τ is not a relevant parameter for describing the charging of a dielectric, rather the relevant parameter is the mobility μ. We shall now look at the problem of ohmic charging with a new approach which takes μ into account. Let us consider a one-dimensional metallised and grounded slab of dielectric of thickness L receiving a uniform flux of charge on its free side (Fig 9.1).

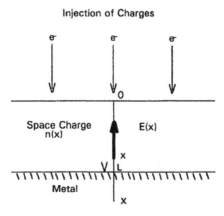

Fig 9.1   Ohmic charging in a slab subjected to an electron beam; $n(x)$ is the space charge density.

In the absence of any trapping, ohmic charging is measured by the potential V which appears on the front side exposed to the flux of charge, and is due to the induced space charge of density n(x). For convenience, the situation is considered where the injection of electrons has a uniform density of jAm^{-2}; furthermore, the grounded electrode is assumed to be neutral (i.e. not hindering the passage of charges) and the intrinsic conduction of the medium is neglected. Assuming that the slab is a perfect parallel-plate capacitor, the potential is given by :

$$V = -\frac{2}{3}\left[\frac{jL^3}{\varepsilon\mu}\right]^{\frac{1}{2}}$$

9.1

and the carrier-transit time t_c from the front side to the grounded electrode is:

$$t_c = \left[\frac{2\varepsilon L}{\mu j}\right]^{\frac{1}{2}}$$

9.2

Numerical calculation of V and t_c is performed taking a current density $j \approx 10$ A cm^{-2}. The sample is supposed to be a very pure αAl_2O_3 oxide slab, 1 mm thick, having a dielectric constant $\varepsilon = 10\varepsilon_0$, a resistivity in the range of 10^{18} Ω cm and an electron mobility at room temperature ≈ 200 cm^2/Vs, which is typical of the magnitude found for metals [18]. With these numerical values we find V $\approx$ 70 V and $t_c \approx$ 1 μs.

These figures clearly show that ohmic charging is extremely low under the very severe conditions mentioned above, and that no durable charging is expected since the space charge will decay in a few μs once the injection of charge is stopped.

Does this work for any insulator? Combining typical electron mobilities μ $\approx$ 10 cm^2/Vs with a relative dielectric constant ranging from 1 to about 1000, and a sample thickness varying from μm to cm, we can find ohmic charging in some extreme cases. However, by a judicious choice of experimental conditions, ohmic charging will be avoided in most cases and, in no way, will a durable charging state be obtained. From this discussion we are bound to conclude that the charging mechanism of insulators is definitely not ohmic.

There is a general agreement that the charging of insulators is attributable to the trapping of charge-carriers at defects present in the material. In a classical band structure model, trapping levels produced by the multiple imperfections of the material are located in the band gap. Trapping is described as in doped semiconductors: an electronic charge-carrier moving in the conduction band is captured on a level located in the band gap. The

334

release of the trapped charge occurs through an activation process which propels the charge in the conduction band. A distinction is made between shallow traps at some tenth of an eV and deep traps at ~ 1-2 eV. In fact the evidence for or against any of these two categories of traps is, for most materials, weak: for example thermally stimulated discharge experiments often lead to the assumption of quasi-continuous distribution of traps extending to a depth of about 2 eV.

Such a quasi-continuous distribution of traps in polymers is justified by the randomness in their structure which in turn, causes a randomisation of the trap depth. However, charging based on the argument of the randomness of the structure is less valid in inorganic crystallised dielectrics and not at all justified in single crystals of stoichiometric composition.

There is no doubt about the fact that structural modifications are responsible for deep trapping sites for which activation energies exceed 1 eV (E' centre for example). Thus, experimental data on deep levels are clear, but on shallow trap levels they are not yet well understood.

9.3.2 Charging Under Electron Irradiation

When insulators are bombarded by an electron beam having an energy of a few keV, the net implanted charge is the difference between the primary current I_p and the total secondary electron current I_σ due to backscattered and secondary electrons emitted from the sample. As a result the charging can be positive or negative depending on the variation of I_σ with the energy of the primary beam, as illustrated in Fig 9.2. Correspondingly, the stabilisation of the surface potential, as observed experimentally under the special conditions when an electrokinetic steady state is attained, requires the conservation of the current. This means that the sample current, given by:

$$I_s = I_p - I_\sigma$$

is flowing to ground through the insulator. Under these circumstances, the stabilisation of the potential has been attributed to the ohmic charging produced by I_σ, as described before. However, this interpretation has not received a quantitative justification and, furthermore, it does not explain why some single crystals such as MgO and Al_2O_3 do not charge at room temperature (Table 9.I).

Charging has also been attributed to charge carriers trapped at defects present in the insulator. According to the rigid band structure model, trapping occurs on energy levels localised in the band gap. This description of trapping is also not satisfactory since, as illustrated in Table 9.1, it does not explain why single crystals, having similar crystallographic structures and very similar band gap widths, charge differently. Furthermore, it does not explain the effect of charging on the internal properties such as the variation of the heat capacity [7], or the change of the crystalline structure [19].

Table 9.1   Charging behaviour of common single-crystal insulating materials

Crystal	Structure	Permittivity	Band gap width in eV	Charging at room temperature
MgO	Cubic	~10	~6	no
Y_2O_3	Cubic	~15	~6	yes
SiO_2	Hexagonal	~4	~10	yes
Al_2O_3	Hexagonal	~10	~9	no

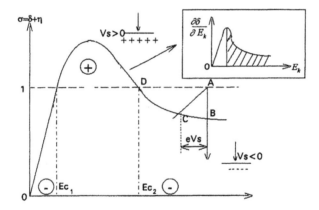

Fig 9.2   The conventional approach of insulator charging is based on the dependence of the total yield curve σ as function of the primary electron energy E_0. When the yield is larger than unity between Ec_1 and Ec_2, it is accepted that secondary electrons are attracted by the positively charged surface. The insert illustrates the decrease of the yield when the surface potential is positive. However, the parallel between a positive charging and a high yield is not a general rule. For example, pure MgO single crystal does not charge and yet the maximum yield is about 20. When the primary beam energy is larger than Ec_2 the sample charges negatively, producing a deceleration of the incident electrons until point D is reached. This is not a general rule because for high energy beam charging increases until the beam is reflected by trapped charges.

A completely different approach to charging has been developed [2], based on the concept of the polaron, as originally used by Mott to describe the conduction properties of insulators [20]. This approach has the advantage of providing a general description of the trapping process which is in better agreement with experimental characteristics.

9.3.3   A New Description of a Charged Medium.

9.3.3 (i)   Charge Trapping.
Let us consider a point charge q moving at a velocity v in a polarisable medium characterised by an optical vibrational frequency ω (dispersion is neglected). Due to the movement of the charge, the polarisability of atoms varies with their distance from the charge, as illustrated in Fig 9.3. At a distance larger than v/ω, the charge appears as static, producing a polarisation with a static dielectric constant $\varepsilon(0)$. At distances less than v/ω, the rapid movement of the charge makes the polarisation purely electronic with a high frequency dielectric constant $\varepsilon(\infty)$.

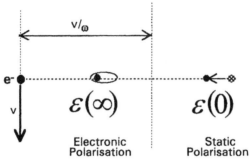

Fig 9.3   The variation with distance of the dielectric response of a medium to the field of a moving electron. Close to the charge, the electronic response dominates (electronic dielectric response $\varepsilon(\infty)$) whilst far from it, the static response dominates (static dielectric response $\varepsilon(0)$).

An important consequence of the variation of the dielectric constant with distance is the presence of a polarisation charge whose density is given by

$$\rho_p = -\mathrm{div}P = -\varepsilon_0 \frac{E_{xyz}}{1+\chi_{xyz}}\mathrm{grad}\chi_{xyz} \qquad 9.3$$

where $P = \varepsilon_0\,\chi_{xyz}\,E_{xyx}$, is the polarisation induced by the Maxwell field E_{xyx} which in turn, is produced by the charge q and the resulting polarisation charge. In equation 9.3 we have introduced the susceptibility χ_{xyz} as a function of space co-ordinates. If the charge is now localised in an orbital of spherical symmetry the field at a distance r from the centre of the orbital is

$$E_r = \frac{1}{4\pi\varepsilon_0}\left[\frac{\varepsilon_0}{\varepsilon(\infty)}q+Q_p(r)\right]\frac{1}{r^2} \qquad 9.4$$

where $Q_p(r)$ is the polarisation charge contained in a sphere of radius r. Combining equations 9.3 and 9.4 we obtain

$$Q_p(r) = \varepsilon_0 \left[\frac{1}{\varepsilon(r)} - \frac{1}{\varepsilon(\infty)} \right] q$$

where $\varepsilon(r)$ is the dielectric function at distance r from the charge. The potential energy between the charge q and polarisation charge contained within two spheres of radius and r_a is r_b will then be

$$W = \frac{q}{4\pi\varepsilon_0} \int_{r_a}^{r_b} \frac{dQ}{r} = \frac{q^2}{4\pi} \int_{r_a}^{r_b} \frac{1}{r} d\left[\frac{1}{\varepsilon(r)} \right] \qquad 9.5$$

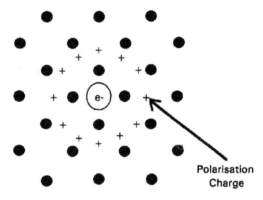

Fig 9.4   Illustrating the binding effect of the polarisation charge that surrounds a charge localised in an atomic orbital.

As a first approximation, the polarisation charge is supposed localised on a sphere of radius r_p as shown in Fig 9.4 so that:

$$\varepsilon(r) = \varepsilon(\infty) \qquad\qquad r < r_p$$
$$\varepsilon(r) = \varepsilon(0) \qquad\qquad r > r_p$$

Using this approximation we finally obtain:

$$W \approx -2E_R \left[\frac{1}{\varepsilon(\infty)} - \frac{1}{\varepsilon(0)} \right] \frac{a_0}{r_p} \qquad 9.6$$

where E_R = 13.56 eV is the Rydberg energy and a_0 the Bohr radius. As $\varepsilon(\infty) < \varepsilon(0)$ always applies, W is negative and a binding effect is predicted from equation 9.6.

338

The quasi-particle formed by the charge surrounded by the polarisation charge is called a polaron. The distance r_p defines the radius of the polaron. Equation 9.6 also clearly shows that the binding effect comes from the effective dielectric constant ε_p defined by:

$$\frac{1}{\varepsilon_p} = \frac{1}{\varepsilon(\infty)} - \frac{1}{\varepsilon(0)} \qquad 9.7$$

which characterises the variation with distance of the dielectric response of the medium to the field of the charge.

Quantum polaron theories are based on the coupling of the charge with the longitudinal phonon, neglecting dispersion. They all use equation 9.7 as the starting point of the binding effect. Their advantage over the very crude electrostatic model outlined here lies in the calculation of the radius r_p of the polaron. Two types of theories have been developed leading to the large and small polaron.

Large polaron: When the effective mass m* of the charge moving in a rigid lattice is not too high, the extension of the orbital of the charge, represented by the radius r_p exceeds the interatomic distance. It follows from equation 9.6 that the binding effect will be extremely small, so that the polaron moves easily, as a free charge [21]. This is the case in semiconductors in which m*<m_0 (m_0 being the electron mass) and the mobility $\mu = e\tau/m*$ is high; typically, $\mu \gg 1000$ cm^2/Vs at room temperature.

Small polaron: When the effective mass m* is high, the radius r_p of the polaron becomes somewhat less than interatomic distance. Then, the binding effect becomes large and the overlapping of the wave functions associated with polaron sites is so small that, as illustrated in Fig 9.5, it leads to the formation of a narrow energy band of width $2J_p$ called a polaron band [20-21]. At low temperatures, the polaron behaves like a heavy particle moving in a narrow band. At high temperatures it is easier for the polaron to move by a thermally activated process over the potential barrier W_H, rather than by the band mechanism of Fig 9.5. At room temperature, the mobility μ is typically ($\mu \approx 10-100$ cm^2/ V.s.) which is one or two orders of magnitude smaller than in the case of semiconductors.

The small polaron theory was used by Mott to interpret the conduction properties of metal oxides [21]. In this approach, conduction is described through a polaron band structure which is not coupled to the ordinary band structure of the rigid lattice. As a consequence, the binding effect can be studied independently of the rigid band structure.

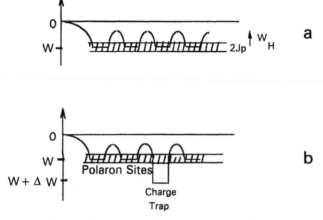

Fig 9.5   Illustration of (a) the potential wells associated with polaron sites and the polaron band produced by the overlapping of polaron orbitals, and (b) the trapping effect that is produced on a polaron site where the local electronic polarisability is smaller than the average electronic polarisability of the surrounding medium.

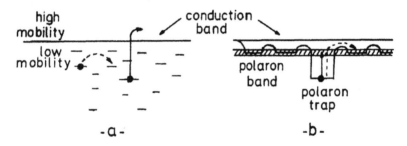

Fig 9.6   Electron trapping-detrapping process as described in reference [33] (a) compared to the process described in the polaron band model, (b) mutatis mutandis for holes.

However, we must speculate how to represent the binding energy due to the polarisation in the classical band energy diagram. The reasoning is simple. If the medium was not polarisable ($\varepsilon(\infty) = \varepsilon(0) = \varepsilon_0$) an electron "at rest" would be at the bottom of the conduction band. Taking the polarisation into account, the electron level is lowered by an amount equal to the binding energy, and consequently falls into the band gap, as shown in Fig 9.6. Let r_j be the position of a polaron site and $\varepsilon(\infty, r_j)$ be the local dielectric function of atoms forming the cell surrounding the charge. The concept of the small polaron can be used to describe the trapping process provided the dielectric constant $\varepsilon(\infty)$ is replaced in equation 9.6 by $\varepsilon(\infty, r_j)$. As $\varepsilon(\infty, r_j) << \varepsilon(0)$ in many materials, the binding effect is essentially governed by $\varepsilon(\infty, r_j)$ that is by the electronic polarisability of atoms attached to the immediate environment of the charge. Let us now consider a medium of uniform static dielectric constant $\varepsilon(0)$. If $\varepsilon(\infty, r_j)$ is independent of r_j, that is independent of the polaron site, the binding energy W given by equation 9.6 is constant. We now have to deal with the classical polaron described by Mott, moving either by hopping at high temperature or by a band mechanism at low temperature.

Suppose now that there is a slight variation $\Delta\varepsilon(\infty, r_k)$ in the electronic dielectric function at a polaron site located at point r_k. The potential well on this site will consequently vary by an amount

$$\Delta W_k = 2E_R \frac{a_0}{r_p} \frac{\Delta\varepsilon(\infty, r_k)}{\varepsilon^2(\infty, r)} \qquad 9.8$$

If $\Delta\varepsilon < 0$, that is, if the electron dielectric function is locally reduced, the binding effect is reinforced. If $\Delta W_k \gg kT$, the trapping of the charge occurs and the site located at point r_k constitutes a polaron trap. By relating $\Delta\varepsilon$ to the local variation of the electron polarisability, through the generalised Clausius-Mosotti expression for example, we can interpret the trapping as due to a local decrease of the electronic polarisability of the surrounding medium. In that sense, these polaron sites, which have an electron polarisability smaller than the average value, can be considered as "defects". Thus, in this approach, "defects" that are responsible for charge trapping take the universal meaning of polarisability defects; that is, defects associated with local variations of polarisability.

This concept of polarisability defects being responsible for charge trapping is not in conflict with the well accepted idea that trapping is due to defects. On the contrary it gives a clear insight into the nature of these defects that were not previously well understood. Furthermore, it is compatible with the following situations typically encountered in insulators.

(i)   Structural defects in crystalline polymers are so numerous and dense that the variation of the electronic polarisability attached to these defects gives a quasi-continuous distribution of polaron traps.

(ii)   In inorganic materials, such as ceramics, point and extended defects are the source of anomalies in electronic polarisability; as are grain boundaries, and vacancies due to non-stoichiometry, concentration gradients, segregation, clustering. All these anomalies can give rise to polaron traps.

(iii)   In stoichiometric single crystals, the cause of trapping must be investigated in the lattice structure itself. At least two sites, having different electron polarisabilities, must be found: a normal site ensuring the conduction of polarons and a polaron trap site of lower electron polarisability. Consequently, (i) MgO, in which all cation sites are equivalent, does not charge, (ii) Y_2O_3, where yttrium atoms are embedded into two types of oxygen octahedrons [22] charges very strongly, (iii) Al_2O_3 where the half of the octahedron sites are filled with Al atom charges at low temperature, (iv) α quartz, which is composed of tetrahedral grouped to form helices, charges very strongly.

It is risky to give a numerical estimate of the trapping energy from equation 9.8, since because most of the parameters of this equation are not known with accuracy. But it is obvious that local variations of polarisability will be small in most cases, from which trapping energies are expected to vary from a few 1/100ths to a few 1/10ths of an eV. This trapping energy range is in agreement with the fact that detrapping of charges occurs in most insulators at temperatures ranging from liquid nitrogen to few hundreds of deg. C.

As equations 9.6 and 9.8 are independent of the sign of the charge, the same argument as that used to explain the trapping of an electron is valid for the trapping of a hole: i.e. a hole is trapped on a polaron site where the electron polarisability is smaller than in the surrounding medium. Thus, polaron traps are capable of fixing positive and negative charges. The fact that polaron traps can fix positive or negative charges equally well allows one to describe, in a very symmetrical way all the charging and polarising phenomena observed in dielectrics, whatever the sign of the electrostatic space charge.

To close this section, the trapping-detrapping process described in reference [33] with the rigid band model is compared in Fig 9.6 with the polaron band model. The main difference is that, in the first one, trapping-detrapping involves electron and hole exchange between the rigid band structure and discrete levels localised in the band gap, whereas in the second one, the trapping-detrapping process is entirely contained in a polaron band which results from the polarisation properties of the medium, independently of the rigid band structure.

9.3.3.(ii)   Energy Stored in a Dielectric

The energy stored in a dielectric polarised by a static applied field comprises two terms:

(i)   A mechanical energy, which results from ion and electron displacements under the applied field, i.e. corresponding to the mechanical deformation of the structure due to polarisation.

(ii)   An electromagnetic energy, which is the consequence of the dipoles formed by the displacement of charges.

342

One might consider that the first term is included in the second one since charges have to be displaced to produce the polarisation field. However, that is not at all the case, as illustrated in the following example, which simulates the situation in a dielectric.

In order to simplify the problem, let us consider the perfect parallel plate capacitor, shown in Fig 9.7 having one electrode fixed and the other hanging on a spring with no damping. With no charging, the mobile electrode is, at rest, at a reference position zero. When the capacitor is charged the mobile electrode is misplaced by a distance x_0 closer the other, with the final separation of the two electrodes being d.

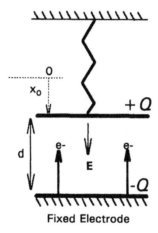

Fixed Electrode

Fig 9.7   Mechanical and electromagnetic energy stored in a parallel-plate capacitor, having one electrode that is suspended from a spring. This device illustrates the two energy terms of a polarised dielectric: the spring simulates the mechanical energy due to the polarisation of the medium, whilst the plate capacitor simulates the electromagnetic energy stored in a polarised medium.

The total energy stored in the system after charging will comprise the following two components:

(a)   The mechanical potential energy, corresponding to the stretching of the spring:

$$w_p = \frac{1}{2}Cx_0^2$$

where C is the constant force of the spring.

(b)   The electromagnetic energy contained in the capacitor

$$w_{em} = \frac{1}{2}\varepsilon E^2 dS$$

343

where E is the field between the conducting plates and S their surface area. Here, the stretching of the spring corresponds to the formation of dipoles in the dielectric, whilst the electromagnetic energy of the capacitor simulates the electromagnetic energy stored in the dielectric. Now we imagine that electrons are released from the fixed electrode by a thermal emission process, and that when they strike the other electrode their kinetic energy is converted into heat. As electrons are emitted, the capacitor becomes discharged and the mobile electrode moves towards its initial reference position. It is supposed that the process is slow enough to prevent the mobile electrode from acquiring kinetic energy. After discharging, the mechanical potential energy of the stretched spring is restored to the system, whereas the electromagnetic energy is converted into heat. A simple calculation shows that the amount of heat just equals the electromagnetic energy w_{em} stored in the capacitor after charging. In other words, the energy dissipated in a real electric circuit (including now dielectric losses) during the discharging of a capacitor filled with a dielectric is the electromagnetic energy available in the capacitor. Thus, the mechanical potential energy plays no role in the process.

The question is "What would happen if charges were instantaneously removed from the two plate electrodes of the capacitor?" To answer this question, it is first necessary to calculate the two energy terms.

(a)   Electromagnetic Energy
The electromagnetic energy per unit volume stored in a medium of dielectric constant ε is given by:

$$w_{em} = \frac{1}{2}E_m D = \frac{1}{2}\varepsilon E_m^2 \qquad 9.9$$

where $D = \varepsilon_0 E_m + P$ is the electric displacement. In order to simplify the presentation, we consider a slab of dielectric with the applied field E_a perpendicular to the surface. From the continuity requirement of the electric displacement between the vacuum and the dielectric $D = \varepsilon_0 E_a = \varepsilon E_m$, we deduce:

$$E_m = E_a - \frac{P}{\varepsilon_0}$$

Consequently equation 9.9 can be re-written as

$$w_{em} = \frac{1}{2}\varepsilon_0 E_a^2 - \frac{1}{2}P E_a \qquad 9.10$$

Equation 9.10 is quite general, as demonstrated in Ref. [23]. The first term in equation 9.10 represents the electromagnetic energy due to the applied field in

344

the vacuum, whilst the second is associated with the polarisation. The minus sign in the second term indicates that the system is more stable when it is polarised that when it is not. The energy contained in equation 9.10 is the electromagnetic energy that is dissipated partly in an electric circuit connected to the capacitor filled with the dielectric, and partly in the dielectric itself (the dielectric loss).

(b)   Mechanical Energy
A dipole is formed by the action of the local field against internal forces. Under static conditions, the mechanical potential energy stored in the medium per unit volume is:

$$w_p = \frac{1}{2} N_0 C x_0^2 = \frac{1}{2} P E_{loc} \qquad 9.11$$

where N_0 is the atomic density and E_{loc} the local field. In this context, attention is drawn to the possible confusion between the term $1/2\ PE_a$ in equation 9.10 and $1/2PE_{loc}$ in equation 9.11 which, though having a similar waveform, involve different fields.

As an application, equation 9.11 has been used to calculate w_p for the case of a non-polar medium [2]:

$$w_p = \frac{1}{2} \varepsilon_0 \chi (1 + \chi \gamma) E_m^2 = \chi \frac{(1 + \chi \gamma)}{(1 + \chi)} w_{em} \qquad 9.12$$

where γ is the Lorentz' coefficient equal to $1/3$ in cubic structures [2].

This indicates that when χ becomes large, that is $\chi \gamma \gg 1$, we have :

$$\frac{w_p}{w_{em}} \approx \chi \gamma \gg 1$$

In other words, for non-polar materials having a high susceptibility the mechanical potential energy stored in the medium can be much higher than the electromagnetic energy. We have, therefore, a reservoir of energy which has not yet been investigated.

Note: a complete analysis of the stored energy would also require taking the entropy into account, that is the degree of disorder. This energy component corresponds to the so-called electrocaloric effect in polar dielectrics (see Ref [23], p.75, and Ref [24], p.9,),but is generally a very small effect that will not be considered in the following treatment.

(c)   Mechanical Energy Around a Charge Due to Polarisation

Let us first approach the subject with the assumption that the Lorentz method is acceptable for calculating the local field around the charge.

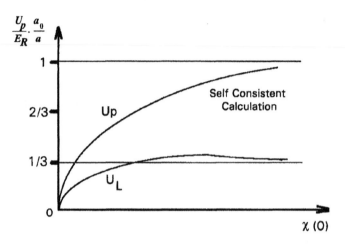

Fig 9.8   Mechanical energy surrounding a charge versus static susceptibility $\chi(0)$. The curve marked U_L is the energy based on the Lorentz method for calculating the local field. The curve marked U_p has been obtained by a direct self-consistent calculation of the field. (After [25].)

By integration of equation 9.11 from the distance a of the first neighbours to infinity we obtain for the mechanical energy U_L stored around a charge (subscript L means calculated using the Lorentz method)

$$U_L = \left(\frac{1+\chi\gamma}{1+\chi}\right)^2 U_0 \qquad\qquad 9.13$$

where $U_0 = \frac{1}{2}\varepsilon_0 N_0 \alpha_d \int_a^\infty E_q(R)4\pi R^2 dR = N_0 \alpha_d E_R \frac{a_0}{a}$ represents the mechanical energy due only to the applied field E_q (R); here α_d is the polarisability. The variation of U_L with χ is qualitatively represented in Fig 9.8 for cubic crystals. When the value of χ is such that the condition $\chi\gamma >> 1$ is fulfilled, equation 9.13 reduces to $U_L = \gamma^2 U_0$.

For non-polar cubic crystals, where $\gamma = 1/3$ and $N_0\alpha_d \approx 3$ for $\chi >> 1$, the maximum of U_L is

$$U_L \approx \frac{1}{9}U_0 \approx 1.5 eV$$

As we know, the validity of the Lorentz method is questionable. Using the direct calculation of polarisation developed above, we can express the mechanical energy w(i) associated with the formation of a dipole i as

$$w_i = \frac{1}{2} p_i E_{loc}(i)$$

Thus the total mechanical energy U_p stored per charge is

$$U_p = \sum_i w(i)$$

which can be rewritten as:

$$U_p = K(\alpha_d)U_0 \qquad\qquad 9.14$$

where $K(\alpha_d)$ is determined by numerical computation.

In cubic crystals we have found that K varies from 1/4 to 1/2 depending on the structure, when $\chi \gg 1$, [25]. Taking the average value $K \approx 1/3$ we obtain

$$U_p = E_r \frac{a_0}{a} \approx 5eV$$

This value of U_p seems to be close to the maximum of the mechanical energy available in cubic crystals, when the Clausius-Mosotti relation is taken into account. The energy U_p is plotted in Fig 9.8 as a function of χ, from which it will be seen that U_p is about 3 times higher than the mechanical energy calculated by the Lorentz method.

Using Onsager's dielectric constant theory in a polar medium, we have derived an expression for U_p of the form

$$U_p = \frac{2}{3}\chi K(\chi)E_R \frac{a_0}{a} \approx 3\chi K(\chi) \qquad\qquad 9.15$$

where preliminary calculations indicate that the energy predicted by equation 9.15 can generally exceed the 5 eV obtained in non - polar cubic crystals.

The conclusion is that a non-negligible quantity of mechanical energy is stored around a charge and therefore a significant change of internal energy has to be expected in a charged dielectric. This change has been confirmed by measuring the variation of the heat capacity as function of the amount of charges [7]. The internal energy stored in a charged dielectric can be at the origin of damage [26] when trapped charges are released. This is why we are now examining the two problems of the space charge dynamics and of the relaxation associated with charge detrapping.

(d)   Space Charge Dynamics

The space charge dynamics resulting from the injection of charges into insulators is controlled by a competition between trapping and detrapping processes. Trapping is characterised by a trap density n_p whereas detrapping is function of the internal field $E(\bar{r})$, which is composed of the applied field plus the space charge field. If $E(\bar{r})$ is smaller than a critical detrapping field E_d, which is function of the trap energy, we can assume that charges cannot be detrapped. In this case a static space charge is formed, perfectly stable in time if the injection of charges is stopped. Continuing the injection of charges results in the expansion of the space charge if $E(\bar{r}) \geq E_d$ and then leads to the establishment of a propagating regime of charges through the dielectric, characterised by a steady state current resulting from an equal number of trapped and detrapped charges per unit time.

A preliminary theoretical description of the space charge dynamics in a parallel-plate capacitor, including the parameters mentioned above, has been proposed [27]. When $E(\bar{r}) \geq E_d$ the detrapping rate is described by a function $\dfrac{f[E(r)]}{\tau_d}$ in which $\dfrac{1}{\tau_d}$ is the detrapping rate for the field E_d and f is a dimensionless increasing function of the total field. Let $J(x,t)$ be the flux of charged particles at point x and time t, and $n(x,t)$ be the density of trapped charges. The continuity equation requires

$$\frac{dn(x,t)}{dt} = -\frac{dJ(x,t)}{dx} \qquad\qquad 9.16$$

and the variation of J is given by:

$$\frac{dJ(x,t)}{dx} = -J(x,t)\big[n_p - n(x,t)\big]\sigma_p + \frac{f[E(x,t)]}{\tau_d}n(x,t) \qquad\qquad 9.17$$

Transport equations are proposed in Ref. [33], p. 118 and [136] to calculate the current emitted during a thermally stimulated discharge experiment. They differ from equation 9.17 in the sense that only trapping-detrapping processes are considered as affecting the mobility of charges, whereas 9. 17 is used to calculate the detrapping rate in order to evaluate the energy released. In the present case equations 9.16 and 9.17 have been used in addition to classic field equations, to calculate (a) the amount of charges detrapped in a parallel-plate capacitor as a function of the applied field and (b) interpret the breakdown process observed near both electrodes as it is described in Refs. [4-29].

(e)   Energy Release Associated with Charge Detrapping

The field around the trapped charges distorts the lattice which, in turn, locally results in an increase in mechanical potential energy. This situation is maintained as long as the charges remain at fixed positions. If, however, the charge distribution is suddenly detrapped, the lattice loses equilibrium locally. This results in a mechanical relaxation process in which the energy of distortion is converted into phonon waves which dissipate the energy. The question that arises is whether the energy is released through dipolar transitions and their coupling to the phonon bath, or through the formation of shock waves. To answer this question we have to estimate the temperature produced by the relaxation process which follows the detrapping of a charge.

More than 80% of the mechanical energy is concentrated around the charge within a sphere of radius R = 5a ≤ 10 Å which contains about one hundred atoms. We have to estimate the temperature increase of this sphere and that of the surrounding medium resulting from a dipolar relaxation process. If we define the characteristic time τ of heat propagation as the time required to cover a distance R outside the sphere we have

$$\tau = \frac{R^2}{\kappa}$$

where $\kappa/\rho c$ is the diffusivity of the material (K being the thermal conductivity, ρ the density and c the specific heat). Taking $R \approx 10$ Å and $\kappa = 6 \times 10^{-2}$ cm^{2}s^{-1}, we find $\tau = 10^{-12}$ s. That means the thermal energy should appear in the sphere of radius R in times $<<10^{-12}$s to produce a temperature increase of several thousands degrees necessary to form a plasma. This result is inconsistent with individual Debye relaxation processes of dipoles in which times range from 10^{-9}-10^{-3} s. It is also inconsistent with a co-operative process of dipolar transitions by tunnelling ($\approx 10^{-12}$ s) [11], followed by the transfer of the energy to the phonon bath within a 10^{-9}-10^{-6} s range.

Thus, it becomes clear that a global increase of the temperature of the surface cannot result from the release of the mechanical energy. If we want to produce some catastrophic event from the release of this energy we have to consider the formation of the thermal shock waves around each source that dissipates energy. Accumulation of theses waves in some regions of the surface should induce the formation of plasmas whose expansion would eventually constitutes the breakdown itself.

Let us mention that calorimetric experiments have clearly demonstrated the exothermic character of the lattice relaxation accompanying the detrapping process [7].

9.3.4   Breakdown as a Consequence of Charge Detrapping

If it is accepted that the breakdown mechanism originates from the destabilisation of the space charge, the problem is to determine whether breakdown must be attributed to the expansion of the space charge itself or to the relaxation of the polarised lattice after charge detrapping. If breakdown is to be attributed to the sudden expansion of the space charge, it is the electromagnetic energy which has to be considered as the source of energy dissipated in the process. If, on the other hand, it is the relaxation of the polarised lattice, we have to consider the mechanical energy stored in the lattice as the cause of breakdown.

The first process can be rejected on the basis of a very simple argument. The electromagnetic energy around a point charge is given by

$$W_{em} = \frac{1}{1+\chi} E_R \frac{a_0}{a}$$

which shows that the smaller this energy, the higher the susceptibility. This suggests that breakdown should preferentially affect dielectrics of low susceptibility, which is not true at all. Since, in addition, we must consider the transfer of the electromagnetic energy to the lattice as a necessity to produce breakdown (transfer ensured by electron-phonon coupling for example), it is quite certain that the process is ineffective. It is therefore to be concluded that the second breakdown process, based on the release of a mechanical energy appears to be the only one possible.

9.3.5   Breakdown by Destabilising a Static Space Charge

The equilibrium of a space charge is very fragile. Many factors can initiate instabilities which eventually lead to the complete destabilisation of the space charge. Some possible causes of instabilities are:

(i)   a temperature increase,
(ii)   a small positive or negative increment of the external applied field,
(iii)   a mechanical stress,
(iv)   a new phase nucleation or a phase transition related to the increase of the material internal energy.

Whatever the cause of the destabilisation, the initiation of a detrapping process in a small region of the space charge produces a local electric field increase which, if strong enough, is capable of detrapping additional charges.

When the destabilisation takes place at the edge of the space charge, a propagation of the detrapping from the edge to the centre of the distribution occurs. If the velocity of propagation of the detrapping process is slowed down, the detrapped charges are trapped again in new sites surrounding the initial space charge distribution. If the destabilisation process is accelerated, all the charge distribution collapses, producing damage of the lattice when a critical energy density is released.

It is interesting to note the analogy that exists between the destabilisation process of the space charge and that of a pyramidal heap of sand. As the grains of sand are removed around the base of the pyramid, just a slight shift of sand along the slope is produced. Conversely, if the pyramid is vertically truncated close to its core, then the whole pyramid collapses. One can conclude then, that breakdown is triggered every time the space charge is rapidly destabilised.

9.3.6 Breakdown Induced by the Trapping-Detrapping of Flowing Charges

Let us consider a parallel-plate capacitor subjected to a total field that exceeds the detrapping field E_d. A steady state current regime is established in which an equal number of charges are trapped and detrapped per unit time. Each detrapped charge produces a local release of energy, and breakdown conditions are achieved when, by increasing the applied field, a critical density of energy w_c is released per unit time. Blaise [27] has shown that the breakdown field E_c which results from these conditions is given by :

$$E_c(s) \approx \frac{1}{\varepsilon s^n} + E_c(\infty) \qquad\qquad 9.18$$

which shows that E_c is inversely proportional to the permittivity ε (which is confirmed by experiment [4]) and depends on the length s of the dielectric spacer through the size factor n related to the space charge distribution. This approach stresses the importance of the trap distribution and energy stored in the vicinity of electrodes, on the field strength.

9.3.7 Is Flashover Relevant to a Dielectric Relaxation Process?

The secondary electron cascade model proposed to interpret the surface flashover of insulators is supported by the correlation between the secondary electron yield δ and the flashover strength: the higher δ, the lower the field strength. It was also observed that a magnetic field [30] improves the flashover voltage. As the yield δ and the influence of a magnetic field are not included in the breakdown process described in the previous section, we must pose the question as to whether there are two different breakdown processes: a bulk breakdown relevant to the dielectric relaxation process and a flashover relevant to the secondary electron cascade model. To resolve this issue the secondary electron emission of insulators must be properly investigated. In the Monte Carlo simulation [31] represented in Fig 9.9, the secondary electron emission due to inelastic scattering processes has been studied by taking into account the effect of the space charge. In this context, the following characteristics have been obtained:

(i)   Charges implanted with a gaussian electron beam having 10 nm diameter, spread into a very large volume whose average cross section greatly exceeds the area covered by the incoming electron beam. The smaller the trapping energy W_p, the more significant is the spreading extension of the space charge.

(ii)   A double layer of charges is obtained, consisting of (a) a positive surface layer, typically 25 nm thick, due to the emission of secondary electrons, and (b) a negative in-depth layer due to implanted charges, whose maximum density is located at a depth that is much smaller than the electron range of primary electrons in an uncharged material. The effect of this charged double layer is to increase the secondary electron yield δ in proportion to the strength of the electric field but, there is no straightforward relationship between δ, W_p and the defect distribution.

There is another electron emission process that is related to the dielectric relaxation of the lattice after charge detrapping and also leads to an increase of δ when it occurs at the surface [32]. This effect of charge detrapping on δ occurs during the expansion of the implanted charge, and can continue during irradiation to regulate the surface potential. The sensitivity to detrapping is controlled by the trapping energy W_p, where the lower is W_p, the easier in the detrapping.

Finally, it has to be pointed out the correlation between δ and the flashover field expresses nothing else but the sensitivity of these two parameters to the charging up phenomenon: the ability for a material to charge and to release its internal energy by detrapping makes δ large and the flashover field low. The conclusion therefore is that flashover is also relevant to dielectric relaxation processes as will be experimentally demonstrated later in this chapter. It is also clear that the measurement of the intrinsic electron yield of a material requires a special attention to the charging effect.

A possible explanation of the effect of a magnetic field on flashover could be as follows. Flashover experiments are usually achieved under low vacuum condition (10^{-6} torr) in which metastable states of molecules of the residual atmosphere, produced by the impact of electrons emitted from the cathode, can be de-excited on the insulator surface with generation of positive charges. Thus, we can suppose that a properly oriented magnetic field will certainly modify the surface charge density and therefore the flashover strength. Once again, it is not sufficient to argue the effect of a magnetic field on the flashover strength cannot be used as unequivocal evidence in support of an electron cascade model.

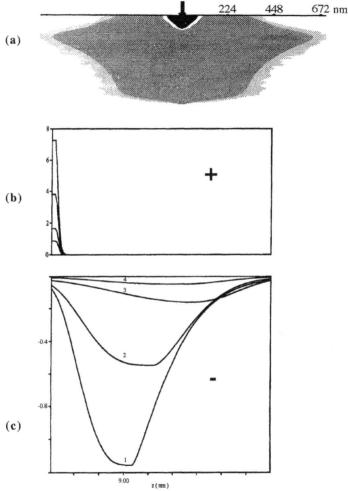

Fig 9.9 (a) A Monte Carlo simulation of the charge distribution in SiO_2. The simulation conditions are: an incident electron beam of energy 5 keV, a beam diameter of 10 nm, and a number of simulated trajectories of 10^6. (b) The variation of the positive charge density as a function of the depth along the normal to the surface. The simulation made with four different numbers of trajectories (10^4-10^6) shows that the depth of positive charges does not depend very much on the number of incident electrons. (c) In-depth distribution of negative charges obtained in the same conditions as in (b). Showing how the simulation reveals that the distribution depends very much on the dose. (After [31].)

9. 4   Applications of Space Charge Physics

In this section we shall review some of the important experimental evidence that reveals how space charge behaves when implanted in an insulating sample.

9.4.1   Space Charge Characterisation

Charge implantation and charge characterisation techniques have been developed for many very different purposes: e.g. the formation of electrets, the investigation of thermally stimulated phenomena (current, luminescence, etc.), the investigation of breakdown phenomena related to space charge formation, the investigation of charging effects related to chemical analysis and research into remedies of charge prevention during analysis. Charge trapping is considered either as a still poorly understood phenomenon by some people or, on the contrary as well understood by others. The reasons for this differing appreciation is reflected not only by the differing responses of insulators to charging, as shown in Table 9.I, but also, a lack of fundamental knowledge.

9.4.1(i)   Choice of Samples
The quality of the sample is an essential prerequisite for verifying the fundamental analysis presented in the previous sections. This is the reason why results reported here are selected from experiments conducted on well defined materials. To complement these, the charging properties of surface-contaminated samples have also been examined. This set of samples has provided the opportunity to investigate the difference of charging properties on i) crystal structure (cubic or hexagonal), ii) crystal properties (piezoelectric), iii) electronic structure (the band gap of SiO_2 is about twice that of Y_2O_3), and iv) type of defects (chemical or structural), including multiphases and internal stresses related to the presence of dopants, interfaces between materials having different permittivities and different conductivities.

9.4.1(ii)   Techniques for Charging an Insulator
Charges can be injected under the influence of an electric field (poling), by using an electrostatic charger or by irradiation with ionising beams. Poling procedures that require the deposition of metal electrodes change the insulator surface characteristics by either chemical reaction at the interface or by interatomic diffusion. Extrinsic Maxwell-Wagner domains, and defects at metal-insulator interfaces, favour the formation of space charge.

Electrostatic chargers are often used under atmospheric pressure by applying a high voltage to a tip that is facing the sample. Provided the voltage is kept below the critical breakdown value of the gas, the sample is charged to a maximum potential equal to the potential of the tip. The charge distribution depends on the sample trap characteristics and on the space charge field. Until now, no attention has been paid to the charge dynamics during implantation.

Irradiation by ionising beams has been used for a long time to form "electrets" [33]. In most cases, samples are polymer sheets, and little attention is paid to the effect of the surface contamination layers or microcracks on the charge distribution. Under electron beams of a few tens of kV, the insulator can reach a maximum potential equal to the implanting beam potential. However, on thin films (20-100 μm for example) mounted on a grounded metal support, a steady state current is often observed. In such a case the maximum potential is lower than the implanting beam potential. As before, no attention has been paid to the charge stability conditions and to the consequence of an energy release due to charge detrapping.

9.4.1(iii)   Charge Detection.

Since it has been recognised that breakdown is related to the formation of a space charge, many space charge characterisation techniques have been developed. Electro-optic methods [34-35] are based on the birefringence of high voltage stressed materials. They are powerful methods requiring transparent insulators.

Charge Distribution Analysis (CDA) [36] is a contact free technique, which consists of measuring the force acting on a dielectric in a static electric field gradient. CDA allows the measurement of charge sign and density, electric field and carrier mobility. CDA appears to be a powerful and promising method for detecting changes in the permittivity of surface layers and large space charge field variations related to the occurrence of phase transitions.

Charge detection by the scanning capacitance microscope [37] is based on the same principle as CDA, and combines the capabilities facilities of the atom dorce microscope technique, allowing the observation of charge distribution with both high sensitivity and high spatial lateral resolution.

Measurements of induced charge variations produced on a reference electrode by the propagation of a thermal wave [38-39] or by a mechanical pressure wave [40], are the methods most commonly used to investigate insulator charge distribution. These methods have some disadvantages, related to heat diffusion and sound velocity; thus, deconvolution techniques are needed to determine the charge distribution. Another limitation of these techniques is the available spatial resolution: there is no possible lateral resolution, and the resolution in the direction of the propagating wave is in the range of micrometer.

In most techniques, the space charge field is deduced from a measurement of the total charge distribution, ignoring the nature of charges (electronic, ionic, polarisation charges, etc.) and their origin (injected from electrodes, produced by electrodiffusion or electrochemistry).

Therefore, in the absence of a physical guideline the interpretation of experimental data is very complex. However, one attempt has been made to correlate the density of charge/cm^2 injected by poling, as a function of the applied field. The variation has sometimes been found linear, the slope having the dimension of a permittivity [41]. However, a comparison of these experimental data with the insulator permittivity has not been achieved. Generally speaking no pertinent space charge parameters and no one-to-one correlation of the space charge characteristics with the breakdown field have been demonstrated. It seems that not enough attention has been paid to the sequence of experimental procedures: charge injection, charge propagation and energy relaxation. Due to the overlapping of these three steps, there is no further possibility of extracting from the experimental results the specific parameters associated with each step.

9.4.2 The Mirror Method for Space Charge Characterisation

An electrostatic method called the Mirror Method (MM) has recently been developed in an SEM [42]. The method is simple and samples of different shapes and sizes are acceptable; furthermore, the sample does not need electrode deposition. It is charged and examined under vacuum or under ultra high vacuum conditions, and can be analysed at a variable temperature.

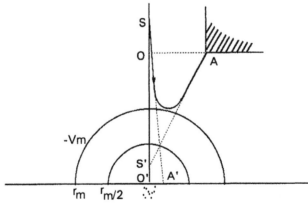

Implanted
Charge

Fig. 9.10   The "Mirror effect" due to negative charge being implanted by a high energy eV_0 electron beam, and observed with an electron beam of lower energy eV_m. Equipotential $-V_m$, formed by the implanted charge, acts on the electron beam as a convex electrostatic mirror. In the gaussian approximation, the mirror gives a virtual image S' of the objective lens aperture S. S' appears as the virtual electron source forming the image of the microscope chamber. For a quasi-punctual implanted charge, the radius r_m of equipotential $-V_m$ is deduced from the objective lens image by the relation $r_m = 4O'A'.O'S'/OA$.

Moreover the method allows the space charge formation to be linked to the secondary electron yield and therefore, offers a way to link space charge characteristics with breakdown parameters. The principle of the technique is to use the electron beam of an SEM at different energies to first charge the sample, and then to work with it as an electrostatic probe to measure the potential distribution around the charge. The electron source of the SEM is at a variable negative potential denoted -V (V > 0), and negative charges are implanted by using a beam of high energy $eV_0 \approx 30$ keV. The charged sample is then scanned with the same electron beam after lowering its energy to $eV_m < eV_0$. In the gaussian approximation, equipotential $-V_m/2$ produced by the implanted charges constitutes a convex electrostatic mirror which, as shown in Fig 9.10, forms a virtual image source S' of the objective lens aperture S. This is why the method is called "Mirror Method" (MM).

In the gaussian approximation, a simple relationship exists between the radius r_m of an equipotential V_m measured along the optical axis of the incident beam, the objective aperture size $\phi_{objective}$, the size ϕ_{image} of the image, the working distance L of the microscope and the image magnification G, namely

$$\frac{1}{r_m} = G \frac{\Phi_{objective}}{\Phi_{image}} \frac{1}{4L} \qquad\qquad 9.19$$

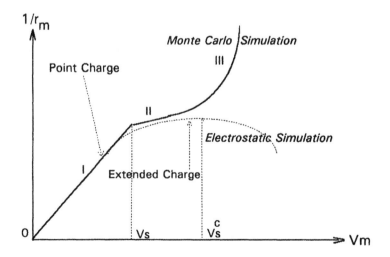

Fig 9.11　　Plot of $1/rm = f (V_m)$. Due to the double layer, charge recombination or charge spreading occurs when the potential reading beam reaches a critical value V_s^c.

The space charge characterisation is deduced from a plot of $1/r_m = f(V_m)$. Such a plot is represented in Fig 9.11 corresponding to the space charge simulation represented in Fig 9.9. The origin of distance r_m is taken at the centre of the negative charge distribution. Three domains of this plot have to be considered:

(i)   Outside the sample (part 1), there is a linear variation of V_m as a function of $1/r_m$ having the form

$$V_m = A_l \frac{Q_i}{r_m}$$ 9.20

Here, Q_i is the implanted charge and A_l the inverse of an "effective dielectric constant" having the form:

$$A_l = A_\infty K(l)$$

where

$$A_\infty = \frac{2}{4\pi\varepsilon_0(\varepsilon_r + 1)}$$

and $K(l)$ is a function of the thickness l of the sample [43] $(K(\infty) \to 1)$.

The implanted charge Q_i is deduced from the difference between the incident charge Q_0, the backscattered charges Q_η and the charge Q_δ of secondary electrons, that is

$$Q_i = Q_0 - Q_\eta - Q_\delta$$

In practice, Q_η is very low for materials of low atomic numbers and will be neglected here, so that the secondary yield δ is:

$$\delta = \frac{Q_\delta}{Q_0} \approx 1 - \frac{Q_i}{Q_0}$$ 9.21

From the plot of equation 9.20 we can deduce the slope

$$\Theta = \frac{1}{A_l Q_i}$$ 9.22

which after substitution in equation 9.21 gives

$$\delta = 1 - \frac{4\pi\varepsilon_0(\varepsilon_r + 1)}{2K(l)\Theta Q_0} \qquad\qquad 9.23$$

(ii)   When the electron beam touches the surface and penetrates slightly into the sample (part II), a change in the slope of the linear variation is observed, due to a change in the constant A_∞. The point of rupture of the slope corresponds to the value of the surface potential V_s.

(iii)   When the electron beam becomes closer and closer to the centroid of the space charge, a Monte Carlo investigation shows that the potential inside the insulator follows part III of the full curve.

The first two domains of variation are well observed experimentally. Of course, the investigation of the potential inside the sample is restricted to the surface layers because when the electron beam penetrates into the sample it is scattered and new positive charges are produced on the surface. As a consequence the field in the double layer increases until it reaches a critical value E_s^c large enough to detrap the negative charges implanted deeply at ≈ 30 keV   Spreading and recombination of charge occur, producing the dashed curve represented in Fig 9.11 beyond the critical potential V_s^c which corresponds to E_s^c. In fact, E_s^c characterises the detrapping field which is related to the trap energy W_p: the higher W_p, the higher E_s^c.

The mirror method has also been applied to investigate space charge dynamics [44]. Work is still in progress but the preliminary results have shown that the size factors deduced from experiment and calculated in the geometry of the mirror experiment agree very well with technological results. The size factor is found to be close to 0.5, but this is not a universal value because it depends both on the permittivity and on the spatial distribution of traps.

Note: the spatial effect of the distribution of the traps on the size factor is also discussed in reference [1] page 500, Fig 5-41.

The coupling of the mirror method with calorimetric experiments [7] gives access to the polarisation energy U_p as a function of the charge concentration, and to u_c the critical energy density released to produce bond breaking and indirectly to the measurement of the trap energy W_p. It is proposed that these three parameters are deterministic of the breakdown strength.

9.4.3 The Mirror Method for Verifying that Flashover is Triggered by Charge Detrapping

The description of the breakdown process proposed in Section 9.3 involves the propagation of two types of waves:

(i) an electric wave associated with the detrapping of the space charge,

(ii) a thermal shock wave attributed to a many-body relaxation process of the polarisation mechanical energy stored in the medium.

The second wave, which must occur a few nanoseconds after the first one, initiates a plasma whose expansion is accompanied by ion, electron and photon emissions. This interpretation is well supported by time-resolved experiments showing that photons begin to be emitted a few nanoseconds before the complete development of breakdown. It is also in agreement with bulk breakdown experiments showing that the zone where the plasma is formed is preceded by a zone where the voltage drops. Finally we should note that the plasma formed at the surface of the insulator can expand in vacuum to trigger a vacuum breakdown event.

Fig 9.12   Treeings formed in contamination layers are characterised by a low contrast of the secondary electron image obtained in a scanning electron microscope.

We have seen in Section 9.3, that the flashover process can be broken into the three steps (i) charging, (ii) rapid destabilisation of the space charge,

(iii) relaxation of the mechanical energy. The mirror method offers the possibility of very carefully controlling these various steps and to verify the various aspects of flashover process in relation to different types of charge distributions [19,32].

When the sample is covered with contamination layers, the flashover propagates in these layers and forms the type of surface treeings shown in Fig 9.12 [45]. An experiment conducted in a scanning Auger microscope has allowed us to show that the number of atoms evaporated during detrapping [19] is in good agreement with the theoretical prediction derived from equation 9.15.

In the absence of surface contamination (or in absence of grain boundary segregation) charges are distributed below the surface and as seen in Fig 9.13(a) and (b), the flashover produces thermal etched pits due to outgassing of gas embedded in the material, the melting of the sample, treeings and mechanical fractures inside or outside the plasma tracks [19].

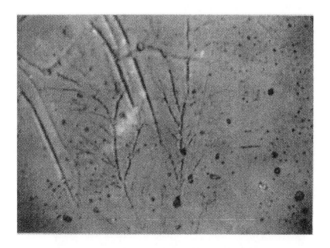

Fig 9.13(a)   An optical micrograph showing treeing tracks left by flashover on the surface of a freshly annealed (1200°C) stoichiometric Y_2O_3 polycrystalline sample.   The sample was charged with a 30 kV electron beam and breakdown was produced at 3 kV.

For polycrystalline ceramics with grain boundary segregation, charges are preferentially trapped at interfaces [19] and detrapping can sometimes produce grain ejection as shown in Fig 9.14 which typically looks like punch-through, and sometimes a combination of grain ejection and surface treeing tracks, as shown in Fig 9.15. The mechanism, and examples of punchthrough have also been proposed by others [46].

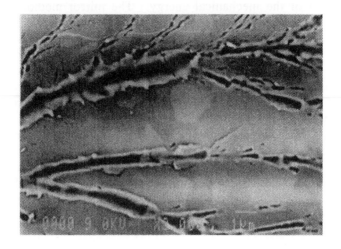

Fig 9.13(b) A scanning electron micrograph treeing tracks left by flashover on the surface of a freshly annealed (1200°C) Stoichiometric Y_2O_3 polycrystalline sample. The sample was charged with a 30 kV electron beam and breakdown was produced at 3 kV. Scanning electron micrograph.

Fig 9.14 Damage left by breakdown of a Y_2O_3 polycrystalline sample stored for three months under air at room temperature. Two grains are pulled away. Charges are trapped in grain boundaries, where segregation of impurities (Ca, K, C) had occurred during the ageing of the sample.

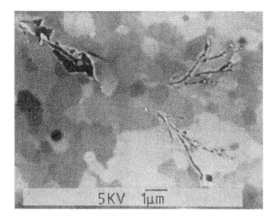

Fig 9.15   Damage left by breakdown of a Y_2O_3 polycrystalline sample stored for three months under air at room temperature. The result of the breakdown is both to pull grains away and to form surface treeing tracks. Also, it is likely that the breakdown was partly developed beneath the surface without producing apparent damage.

9. 5   Applications of the Mirror Method to Technological Problems

In this section, we shall illustrate how the mirror method described above can be used to investigate technological problems.

9.5.1   Surface Finishing Effect on Alumina Flashover Strength

The mirror method has been applied to demonstrate the effect of machining on the trapping properties of alumina. Thus, machining generates defects (F^+ centres) which have a trap energy a little higher than that of intrinsic defects. However, it is probable that in relation to the insulator structural modification which occurs under electrical stress (F^+ centres transform in F^0 centres [47]) machined samples have a lower breakdown strength than high temperature annealed samples.

9.5.2   Surface Contamination Effect on the Flashover Strength

Contamination layers on alumina, due either to machining or to exposure to atmosphere (hydroxide and carbonated layers) form Maxwell-Wagner domains where charges are preferentially trapped. Such layers are decomposed at a temperature higher than 400°C, which explains the effect of vacuum tube outgassing on the flashover strength.

363

9.5.3   Effect of Doping on the Ceramic Flashover Strength

Dopants can produce high energy traps depending on their electronegativity and size. For example, a low concentration of transition metal dopants in single crystal alumina increases w_p by about a factor of two. Also, $\delta(30 \text{ keV})$ is decreased by doping [48].

9.5.4   Flashover on Metal Covered with a Contaminated Dielectric Layer

The interpretation of breakdown on the basis of space charge relaxation also finds an application in vacuum breakdown and its relation to electrode contamination. In this case, the electrode is covered with a dielectric film where charges can be trapped. In the following example (Fig 9.16) the electrode was a copper-zinc alloy where the anode was bombarded with very high energy electrons and very high current (1000 A). Treeings are formed during high power pulses [49].

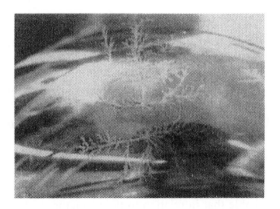

Fig. 9.16.   An optical microscope image of the anode of a high power pulse generator. Published by courtesy of M. Devin, Centre d'Essais Techniques d'Aquitaine, Commissariat à l'Energie Atomique, Le Barp, France.

9.5.5   Effect of Ionising Radiation on the Flashover Strength

Finally, it should be noted that glass and alumina subjected to X rays or γ radiation, have shown a lowering of w_p [50]. Also, it is known that any ionising radiation lowers the breakdown strength [51].

9. 6   Conclusions

Experiments have confirmed the existence of a breakdown process in which the energy is supplied by the charged medium itself, rather than by any external energy source. This central feature is not relevant to classical theories of breakdown proposed up to now, which generally attribute the motive force to the action of an external applied field on electrons.

The new interpretation of breakdown is forged on the idea that the movement of space charge in a dielectric medium is accompanied by a mechanical relaxation of the lattice. Under certain circumstances this, produces a local heating that is high enough to melt or to sublimate the solid.

The first fundamental point to be discussed regarding this process was to elucidate the nature of defects acting as trapping centres. In place of the classical interpretation of trapping based on levels located in the band gap, we have substituted an approach based on the polarisable property of the material. The concept of defects, related to local variations of the electronic polarisability, gives a universal explanation of charge trapping in pure and crystallised oxides, as well as in any heterogeneous and chemically complex material. This description of trapping is not in conflict with the classical trapping model, but is different in the sense that polarisable defect trapping centres are independent of the band structure and are not specific to the species of charge carrier.

The second point of importance in the description of the process is the introduction of the mechanical energy that is related to the polarisation produced by trapped charge. In this respect, a clear distinction is made between electromagnetic and mechanical energies. Both are a consequence of polarisation, but the energy source of breakdown lies in the relaxation of the mechanical energy.

The third point deals with the relaxation mechanism of the stored mechanical energy. It is likely that the relaxation occurs via a many-body process, but its theoretical description is still a problem requiring much further work. However, from a qualitative point of view, the mechanism is described as follows. The polarisation produced by the trapping of a charge involves mechanical energy to be "pumped" to the medium in a quasi-reversible way, whereas the energy released during the detrapping process is returned to the medium in an irreversible way, producing a local heating.

Having addressed, in the Introduction, the question of whether space charge is the cause or the consequence of breakdown, experiments reported here demonstrate with no ambiguity, that a breakdown process exists whose cause is the presence of space charge. Thus, it is proposed that the primary cause of breakdown in dielectrics subjected to an electric field applied between the electrodes, can be attributed to the formation of a space charge. From this perspective, the applied field is seen as just a possible cause of space charge build up or of space charge destabilisation, i.e. it is not seen as the driving force, as has been proposed up to now. Furthermore, any particle emission from insulators subjected to electric or mechanical strains can be interpreted as a consequence of the relaxation of the mechanical energy stored in the charged medium.

In the course of this work it has also been found that the secondary electron emission from the insulator may be a complex process. The measurement of the yield requires careful experiments and there is not a simple relationship between the yield and the breakdown strength.

In conclusion, the relaxation of the mechanical energy in a charged medium allows one to give a consistent description of a number of phenomena observed in dielectrics. Thus, it is proposed that studies of the properties of dielectrics are broader based, to include studies of the charge mechanisms involved in electromagnetism, fracture mechanics, and tribology.

9.7   References.

1. Nelson, JK., *In* "Engineering Dielectrics", Vol. II-a, 445-520, Bartnikas/Eichhorn Editors, ASTM STP 783, 1990.
2. Blaise, G., Interdisciplinary Conf. on Dielectrics, Supl. Revue "Le Vide, les Couches Minces", 260, 1-27, 1992.
3. Blaise, G., Le Gressus, C., Conf. on Electr. Insul. and Dielectric Phenomena, CEIDP 1990, IEEE Service Centre, 445 Hoes Lane, Piscataway, N.J., 08854.
4. Miller, HC., Interdisciplinary Conf. on Dielectrics, Supl. Revue "Le Vide, les Couches Minces", 260, 250, 1992.
5. Pauwels, JC., *XV-ISDEIV*, 94, Moscow, Russia.
6. Boersch, H., Hamisch, H., Erhlich, W., *Angew, Z. Physik*, **15**, 518, 1963.
7. Moya-Siesse, D., Moya, G., Le Gressus, C., Proc. CEIDP, 1993.
8. Eberle, G., Schmidt, H., W. Eisenmenger, Proc., 1993.
9. Menzel, D., Gomer, R., *J. Chem. Phys.*, **40**, 1164. 1964.
10. Knotek, ML., Feibelman, PJ., *Phys. Rev. Letters*, **40**, 964, 1978.
11. Jonscher, AK., Dielectric Relaxation in Solids, Chelsea Dielectrics, London, 1983.
12. Hagstrum, HD., *Phys. Rev. Letters*, **43**, 1050-3, 1979.
13. O'Dwyer, JJ., The Theory of Electrical Conduction and Breakdown in Solid Dielectrics, Clarendon Press, Oxford, 235-292, 1973.
14. Bommakanti, RG., Sudarshan, TS., *J. Appl. Phys*,. **66**, 2091-8, 1989.
15. di Maria, D, J., Arnold, D., Cartier, E., *Phys. Rev.*, **B 45**, 1991.
16. Bommakanti, R.G., Sudarshan, TS., *J. Appl. Phys*.., **71**, 2181-90, 1992.
17. Jardin, C., Durupt, P., Robert, D., Le Gressus, C., Proc. CEIDP, 1993.

18. Will, FG., de Lorenzi, HG., Janora, KH., *J. Am. Ceram. Soc.* **75**(é), 295-304, 1992.
19. Le Gressus, C., Blaise, G., *IEEE Trans. Electr. Insul.*, **27**, 472-479, 1992.
20. Austin, IG., Mott, NF., *Advances in Physics*, **18**, 71, 41-102, 1969.
21. Fröhlich, H., *In*, "Polarons and Excitons", Ed. C.G. Kuper and G.D. Whitfield, Oliver and Boyd, Edinburgh and London, 1-44, 1963.
22. Blaise, G., *IEEE Trans. Elec. Insul.*, **28**, (4), 437, 1993.
23. Landau, L., Lifchitz, E., Electrodynamique des Milieux Continus, Ed. Mir, Moscou, 76-79, 1969.
24. Scaife, BKP., "Principles of Dielectrics", Clarendon Press, Oxford, 178, 1989.
25. Coudray, C., Private Communication.
26. Le Gressus, C., Blaise, G., Insulator Damage Related to Charge Detrapping, Proc. CEIDP, 1993.
27. Blaise, G., Proc. CEIDP, 1993.
28. Sessler, GM., Proc. CEIDP, 1992.
29. Coaker, BM., Xu, NS., Jones, F., Latham., RV., Proc. XV, ISDEIV 1993.
30. Hergeler, F., Masten, G., Krompholz, H., Hatfield, L.L., Proc. CEIDP, 1992.
31. Vicario, E., Rosenberg, N., Renoud, R., Proc. CEIDP, 1993.
32. Blaise, G., Le Gressus, G., *C.R. Acad. Sci. Paris*, **t.314**, Série II, 1017-1024, 1992.
33. Sessler, GM., *In*, "Electrets", 2nd Edition, Ed. G.M. Sessler, Topics in Applied Physics, Springer Verlag, 33, 1987.
34. Zahn, M., Hikita, M., Wright, KA., Cooke, CM., Brennam, J., Kerr *IEEE Trans. on Elec. Insul.*, **EI-22(2)**, 181-185, April 1987, EI-**23(5)**, 861-885, October, 1988.
35. Kawasaki, T., Arai, Y., Takada, T., Proc. CEIDP, 1993.
36. Freund, MM., Freund, F., Battlo, F., Highly Mobile Oxygen Holes in Magnesium Oxide. *Phys. Rev. Letters*, **63**, 2096-2099,1989.
37. Barret, RC., Quate, CF., *J. Appl. Phys.*, **70**,(5), 2725-2733, 1, 1991.
38. de Reggi, AS., Guttman, CM., Mopsik, I., Davis, GT., Broadhurst, M.G., *Phys. Rev. Letters*, **40**, 413, 1978.
39. Toureille, A., Reboul, JP., Merle, P., *J. Phys. III*, **1**, 111-123, 1991.
40. Laurenceau, P., Dreyfus,G., Lewiner, G., *J Phys. Lett.*, **38**, 46, 1977.
41. de Reggi, AS., Interdisciplinary Conf. on Dielectrics, Supl. Revue "Le Vide, les Couches Minces", **260**, 260-8, 1992.
42. Le Gressus, C., Blaise, G., *J. of Electron   Spectroscopy and Related Phenomena*, **59**, 73-76, 1992.
43. Vallayer, B., in press.

44. Oh, KH.; Ong, CK.; Tan, BTG., Le Gressus, C.; Blaise, G. *J. of App Phys.*, **74**, (3), 1960-7. 15, 1993

45. Henriot, F., Gautier, M., Duraud, JP., Le Gressus, C.Valin, TS. Sudarshan, Bommakanti, RG., Blaise, G., *J. of App.Phys.*, **69**,   no.9, 6325-33, 1991.

46. Shea, JJ., Proc. CEIDP, 1990/

47. Saito, Y., Michizono, S., Anami, S., Kobayashi, S., Flashover on RF Alumina Wundows for High Power Use, *IEEE Trans. on Elec. Insul.*, **28**, (4), 566, 1993.

48. Vallayer, B., in press.

49. Devin, CESTA, Le Barp, F. Private communication.

50. Berrough, A., Fayeulles, S., Hamzaoui, B., Tréheux, D., Le Gressus, C., *IEEE Trans. Elec. Insul.*, **28**, (4), 528, 1993.

51. Kalbreier, W., Goddard, B., Radiation Triggered Phenomena in High Energy e^+e^- colliders, *IEEE Trans. Elec. Insul.*, **28**, (4), 444-56, 1993.

10

Pulsed Power

A Maitland

10.1 Introduction

10.2 Breakdown-Initiating Processes

10.3 Pulsed Power Devices and Technologies

10.4 References

10. 1   Introduction

The end of the cold war brought with it intense pressure to turn swords into plough shares. How do we get a peace dividend from the enormous progress that has been made in pulse power technology developed to meet defence specifications over the last 40 years or so? How do we exploit the opportunity for technology transfer from defence to industrial/commercial applications? The answers to these questions are not easy to find and they motivate much of the current research and development effort in pulse power technology. Commercial applications require levels of cost effectiveness, reliability and specifications as demanding as any of those established in defence applications. Indeed, many envisaged commercial applications require even greater powers and energies to be delivered to the load than has been the case in defence. The requirements of particle accelerators for higher and higher energies at lower and lower costs also push hard for greater efficiencies and effectiveness of the pulse power technologies used. Shift from National Laboratories and Defence Research Laboratories to Industrial Research Laboratories inevitably enlarges the pool of pulse power users and brings the necessity for a wider and deeper understanding of pulse power engineering and the processes of the transfer from vacuum insulation to conduction. Newcomers to pulse power engineering may find R J Adler's 1989 "Pulse Power Formulary" useful [1].

10.1.1   General Concepts

Pulse power is a subset of power electronics in which the terms *pulsed power* and *pulse power* are synonymous, as are the terms repetitive rate *pulse power* and *power modulator*. Hence, the principal sources of theory, technology and applications are the published volumes of the biennial series of the IEEE Pulsed Power Conferences and the IEEE Power Modulator Symposia which are held in alternate years. Other valuable sources are the many published proceedings of Accelerator Conferences sponsored by IEEE and others from time to time.

Pulsed power systems start with an energy source and end at an interface with a load (laser, microwave, radio frequency, x-ray, particle beam) which is where vacuum breakdown problems generally arise. Thus, pulse power is not an end in itself, it is an "enabling technology" in that it enables particular loads to be exploited. Many applications require the load to be pulsed for optimum operation, indeed many lasers will only operate at all if they are excited by pulses.

The purpose of a pulse power source is usually to produce very high powers at very high voltages by using a source of much lower voltage and power. To do this, the energy to be delivered by a pulse must be stored over a time which is much longer than the duration of the pulse. If the energy is

370

stored capacitively, the switch used is a "closing" switch (this is the most common); if the energy is stored inductively, the switch must be suitable to act as an "opening" switch. To control the pulse repetition rate and the timing of the transfer of energy from the "store" to the load, a switch is necessary; thus, switches feature as key components in all pulse power systems. From an operational point of view, Levy et al [2] identify four advantages of pulsed power over DC power.

(i) Nonlinearity
High peak powers can drive the normally linear response of a material into a strongly nonlinear regime, which often leads to saturation, such as with the ferromagnetic materials used in magnetic switching. Combined non-linear dielectric and magnetic effects are exploited in lines used for pulse compression which aim to decrease the rise-time of a propagating pulse.

(ii) Time isolation
Pulses with a rise-time shorter than the response times of charge carriers can establish conditions which are independent of charge carrier mobility (but for a short time only); for example, intense electron beams are often established, and produce their desired effects before the plasma which leads to gap closure develops in the electron gun. A classic case occurs in the laser process which enables an eyeball to be "machined" to correct for myopia, etc. For this process, the pulsed laser radiation is of such power that each pulse evaporates, or ablates a surface layer so quickly that there is no heat transferred to the underlayer to cause "burn" damage.

(iii) Efficiency
A high power modulator can have higher efficiency than a DC power supply of the same average power. This is largely due to the nonlinear properties of most pulsed power switches, gaseous or solid-state. Relative switch losses are less as the power per pulse is increased, so switch losses are less for the same average power.

(iv) Adiabatic Heating
The basis of this advantage lies in the relatively slow thermal propagation velocities. Thus, the advantages of high peak powers may be enjoyed without the penalties created by thermal effects if low duty cycles are used. Without careful thermal management, the size of a high power unit can be prohibitive. In practice, the usual size limiting factor for pulsed systems is insulator flashover, rather than overheating.

10.1.2   Pulsed Power and Vacuum Breakdown

Components with which pulsed power systems are implemented include the following: capacitors, pulse transformers, inductors, resistances, diodes, varistors. For each of these devices, the problem of vacuum breakdown can severely limit performance. The all-important switches include thyratrons (the principal work-horse), ignitrons (fast being replaced by solid state devices except at the very highest energies), spark gaps of one form or another, crossatrons, pseudospark devices, backlit-thyratrons (BLTs), magnetic switches based on the saturation of ferromagnetics, and many other new solid state and optically triggered switches. Vacuum switches and triggered vacuum gaps continue to give sterling service in routine and esoteric applications. Vacuum switches and some of the new solid state switches can act as both closing and opening switches. The "product" of a pulsed power system is generally a plasma, a particle beam, a microwave beam, or electromagnetic radiation.

The properties of vacuum-surface combinations, as they occur in the above range of devices, are sufficiently diverse to make them exploitable in both insulating and conducting contexts. Thus, the same processes, regarded as breakdown in one case, may be cultivated deliberately in another. For example, field emission processes must be suppressed when vacuum is required to be an insulator, but much effort has been devoted to ways of enhancing field emission when a high power electron beam is required. Many pulse power applications rely on the ease and speed with which a transition from the insulating state to the conducting state can be induced, and vice versa. Included in these applications we have vacuum switches, triggered vacuum gaps, field emitting vacuum electronic devices, and field switched ferroelectric cathodes.

10.1.3   Applications of Pulsed Power

The principal thrusts of pulsed power during the past 20 years or so have been associated with the simulation of nuclear generated EMP, and the development of accelerators and pulsed radar. Significantly, the technology and information processing developed for pulsed radars is now exploited in such commercial systems as Computer Aided Tomography (CAT) scanners. Indeed, a comprehensive review of commercial applications for modulators and pulse power technology has been given by Levy et al. [2]. Other areas of research and development of applications for pulse power include treatment of flue gas, disinfestation of food, food preservation and processing, surface processing, metal forming, joining and cutting, sterilisation of infectious waste, toxic waste processing and other matters of environmental clean-up, and a wide range of particle and photon based radiation therapies for medical purposes. Several companies already have commercial products in these areas. In assessing the viability of a given process in a commercial context, it is found to be useful to place the process in one of the following categories: A-Commercially used (Established); B-Proof of principal obtained (verified, but not commercially developed); C - New ideas which do not violate the laws of physics (possible).

To make the next generation of microchips even smaller and faster, X-ray photolithography, based on intense pulses of X-radiation, is being developed for commercialisation. With their status of enabling technologies, "pulse power" and vacuum insulator properties will play an increasingly crucial role in improving and maintaining our standard of life.

Pulsed power systems of the highest energies and powers operate in accelerator and beam fusion programmes which respectively aim to produce beams at 10^{12} eV and 10^{14} W. A general system for particle beam fusion, Fig 10.1, described in a review by Toepfer [3], consists of a Marx generator, a pulse forming network followed by a magnetically insulated vacuum transmission line to a load which may be an e-beam, an ion-beam, or an imploder to deliver the power/energy to the target. The so-called vacuum transmission line usually consists of a coaxial, water-insulated transmission line which leads to a solid transition region and thence to a vacuum diode.

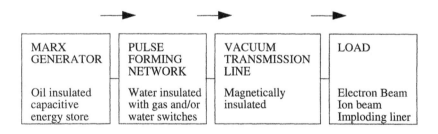

Fig 10.1   Block diagram illustrating the general requirements for a particle beam fusion driver system.

As an example of how pulse power technology may be implemented in the design of a high power microwave facility with minimum microwave breakdown problems, we cite some recent work by Moreland et al. [4]. They describe an investigation of non-uniform-amplitude backward wave oscillators (BWOs) in which a high current, high power, relativistically repetitively-pulsed electron beam accelerator (Sinus-6) injects an e-beam into a variety of BWO structures over a wide range of input parameters. Sinus-6 starts with a Tesla transformer and pulse forming line which steps up the voltage from 300 V to 700k V. A nitrogen spark gap (18 atmos) then switches the voltage to an oil-filled adiabatic line which matches the 22 ohm input to a 100 ohm output consisting of a magnetically insulated coaxial vacuum diode with a cold explosive-emission graphite cathode. The emitted electron beam is confined by a magnetic field of 2.6 T produced by a pulsed solenoid system. The e-beam produced is 12 nsec FWHM at a current up to 5 kA. The tube is overmoded and the extraction of power through the volume mode minimises microwave breakdown problems. The above exemplifies a general method whereby an accelerator injects an e-beam into the microwave system to be investigated.

10. 2   Breakdown-Initiating Processes

The technological problems associated with insulating high voltages or fields in high frequency (HF) or pulsed field (PF) devices feature prominently in this text. Thus, Chapter 11 focuses on the design criteria of travelling wave tubes (TWTs), whilst Chapter 12 discusses the operational requirements of superconducting RF cavities. In this section, consideration will be given to a range of breakdown initiating processes that arise to a greater or lesser extent in all such HF and PF devices.

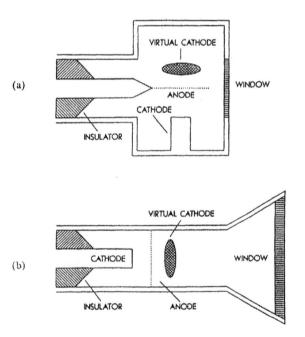

Fig 10.2   Geometries of two principal types of virtual oscillator: (a) reflex triode. (b) vircator with axial-extraction. (After Benford and Swegle [5].)

An ideal vehicle for focusing this discussion is the "vircator", or virtual cathode oscillator, since this simple type of microwave tube may be used to visualise the fundamental generic phenomena to be described in the following paragraphs.

The vircator is a device which exploits space charge effects. Thus, referring to Fig 10.2, when an electron beam is injected into a drift space and its current exceeds the space charge limited current appropriate in the drift space, a so-called "virtual cathode" develops [5]. As the beam travels away from the anode through which is has passed, it loses kinetic energy and gains electrostatic potential energy so that, ultimately, the beam is reflected by the virtual cathode and an oscillation of charge occurs within the potential well developed between the virtual cathode and the anode which transmitted the beam. The oscillation of charge generates microwaves of the same frequency, given by

$$\omega_p = [4\pi n_{e\text{-}beam}\, e^2/\gamma m]^{1/2} \qquad\qquad 10.1$$

where $n_{e\text{-}beam}$ is the electron density of the beam at the point of injection, γ, is the relativistic factor at injection, and m is the electron mass. It be further noted that the microwave emission has also been explained by Mahaffey et al [6] in terms of phase bunching of electrons reflecting inside the self-potential well.

10.2.1 Gap Closure

Ideally, in most anode-cathode vacuum devices, the emission of particles should be confined to the anode and cathode surfaces; all other surfaces should be non-emissive. However, even in these ideal conditions, we find that the anode-cathode conduction processes can leave their desired parameter space and develop new characteristics which can have disastrous consequences. Such developments are inherently time dependent as, for example, is the case of gap closure which occurs in relativistic magnetrons and other microwave devices when plasma from either or both electrodes expands into the interelectrode gap. In a relativistic magnetron, the duration of the microwave pulse is limited by gap closure. Typical velocities of closure are about 5 cm/μsec. A form of the Child-Langmuir equation for electrons given by Toepfer [3] is

$$I/A = 2.34 \times 10^3\, V^{3/2} / d^2 \qquad\qquad 10.2$$

where V is in megavolts, d is the geometrical interelectrode gap in centimetres, I is the current in amperes and A is the cathode area. The ratio $I/V^{3/2}$ is known as the perveance, P of an electron gun of given geometry

A/d^2. If A and d remain constant, then P is a fixed property of the gun. However, if d or A change, then the value of P changes. When plasma forms in a vacuum breakdown process, it expands into the gap (gap closure) with velocity v_{exp}, so that the effective gap d_{eff} at time t after the plasma formation by the explosion of a field emitting point (say), becomes

$$d_{eff} = d - v_{exp}t \qquad\qquad 10.3$$

For this case, d_{eff} should replace d in equation 10.1 and the perveance becomes a function of time. Parker et al. [7] have measured the perveance of an electron gun at times covering the range 20-100 nsec and produced the data shown in Fig 10.3.

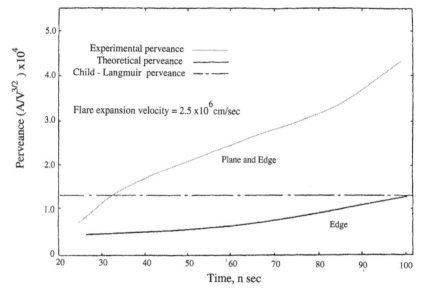

Fig 10.3   Perveance comparison for the diode formed by positioning a 5.08 cm-diam graphite cathode a distance of 6.05 mm from a planar anode (After Parker et al [7].)

The gun consisted of a 5.08-cm-diameter graphite cathode located 6.05 mm from a planer diode. They deduced that the plasma velocities were 2-3 cm/μsec. For the interested reader, the properties of electron guns used in microwave devices are further discussed in Chapter 11.

10.2.2   Explosive Emission

As noted by Parker et al. [7], and later by Mesyats [8], pulsed vacuum discharge phenomena, and possible gap closure, develop in the diodes of high current electron accelerators. As previously discussed in Chapter 6, the basic process of explosive electron emission is where Joule heating of an emitting micropoint produces a blob of plasma at the exploding point. As the field emission process at a cathode point develops to explosion of the point and plasma formation, the current moves from pure field emission to a space charge limited flow governed by the Child-Langmuir equation. However, the ions produced, together with those from the anode or elsewhere, act to reduce the space charge in the gap, and hence a greater current, which further heats the electrodes to cause yet more plasma with its ions; i.e. a "runaway" situation is created. A potential difference of several tens of volts then develops over the Debye length between the metal of the cathode emitting site and the plasma. The resulting electric field is sufficient to produce further intense electron emission from the metal to the plasma.

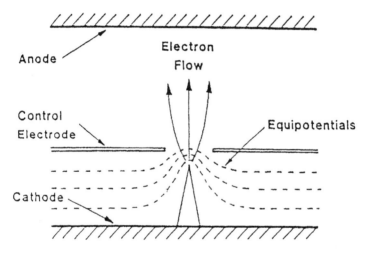

Fig 10.4   Illustration of field enhancement and electron emission from a needle-shaped emitter.

Whilst each explosive emission site, with its plasma, is a source of an intense electron beam, the emission sites are not uniformly distributed, so that the individual beamlets together form a somewhat non-uniform total electron beam. However, as will be discussed in a later section, the exponential increase of the field emission current with applied voltage, coupled with the capability of achieving current densities up to about 10^8 amp/cm^2 from a single point emitter, makes an array of such emitters a very attractive proposition for cathodes in large e-beam systems, including accelerators and microwave tubes. A major effort is also being made to exploit this type of emitter in microelectronics applications, based on the principle shown in Fig 10.4. It should be noted however that the review by Toepfer concludes that high current electron beams appear to be limited to about 10 TW/cm^2.

A diode with a plane cathode and a plane anode separated by distance d has a space charge limited current given by the Child-Langmuir relation of equation 10.2. However, the assumptions on which the equation is based include no account of either the expanding plasma from the explosive emission process or the non-uniform distribution of emission sites. To allow for the expanding plasma (so-called "gap closure"), d is replaced by d - vt, where v is the expansion velocity and t is the elapsed time from the initial formation of the plasma. With regard to non-uniformity of emission site distribution, it has been shown by Mesyats et al. [8] that, when an emission site is formed, the electric field at the cathode surface is rapidly decreased in the neighbourhood of that emission centre; this therefore has the effect that micropoints close to the emitting site do not explode to form another site. This so-called "screening effect" is attributed to the space charge of the electron beamlet from the emission site. In this context the non-relativistic equation for the screening radius r_s due to a beamlet current i, is given by Mesyats [8] as

$$r_s = 5 \times 10^2 \ i^{1/2} \ V^{-3/4} \ d \qquad\qquad 10.4$$

where i, d, and V are, respectively, in amperes, centimetres and volts. The relativistic version of this equation is

$$r_s = [8mcd^2i \ (\gamma^2 - 1)^{1/2}/(eV^2)]^{1/2} \qquad\qquad 10.5$$

where $\gamma = 1 - eV/(mc^2)$.

10.2.3   The Multipactor Effect (Classical)

Multipactoring is an important mechanism that can lead to the development of vacuum breakdown in high power RF and microwave systems. The phenomenon was first described by Farnsworth [9], the word deriving from "multiple electron impact".

The strong RF fields present in many vacuum systems accelerate charged particles (e.g. field and photoemitted electrons, ionised residual gas molecules etc.) towards dielectric surfaces (walls or windows) which then deteriorate under the bombardment. The number of charged particles involved depends upon the gas pressure within the system. If the bombarding particles acquire enough energy to produce secondary emission, the secondary particles (usually electrons) may find the RF field normal to the dielectric surface and in the correct phase to accelerate them to bombard another surface which again emits secondary electrons. This "bouncing" back and forth between surfaces is the multipactor effect. When the secondary emission coefficient is greater than unity, the electron density and the power delivered to the surfaces increase exponentially, usually ending with catastrophic failure such as melting or fracture. Even if the power transferred by the multipactor process is too small to cause thermal damage or arcing, the process may cause surface charging which sets up local electric fields which may in turn exceed the dielectric strength of the material. To mitigate this possibility, dielectric surfaces are sometimes coated with a conducting film (e.g. graphite).

In summary then, multipactoring develops by secondary electron multiplication between two opposed surfaces when a cyclic voltage is maintained between them, and if (a) the primary electron energy at collision with the wall results in a secondary emission coefficient greater than unity, and (b) the RF electric field is in the correct phase.

Basically, the same processes as have been described above can also occur between metal surfaces, especially if the surfaces have patches of thin dielectric films such as oxide or pump oil. For the multipactor effect to develop, we require a non-uniform RF field of high intensity. A particle always gains more energy during the half cycle in which it is travelling towards the region of weaker field than it loses in travelling towards the region of stronger field during the reverse half cycle. An equation given by Gittins [10], previously derived by Boot et al. [11], for the time averaged force acting on a particle is

$$F = (e^2/4\omega^2 m) \, \Delta \, |E^2|$$
\hfill 10.6

where ω is the angular frequency of the oscillating field.

As a multipactor discharge develops between two opposing surfaces, a thin electron cloud forms as a sheet approximately parallel with the two surfaces. The current carried by the electron cloud at each half cycle increases exponentially to a limit set by the perveance of the gap. This perveance limit is probably due to spacecharge repulsion which causes some electrons to fall out of phase with the applied RF field, thereby limiting the maximum electron density attainable by the electron cloud. A simple analysis of the classical multipactor effect has been given by Gallagher [12] whose work we shall now follow. The equation of motion for a particle of mass m and charge e moving in a gap d in an RF field given by $E = (V_0/d) \sin wt$ is

$$\frac{d^2 x}{dt^2} = \frac{eV_0}{md} \sin(\omega t + \phi)$$
\hfill 10.7

where ϕ is the field phase at electron emission. In practice, the relativistic equation of motion is more appropriate when field strengths are as high as they are in some high energy accelerators, klystron cavities, superconducting cavities, and the like. Thus, relativistically, Gallagher [13] has shown that

$$\frac{d(\delta x)}{dt} = \frac{eV_0}{md}\sin(\omega t + \phi) \qquad 10.8$$

where $\qquad \delta = \left[1 - (dx/cdt)^2\right]^{\frac{1}{2}}$

This equation is most conveniently solved by a binomial expansion of

$$\left[1 + \left(eV_0/\omega mcd\right)^2\left\{\cos\phi - \cos(\omega t + \phi)\right\}^2\right]^{-\frac{1}{2}}$$

which appears in the equation for dx/dt. The expansion is followed by term-by-term integration rather than by using integral tables, such as those by Grobner and Hofreiter [14].

Continuing classically, we may integrate equation 10.7 to obtain

$$x = \frac{eV_0}{\omega^2 md}\left(\omega t \cos\phi - \sin(\omega t + \phi) + \sin\phi\right) \qquad 10.9$$

To have gain, the transit time must be an odd number of half-cycles, so we have

$$\omega t = 2n + 1$$

with equation 10.9 becoming

$$x = \frac{eV_0}{\omega^2 md}\left[(2n+1)\pi\cos\phi + 2\sin\phi\right] \qquad 10.10$$

Since the particle motion must be within the range 0 < x < d, the phase range in equation 10.10 must be restricted to

$$0 \le \phi \le \arctan\frac{2}{(2n+1)\pi} \qquad 10.11$$

Finally, with x = d, equation 10.10 gives

$$\left(\frac{d}{\lambda/2}\right) = \frac{eV_0}{mc^2}\left[\frac{(2n+1)\pi\cos\phi + 2\sin\phi}{\pi^2}\right] \qquad 10.12$$

380

The strongest multipactor effect occurs in the first order, so we have

$$\left(\frac{d}{\lambda/2}\right) = \frac{eV_0}{mc^2} \qquad\qquad 10.13$$

If we now assume that the primary electron with an impact velocity of v_i produces a secondary electron with an emission velocity v (noting that a high secondary emission coefficient implies a low emission velocity), appropriately specified boundary conditions change the coefficient $(2n + 1)\,\pi$ to

$$\frac{K-1}{K+1}(2n+1)\pi \qquad\qquad 10.14$$

where K is the ratio v_i/v. It follows that the phase range given by equation 10.14 changes similarly. Finally, it should be noted that experimental results obtained by Hatch and Williams [15] agree well with the predictions of the above classical theory.

10.2.4   Surface Flashover

As discussed at length in Chapters 8 and 9, there is always a serious risk of breakdown occurring along the surface of insulators separating conductors or electrodes. Frequently, it is initiated by the bombardment of insulators by charged particles such as can occur in space, in RF equipment (multipactoring), and in beam machines.

Technological experience indicates that surface flashover generally involves the presence of surface inhomogeneities in the structure of the pure dielectric occurring at sites on the surface, or in the solid close to the surface. Possible inhomogeneties have been identified as lattice defects, impurities, and absorbed gases. The traditional interpretation assumes that when a sufficiently high electric field is applied between the electrodes bridged by the insulator, electrons are emitted at the triple junction formed by the cathode-insulator-vacuum combination. These are then assumed to "hop" along the insulator surface towards the anode following paths of least resistance between the randomly distributed sites, changing the solid state, surface state and vacuum state characterising each site with which the electrons interact [16,17]. Definitively detailed models of exactly how the various stages of a complete flashover develop from prebreakdown currents, through circuit limited discharge current, to full recovery have yet to elucidated. However, recent ideas discussed in Chapter 9 indicate that the relaxation of trapped charge could play a crucial role.

In a recent survey, Asokan et al. [18] used streak photography to record surface flashover caused by pulsed electric fields across an insulator-vacuum interface at a streak speed of 1 nsec/mm. An interesting feature of this detailed study is the planar geometry of the electrode-insulator configuration, as shown in Fig 10.5. The more usual geometry for flashover studies is cylindrical, whereby a cylinder of dielectric bridges the inter-electrode gap.

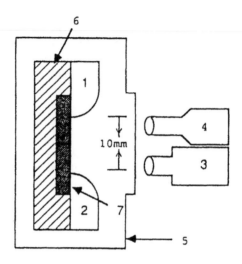

Fig 10.5   Side view of electrode configuration used for surface flashover studies [16]. 1 and 2 - electrodes, 3 - streak camera, 4 - phototube. (After Asokan et al. [18].)

However, a cylindrical geometry is quite unsuitable for photographic studies because only the near half of the cylindrical area is observable by the camera, the far half is hidden by the dielectric itself. Also, the aspect of a flashover channel which is observed by the camera changes according to the location of the channel on the semi-cylindrical face presented to the camera. Such changes can lead to interpretational ambiguities. The planar geometry of Asokan et al. [18] avoids these problems. Their electrodes are clamped onto one of the flat surfaces of a dielectric disc (crystal quartz, 25.4 mm diameter × 5 mm thickness) so as to give a uniform inter-electrode spacing of 1 cm. The plane to which the flashover is thereby constrained is arranged to be normal to the optical axis of the streak camera. A typical sequence of events observed during flashover by Asokan et al. [18] is as follows. Visible light of high intensity originates at the cathode followed, about a nanosecond later, by visible light of high intensity originating independently at the anode. The anode and cathode light emitting regions then propagate towards the opposite electrode at about 3×10^8 cm/sec. They also observed that, in two-day tests, the greater flashover voltages were achieved on the second day and attributed this to the formation of negative charges at trap sites on the insulator surface during the first day test.

At voltages above a few tens of kV, the voltage hold-off of insulators is proportional to the square root of the insulation thickness; thus a ten-fold increase of voltage requires a hundred-fold increase in thickness of insulation for hold-off. By using n metal grading rings, separating insulating discs of thickness x, to create a uniform potential gradient between the anode and cathode, the hold-off for each stage can be such that the total hold-off voltage can become linearly proportional to total insulator thickness. In general we can write

$$V_h \ll nV_n$$

where V_h is the voltage hold-off of a single ungraded insulator of thickness X, and $nx = X$. As a practical illustration, Fig 10.6 shows how a vacuum insulator is built into a typical system for generating an e-beam for injection into a microwave generator. Each insulator annulus is separated from its neighbour by a metal annulus. The slope of the inner surface of each insulator is cut at the angle which gives maximum hold-off voltage.

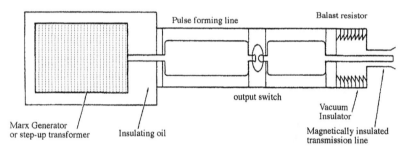

Fig 10.6   Schematic of a typical pulse forming line system employing a graded insulator to isolate the load.

More recently Elizondo [19] has reported how much greater vacuum surface flashover voltages can be achieved by forming a so-called microstack of thin metal discs alternating with thin insulator discs. Samples with the optimum configuration failed by dielectric bulk breakdown at 400 kV/cm rather than by surface flashover.

The voltages at which surface flashover occurs with pulses of duration greater than microseconds appears to be independent of the pulse duration. If t_{if} is the longest formative time necessary for the completion of an individual process necessary to establish a full breakdown, the energy of any pulse of duration greater than t_{if} will be dissipated in other conduction processes which maintain the discharge rather than established it. However, for pulse durations less than 100 nsec, surface flashover voltages become time-dependent according to which processes have enough time to develop to full breakdown in the applied field. Factors which become influential are insulator relaxation times and field enhancement at triple points. Generation of surface plasma moves time scales from the purely electronic to the longer ones associated with heavy particles such as neutrals and ions.

With the growing importance of high voltage semiconductor switches, many research projects have been aimed at understanding the flashover mechanism of a semiconductor-vacuum interface. These studies have revealed that this type of regime exhibits many significant differences from the corresponding insulator-vacuum behaviour. For example, Nam and Sudarshan [20] have found that surface breakdown (flashover) across the semiconductor/vacuum interface occurs at fields much less than the theoretical value corresponding to the breakdown field of the pure semiconductor bulk material, and much less than the breakdown field of pure vacuum. In summary, surface flashover appears to involve interactions between semiconductor conduction processes within the surface layers and gas conduction processes in the vacuum ambient close to the semiconductor surface. Clearly, much work will be needed to understand the physical mechanisms that control the performance of these advanced systems.

10.2.5   Magnetic Insulation

As indicated in the previous section, a solid insulator becomes vulnerable to surface charging processes, and consequently flashover, if it has to operate in a plasma environment. Thus, in low earth orbit, where the typical plasma density is $\sim 10^7$ cm^{-3}, satellite-based devices are prone to instability. However steps can be taken to inhibit flashover when the region of insulator where it is likely to occur is defined by electrodes. In this case, a magnetic field (magnetic insulation) can be applied precisely to alter inter-electrode particle trajectories and thereby increase the flashover voltage by significant factors. An accepted model of flashover development, due to Boersch et al. [16], involves a combination of electron avalanche and electron-induced gas desorption. Magnetic insulation is most effective when the E × B vector forces the avalanche electrons away from the surface (lift-off of electrons); where a number of authors, including Bergeron [17], Van Devender et al. [21], and Bergeron and McDaniel [22], have discussed the conditions for lift-off. The critical magnetic field given by Bergeron and McDaniel [22] is

$$B_c = 1.6\,(v_0/v_1{}^2)\,E \qquad\qquad 10.15$$

where E is the average electric field applied parallel to the surface, v_0 is the average emission velocity of secondary electrons, and v_1 is the impact velocity of electrons for saturated avalanche conditions. Korzekwa et al. [23] have found that magnetic fields of several tenths of a tesla are sufficient to double flashover voltages. Materials studied in the presence of DC or pulsed magnetic fields were an epoxy-fibreglass composite (G-10), pyrex, lucite, and teflon. Results indicate that significant magnetic insulation can be obtained from small (several cm^3) lightweight (several tens of grams) permanent magnets which apply fields up to 0.2 T to the cathode region only.

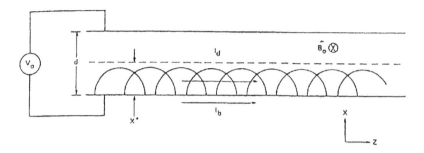

Fig 10.7   A magnetically insulated parallel plate line. (After Toepfer [3]).

As a general rule, electromagnetic energy is converted to particle beam energy by an induction process or by an emission process in vacuum. Magnetic insulation was proposed by Winterberg [24] to increase the breakdown voltage of a vacuum gap. As a practical example of its applications, Fig 10.7 shows a magnetically insulated parallel plate line [3]. In this context, Lovelace and Ott [25] have derived the following equation for the minimum magnetic field B_0 which will insulate a vacuum coaxial transmission line,

$$B_O = \frac{mc^2}{e}\left(\frac{2r_+}{r_o^2 - r_i^2}\right)\left(\gamma_o^2 - 1\right)^{1/2}$$

10.16

where we have

$$\gamma_o = 1 + \left(eV_o / mc^2\right)$$

and r_+ is the anode radius, $r_o(r_i)$ is the radius of the outer (inner) cylinder and m, c, and e have their usual meanings.

An important case of magnetic insulation is where the insulation is not provided by an external magnetic field, but by the self-magnetic field due to the line current flow itself. For example, transmission lines feeding power to high current vacuum diodes are often magnetically insulated by such a mechanism. The modelling of this type of system has been based on cold fluid equations [26] and particle-in-cell simulations [27].

For the application of magnetic insulation in the space environment, the reader is referred to the review of Krompholz et al. [28].

10.2.6 Stability of Microwave Transmission Windows

When millimetric microwave generators operate at powers approaching the gigawatt level they do so with pulses of 10^{-8} - 10^{-6} sec at repetition rates which are generally slower for the longer pulses. Field gradients can be greater than 10^6 V/cm and energy densities can cause materials problems, particularly at windows which enable the generated microwaves to be coupled to the outside world.

An equation which is useful for rough calculation/estimation of the magnitude of the electric vector in electromagnetic radiation is

$$w = E^2/(\mu_o/\varepsilon_o)^{1/2} \sim E^2/77 \text{ W/m}^2 \qquad 10.17$$

where w is the incident power per unit area and the other symbols have their usual meaning. For a diffraction limited spot of area λ^2, we have

$$w = P/\lambda^2 \qquad 10.18$$

where W_T is the total beam power. Equations 10.17 and 10.18 then give

$$W_T = E^2 \ \lambda^2/377 \qquad 10.19$$

The critical field for the Kilpatrick limit at 35 GHz (8 mm) is about 10^8 V/m, so that from equation 10.19 we find the critical power to be $W_T = 1.7$ GW.

An electron beam device for generating microwaves, such as a travelling wave tube, usually consists of a beam source which is often a diode, an accelerating region and a wave-beam interaction region where the electromagnetic radiation is generated. After the electrons have passed through the interaction region, the beam and its unused energy is deflected by a magnetic field and upped into a collector which must cope with the thermal load, and high energy electron bombardment which may cause outgasssing and surface charging leading to plasma formation and/or surface flashover. The microwaves generated pass along a vacuum waveguide which terminates at a window which separates the vacuum of the waveguide from the atmospheric pressure of the antenna (or load). When the peak powers of the microwaves reach values such that their associated electric fields are greater than those necessary to produce plasma at the window surface, catastrophic window failure may occur if the window design has not been very carefully considered.

Factors to be taken into account include material selection, dimensions, and geometry. Low microwave absorption, thermal shock resistance, and high stability at the anticipated working temperatures are important properties of the window material. In selecting material for the windows of high power RF/microwave systems, high values of the ratio

$$\frac{\text{Thermal conductivity}}{\text{Dielectric constant} \times \text{Loss tangent}}$$

should be sought. Geometrical considerations may include enlargement of wave guide section for the window installation so as to reduce the power flux/electric field at the window surface and to improve thermal management. There is plenty of scope for exploitation of engineering ingenuity and "know-how" to shine in window design for high power, high repetition rate microwave systems.

As a practical illustration, Fig 10.8 shows a transition from water to vacuum [3]. Respective breakdown strengths given by Toepfer [3] for water and solid insulator are

$$E_{BW} t_{eff}^{1/3} A^{0.058} = k^{\pm} \qquad\qquad 10.20$$

and $\qquad E_{BS} V_S^{0.1} = k^1 \qquad\qquad 10.21$

where E_{BW} and E_{BS} are the breakdown fields in megavolts per centimetre, t_{eff} is in microseconds, and A is the electrode area in square centimetres. In a coaxial system the constant k is a function of the polarity of the centre electrode, k^+ $(k^-) = 0.3$ (0.6). The volume of solid (cm^3) is V_S and k is a constant of value depending on the material. The maximum power flow sustainable by the system, is not limited by the breakdown strengths of water or the solid but by flashover at the solid-vacuum interface, which is usually initiated by events at the "triple point" where metal, dielectric and vacuum meet. Toepfer [3] quotes a limiting power flux of 2 TW/m² at the vacuum interface for flashover at 270 kV/cm for the system he describes. As noted previously in the case of microwave systems, by gradually flaring the line/guide to increase the cross-sectional area up to that of the solid transition region, the power per unit area arriving at the solid-vacuum interface is reduced to levels which enable the peak powers delivered safely to the diode to be significantly increased [29]. The peak power W_L deliverable to a load of impedance Z_D is

$$W_L = \alpha V_0^2 / Z_D \qquad\qquad 10.22$$

where α is defined as

$$\alpha \equiv [4(Z_L/Z_D)]/[(Z_L/Z_D) + (2\tau_D/\tau_R)]^2 \qquad\qquad 10.23$$

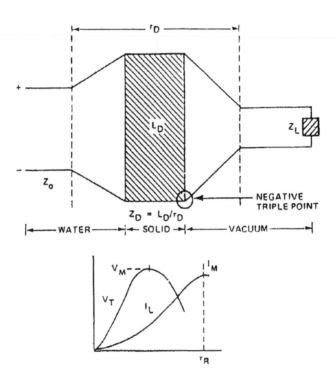

Fig 10.8   The transition from water → dielectric → vacuum. Line is flared in the region close to solid dielectric-vacuum interface. (After Toepfer [3].)

The maximum voltage across the interface is V_0 and $Z_D = L_D/\tau_D$ is the impedance of the transition region. The inductance of the diode is L_D and τ_D is its electrical length. The rise time to peak current at the diode output is τ_R.

10. 3   Pulsed Power Devices and Technologies

In the previous sections, we have considered some of the key physical processes that limit the operation of HF and PF systems. We shall now describe a range of devices that have been developed to facilitate practical applications of pulsed power technology.

388

10.3.1   Electrically Triggered Vacuum Switches

While surface flashover in vacuum is generally undesirable, it is necessary for the operation of triggered vacuum gaps, i.e. the "switch" for many pulse power systems. Much has been written about the mechanisms by which triggered vacuum gaps operate and many ingenious designs have been described [30-33]. By way of illustration, we shall describe a recent cathode design by Sampayan et al. [34], shown in Fig 10.9. This produces a uniform array of surface flashover sparks to trigger the vacuum gap uniformly so that current densities in the main discharge are much reduced. Their cathode consists of an insulating substrate with a high voltage bottom plane and isolated metal discs surrounded by a ground plane on the top plane. Each disc was isolated from the pulsed trigger source by a resistance. Essentially, the overall geometry of the cathode, plus its trigger electrode structure, was made from a standard 1/16 in-thick, copper clad double sided fibreglass-epoxy circuit board. The bottom copper plate was etched so that identical circular areas of epoxy were exposed as an array. The remaining copper surrounding each and every circular epoxy patch was used as the trigger pulse connection (equipotential because of the high conductivity of copper). The top copper plate was etched so that it was covered by an array of identical rings (annuli) of exposed epoxy in such a way that no rings touched. In this way, the centre of each epoxy ring isolates its central copper disc from the copper which surrounds each epoxy ring. The surrounding copper was used as a ground plane (again an equipotential). The top and bottom arrays were sized and aligned so that each and every top copper disc was concentric with a bottom circular epoxy patch of equal area to form a "mirrored" pair. To provide a resistive path (found to be essential for reliable triggering) between the high voltage bottom plane (trigger pulse connection) and the isolated copper discs in the top plane, a small connecting hole was first drilled between the centres of each and every "mirrored" pair and then the walls of the connecting holes were plated. The spacing between an isolated copper disc and the ground plane was about 0.25 mm and it was across this insulator (epoxy) surface that the trigger plasma was formed.

In the triggering operation, a high voltage pulse applied to the bottom plate is resistivly coupled to each and every copper disc on the top plate through the plated holes. The effective isolation resistance between the bottom (trigger) plane and the isolated copper discs of the top plane was of the order of 10^4 ohm. Triggering results in surface flashover sparks between most of the copper discs and the ground plane which effectively surrounds them. The multiplicity of flashover plasmas initiate a discharge between the anode and cathode of the vacuum spark gap. Other materials used instead of copper double sided epoxy circuit board include platinum-palladium on ultra-pure alumina ceramic.

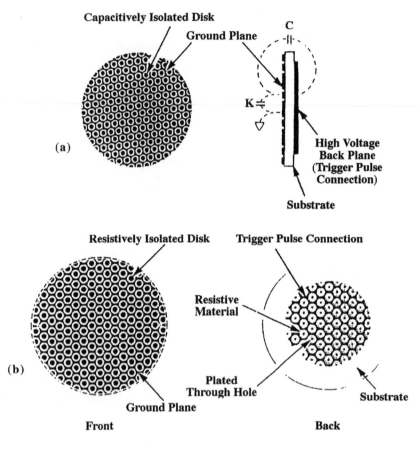

Fig 10.9   (a) Capacitively coupled plasma source. (b) Resistively coupled plasma source. (After Sampayan et al. [34].)

10.3.2   Optically Triggered Photoconductive Switches

Pulse power applications for e-beam and laser controlled semiconductor switches are increasing apace as both closing and opening switches. As their commercial potential grows, so does the need and motivation to understand their basic operating principles and failure modes. As indicated in Section 10.2.4 the principal factors limiting the operating voltages is the flashover of both the dielectric-vacuum and semiconductor-vacuum interfaces. Much research therefore remains to be done on this topic if the performance of this type of switch is to be optimised.

With contemporary switches, the current densities reached by conducting filaments in the GaAs photoconductive switches used in pulsed power applications can be 10^6 A/cm^2 or more. At these densities, damage accumulates near the contacts until flashover occurs across the semiconductor surface between the electrodes. Baca et al. [35] describe results which show that accumulated damage at contacts can be significantly reduced by modifying the alloying procedure. It is also noted that Au-Ge is better for a negative contact and Au-Be for a positive contact. The above observations are in line with other studies which have concluded that the development of surface flashover conduction processes at a semiconductor-vacuum interface begins in the bulk semiconductor.

For high average power switching, industrial and government laboratories have identified laser and electron beam triggered photoconductive solid-state switches as prime candidates for the next generation of high performance switches in advanced pulsed power applications.

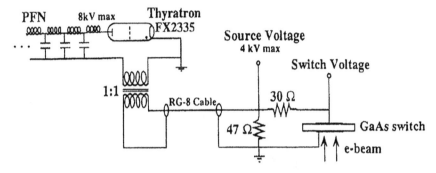

Fig 10.10   Schematic diagram of a PFN-based test circuit for studying the closing and opening behaviour of an electron beam CSS module. The thyratron-switched PFN is impedance matched at the load. A 30 Ω resistor is used to limit the switch current. (After Kirkman et al. [36].)

The package of important operating characteristic features of solid-state photoconductive switches includes high hold-off fields, high switched currents, sub-nanosecond jitters, high repetition rates, long operating life, and compactness. The combination of compactness and comparable (or superior) performance makes these new switches the preferred design option for many new pulse power systems. The most important materials for photoconductive switches appear to be GaAs and bulk silicon, either doped or intrinsic. Pulse specifications attainable by using GaAs photoconductive switches are 10^7 W with pulse durations of many nanoseconds at repetition rates greater than 1 kHz and sub-nanosecond rise times and jitters. Diamond in natural or CVD grown forms is also attracting much research attention. For these, and other materials, hold-off fields greater than 100 kV/cm can be switched to pass currents in excess of 10^4 amp/cm^2 by either laser or electron beams, depending on conditions. Figs 10.10 to 10.12 show various aspects of e-beam switched photoconductive devices.

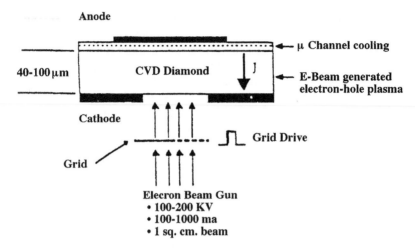

Fig 10.11   Electron beam controlled diamond switch. (After Hofer et al. [37].)

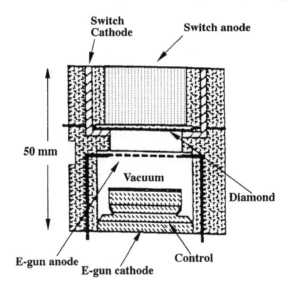

Fig 10.12   Diamond switch and electron beam assembly concept. (After Hofer et al. [37].)

The level of sophistication of solid-state photoconductive switches has been raised to include both the closing and opening operation in the one device known as a Bulk Optically Controlled Semiconductor Switch (BOSS). This is a switch which can be closed on command with one laser pulse (1 μm, say) and opened on command with a second laser pulse with a different wavelength

(2 μm, say). Clearly, such control enables BOSS devices to compete with traditional hard tubes such as the triode. A semiconductor material which operates as a BOSS is copper-compensated, silicon-doped gallium arsenide (GaAs:Si:Cu), [38-41]. For faster (sub-nanosecond) opening, chromium doping is of particular interest [42]. Closing and opening has also been achieved with a GaAs bulk switch which has a built-in e-beam of 50 keV from a dispenser cathode. Only one beam is necessary because the current passed by the switch follows the profile of the e-beam. The device is compact and megahertz repetition rates have been demonstrated with applications including pulse length modulation.

Vacuum is often the chosen ambient within the capsule of the switch device, especially if it is e-beam triggered. The hold-off fields attainable are thus limited by the processes of surface breakdown in vacuum. To date powers greater than 400 kW have been switched into a 50 Ω load

10.3.3   Electron Beam Applications

Applications identified for high average power electron beam systems include the treatment of medical and chemical waste, sterilisation of food, and clean-up of flue gases. Accelerators using multiple stages of pulse compression are now available to give output pulses of tens of nanoseconds durations at voltages up to 20 MV and average powers up to MW. e-beam irradiation can be switched off and has none of the problems associated with radioactive demand sources such as cobalt-60. Some US regulations limit acceleration voltages to 5 MV. For cost effective irradiation, large volumes must be processed rapidly, so an important emphasis of research and development is to create uniform e-beams, or X-ray beams, of large cross-sectional area ($\sim$ m^2), with average powers of several hundred kW to several MW, and which are good for upwards of 10^9 shots. For this level of performance, the cathodes should emit current densities greater than 100 A/cm^2. In general, electrons for such beams can be derived from a number of well defined processes or combinations of them. These are field emission, thermionic emission, photoelectric emission, secondary emission, ferroelectric emission, or plasma. Ultimately, the quality of an electron beam is very dependent on the quality of the pulse power engineering.

For the above applications however, pulsed e-beams to deliver a few kJ in about 50 nsec FWHM at rates of 100 pps are required. Reliability considerations require the e-beam diode to operate for at least 10^8 shots without maintenance. The cathode of such a diode is often a hollow graphite cylinder which develops an almost uniform distribution of electron emitting sites by the familiar vacuum breakdown processes; the anode may be a metal such as tantalum. McClenahan et al. [43] have reported such a system, where the design parameters of their diode were

Cathode graphite, OD 1.9 cm, wall thickness 0.3 cm
Anode: tantalum, 0.76 mm thick
Anode-cathode gap; 0.6 cm - 1 cm

The impedance of their diode varied between 100 Ω and 200 Ω depending on the gap. The typical gap closure velocity was 5 cm/μsec.

In practical terms the maximum power available from an electron beam is essentially limited by self-space-charge effects (see Child-Langmuir equation 10.2). A voltage limitation exists but cannot be specified precisely because the electric fields involved may initiate breakdown by any of a wide variety of processes ranging from multipactoring to deterioration of vacuum conditions by outgassing. An electron beam device for generating microwaves will typically have three vulnerable sectors; namely, the beam source, which is often a diode, an accelerating region and a wave-beam interaction region where the electromagnetic radiation is generated. It can however be assumed for practical purposes that the output powers of electron beam driven radiation sources are limited by the input powers. Performance specifications now demand greater e-beam currents with higher brightness and lower emittance.

10.3.4   X-Ray Lithography

The microelectronics industry is driving towards higher densities of circuit elements on circuit chips and faster circuits. The feature size, which will result in an order of magnitude improvement in both these goals, is about 0.25 μm, or less. To achieve this, X-ray photolithography is the preferred route. For this route, soft X-ray sources of high brightness and stability are needed for photolithography of large-scale integrated circuits. Sources considered include plasma focus [43], gas-puff Z-pinch [45] and a vacuum spark [46,47]. No single source seems to have such dominant advantages over the other sources that they would make it the "obvious" choice, but vacuum sparks are strong contenders. The X-ray intensity and wavelengths emitted by vacuum sparks depend on electrode configurations and materials, so optimisation of these has become the main thrust of many investigations.

One of the options being seriously considered as a source for X-ray photolithography, because of its simplicity and potential for repetition rates of 50 pps or so, is the X-pinch. The X-pinch consists of two wires (or more) arranged in an X-configuration so that they touch at the intersection point. The wires are arranged stretched between the output electrodes of a source of current pulses (10^5 amp). A current pulse causes a plasma, which is essentially a vacuum arc, to form at the wires and pinch towards the single point where they touch under the influence of the self-magnetic field of the current. Intense X-ray radiation is generated in the plasma surrounding the "intersection point" of the "X". The recovery properties of this arc naturally influence the pulse repetition rates attainable. According to Glidden et al, [48], magnesium wires have given radiation between 0.84 nm and 0.94 nm (mostly for K-shell lines). The design of the vacuum insulator, which is the interface between oil and vacuum regions, is crucial to the successful operation of this pulsed system. This is mainly because of the need to dissipate the excess inductive energy after X-ray pulse, and the fact that the flashover strength of insulators in vacuum depends strongly on the direction of the applied electric field, whatever its origin. The design recommendation of Glidden et al. [48] for the vacuum region is to minimise the inductance while maintaining the vacuum electric fields on the cathode surfaces at values no greater than 200 kV/cm.

Complete exposure of a proprietary resist with soft X-ray sensitivity of 50 mJ/cm^2 in one second requires a pulse current of 500 kA at a repetition rate of about 40 pulses per second [49]. This yields an average X-ray power output greater than 1 kW in the 0.7 to 1.0 nm range. Repetition rates should generally be above 10 pps to avoid mask damage.

A vacuum spark itself can be an efficient source of X-rays [44,46] in both line and continuum form [47,49-52]. The X-rays can be produced by e-beam interaction at the anode or beam-plasma interactions and plasma processes from almost any region of the inter-electrode space according to electrode geometry and mutual location. For X-ray photolithographic purposes, the emission site must be stable during the exposure time.

In a study by Arita et al. [53], a 30 kV, 10.6 μF, 7.2 kJ capacitor was used to produce currents of about 300 kA in a half sinusoid of about 5 μsec duration. Arc and X-ray emission spot sizes were recorded by an optical camera and a pin-hole camera, respectively. A Be foil covered the pin-hole to block wavelengths longer than 1.5 nm. Flat-field grazing incident XUV spectrography [54] was used to record the soft X-ray spectra (line continuum). The time delay to X-ray line emission was observed to be about 1 μsec. The studies of Arita et al. [53] relate to Mo, Al, C and cathodes and include X-ray spectra from Z-pinched vacuum sparks. Of the various electrode geometries studied, the greatest X-ray intensities were observed when using a spherical anode and a conical cathode with a gap of 2 mm. X-ray emission wavelengths were in the range 0.52-1.2 mm - quite close to the range 0.4-1.4 nm considered to be suitable for X-ray lithography. In considering the optimum/efficient production of X-rays from a pulsed source, it is important to remember that both the electrode-plasma combination and the circuit configuration are ultimately related.

In summary, we note that for X-ray photolithography, the X-ray source can be a vacuum spark in systems configured to exploit the well known properties of X-ray emission, current carrying capabilities, and recovery time of vacuum arcs. The research issues now are focused on simplicity of construction, high repetition rates and consistency of X-ray output with respect to both spectral content and location of the emitting region. As the references show, formal investigations over the last 30 years or so have provided a substantial background of data, phenomena and understanding which are now forming the basis of what promises to provide the microelectronics industry with yet another order of magnitude advance on product performance.

10.3.5   Photomultipliers and Image Tubes

Image tubes, photomultipliers, and the like have a photocathode selected from a large number of types (S1, silver-oxygen-caesium; S11, antimony-caesium, etc.), according to the desired spectral range and sensitivity. Both types of tube are widely used for recording transient optical events with time resolutions of nsec (photomultipliers) or tens of picoseconds (image tubes). Fine wires and thin plates are generally used for electrical, electro-optical, and structural purposes; insulator-metal seals abound. At the high voltages (a few hundred for photomultipliers and a few kV for image tubes) commonly used to obtain optimum system sensitivity, field emission effects can occur. In this context, Fig 10.13 presents analytical formulae for electric field enhancement factors at metal surfaces of various geometries [55]. Fortunately, for the designer, the field emission characteristics are, more or less, independent of the type of cathode (e.g. S1, S11). Ideally, all electrons arriving at the anode of an electro-optic device should originate as "signal"; any other electrons, unrelated to the signal, merely generate noise.

Of course, field-emitted electrons can arrive at almost any surface within an electro-optic device and their presence can often be deduced by the fluorescence or changes of potential and current they produce. Spurious photons from fluorescence or plasma at field emission sites can reach the photocathode surface by scattering and reflection processes, and so increase the background noise level.

Many of the sites of field emission are fixed by the prevailing fields but others may come and go randomly according to the location of small particles of loose debris which may be present on assembly of the tube, or may develop during the tube life, especially under conditions of high vibration energies and frequencies. Very thorough cleaning and assembly procedures can eliminate debris problems almost entirely. Other causes of field emission may have their origin in the migration of material of low work function from the photocathode to other surfaces subjected to high electrical fields. Such unwanted low work function surfaces can also emit photoelectrically into high field regions. A good account of field emission in image tubes has been given by Essig [56].

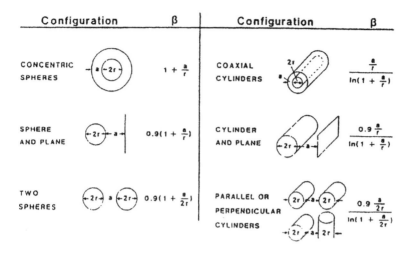

Fig 10.13 Maximum field-enhancement factor, β, for various electrode configurations. (After Gilmour [55].)

10.3.6 Space Applications

The major applications for pulse power in defence lie in the areas of high resolution radar, kinetic energy weapons and beam weapons. However, as indicated earlier in Section 10.2.4, when any of these applications are considered for a space environment, vacuum breakdown problems arise to a greater or lesser degree. All components to which high voltages are applied and which perform perfectly in earth surface applications may have to be redesigned to avoid surface flashover if they are to be used in space systems.

The vacuum of space appears at first sight to be a ready made insulator for high voltages. Indeed, it would be were it not for all the gas which is steadily evolved from all the surfaces of a space vehicle and the high energy charged particles arriving during solar flares. The Strategic Defence Initiative (SDI) provided very powerful incentives to operate high energy pulsed power systems from space platforms. The systems seriously considered ranged from megavolt particle beams to kilovolt electromagnetic launchers (railguns). Gordon and Logue [57] have summarised possible effects of gas contamination on the performance of high power systems aboard orbiting spacecraft and discussed observations of contamination effects on the high voltage space experiments BEAR, SPEAR I and SPEAR II. (SPEAR - Space Power Experiments Aboard Rockets, BEAR - Beam Experiments Aboard Rockets).

A major conclusion of SPEAR I was that outgassing resulted in local pressures up to three orders of magnitude higher than the space ambient pressure.

Among the objects of SPEAR I were studies of the altitude dependence of the I-V characteristic of a metallic conductor exposed to the space environment at biases up to tens of kilovolts, and the interaction of kV biased conductors within each others charge sheaths. Recommendations resulting from the data collected from SPEAR I include the following.

(i)   Materials used in a payload must satisfy a minimum mass loss criterion.
(ii)   After vacuum bake-out, high voltage components should be tested for evidence of trapped gas or incomplete outgassing.
(iii)   Exposed dielectric surfaces should be shielded from the space environment of particles and photons so as to avoid surface charging by direct charge trapping, secondary emission, or photo-emission.
(iv)   Materials and surface treatments should be selected for their low outgassing rates, sputtering coefficients and secondary emission coefficients.
(v)   Triple points should be shielded/covered so as to reduce the probability of surface flashover at a given inter-electrode voltage.

The main objective of SPEAR II was to test components and systems for operation at high-voltages and powers in the environment of low earth orbits (LEO). The systems chosen for study were a klystron (high voltages) and a plasma accelerator (high current). Also in the payload were a turbomolecular pump and a cryogenic pump. Other issues considered included the behaviour of surface insulation in the presence of plasma, short pulse phenomena, conditioning and the effects of high magnetic fields. The following additional important conclusions relevant to the design of space-based HV equipment were reached by the SPEAR II programme.

(vi)   Vacuum insulators must be designed to survive several surface flashovers since conditioning is inevitable.
(vii)   It must be possible to pump evolved gas quickly away from regions where flashover is likely.
(viii) Charged particles reaching dielectric surfaces distort surface fields and cause breakdown.

The principle objective of the BEAR project, as detailed in its Final Report [58], was achieved by operating a neutral particle beam system in space for the first time. The neutral particle beam consisted of neutral hydrogen atoms and hydrogen ions of energy 1 MeV emitted by an accelerator in 50 μsec pulses. Among subsidiary objectives was the measurement of spacecraft charging, again emphasising the importance attached to the problem of surface flashover at high voltages aboard spacecraft at LEO. In this context, particular note should be taken of our earlier observations in Sections 10.2.4 and 10.2.5, where we highlighted the adverse effects of surface charging on insulator hold-off performance.

For further information about space-based HV applications, the reader is referred to Chapter 13.

10. 4 References

1. Adler, RJ., Pulse Power Formulary, North Star Research Corporation, August, Supported by US Air Force Office of Scientific Research under contract F 49620-89-C-0005, 1989.

2. Levy, S., Nikolich, M., Alexeff, I., Rader, M., Buttram, M.T. and Sarjeant, WJ., Conference Record of 1992, 20th Power Modulator Symposium, 23-25 June, 1992, Myrtle Beach, South Carolina, pp 8-14. Copyright 1992, IEEE Inc, 345 East 47 St., New York, NY 10017.

3. Toepfer, AJ., Advances in Electronics and Electron Physics, **53**, 1-45, Academic Press, ISBN 0-12-014653-3, 1980.

4. Moreland, LD, Schamiloglu, E., Lemke, R., Gahl, J. and Shiffler, D. Proc 9th IEEE Pulsed Power Conference, 1993.

5. Benford, J. and Swegle, J., *In* "High Power Microwaves", Artech House, (ISBN - 089006-415-6), 1992.

6. Mahaffey, RA., Sprangle, P., Golden, J. and Kapetanakos, CA., *Phys. Rev. Lett.*, **39**, 843-7, 1977.

7. Parker, RK., Anderson, RE. and Duncan, CV., *J . Appl. Phys.*, **45**, 2463-70, 1974.

8. Mesyats, GA., *IEEE Transactions on Plasma Science*, **19**, 683-689, 1991

9. Farnsworth, P., *J. Franklin Inst.*, 2, 411-16, 1934, and *Z. Phys.*, **129**, 491-6, 1951.

10. Gittins, JF., in "Power Travelling-Wave Tubes", English Universities Press, 1964.

11. Boot, HAH., Gittins, JF. and Shersby-Harvie, RBR, *J. Electronics and Control*, **4**, 434-41, 1958.

12. Gallagher, WJ., *IEEE Trans. on Nucl. Eng.*, **NS-26**, 4280-4282, 1979.

13. Gallagher, WJ., IEEE Particle Accel. Conf. CH2669-0/89, 2008-2009, 1989.

14. Grobner, H and Hofreiter, W., Integraltafel, Unbestimmte Integrale, Wien, No. 867, sec 244, 88-95, 1957.

15. Hatch, AJ. and Williams, HB., *J . Appl. Phys.*, **25**, 417-23, 1954.

16. Boersch, H., Hamisch, H. and Ehrlich, W., *Z. Agnew Phys.*, **15**, 534-42, 1963.

17. Bergeron, KD., *J. Appl. Phys.*, **48**, 3073-3080, 1977.

18. Asokan, T., Morris., G. and Sudarshan, T. S, 20th Power Modulator Symposium, pp. 336-9, CH 3180-7/92, IEEE, 1992.
19. Elizondo, JM., 9th IEEE Pulse Power Conference, Albuquerque, NM, 1993.
20. Nam, S.H. and Sudarshan, T.S., *IEEE Trans. Electron Devices*, ED-37, 2466-2471, 1990.
21. Van Devender, J. P, et al., *J. Appl. Phys.*, **53**, 4441-4447, 1982.
22. Bergeron, KD. and McDaniel, KD., *Appl. Phys. Lett.*, **29**, 534-536, 1976.
23. Korzekwa, R., Lehr, FM., Krompholz, HG. and Kristiansen, M. *IEEE Transactions on Plasma Science*, **17**, 612-615, 1989.
24. Winterberg, F, *Phys Rev,* **174**, 212-218, 1968.
25. Lovelace, RV and Ott, E., *Phys. Fluids,* **17**, 1263-71, 1974
26. Creedon, JM., *J . Appl Phys.,* **48**, 1070-8, 1977.
27. Poukey, JW. and Bergeron, KD., *Appl. Phys. Lett.,* **32,** 8-12, 1978.
28. Krompholz, H., Korzekwa, R., Lehr, M. and Kristiansen, M., Proc. of SPIE, Vol. 871, Space Structures and Power Conditioning, pp.341-347, 1988.
29. Van Devender, JP. and McDaniel, DH., Proc. VII-ISDEIV, Paper E1, 1978.
30. Lafferty, JM., *Proc. IEEE.*, **54**, 23-32, 1966.
31. Kellogg, JC., Boller, J.R., Commisso, RJ., Jenkins, DJ., Ford, RD., Lupton, WH and Shipman, JD., *Rev-Sci. Instrum.*, **62**, 2689-94, 1991.
32. Arita, H, Suzuki, K and Kurosawa, Y., *IEEE Trans. Plasma Sci.*, **20**, 76-79, 1992.
33. Coaker, BM., Xu, NS., Jones, FJ., and Latham, RV., *IEEE Trans. Plasma Sci.,* **21**, 400-406, 1993.
34. Sampayan, SE, Gurbaxani, SH. and Buttram, MT., *IEEE. Trans. on Plasma Science*, **17**, 889-897, 1989.
35. Baca, AG., Jalmarson, HPH., Loubriel, GM., McLaughlin, DL. and Zutavern, F.J., 8th IEEE Pulse Power Conf., 1991.
36. Kirkman, G., Hur, J., Jiung, B., Reinhardt, N. and Schoenbach, K., 20th Power Modulator Symposium, IEEE, CH3180-7/92, 1992.
37. Hofer, WW., Schoenbach, KH. and Joshi, RP., 20th Power Modulator Symposium, IEEE, CH3180-7/92 pp241-244, 1992
38. Mazzola, MS., Schoenbach, KH., Lakdawala, V.K., Germer, R., Loubriel, GM. and Zutavern, F. *J. Appl. Phys. Lett*, **54**, 742-744, 1989.
39. Mazzola, MS, Schoenbach, KH., Lakdawala, VK. and Ko, ST., *Appl. Phys. Lett.*, **55**, 2102-2104, 1989.
40. Mazzola, MS., Schoenbach, KH., Lakdawala, VK. and Roush, RA., *IEEE Trans. Electron Devices*, **37**, 2499-2505, 1990.
41. Stoudt, DC., Roush, RA., Mazzola, MS. and Griffiths, SF., 8th IEEE Pulsed Power Conference, 1991.
42. Gupta, S., *Appl. Phys. Lett.*, **59**, 3276-78, 1991.

43. McClenahan, CR, Martinez, LE, Pena, G.E, Weber, GJ., 9th IEEE Pulse Power Conference, Albuquerque, NM (Digest of abstracts, PI-20), 1993.
44. Lebedev, SV, Mandel'shtam, SL and Rodin, GM., *Sov. Phys.*, *JETP*, **37**, 248-252, 1960.
45. Okada, I., Saitoh, Y., Itabashi, S and Yoshihara, H., *J. Vac. Sci. Technol* **B4**, 243-247, 1986.
46. Handel, SK. and Berg, JM., *Ark. Fys.*, **31**., no. 1, 1-18, 1965.
47. Cohen, L, Feldman, U., Swartz, M and Underwood, J.H., *J. Opt. Soc. Amer.*, **58**, 843-846, 1968.
48. Glidden, SC., Hammer, DA. and Kalantar, DH., p.40, Conference Record of 1992, 20th Power Modulator Symposium, Myrtle Beach, South Carolina, Copyright 1992, IEEE Inc. 345 East 47 St., New York, Ny 10017. Lib. Cong. No., 92-53-252.
49. Chillers, WA., Datla, RU., and Griem, HR., *Phys. Rev. A.*, **12**, 1408-1418, 1975.
50. Turechek, JJ and Kunze, JH., *Z. Phys.*, **A273**, 111-121, 1975.
51. Negus, CR and Peacock, NJ., *J. Phys.*, *E: Nucl. Instrum.*, **22**, 91-111, 1979.
52. Morita, S. and Fujita, J., *Appl. Phys. Lett.*, **45**, 443-445, 1983.
53. Arita, H., Suzuki, K., Kurosawa, Y. and Hirasawa, K., *IEEE Trans on Plasma Science*, **18**, 695-697, 1990.
54. Nakano, N et al., *Appl. Opt.*, **23**., 2386-94, 1984.
55. Gilmour, AS., in "Microwave Tubes", Artec House, (ISBN 0-089006-181-5), 1986.
56. Essig, J., *Adv. in Electron Phys.*, **20**, 73-85, 1990.
57. Gordon, LB. and Logue, AC., Proc. 8th IEEE Pulsed Power Conference, p 1046-1049, 1991.
58. BEAR Project Final Report, Vol 1: Project Summary, Los Alamos National Laboratory, LA-11737-MS, Vol1 BEAR-DT-7-1, 1990.

11

High Voltage Breakdown in the Electron Gun of Linear Microwave Tubes

AJ Durand & AM Shroff

11. 1   Introduction

The electron gun in a microwave tube is used to shape the electron emission from the cathode into a beam suitable for interaction with the microwave circuit. A typical example of a high power microwave tube (a multi-cavity klystron) is presented in Fig 11.1. The lower part is the gun; above are located the cavities where the beam interacts with the field and finally the collector where the electrons dissipate their energy.

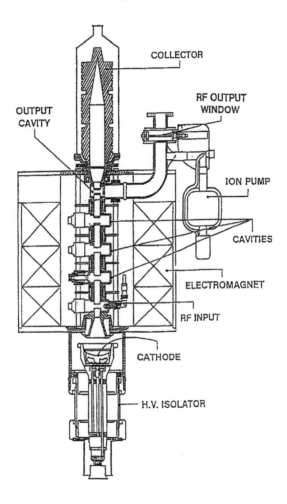

Fig 11.1   Section through a multi-cavity klystron.

The construction of very high peak-power electron tubes (i.e. few tens to hundreds of megawatts under very short pulse operation), or electron tubes with lower peak power operating under longer pulse, or DC conditions, is rapidly faced with the problem of voltage breakdown in the gun structure. Because of the many parameters involved, the problem can be very complex. For example, it is influenced by the voltage, the electric field, the presence of a high current electron beam, the material used to manufacture the electrodes, the presence of a hot cathode that generates a range of pollutants, and the physico-chemical preparation treatments.

The particular application domain to be discussed in this review is defined in Section 11.2. Subsequently, in Section 11.3, both published and non-published data on a variety of electron guns will be presented. These data are then analysed to bring out the most significant parameters. This has led to the data being plotted as a function of four parameters: the cathode voltage V_k, the cathode diameter $\emptyset_k$, the cathode current I_k, and the ratio of the maximum electric field on the focusing electrode to the cathode voltage. In the same section, the experimental data will be analysed in order to highlight the influence of the geometry of the focusing electrode on the electric field.

To evaluate the quality of a given electron gun, it is interesting to have an estimate of the minimum electric field which is achievable by optimising the shape of the electrodes. Accordingly, Section 11.4 presents a "ranging" computation, based on a computer program, that leads to a simple formula for $(E_{max})_{min}$. In Section 11.5, the various data known on electron guns are compared with results obtained by Cranberg on voltage limitations for electrodes in vacuum. The most interesting consequence is the reformulation of the limit of electric fields as a function of pulse length.

11. 2   Specification of Selected Class of Tubes

The subject area to be considered is that of "O" type electron tubes which essentially includes two families:

- travelling wave tubes (TWTs),
- klystrons.

both of these groups of tubes usually employ Pierce type electron guns, in which the cathode has the shape of a spherical concave surface.

The electron gun is characterised by two electrical parameters, the electron current I_k emitted by the cathode, and the accelerating voltage V_k. According to the user's needs, the current may be DC or modulated under pulse field conditions, as illustrated in Table 11.1.

Table 11.1 Illustrating the possible operating regimes of Pierce-type electron guns.

		Cathode Voltage (V_k)	
		DC	**Pulse**
Electron emission current I_k	DC	Diode electron gun	
	Pulse	<u>Types of modulation</u>: Electron gun with anode Electron gun with focusing electrode Electron gun with grid	Cathode modulation

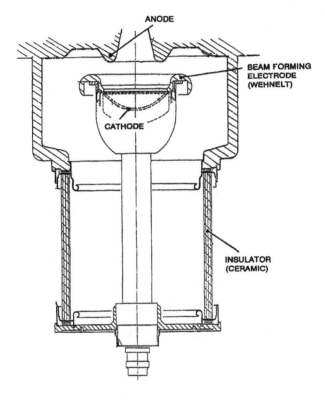

Fig 11.2   Structure of diode-type electron gun.

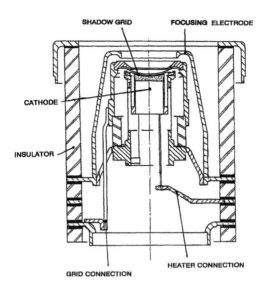

Fig 11.3 Structure of a grid modulated electron gun.

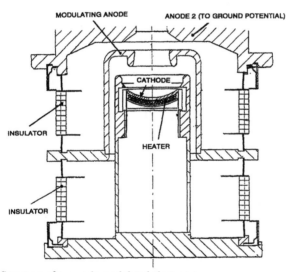

Fig 11.4 Structure of an anode modulated electron gun

Figs 11.2 to 11.4 present the various types of electron gun structures referred to in Table 11.1. For these, the applied voltage covers the following range:

- DC voltages up to 100 kV,
- Pulsed voltages up to 5-600 kV.

407

For all types of gun, the electron emission is related to the voltage through the equation:

$$I_k = PV_k^{3/2} \qquad\qquad 11.1$$

where P is the perveance of the gun structure, which is usually in the range

$$10^{-7} < P < 2.5 \times 10^{-6} \qquad [A/V^{3/2}]$$

The above limits on current and voltage lead to corresponding limitations on the electron beam power generated by the gun and, in consequence, the output power. This is clearly shown on Fig 11.5 which presents the peak and mean power limitation as a function of the frequency.

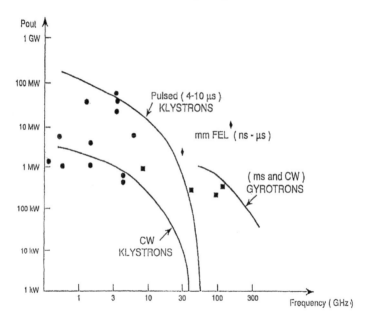

Fig 11.5   Peak and mean power limits of electron guns.

Among the wide ranging and varied reasons leading to arc generation is the effect of the electric field on the electrodes at cathode potential. Typically, this can range from around 1 kV/mm to 10-20 kV/mm; indeed, this field is one of the principal causes of arcing. From this, the beam area convergence (defined as the ratio of the surface area of the cathode to the cross-sectional area of the beam) has to be < 100 in order to obtain good

408

quality electron optics and an adequate current density; the latter being related to the cathode temperature (800 < T < 1100°C). If one tries to exceed the operational limits imposed by the state of the art, one is rapidly faced with the problems of voltage breakdown phenomena (arcs and microdischarges) in the gun.

11. 3   Experimental Data on Pierce Type Guns

This section is divided into two sub-sections. In the first, a set of parameters are selected that provide the best description of the electrical phenomena that occur in an electron gun; in the second, the values of these parameters, and the conclusions to which they lead, are presented for a large number of travelling wave tubes and klystron gun structures.

11.3.1   Choice of Parameters

These have been selected according to three important physical phenomena that occur in electron guns.

a)   *Electron Emission from the Cathode*

In the simplest case, namely the diode, the electron emission follows the Child-Langmuir law of equation 11.1. For such a regime, the perveance is characteristic of a given geometry: it does not depend upon the applied voltage, or the geometric scale. In any case, if the voltage becomes very high, a slight correction is necessary to take into account relativistic effects. The perveance of the electron gun tubes considered here is around 10^{-6}. (N.B Microperveance (μP) is defined as equal to 10^6 P).

b)   *Field Emission and Conduction*

If a voltage is applied to a gun with a cold cathode (i.e. unheated), a small electron current is measured which is called the leakage current. This current is due to either conduction along the insulating parts or electron emission from electrodes that are at a negative (cathode) potential. In this latter case, the emitted current in general follows the Fowler-Nordheim law which is characteristic of field emission (see Chapters 2 and 4) i.e.

$$I = A\ E^2 exp(-B/E)$$ 11.2

where E is electric field at the emission site.

For a given geometry, the electric field is proportional to the voltage, so the relation becomes

$$I = A'V^2 exp(-B'/V) \qquad 11.3$$

or

$$I/V^2 = A' exp(-B'/V) \qquad 11.4$$

and

$$log\ (I/V^2) = log\ A' - B'/V \qquad 11.5$$

Thus a plot of log (I/V^2) versus (I/V) (known as Fowler-Nordheim or F-N plot) gives a straight lines with a negative slope when field emission is responsible for the phenomenon.

In contrast, when conduction along insulating surfaces is responsible for the phenomenon, the current I is proportional to V, i.e.

$$I = C\ V \qquad 11.6$$

where C is the conductance. This relation can alternatively be written as

$$I/V^2 = C/V \qquad 11.7$$

or

$$log(I/V^2) = log\ C + log\ 1/V \qquad 11.8$$

It follows that a F-N plot in the case of insulator conduction will be characterised by a logarithmic curve.

An example of this type of plot is presented in Fig 11.6 for a travelling wave tube operated at 20 kV. For low voltages (I/V large), the conduction regime is predominant, while for large values of V, it is the field emission process that dominates.

c)   *Electric Arcs*

There is no simple relation between current and voltage when an arc occurs. In addition, these parameters vary very rapidly with time, and are a function of the characteristics of the power supplies and the circuit components. Generally, it is a perturbing phenomenon that interrupts the tube operation and causes quite severe damage due to the energy developed during the arcing. It is mainly this phenomenon that one wishes to avoid. Among the many possibilities of breakdown initiation, two cases have been chosen:

-     field emission from the cathode,
-     clumps, as postulated by Cranberg.

410

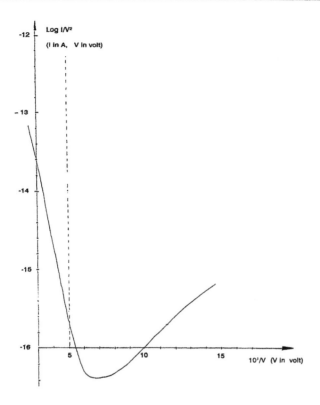

Fig 11.6 Fowler-Nordheim type plots for the electron gun of a 20 kV travelling wave tube.

In the case of field emission, arcing is assumed to be initiated at the site that emits the highest current density. Normally, this emission site will be located in the zone where the electric field is maximum.

As previously discussed in Chapter 7, Cranberg [4] postulated in 1952 the existence of small aggregates, clumps, or microparticles, that have a weak adherence to the electrode on which they are deposited. These microparticles acquire a charge Q proportional to the surface charge density, which is, in turn proportional to the local electric field E. The mutual repulsion between charges having the same polarity results in the ejection of the microparticle which, after acceleration, bombards the opposite electrode with an energy QV proportional to the product EV. If the energy is above a certain level an arc is produced. This mechanism is thought to be particularly important in the case of devices employing dispenser cathodes where the possibility exists for particles of cathode material (e.g. barium) to be injected into the high voltage gap.

If one accepts this theory of arc initiation by clumps, the product EV has to be higher than a given limit, if an arc is to be produced. As the clumps may originate from either the cathode or anode, consideration has to be given to both the cathode and anode electric field. Many experiments [5,6] prove that the field referred to in the product EV is the cathode field. On all electrodes that are at cathode potential, the value of V is the same. Hence, to keep the product EV lower than a given limit is equivalent to limiting the electric field. As a result, the probability of arcing will be highest in the area where the electric field is the maximum. So, for both cases, the conclusion is the same. If, then, the three electric phenomena occurring in electron guns are taken into account, four parameters may be retained

V_k - the cathode voltage (referenced to the anode)
I_k - the cathode current
P - the perveance
E_{max} - the maximum electric field on the electrode at cathode potential.

At this stage, it should be noted that two of the parameters, P and E_{max}, depict the gun geometry: however, they vary differently according to the following scaling transformations.

Scaling of the Voltage
If the voltage is multiplied by a factor m, while the dimensions of the gun remain constant, the electric field E is multiplied by m and the perveance remains constant. The ratio E/V is independent of V and depends on the geometry.

Scaling of the Dimensions
If all the dimensions of the gun are multiplied by h, and the applied voltage and remains constant, the perveance and the convergence of the gun will remain unchanged, and the electric field will be multiplied by 1/h.

In order to compare guns, a reference length has been introduced: i.e. the cathode diameter $\emptyset_k$ is related to I_k and the current density J_k through

$$I_k = (\pi/4)\emptyset_k.^2 J_k \qquad\qquad 11.9$$

The seven parameters I_k, V_k, P, E_{max}, $\emptyset_k$, j_k and the ratio E_{max}/V_k are then connected by the three relations 11.1, 11.9 and the following:

$$E_{max} = (E_{max}/V_k)V_k \qquad\qquad 11.10$$

In fact only four parameters are independent. Thus, if the geometry of the

412

electron gun is to be characterised, the independent parameters (E_{max}/V_k) and $\emptyset_k$ will be chosen. The two other independent parameters will be I_k and V_k, and the linked parameters E_{max}, P and I_k, from which the diagram of an electron gun can be obtained, as shown in Fig 11.7. In this figure:

- the first quadrant gives I_k versus V_k (curves of constant perveance having a parabolic shape),
- the second quadrant gives I_k versus $\emptyset_k$ (curves of constant current density J_k also having a parabolic shape),
- the fourth quadrant gives (E_{max}/V_k) versus V_k (plots of constant field E_{max} are hyperbolic).

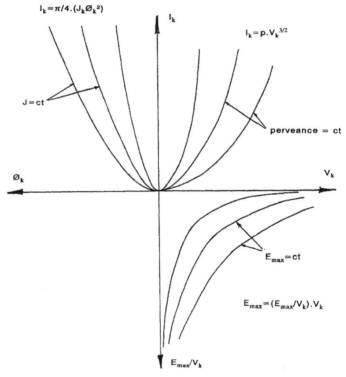

Fig 11.7   Electron gun diagram (linear plot).

From what has been said previously, E_{max} and E_{max}/V_k vary as the reciprocal of $\emptyset_k$ for guns having the same perveance. In the third quadrant, curves of constant perveance may be plotted that are hyperbolic and characterise families of similar guns.

For a better presentation of the gun diagram, logarithmic scales have been used on Fig 11.8, 11.9 and 11.10. As a result, the previous curves become straight lines. To illustrate how a gun diagram is used, the values of three tubes are presented in Fig 11.8:

- a travelling wave tube operated at 70 kV
- a SLAC klystron delivering a peak power of 150 MW at 3 GHz [8]
- an electron gun used in a free electron maser operated at 580 kV [3].

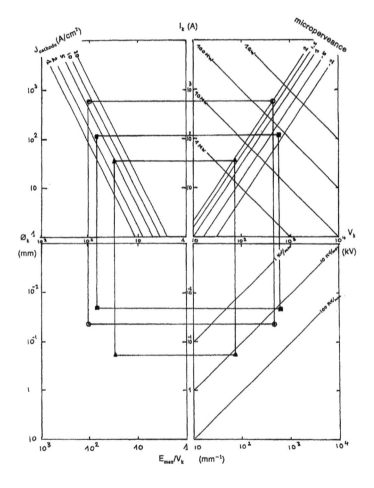

Fig 11.8   Electron gun diagram using log-log scales.

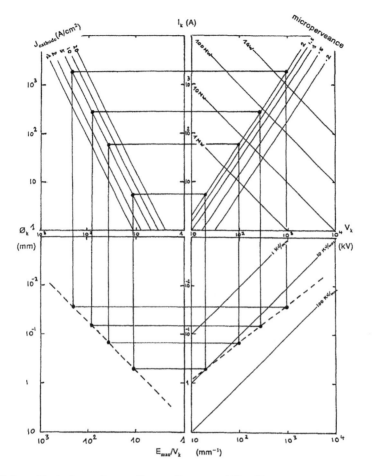

Fig 11.9   Typical use of a gun diagram (increase of V_k with constant perveance and current density J_{k}.)

This type of diagram (Fig 11.8) allows one to follow the evolution of the parameters as one of them is varied, while keeping the others constant. Thus, assuming the electron current density (and therefore the set cathode operating temperature) and the perveance are kept constant, it will be seen from Fig 11.9 that:

-     the cathode diameter $\emptyset_k$ increases as $V_k^{0.75}$
-     the ratio E_{max}/V_k varies as the reciprocal of $\emptyset_k$
-     the electric field increases as $V^{0.25}$ from 10 kV/mm at $V_k = 20$ kV to 20 kV/mm at $V_k = 300$ kV.

415

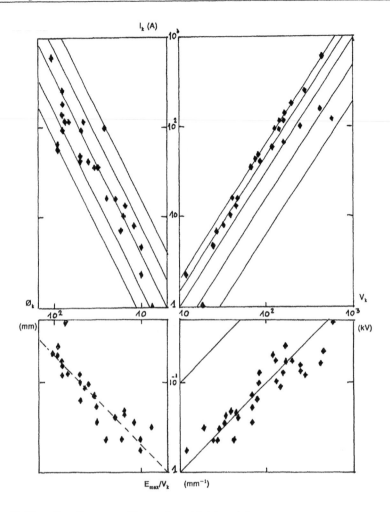

Fig 11.10 Gun diagrams with experimental data points.

This clearly shows that the electric field increases rapidly and reaches electrical breakdown levels. To avoid such a breakdown, it is possible to operate at constant electric field, while keeping the same perveance as in the previous example. Then,

- the current density varies as $V^{-0.5}$
- the cathode diameter and voltage increase rapidly, which results in an increase of the gun convergence.

These two examples show that the choice of the cathode diameter, from among the other parameters, results in a compromise between the

416

convergence and the value of maximum electric field at the focusing electrode.

11.3.2   Use Gun Parameters

The values of the parameters V_k, I_k, $\emptyset_k$ and E_{max}/V_k, corresponding to a large variety of travelling wave tubes and klystrons, have been plotted on the gun diagram of Fig 11.10.   These data will now be used to highlight a number of important design considerations.

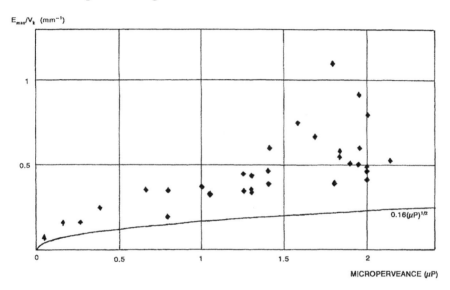

Fig 11.11   (E_{max}/V_k) versus the perveance, for a cathode diameter of 10 mm.

a)   *The Third Quadrant*
If a series of similar guns having the same perveance are considered, the electric field varies as the reciprocal of the cathode diameter. In the third quadrant, these points are located on a straight line having a negative slope of -1. However, Fig 11.10 shows that all the electron guns having a perveance of 2×10^{-6} are not located on a straight line of negative slope -1. While some are located on the line, others are below it, i.e. corresponding to higher electrical fields.   Moreover, other electron guns, having low perveances, are also located below the straight line of slope -1. In order to compare all possible electron guns, a preliminary scaling has to be implemented in order to reduce them to a common cathode diameter. Thus, Fig 11.11 presents a plot of E_{max}/V_k for $\emptyset_k = 10$ mm as a function of the perveance. It shows that E_{max}/V_k increases with the perveance, and that for a given perveance, the dispersion can be important.

417

Two questions arise:

(i)   *Is there a lower limit for E_{max}/V_k for a given perveance?*
      The answer to this question is in the affirmative, where an estimate
      based on computation is given in Section 11.4.

(ii) *What is the origin of the observed dispersion?*
      This point will be treated in the following section.

b)   ***Influence of the Electrode Geometry on the Maximum
      Electric   Field***

Fig 11.12 presents a plot of the equipotentials of a high power klystron
electron gun having a perveance of 2×10^{-6}. The electric field is maximum
in the vicinity of the focusing electrode, and increases as the radius of
curvature of the focusing electrode R_w decreases.

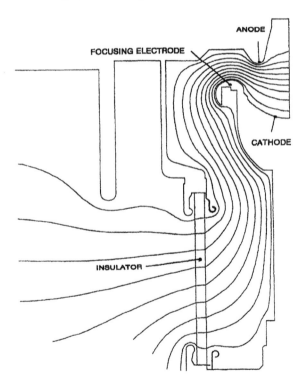

Fig 11.12   Equipotentials of a high power klystron gun.

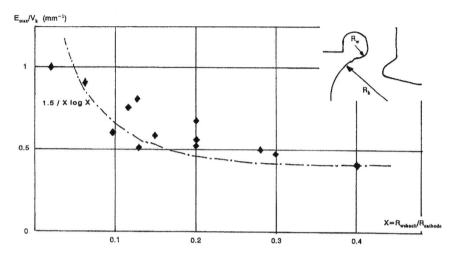

Fig 11.13 E_{max}/V_k versus the ratio $X = R_{wehnelt}/R_{cathode}$.

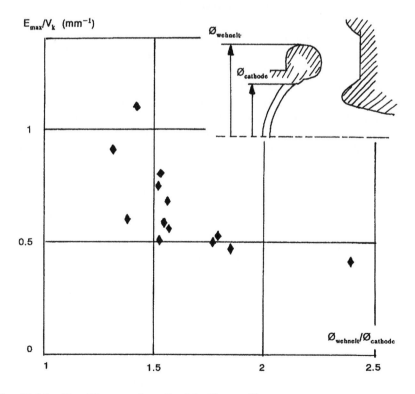

Fig 11.14 E_{max}/V_k versus the ratio of the $\varnothing_{wehnelt}/\varnothing_{cathode}$.

419

Fig 11.13 presents the variation of E_{max}/V_k versus $X = R_w/R_k$, for $\emptyset = 10$ mm (R_k is the radius of curvature of the cathode). This shows that the smaller is R_w the greater is the ratio E_{max}/V_k. These results can be compared with the electric field occurring between two coaxial cylinders. If the radius of the inner cylinder is b and the radius of the outer cylinder is a, the maximum electric field is located on the central conductor and its value is $E_{max} = V/b \log(a/b)$. Whilst the focusing electrode of radius R_w can be simulated to b, it is difficult to find an equivalence to a, i.e. since the distance between the focusing electrode and the other electrodes, which are at ground potential, varies strongly from one gun to another. However, if R_w is normalised against the cathode radius R_k, it can be considered as a variation of the type k/x log x. In Fig 11.13, 1.5/x log x has been plotted against x. These considerations are useful in the design of an electron gun.

N.B. If a large radius of curvature R_w is needed in order to obtain a low field at the edge of the focusing electrode, this leads to a greater overall diameter of the focusing electrode $\emptyset_w$, i.e. as illustrated in Fig 11.14. Thus, a ratio of $\emptyset_w/\emptyset_k$ of the order of 1.8 would be a conservative choice in this case.

c) *Influence of the Pulse Length*
It is known from experiments [9,10] that the breakdown limit of an electron tube operated under pulse conditions is improved when the high voltage pulse length is decreased. As an example, Fig 11.15 presents the maximum electric field at the edge of the focusing electrode as a function of the pulse length. For the tube referred to as **A** in Fig 11.15, the decrease from 600 μs to 10 μs allows an increase in the cathode voltage from 130 kV to 200 kV, and hence an increase of the electric field on the edge of the focusing electrode from 7.56 to 11.6 kV/mm.

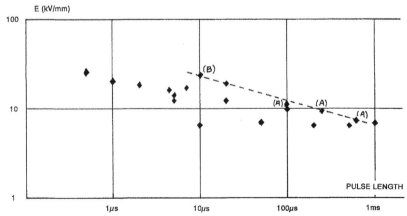

Fig 11.15   The dependence of E_{max} on pulse length.

It will also be seen from Fig 11.15 that a limit can be drawn which links tube **A** and tube **B**, above which the breakdown occurs. However, if the limit is extrapolated towards small pulse lengths, the experimental points remain well below this limit and the field in general does not exceed 20 kV/mm, even when very small pulse lengths are considered. To explain this limitation, it should be noted that small pulse lengths correspond to electron tubes operated at very high voltages, and the maximum permitted electric field decreases when the voltage is increased. This point will be developed in Section 11.5 where the Cranberg results are presented, and a reformulation of the limit of the field E versus τ is proposed.

11. 4　Mathematical Computation of the Minimum Electric Field in a Pierce Gun

Fig 11.16 represents the geometry of a Pierce gun. If the effect of the radius of curvature of the focusing electrode is neglected, and the shape of the truncated cone is adopted, the maximum electric field on the focusing electrode is obtained at the point where the distance d' to the anode is a minimum.

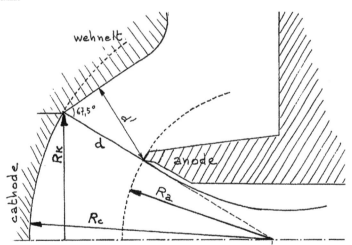

Fig 11.16　Pierce type gun geometry.

R_k　　　　　Cathode radius ($\emptyset_k = 2R_k$)
R_c　　　　　Spherical radius of curvature of the cathode
R_a　　　　　Spherical radius of the anode
d　　$=$　　$R_c - R_a$
d'　　　　　Minimum distance between the focusing electrode and the anode
($d' = d \sin 67.5° = 0.92d$)

The minimum value of the maximum electric field on the focusing electrode is obtained when d' is minimum, i.e. when

$$d' = d.\sin(67.5°) = 0.92\ d \qquad\qquad 11.11$$

with $\qquad d = R_c - R_a$

and

$$(E_{max})_{min} = V_k/d' = V_k/0.92d = 1.08V_k/d$$

or

$$(E_{max}/V_k)_{min} = 1.08/d \qquad\qquad 11.12$$

In order to determine the lower limit of E_{max}/V_k, it is sufficient to compute

$d= R_c - R_a / R_k$

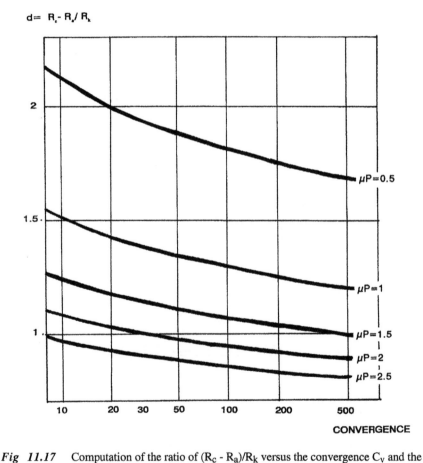

Fig 11.17 Computation of the ratio of $(R_c - R_a)/R_k$ versus the convergence C_v and the microperveance μ_p.

d versus the electron gun parameters (perveance, area convergence). This computation has been made using the computer program "**Pierce gun synthesis**" of Vaughan [12], where Fig 11.17 presents a plot of d/R_k versus the convergence, for microperveances ranging from 0.5 to 2.5.

With the exception of millimetric wave tubes, the electron guns have a convergence lower than 100; indeed, with the exception of UHF and VHF tubes, convergences are for the most part below 40 to 50 and above 10. Fig 11.18 presents a plot of the reciprocal of d versus perveance for convergences between 10 and 100, and a cathode diameter equal to 10 mm.

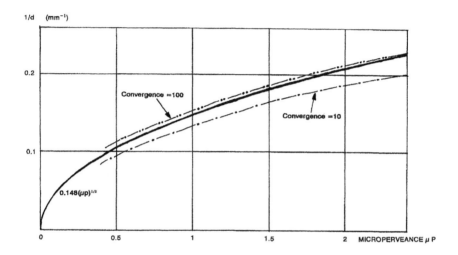

Fig 11.18   Computation of 1/d versus the perveance for a cathode diameter of 10 mm.

This verifies that the reciprocal of d increases with increasing perveance, i.e. in agreement with the results of Fig 11.11. It should also be pointed out that, for values of the microperveances (μ_p) between 0.5 and 2, the expression $0.148 \, (\mu P)^{1/2}$ is a good approximation for the reciprocal of d. Indeed, the discrepancy with calculated values for convergences of 10 and 100 is lower or equal to 10 %. This result is not caused; for a planar diode, having two surfaces Σ separated by a distance d, the perveance is

$$p = 2.33 \times 10^{-6} \, \Sigma/d^2 \qquad\qquad 11.13$$

and $E = V/d$

Hence $E/V = 1/d$ varies as $(p)^{1/2}$. It is therefore natural to look for an approximation of E_{max}/V_k which varies as $(p)^{1/2}$.

For the Pierce gun, the equation is

423

$$(E_{max}/V_k)_{min} = 1.08/d = (1.08)\ (0.148)\ (\mu P)^{1/2} = 0.16\ (\mu P)^{1/2} \qquad 11.14$$

which is plotted on Fig 11.11. From this plot it can be seen that the curve constitutes a lower limit. To assist the comparison between two guns having different perveances, the ratio between (E_{max}/V_k) for $\varnothing_k = 10$ mm and the calculated limit $0.16\ (\mu P)^{1/2}$ has been plotted versus the microperveance in Fig 11.19.

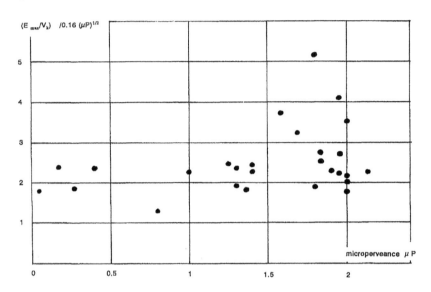

Fig 11.19   Ratio of E_{max}/V_k over $0.16(\mu P)^{1/2}$ versus microperveance μP

Apart from a few exceptions, the ratio is in general greater than 2. The ratio (E_{max}/V_k) for a cathode of $\varnothing_k = 10$ mm can be written

$$(E_{max}/V_k) = \alpha\ (\mu P)^{1/2} \qquad 11.15$$

where computed value of $(E_{max}/V_k)_{min}$ have been obtained for $\alpha = 0.16$. Here, one must keep in mind that $a_{min} = 0.16$ is also a computed value with an accuracy that is no better than 10% for a cathode having a diameter $\varnothing_k$ (mm); hence E_{max}/V_k is given by

$$E_{max}/V_k = \alpha\ (\mu P)^{1/2}(10/\varnothing_k) \qquad 11.16$$

This relation characterises the third quadrant of the diagrams of Figs 11.7 to 11.10. Now, E_{max} can be estimated through the elimination of the various parameters between the four equations 11.1, 11.9, 11.10 and 11.16, and leads to

$$E_{max} = 0.886 \; \alpha(J_k)^{1/2}V_k^{1/4} \qquad\qquad 11.17$$

where E_{max} is in kV/mm, J_k is Acm2, V_k in volts, and $(E_{max})_{min}$ is obtained for $\alpha_{min} = 0.16$. The parameter α is characteristic of the geometry, and so $\alpha > 0.16$ can be explained by three possibilities:
- the distance between the anode and the focusing electrode is too small,
- the radius of curvature of the latter is too small,
- the external envelope is too close to the focusing electrode.

The formula points out the importance in the choice of J_k when a gun is designed. It is not only the choice of the cathode current density that determines its operating temperature, and consequently its life, but also the cathode dimensions since $(J_k^{1/2})$ varies as the reciprocal of $\emptyset_k$, and therefore the electric field.

11. 5   Comparison with Cranberg Results

In Section 11.4 the estimated minimum value of the maximum electric field in an electron gun was computed. It is now important to compare these data with the electric field limits at which arcs occur; the limiting electric field being either deduced from experiment or computed from theoretical models. Thus Fig 11.15 shows that there is a tendency towards increasing the electric field when the pulse length is decreased, but without a clear limit. It can be also noted that the combined effect of increasing the high voltage and shortening the pulse length are partially compensating. However, it remains important to have the possibility of separating these two effects in order to better understand these dependencies.

It is known from earlier experience [1] that, when the distance between two electrodes is increased, the breakdown voltage increases and the breakdown electric field E_{lim} decreases according to the relationship [4]

$$E_{lim} = 10/V \qquad\qquad 11.18$$

where E is in MV/m (or kV/mm), and V is in MV. This limiting value of E can be rewritten by introducing the product EV, i.e.

$$E_{lim}V = 10^4 \qquad\qquad 11.19$$

where E is in kV/mm and V in kV. Figure 1 in Cranberg's original article [4] presents the breakdown voltage versus the distance between electrodes having geometrical shapes such that the field is approximately uniform; i.e. where the electric field is given by the ratio V/d. This figure has been re-plotted in Fig 11.20 with modified co-ordinates: namely, the product EV plotted against the high voltage V. The results from electron guns can now be plotted on such a diagram. Thus, in Fig 11.21, the close circles correspond to cathode modulated tubes and the open circles to tubes operated under DC conditions.

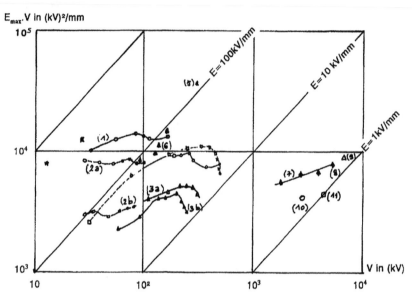

Fig 11.20 Plots of the product EV versus V. (Values after Cranberg [4].)

The Fig 11.21 shows that, for tubes operated under pulse conditions, the values of V_k and of the product $V_k E_{max}$ are higher than that obtained with tubes operated under DC conditions, where the high voltage does not exceed 100 kV and the product EV is lower than 10^3 (kV)²/mm. The evolution of EV as a function of V can be described by equation 11.17, and reformulated as

$$EV = 0.886 \, \alpha \, J_k^{1/2} V^{1.25} \qquad\qquad 11.20$$

It can also be noted from Fig 11.21 that, when the voltage increases, the field is not constant.

To specify the influence of the pulse length, $E_{max}.V$ is plotted against pulse length τ, as shown in Fig 11.22. This indicates that when τ decreases, the value of EV increases; however from Fig 11.15, it will be seen that when τ decreases, the electric field does not increase significantly.

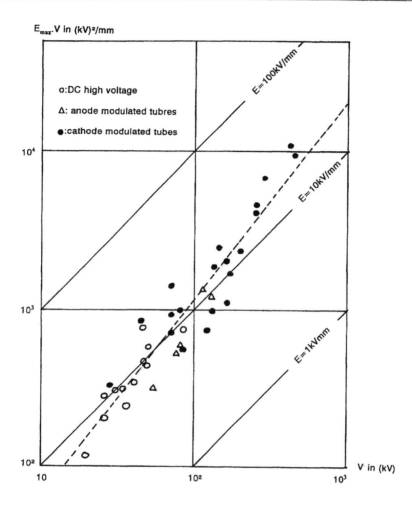

Fig 11.21 $E_{max}V$ versus V for a Pierce type gun.

If the points corresponding to high values of E.V are linked together, almost all guns are below a limit defined as

$$EV \sim 100 \, \tau^{-0.34} \qquad\qquad 11.21$$

or its equivalent $\tau \sim (E.V/100)^{-3}$

Such a relationship has to be taken into account for electron tube design. The points **A** In Fig 11.22 correspond to breakdown experiments on high voltage electron guns from X-band klystrons at SLAC.[7]; clearly, these points are beyond the limit $EV = 100 \, \tau^{-0.34}$. It should also be noted that

427

Staprans [9] obtained a good fit of his experimental data with the empirical relation

$$V_{max} = K_2 \, L^{0.8} \qquad\qquad 11.22$$

where V_{max} is the breakdown voltage of the gun, L the electrode spacing, and K_2 a parameter depending on pulse length. His results are coherent with equation 11.20.

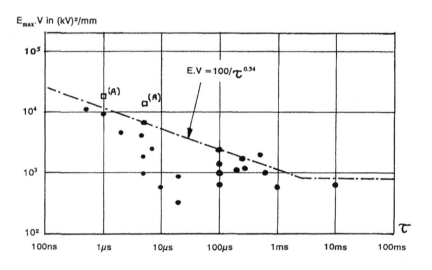

Fig 11.22 E_{max} V versus pulse length.

For guns having the same perveance and area convergence, the ratio Ø/L is constant (Ø being the cathode diameter); also, from equation 11.9, $J^{1/2}$ varies as 1/Ø, or 1/L, and the product EV, given by equation 11.20, varies as $V^{1.25}/L$. Thus, if EV is constant, it follows that $V^{1.25}/L$ is constant and likewise $V/L^{0.8}$.

For a pulse duration greater than 1 ms, and for DC voltages, the limit does not follow equation 11.21, but becomes approximately

$$EV = 8 \times 10^2 \ (kV)^2/mm \qquad\qquad 11.23$$

Compared to the value of EV given by equation 11.19, there is a difference of approximately 11dB. This is probably due to the presence of a hot cathode, hot electrodes in the gun, or electrodes polluted by barium deposited from the cathode. This EV limit may be increased by attention to the physico-chemistry of the gun electrodes and in particular by:

- lowering the electrode temperature,
- having well cleaned componant parts (physically and chemically treated),
- having well degassed parts through firing and baking.

Improvements can also be obtained by modifying the gun structure by way of introducing intermediate electrodes between the cathode and the anode; i.e. which corresponding to a transposition of the classical results of Van de Graaf for insulators to the gun structure. In pursuance of this approach, a recent calculation has been made [11], and experiments are under way with the testing of a 1MV ku band klystron [2].

11. 6   Conclusions

Breakdown in the electron gun is one of the major limitations on the performance of microwave tubes. If continuous voltages are applied to the tube, the limitation of the product EV results in a limitation of the CW output power. However, peak output powers, higher than the limitation of CW tubes, can be obtained with pulsed voltages: the shorter the pulse length, the higher the output power.

Acknowledgements

The authors wish to thank Dr. B. Epsztein for helpful suggestions and criticisms during the preparation of the manuscript.

11. 7 References

1.   Anderson, HW., *Electrical Engineering*, **54**, 1315-1320, 1935.
2.   Balakin, VE. et al., Int. Workshop on Pulsed R.F. Power Sources, Dubna, 1933.
3.   Chu, TS. et al., *Nuclear Instr. Meth.*, **A289**, 1-7, 1992.
4.   Cranberg, L., *J. Appl. Phys.*, **23**, 518-522, 1952.
5.   Farral, G.A. et al., *J. Appl. Phys.*, **36**, p 2966, 1965.
6.   Kaljatsky, I. et al., *IEEE*, Trans. Elec. Instrum., **20**, 701-703, 1985.
7.   Koontz, R. et al., *SLAC Pub 5257*, 1990.
8.   Lee, TG. et al., *IEEE. Plasma Science*, **13**, 545-552, 1985.
9.   Staprans, A., *Proc . II-ISDEIV*, p 293-303, 1966.
10.   TTE internal report, 1993.
11.   True, R., I.E.D.M.,Trans. Inst. Elec. Dev. Man., 403-406, 1991.
12.   Vaughan, JR., *IEEE,* Trans. Elec. Dev., **28**, 37-41, 1981.

12

RF Superconducting Accelerating Cavities

H Padamsee

12. 1   Superconducting Microwave Cavities for Particle Accelerators

Superconductivity is at the heart of new accelerator technology. Both electron and proton accelerators are reaping benefits from this phenomenon. Whereas superconducting (SC) magnets provide the high magnetic fields needed to guide charged particles in circular orbits, microwave cavity resonators provide the high voltages needed for acceleration. The resonant frequency of such cavities is usually between 100 and 3000 MHz.

Most accelerators today are operated with copper cavities. For continuous operation, as is required for many applications, the power dissipation in the walls of a copper structure is quite substantial, e.g. 0.1 megawatts per metre of structure operating at an accelerating field of 1 million volts/metre (MV/m). Since losses increase as the square of the accelerating field, copper cavities become severely un-economical as demand for higher fields grows with the higher energies called for by experimenters who wish to probe ever deeper into the structure of matter.

Here superconductivity comes to the rescue. The microwave surface resistance of a superconductor such as niobium is 5-6 orders of magnitude lower than that of copper. The quality factor of an SC resonating cavity is of the order of 10^{10}. However, not all of this phenomenal gain is available for practical exploitation, since the wall losses of SC cavities are dissipated at low temperatures (2-4 K). Nevertheless, after allowing for the refrigerator power, the net gain is still a factor of several hundred. In fact, it becomes more attractive to use higher accelerating fields. Typically, for an accelerating field of 5 MV/m, which is readily accessible with today's SC structures, the losses are a few watts/metre. This therefore is the driving incentive for the use of an SC RF system in accelerators.

Apart from the reduced operating power and higher operating fields, there is yet another area of benefit offered by SC cavities to particle accelerators. The presence of microwave cavities in accelerators has a disruptive effect on the beam, limiting its quality and the maximum current available. With the higher voltage capability, an SC system can be made shorter, thus reducing overall beam-cavity interaction. To minimise power dissipation for a desired accelerating field, copper cavities are forced to have geometries that intensify the harmful beam-cavity interaction, i.e. small beam holes. With SC cavities on the other hand, large beam holes become affordable. In the long run, this translates into superior beam quality or higher beam current, qualities that make new applications attractive.

Sustained progress in achieving higher operating field gradients with SC niobium cavities has given impetus to wide-ranging applications in accelerators for high energy physics and nuclear physics research, as well as to drivers for free electron lasers (FEL).

432

Excellent accounts of the various projects, and review articles on all aspects of the technology can be found in the proceedings of five Workshops on RF superconductivity [1,2,3,4,5]. All together, about 150 metres of structures have been installed and operated at accelerating fields up to 6 MV/m. In tests of the same structures before installation into the targeted accelerator, gradients around 10 MV/m are regularly achieved. In many cases, these pre-installation tests are performed without all the components necessary for accelerator operation, such as power couplers and tuners.

Exciting new applications are forthcoming in the areas of storage rings for high energy physics, light sources, and linacs for FELs. To realise these, higher gradients must be reliably achieved. In the most optimistic case, 20 kilometres of SC cavities, with gradients between 20-30 MV/m, will be needed by the early part of the next century for colliding electron and positron beams with energies in the trillion electron volt (TeV) range. Such an accelerator is regarded as an important probe into the ultimate structure of matter, complementary to the LHC (Large Hadron Collider). The LHC will be the largest application of superconducting technology using several thousand SC magnets.

12. 2   Present Performance Compared to Ultimate Limits

To reach the present level of performance in SC niobium cavities, it was necessary to overcome limitations in several categories: weld defects, resonant electron multiplication (multipacting), and thermal breakdown of superconductivity from sub-millimetre sized, non-superconducting regions. Reliable techniques to surmount these difficulties have been invented and implemented. Selecting the proper curvature for the shape of the cavity wall eliminates multipacting. Defocused electron beam welding eliminates weld defects. High purity, high thermal conductivity niobium stabilises the RF surface against thermal breakdown.

The ultimate potential of SC cavities continues to elude us, however; present theoretical estimates set the accelerating field limit at 50 MV/m for structures designed to accelerate particles travelling at near the velocity of light. At this field, the surface RF magnetic field in a well-designed cavity would reach 2000 oersted, equal to thermodynamic critical magnetic field of niobium. Above this level, superconductivity is expected to quench. At the theoretical limit of 50 MV/m accelerating, a well designed accelerating structure (for electrons) would have to support not only a surface magnetic field of 2000 Oe, but also a surface electric field of 100 MV/m. Although the expected magnetic field limit has never been surpassed (the record for the surface RF magnetic field in an SC niobium cavity being 1600 Oe [6]), the electric field counterpart (100 MV/m) has been exceeded on several occasions.

433

In specially designed cavities, a surface RF electric field of 145 MV/m was reached in continuous wave operation [7], and 220 MV/m in pulsed operation [8]. Accordingly, there is no fundamental limit to reaching a surface RF electric field corresponding to 50 MV/m, accelerating.

However, in regular accelerating structures, field emission takes over as the overriding performance limitation above 20 MV/m surface field (about 10 MV/m, accelerating). Excessive heating from field emission currents increases exponentially with field, making SC cavities unattractive above 10 MV/m accelerating. Progress is being made towards a better understanding of field emission in SC cavities, as well as toward inventing techniques to suppress emission. With these new approaches, it is now possible to prepare SC accelerator structures that reach gradients of 15-25 MV/m accelerating in laboratory tests. This chapter will describe the nature of field emission in SC cavities, independent studies of the emission phenomenon and successes in reducing emission.

12. 3   Nature of Field Emission in Superconducting Cavities

A typical SC structure for accelerating charged particles, travelling at near the velocity of light, is shown in Fig 12.1. Electrons from field emission sites located on the cavity surface are accelerated and strike the cavity wall, causing heating and bremstrahlung X-rays. If sufficiently intense, heat deposited from electron bombardment can initiate thermal breakdown, as the temperature of the surface rises above T_c, the critical temperature for superconductivity. Within a single cell 1.5 GHz cavity, Fig 12.2 shows the calculated trajectories in the electromagnetic fields traced by electrons emitted during an RF cycle from a hypothetical emitter. The peak surface electric field chosen is 24 MV/m. Due to the cylindrical symmetry of the accelerating resonant mode, electrons are confined to travel in a plane. The impact energy of the electrons, plotted in Fig 12.3 reaches a maximum of 0.6 MeV in this case. Before 1980, emission studies in niobium RF cavities had been correlated to measurements of (a) the current collected by a pick-up probe (see Fig 12.1) located near the axis of the beam tube of a test cavity, (b) the X-ray intensity outside the cavity, and (c) the increased power dissipation from the bombarding electrons.

These indicators are all observed to independently follow the functional dependence of the Fowler-Nordheim (FN) field-emitted current relationship [9], previously discussed at length in Chapter 7, namely $I = (A\beta^2 E^2 S/\phi) \exp(-B\phi^{3/2}/\beta E)$. Here I is the field-emitted current in amps, E is the applied electric field in V/cm, β is the customary "field enhancement factor", S is the emitter area in cm^2, ϕ is the work function of the emitter

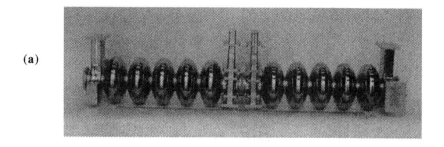

(a)

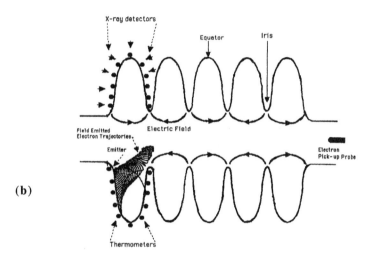

(b)

Fig 12.1   (a) A typical superconducting multicavity structure for accelerating electrons. The wave guides are input and output power couplers.
(b) The electric field lines of the accelerating mode. Thermometers are often placed in contact with the outer cavity wall to detect heating from areas of increased resistance, or impacting electrons from field emitters. X-ray detectors are also used. The equator and iris regions of the cavity are identified.

(4 eV for Nb), whilst A and B are constants having the respective value of A = 1.54×10^{-6}, and B = 6.83×10^7 in the present formulation of the FN relation.   There is also a good correspondence between the different indicators, when their increase with field level is studied [10].   In fact an examination of any of these quantities by means of an FN plot yields the β-values of the emission regime.   For example, an FN plot of the collected current (I), against the applied field E, is a semi-log plot of the quantity I/E^2 vs $1/E$; β is found from the inverse of the slope of the resulting straight line (as reviewed by Lyneis [11] Noer [12], and Weingarten [13].)

435

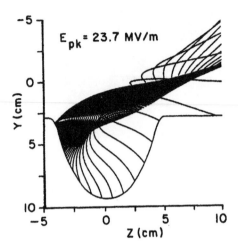

Fig 12.2   Calculated electron trajectories in one RF period emanating from a hypothetical emitter. The maximum surface electric field is chosen to be 24 MV/m for this 1-cell 1500 MHz cavity. The trajectories curve back and hit the wall because of the magnetic field. When tested in the laboratory a pick-up probe placed on the beam axis can collect field-emitted electrons or secondaries.

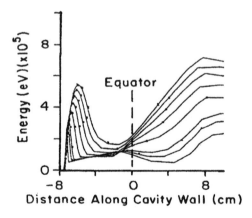

Fig 12.3   Impact energy of electrons plotted as a function of the impact position for the emitter of Fig 12.2. Here 0 is the equator and -8 cm is the emitter location as measured along the cavity wall.

Field emission studies in cavities that span RF frequencies between 0.5 and 10 GHz show β-values in the range from 80 to 1700. It should be recognised that the FN parameters, β-factor and areas, are traditionally used only as parameters to express the emitted current and its field dependence.

Their physical significance is still a matter for debate. For this reason, throughout this chapter we use the simplified Fowler-Nordheim expression which ignores the rounding of the potential barrier due to image charge effects.

Apart from these global indicators, individual emission sites in SC RF cavities can be localised and characterised by measuring in detail the heating of the cavity wall from the spray of emanating electrons [14]. For this purpose, the outer wall of the cavity is covered by a dense array of sensitive thermometers [15] (schematically shown in Fig 12.1). Another method is to mechanically scan the outer surface with a suitable movable arm, bearing a smaller number of thermometers [14]. Using either technique, a series of temperature "maps" may be recorded at increasing field levels, which as illustrated by Fig 12.4, reveal any localised heating of the cavity wall due to emission from one or several emission sites. Information from companion X-ray mapping systems are often used to corroborate the heating patterns observed: thus Fig 12.5 presents an example of correlated temperature and X-ray maps [16].

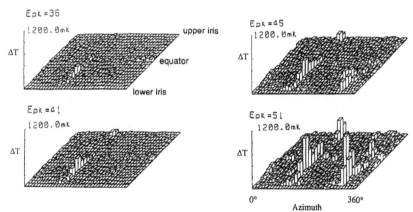

Fig 12.4   A series of temperature maps from a heat treated cavity, taken at increasing peak field levels from 36 to 51 MV/m. Heating is visible from impacting electrons that originate from several field emitters.

From the pattern and amplitude of the observed temperature rise at increasing field levels, it is possible to deduce the Fowler-Nordheim properties of the emission site. The calculated impact current density and impact power density for the candidate emitter of Figs 12.2 and 12.3 are shown in Fig 12.6.

In fact, this is a particularly strong emitter where the maximum impact current density is a few microamps/cm. The lineal power density from electron impact is near 1 watt/cm, and is substantially higher than the power loss of 10^{-2} watts/cm^2 from the background microwave surface resistance

437

(10⁻⁷ ohms). At the outer wall of the cavity, the corresponding temperature rise can be calculated by simulating heat flow through the niobium wall. Because of the thermal transport properties of niobium at 2 K, heat from a point source spreads with an FWHM of $\pm$ 3 mm. The simulated temperature rise of the outer wall is shown in Fig 12.7.

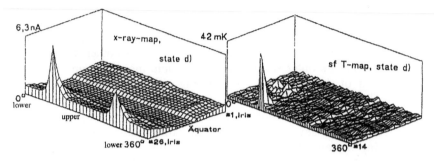

Fig 12.5 Side by side X-ray (left) and temperature (right) maps showing how the heating from impacting electrons and bremstrahlung X-ray emission are strongly correlated. The second X-ray peak, where there is no temperature rise, is from back-scattered X-rays [16].

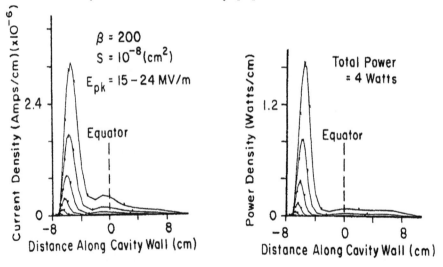

Fig 12.6 Impact current (left) and power density (right) for the emitter of Fig 12.2 at various field levels from 15 to 24 MV/m. The emitter properties selected are $\beta = 200$, area = 10^{-12} m².

The expected 3 K temperature rise from electron impact heating is orders of magnitude larger than the 20 millikelvin increase expected from the uniform RF wall losses at the same field level. Indeed, heating from such a strong emitter can easily be detected, since low temperature thermometry

438

techniques permit detection of millikelvin temperature increments. This corresponds to heat flux densities of milliwatt/cm^2, which in turn corresponds to lineal impact current density of nanoamps/cm. A good temperature mapping system is therefore capable of picking up many emission sites.

Heating profiles observed with the thermometry diagnostic system correspond well to calculations [14,15]. The β-factor and areas of an emitter may then be deduced from a comparison between the observed and simulated heating patterns. As analysed from temperature maps, the β and S values of emitters found in RF cavities are shown in Fig 12.8 for 500 MHz [13], 1500 MHz [15] and 3000 MHz [17] single cell test resonators.

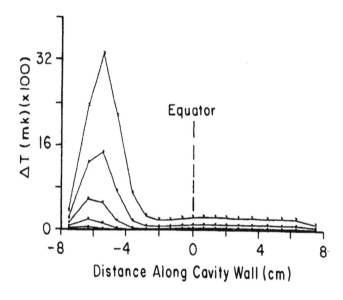

Fig 12.7   Simulated temperature rise at the outer wall of the cavity from the emitter of Figs 12.2 and 12.6. The temperature increase is plotted for several field levels from 15 to 24 MV/m.

It is clear, that to reach high fields, low β values must be attained. On the other hand, there appears to be no strong correlation between emitter area and the maximum field reached. Temperature maps are also used to count the number of emitters found on a cavity surface [18]. Thus, the typical density of significant emitters, as a function of surface electric field, is given Fig 12.9 (solid triangles).

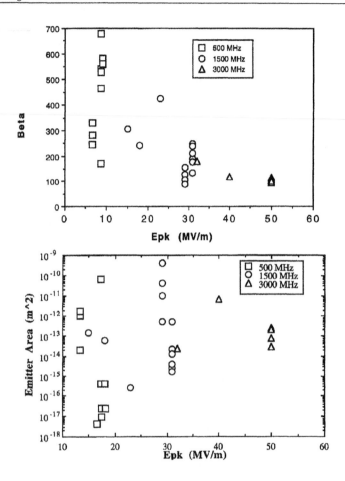

Fig 12.8 Beta (upper) and area (lower) values for various emitters detected in 1-cell SC cavities at three frequencies. The FN properties were obtained from temperature maps.

Here, the effective area is taken as that part of the surface where the field is > 80% of its peak value. It will be seen that there is a rapid increase observed in density of significant emission sites with increasing field.

For comparison, typical emitter densities found on Nb samples of several square centimetres, scanned with a DC needle, are also given [19]. Other information given in Fig 12.9 will be discussed later. These findings make understandable the difficulties in reaching higher operating field levels. There is a clear need for better surface preparation techniques to reduce emitter density and emissive properties i.e. the β- and S-values.

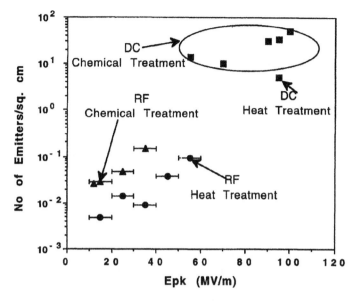

Fig 12.9   Emitter density vs. increasing field level observed in tests of chemically treated (solid triangles) and heat treated (solid circles) cavities. At much higher fields emitter density data from DC studies are included (solid squares).

12. 4   Methods to Reduce Field Emission in SC Cavities

12.4.1   Standard Preparation Techniques

The emitter properties described above were derived from tests on RF cavities prepared by the standard chemical etching procedure, in which a 1:1:1 mixture of HNO_3, H_3PO_4, HF was used. In some cases, electropolishing with a H_2SO_4/HF electrolyte was employed, but there is no clear indication that one chemical treatment method is superior. After etching, the cavities were rinsed in ultra pure water (18 MΩ cm) filtered with 0.2 μm filters for one to several hours, and assembled in a dust-free clean room (Class 10-100). Long rinsing times and clean room assembly have proved to be effective in reaching a good baseline performance. Deviations from these good practices are known to cause excess emission.

To reduce emission in cavities prepared by these standard techniques, three methods have been effective: helium processing, Heat Treatment (HT) in UHV, and High Power Pulsed RF Processing (HPP). The first of these methods has been in use for some time. The last two are more recently developed procedures.

12.4.2   Helium Processing

In He processing, He gas at low pressure (10^{-5} torr, or just below discharge threshold) is admitted into a cold SC cavity [11,13]. The cavity is operated near the maximum possible field level for several hours. Several forms of benefit are realised. Improvement in achievable field levels have been as high as a factor of 2. Orders of magnitude reduction in X-radiation along with factors of 2 reduction in β-values have been reported. Emission heating from individual sites observed by temperature maps were reported to disappear. The most rapid progress in field improvement occurs during the first few hours, where processing typically extends over periods between 1-50 hours.

He processing has also been found to be effective in suppressing field emission from an artificial emitter source, e.g. such as carbon flakes, deliberately placed on the SC cavity surface [20]. In this context, it should be noted that carbon had been shown to be an excellent field emission source as under DC field emission conditions [19].

It has been established that at least part of the benefit of He processing is derived from removal of gas condensates. Emitters activated by deliberately condensing gas were identified by thermometry and subsequently removed by He processing [21]. The remainder of the improvement from He processing has traditionally been interpreted as sputtering of the bulk emitter over longer periods of time. Recent gas conditioning studies on DC field emission from Cu surfaces suggest that gas ion implantation may be an alternative explanation [22].

The processing has some important disadvantages. There is a danger of sputtering metal coatings on to ceramic insulators present in coupling devices attached to the cavity, and thereby damaging them. If all the He gas is not removed, large shifts in frequency of structures during operation are a serious control problem for use in an accelerator [11].

12.4.3   Heat Treatment of Nb Cavities [18,23]

In DC studies, Nb samples scanned for emission sites were high temperature annealed and subsequently re-scanned at ambient temperature, but without removal from the UHV system [19]. Above 1200°C, the density of emitters was drastically reduced. Surfaces several cm^2 in size, which do not emit up to 100 MV/m, were repeatedly obtained after heating to above 1400°C. Both surface particles and emission were observed to disappear. It is reasonable to surmise that emission sources are cleaned up by dissolution and/or evaporation at high temperature.

Encouraged by these results, the influence of high temperature annealing in the final stages of RF cavity surface preparation has been studied. An important difference from the DC studies was that the cavities had to be let up to filtered air after removal from the furnace. The most significant reduction in field emission was observed for 4-8 hour heat treatments at 1500°C. Thus Fig 12.10 shows the surface field reached in several tests of 1-cell 1500 MHz cavities, compared with a large number of tests on chemically treated cavities of the same type.

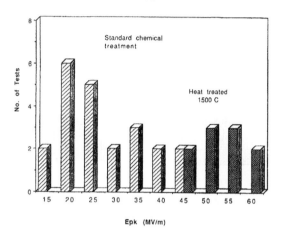

Fig 12.10   Increase of the maximum surface field achieved by heat treatment at 1500°C of many 1.5 GHz, 1-cell cavities.

Fig 12.11 compares temperature maps obtained from heat-treated and non-heat-treated cavities at 30 MV/m peak surface field. Many strong emitters are present when standard treatment is used, whereas the heat treated surfaces are virtually free of emitters at 30 MV/m. Lower temperature treatments (1100-1250°C) were also found useful in reducing field emission, but less substantially.

From the temperature maps, statistics have been compiled on the number of emitters detected as a function of peak field level. Thus, if the effective area of the cavity is taken as that part of the surface over which field is > 80% of the peak field Fig 12.9 shows that there can be a factor of 10 reduction in emitter density as a result of heat treatment. However, it must be stressed that, even though there is strong evidence to show that HT reduces the density of emitters, it remains important that additional emitters should not be introduced in subsequent preparation steps. In many cases, the best results were obtained for HT cavities that were rinsed after HT with dust free methanol, prior to final assembly for cold RF tests.

Heat treatment was also observed to reduce the emissivity of sites as characterised by the β-factor and the emitting areas. Emitters with β-values of 100 show two orders of magnitude lower emission current than emitters with the same β-values seen on chemically prepared surfaces.

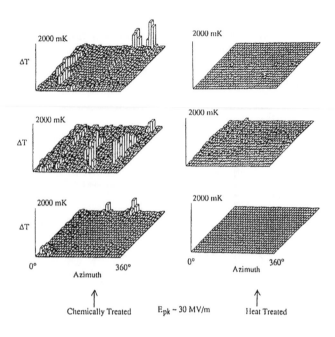

Fig 12.11 A comparison of temperature maps at 30 MV/m between heat treated (right) and chemically treated (left) cavities. Almost no emitters are present in the heat treated cavities. All together the results of six separate cavity tests are presented here.

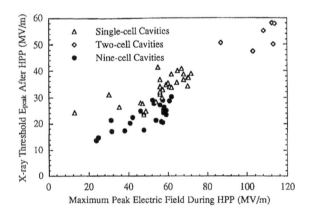

Fig 12.12 CW field level at which first X-rays are observed (y-axis) after processing cavities with pulsed high power (HPP) at field levels shown on x-axis. Absence of X-rays is one of the several signs indicating that the cavities are emission free. Data for 1-cell, 2-cell and 9-cell 3 GHz cavities are shown.

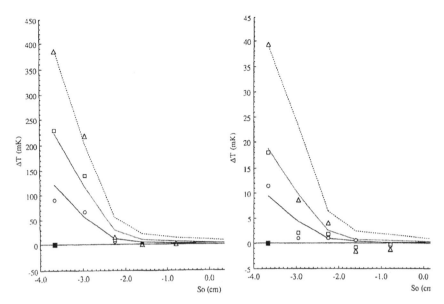

Fig 12.13 Temperature maps show successful processing of a field emission site (left) before processing and (right) after. Open points are data. Lines are calculations after selecting emitter properties. The solid square is the location of the site. Note that after HPP the temperature rise is lower by a factor of 10.

The benefits of HT are being extended to multicell structures, where accelerating fields of between 15-20 MV/m have been reported [24]. With temperature maps, a reduced density of emitters has also been observed in 9-cell cavities at 3 GHz [25]. However, maintaining the cleanliness of HT cavities still remains a problem [25].

12.4.4   RF Processing

When the RF power is increased for the first time on a freshly prepared cavity surface, emission is occasionally observed to decrease, sometimes abruptly. Eventually however, the emission becomes stable, and with the ten to few hundred watts of RF power typically used in Nb cavity tests, no further progress is realised. A comparison of the emissive properties of a few processed emitters with the more stable emitters showed that the emissive properties of the unstable emitters are much stronger, i.e., they have a larger β-factor and emissive areas [18]. As discussed later, it is presumed that higher field emission current plays an important role in the reducing the stability of the emitter.

Following early indications that higher RF power levels in the pulsed mode would continue to be effective [26], there has recently been a thorough

445

and systematic exploration of pulsed high power RF processing (HPP) of SC cavities. With power levels between 5 and 200 kwatts, and pulse lengths between 5 μsec and 2 msec, several 1-cell, 2-cell and 9-cell Nb cavities at 3 GHz have been tested [17]. At first, each cavity was tested with about 50 watts of power in the continuous wave (CW) mode. In most cases, field emission loading limited the maximum field achievable. However, after several doses of HPP, field emission was substantially reduced. During each high power pulse, the Q_0 of the cavity was estimated to fall from 10^{10} to between 10^6 and 10^7, so that the power coupled into field emission was between 3 - 30 kwatts, shared between all the active emitters.

The predominant factor in determining the degree of improvement was always the maximum electric field reached during the HPP stage. Accordingly, the processing power level, the pulse length and the input power coupling strength were adjusted to progressively increase the HPP peak field level. Thus, Fig 12.12 shows how the field emission threshold field level was increased by HPP for 1-cell, 2-cell and 9-cell SC cavities at 3 GHz. Here "threshold" is defined as the lowest field level at which there was the first sign of X-rays. To eliminate field emission in single cell cavities, it was necessary for the HPP field to reach 55 - 70 MV/m. After processing, there were no signs of X-rays up to 40 MV/m. Without any HPP the X-ray thresholds were considerably (50%) lower.

In trying to extend the benefits of this method to go above Epk = 70 MV/m, it was found that the surface RF magnetic field (Hpk) in the equatorial region of the cavity exceeded the limit for thermal stability. By changing the cell geometry to reduce the ratio of Hpk/Epk by 64%, a 2-cell cavity could be processed to 115 MV/m, so that the benefits of HPP could be continued. By this approach, field emission was completely eliminated (absence of X-rays) to 60 MV/m CW.

In 9-cell accelerating structures, using processing fields as high as 60 MV/m, field emission was completely eliminated for CW surface field levels up to 30 MV/m. By tolerating 50 watts of field emission deposited power, the accelerating field reached in 7 tests on two 9-cell cavities ranged between 15-20 MV/m. Nine cell cavities could not be processed much further because of the maximum available 200 kwatts of RF power.

12. 5   Microscopic Examination of Emitters that Process

Electron microscopy studies have yielded direct information about emission sites that process; i.e. exhibit a reduction in emission, or altogether cease to emit [17,27]. For these studies, several 1-cell 3 GHz cavities were dissected after thermometer maps had previously located discrete field emission sites that successfully responded to processing. In one such study, a strong field emitter, identified by thermometry was processed with HPP at 54 MV/m, where Fig 12.13 shows the "before" and "after" temperature maps. On dissecting the cavity, the feature shown in Fig 12.14 was found at the processed emitter location.

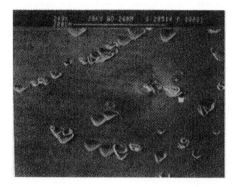

Fig 12.14   SEM photograph of area located by the maps of Fig 12.13. Note the 200 μm size starburst, the two central molten craters near the centre (about 10 μm in size) and the debris field at the periphery of the starburst. The triangular shaped etch pits were present over the entire cavity and are not believed to be related to emission.

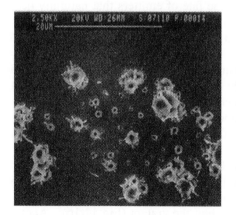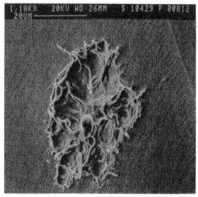

Fig 12.15   Typical clustered craters (left) and overlapped craters (right) of molten Nb found at the centre of starbursts.

This is seen to consist of a 200 μm spot that has a starburst shaped periphery with a core that is a molten crater-like region. At the periphery of the starburst, Ti dust was found. The triangular features are etch pits and were found over the entire surface of the cavity. These are not believed relevant to field emission.

Starbursts and molten craters were found at the high electric field regions of other dissected cavities. Nearby, evidence was found of the molten remnants of the contaminant particle presumed to be originally responsible for the emission, e.g. In, Cu, Fe, stainless steel, Ti and C. Many of these elements are traceable to materials used in cavity assembly and the RF test set-up. This therefore calls for an increased vigilance in cleanliness. In many cases, however, no foreign elements were found to be associated with the starburst feature, only molten Nb.

The presence of the 5-10 μm molten craters does not degrade the performance of SC cavities. In a 1-cell cavity, 40 starburst/molten crater sites were found, but the Q value of the cavity remained above 10^{10} at 40 MV/m surface field.

12. 6   Starbursts, Molten Craters and Other Features [17, 27]

A special shaped SC cavity, called the "mushroom" cavity, was developed in which a small region can be exposed to a very high RF electric field at 6 GHz. This region, the "dimple", is part of a demountable end plate. The total area exposed to fields within 50% of the maximum is 80 mm^2. With this vehicle, cavities do not have to be destroyed. The entire plate fits into the SEM, and the central dimple can be examined. The fate of emitters that naturally occur in the dimple region was studied after the region was exposed to a high RF electric field. In one series of tests, artificial emitters were placed at a single high field location.

In the many of these studies, surface fields at the dimple of between 30 and 90 MV/m were reached. SEM examination revealed starbursts with molten craters and many other interesting features, a few of which are described below. Actually, the mushroom cavity study preceded the dissection of 1-cell 3 GHz cavities, and so the first starbursts and molten craters were found in the mushroom cavity.

At the core of the starbursts, besides the individual craters already shown, clusters of craters and overlapped craters were observed, as shown in Fig 12.15. The overlapped craters apparently formed a moving path. Very often there are ripples or rings surrounding the craters (Fig 12.16). Another observed feature, shown in Fig 12.16, is tracks emanating from starbursts. Foreign elements were frequently found in particles close to the centre of a starburst. The morphology indicates that the particles are partially or fully melted. As an example Fig 12.17 shows a molten stainless steel particle found inside a starburst, near a molten crater. In one dissected 1-cell cavity, a large number of starbursts were found containing molten indium debris (Fig 12.17) at the core. In the same cavity, two partially molten indium flakes <u>without</u> starbursts were also found (see Fig 12.18).

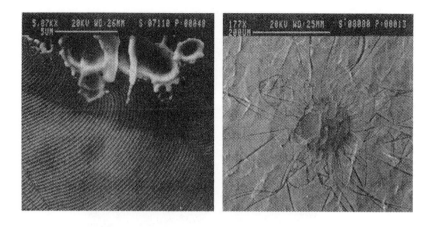

Fig 12.16 Ripples (left) surrounding a molten crater and tracks (right) emanating from the periphery of a starbust.

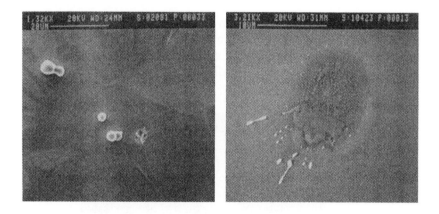

Fig 12.17 (Left) Molten stainless steel particles outside a crater. (Right) Splashed remains of an indium particle. Both features were found inside starbursts.

However, because only a small region of these two particles was observed to be molten, it is reasonable to surmise that the melting did not take place by RF heating. Field emission current heating is the most likely explanation.

The starburst feature is discernible only using the SEM. It is not visible using optical microscopy. In addition, its contrast tends to fade away after a few hours of exposure to air, but the molten core remains unaltered. The ripples also survive exposure to air.

449

Fig 12.18 An indium particle only partially molten. There was no starburst associated with this particle.

Studies of cathode spots from DC arcs [28,29] reveal many similarities to the results discussed here. Photographs show the plasma, whilst post-mortem pictures show the molten eroded areas. But the starburst features have never been reported in DC studies.

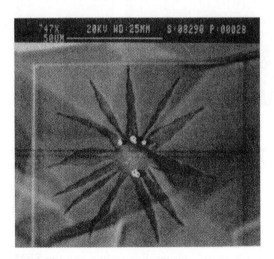

Fig 12.19 Starburst found on a DC field emission Nb cathode. A spark was observed at this site at 110 MV/m.

450

Pursuing these close similarities, experiments were carried out to look for starbursts by initiating a spark with high DC voltage across a small gap between two Nb electrodes in UHV; the Nb surface having been treated in the same fashion as cavity surfaces. Referring to Fig 12.19, subsequent SEM studies of the Nb cathode in the sparked area showed a starburst with molten cores, i.e. similar to those found in RF cavities. It is therefore reasonable to surmise that sparks or microdischarges must also take place in the SC RF cavities.

12. 7   A Model for Emitter Processing and Associated Microscopic Features

Microscopic features found at the processed emission sites after the occurrence of an RF spark indicate that emitter extinction takes place by an explosive process. A superficial contaminant particle, such as the In flake of Fig 12.18, field emits from a small region. The emission area heats up by Joule heating from the emission current, possibly assisted by ion bombardment from the residual gas or desorbing gas. With increasing surface field, and when the field emitted current becomes high enough, temperatures can approach the melting point of the superficial particles, and even of the underlying Nb. A rudimentary estimate [27] shows that ohmic heating from field emission current at a β-value E value of 5000[*] is sufficient to melt a cubic micron of Nb within 10 nanosec. DC studies [29] indicate the time scale for the formation of the arc is also in the nanosec range. In both cases, the vapour resulting from this intense heat is assumed to provide the gas media for initiating a spark or microdischarge. Plasma pressure during the discharge could be one way in which the molten zone is excavated and droplets are ejected. Possibly the plasma cloud associated with the spark is responsible for the starburst shape feature. Thus, there could be etching of the surface oxide layer by the plasma cloud, or perhaps a cracking of the residual surface hydrocarbon layer.

At the higher surface fields accessible with high pulse RF power (HPP), more emitters can be eliminated by the explosive process. The RF pulse length does not have to be very long, judging from the spark formation times of nsecs. Hence, the average power can be kept low for this type of processing to be useful in SC cavities.

---

[*]Here the βE product is again referred to the simplified FN expression quoted earlier in section.

12. 8   A Statistical Model for Field Emission in SC Cavities

When comparing the performance of SC cavities of various sizes, it has frequently been noted that the high field performance of large area cavities is more severely limited by field emission. At one point there was a suspicion that there may be an intrinsic RF frequency dependence to the emissive properties of field emitters, since the large area cavities were invariably also of low RF frequency. This possibility has been carefully excluded in an experiment when an RF cavity was excited at several different resonant frequencies from 500 to 3500 MHz, and the field emitted current from a tip in a high field region was studied. The β-value was found to be independent of frequency [30].

A simpler explanation is that the number of emitters increases with cavity area. Indeed with an elementary statistical model, it is possible to closely mimic several features of performance of an ensemble of SC cavities [31]. In 100 tests of 1-cell 3 GHz cavities [32], the fraction of tests that reached the maximum field Epk is given in Fig 12.20, together with results from simulations.

The maximum dissipated power allowed for a successful test was 20 watts, which was typical for the size of the laboratory RF power source. Similarly, experimental and simulation results from 100, 5-cell, 1500 MHz cavities [33] showed the distribution in Fig 12.20, where the maximum dissipated power allowed was 100 watts. The surface area of a 5-cell, 1500 MHz cavity is 20 times higher than a 1-cell, 3000 MHz cavity. In both cases, the statistical model closely approximates the test results.

The statistical model was based on the following assumptions.

1.   The distribution function for β-values is: $N(\beta) = \exp(-0.01\beta)$, with β-values from 40 to 540.
2.   The log of the emitter areas in m^2 is distributed randomly between -18 and - 9, as in Fig 12.8.
3.   The starting emitter density is a random number between 0 and 0.3 emitters/cm^2.
4.   Very strong emitters (those that deposit more than 5 times the available RF power) are assumed to process successfully.

Also, the power deposited by emitters is calculated from the Fowler-Nordheim field emission current and the electron impact energy. The high power originates predominantly from the large field-emitted currents.

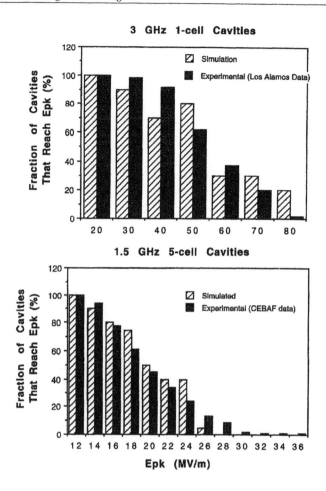

Fig 12.20 In an ensemble of 100 cavities, the fraction (%) that reach Epk with the available RF power. Two sets of data are shown: (upper) for 1-cell, 3 GHz cavities tested at Los Alamos and (lower) 5-cell, 1.5 GHz cavities tested at CEBAF. The fluctuations are statistical. The CEBAF cavities have a surface area 20 times larger than the Los Alamos cavities.

12. 9   Conclusion

Over the last decade there has been substantial progress in both understanding the origin field emission, as well as in reducing it to improve the performance of SCRF cavities. The best surface fields obtained in cavities, as a function of the cavity surface area subjected to high fields, are shown in Fig 12.21.

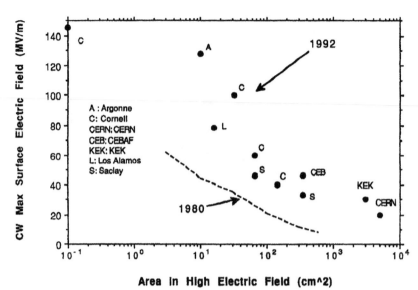

Fig 12.21 A decade of progress in surface RF electric fields reached with SC cavities. Only the highest field from each laboratory is shown.

In 1980, the highest RF electric field ever reached in an SC cavity was 70 MV/m [34]. The record electric field is now 145 MV/m (CW) and 220 MV/m pulsed. Surface fields of 20 MV/m are regularly achieved today in RF cavities with an area of nearly one square metre in high electric fields.

Improved cleanliness is the key to higher fields. Molten debris found at the processed sites clearly show the source of some of the troublesome emitters in SC cavities. By heat treating in UHV, niobium surfaces can be rendered even cleaner. Nb surfaces can be conditioned with high pulsed power to destroy remaining emitters. The mechanism of processing emitters is now better understood and much new effort is being devoted to realising the full benefits of this approach.

12.10   Updates on Recent Developments

Since the time this chapter was written, there have been several advances in the understanding of field emission from RF cavities, in the processing of field emission in superconducting RF cavities, as well as an improved performance of superconducting cavities by suppressing field emission.

As mentioned in section 12.5, for the examination of starbursts (Fig 12.19) and craters (Fig 12.15) on niobium surfaces which sustained high electric fields, we (at the Cornell Laboratory of Nuclear Studies) used a scanning electron microscope (SEM) equipped with an energy dispersive

X-ray spectroscopy (EDS) system. With EDS we were able to detect impurities (presumably responsible for the enhanced field emission) within the starbursts in only 50% of the processed field emission sites. This raised some important questions. Are there impurities present in field emission sites but which are not being detected by the EDS because of its sensitivity limitations, or are there indeed field emission sites which have no impurities?

To address these questions, we conducted more surface sensitive Auger Electron Spectroscopy (AES) analysis on 29 central craters on niobium plates from two RF tests, during which the maximum surface fields reached were between 80-95 MV/m [35]. With AES, impurities were in 100% of the central craters, supporting the idea that field emission is caused predominantly by the presence of some impurity. The dominant impurities found (in order of decreasing frequency) were Fe, Cr, Si. When iron was found, chromium was often found as well suggesting that stainless steel was in the emission source. In less than 20% of the cases we also found excess Ni, Mn, Cu, C, F, Cl and Mg. Since many of the sites examined under AES did not reveal impurities, with the EDS we surmise that the depth of the film covering craters is less than 100 nm (the approximate depth limit of X-ray production in the EDS technique).

The elemental impurities found are consistent with the contaminants one may find in the typical clean room cavity assembly environment. The metals may come from stainless steel bolts and nuts or assembly tools. Possibly the silicon may come from dust or from the borosilcate glass found in High Efficiency Particulate Air (HEPA) filters, which are generally used to provide the clean room environment.

The new AES study also shed light on the nature of the starburst feature accompanying processed field emission sites. After comparing Auger spectra inside and outside a starburst, we found fluorine outside but no fluorine inside. A 2-D fluorine map of the surface near the starburst clearly showed a good overlap between the geometric shape of the starburst and the region of fluorine depletion. After two argon sputterings of the surface, we determined that the fluorine layer present everywhere on the niobium surface was between 50 to 1500 angstroms deep. Fluorine presumably comes from the HF in acid etching solutions used for preparing the niobium surface for RF tests.

As before, it is reasonable to conclude that the electron or ion activity within the plasma of the RF spark is sufficient to "clean" the surface to remove the fluorine and other surface layers so as to lower the secondary emission and reveal a starburst shaped pattern.

Atomic Force Microscopy (AFM) images were obtained from cratered region surrounded by ripples (Fig 12.16). With the topological imaging capability of the AFM, the large craters are found to be depressions 1 μm deep and a raised edge equally high. The ripple patterns are clearly topographical features which generally seem to emanate from a point near the central crater and propagate on the order of 100 μm. The ripple wavelength is several hundred nanometres. Crest to trough amplitudes are in the range of 10-35 nm [35].

Encouraged by high pulsed power processing of field emitters in 3 GHz cavities with 150 kwatt power (Section 12.4.4), we have now successfully extended the application of the technique to processing field emission in 1.3 GHz superconducting cavities and accelerating structures (5-cell units). The RF power levels for processing were increased to 1 Mwatt and the surface fields during pulsed processing were increased to 90 MV/m. After the conditioning several cavities and structures with high power, the units could be operated in continuous wave mode at accelerating fields of 26-28 MV/m (surface field = 50-70 MV/m). These results [36] open the application of superconducting cavities to higher energy particle accelerators being contemplated for the next century.

Acknowledgements

I am deeply grateful to my colleagues at Cornell: Phil Barnes, Joel Graber, Walter Hartung, Joe Kirchgessner, Jens Knobloch, Dave Moffat, Richard Noer, James Sears and Quan Sheng Shu. Without their work, the results reported here would not be possible. Discussions with H. Piel and W. Weingarten are much appreciated. I am also grateful to Dave Moffat for reviewing the manuscript and providing helpful criticism.

This work was supported by the National Science Foundation with Supplementary Support from the US-Japan Collaboration.

12. 11   References

1. *Proc. 1st Workshop on RF Superconductivity*, Karlsruhe, Germany, KFK Report No. 3019, Kuntze, M., Editor, 1980.
2. *Proc. 2nd Workshop on RF Superconductivity*, Geneva, Switzerland, H. Lengeler, Editor, CERN, 1985.
3. *Proc. 3rd Workshop on RF Superconductivity*, Argonne, Ilinois, ANL-PHY-88-1- Argonne National Lab, Shepard, K., Editor, 1988.
4. *Proc 4th Workshop on RF Superconductivity*, KEK, Tsukuba, Japan, KEK Report No. 89-21, Kojima, Y., Editor, 1990,
5. *Proc 5th Workshop on RF Superconductivity*, DESY, Hamburg, Germany, Report No. DESY M-92-01, D. Proch, Editor,1992.
6. Schnitzke, K., *Phys. Lett.* **45A**, 241, 1973.
7. Moffat, D., in Ref. 4, p. 445.
8. Delayen, J., in Ref. 5, p. 376.
9. Fowler, RH. and Nordheim, L, Proc. Royal Society, Ldn, **A119**, 173, 1928.

10. Sh. Noguchi, *Nuclear Instruments and Methods*, **179**, 205, 1981
11. Lyneis, C, in Ref. 1, p.119.
12. Noer, R.J., *Applied Physics*, **A28**, 1, 1982
13. Weingarten, W., in Ref. 2, p. 551.
14. Piel, H., in Ref. 1, p.85.
15. Padamsee, H., in Ref. 3 p. 251.
16. Roth, R.W., in Ref. 5, p. 599.
17. Graber, J., PhD Thesis, Cornell University (1993). See also, Graber, in Ref. 5, p. 758.
18. Padamsee, H., in Ref. 4, p. 207. *Also Proc. 1991 Particle Accelerator Conference*, IEEE 91CH3038-7, p. 2420, 1991
19. Ph. Niedermann, PhD Thesis No. 2197, U. of Geneva, 1986.
20. Athwal, C. and Weingarten, W., CERN/EF/RF 84-7, 1984.
21. Shu, Q.S, *IEEE Trans. Mag.,* **25**, 1868, 1989.
22. Bajic, S., *Proc. XIII ISDEIV*, Paris, p. 8, 1988.
23. Shu, Q.S, *Proc. 1990 Applied Superconductivity Conference, IEEE Trans. Magnetics*, Sno-Mass, Colarado.
24. Padamsee, H., *Proc. 1991 Particle Accelerator Conference, IEEE 91CH3038-7*, p.2042, 1991.
25. Reschke, D., *Proc. 1992 European Particle Accelerator Conference, Editions Frontiers*, p. 1283.
26. Shepard, K., Argonne National Laboratory, private communication. Also I. Campisi, *IEEE Trans. Mag.*, **21**, 134, 1985.
27. Moffat, D., in Ref. 5, p. 245 (1992) & *Particle Accelerators*, **40**, p. 85, 1992.
28. Jüttner, B., *Physica,* 114C, 255, 1982. Also E. a. Litvinov, *Sov. Phys. Usp.,* **26** (2), 138, 1983.
29. Mesyats, G.A., *IEEE Trans. Electrical Insulation*, **EI 8** (3), 218-24 1983. Also Jüttner, B., *IEEE Trans. Plasma Science,* **PS-15** (5), 474-80, 1987.
30. Klein, U., PhD Thesis, Wuppertal University, WUB-DI 81-2, 1981
31. Padamsee, H. and Wright, B., Cornell Univ. Internal Report SRF 850501, 1985. Also W. Weingarten, Proc. XIII ISDEIV, Paris, pp 480-85, 1988.
32. Rusnak, B., *Proc. 1992 Linear Accelerator Conference, AECL-10728*, p. 728, 1992.
33. Schneider, W., CEBAF, private communication, 1992.
34. Citron, A., in Ref. 1, pp. 3-8, 1980.
35. Hays, T., et al., Proc 6th Workshop on RF Superconductivity, CEBAF, Oct, 1993.
36. Crawford, C., et al., Proc 6th Workshop on RF Superconductivity, CEBAF, Oct, 1993.

13

HV Insulation in Space-based Device Applications

MF Rose

13.1　　Introduction

Vacuum breakdown is important to spacecraft for two reasons. First, the natural environment is conducive to spacecraft charging and subsequent breakdown, somewhat independent of the power system. Secondly, the voltage level at which an exposed spacecraft power bus system can operate is determined by the properties of the "space vacuum" and the materials used in the spacecraft construction. While there have been numerous instances of spacecraft anomalies which are attributed to breakdown/discharge phenomena associated with environmental effects, it has not been possible to examine the spacecraft in its environment to determine the exact nature of the phenomena. Consequently, simulation of postulated mechanisms in a terrestrial laboratory is important. There are three distinct mechanisms which can be postulated for breakdown induced by the space environment. "Breakdown" can occur when the potential exceeds that necessary for significant flow of current to the space plasma. Charge balance will eventually be achieved and current flow terminated. If there is a dielectric layer, either intended or as a result of interaction with the space environment, charge accumulates and when the field is high enough, "punch through" occurs and local currents flow through the rupture. Flashovers and punctures can occur as a result of large potential differences due to differential charging or as a result of embedded charge from impact by high energy particles. Each of these mechanisms, or combinations of them, must be taken into account when designing a spacecraft.

In a more generic sense, electrical insulation is a critical design parameter in the field of space power. As the demand for higher power levels increases, joule heating also increases if the voltage levels must remain below that which is known to be safe from operating experience. If higher power levels are obtained by increasing the current only, there are additional demands placed upon the thermal management or alternatively, more massive electrical conductors must be used. Increased mass immediately translates to less payload or additional cost due to the need for larger booster rocket systems. If higher voltages and improved vacuum insulation techniques could be employed, there is the potential for significant savings in the spacecraft mass. In this sense, high voltage may mean voltages less than 1000 volts and associated with the power bus rather than a part of the spacecraft complement of instruments. For the bus, it may be difficult to shield from the detrimental effects of the space environment due to its distributed nature (large solar arrays for example).

There are numerous practical specialised systems such as X-ray tubes, accelerators, electron microscopes and vacuum switch gear which employ vacuum insulation. Conduction in a well conditioned vacuum line is essentially zero until a threshold value is reached. Ultimately, due to several mechanisms, breakdown does occur (see Chapters 5 to 7). The processes involved are complicated and depend upon such parameters as electrode separation, electrode materials, temperature, and the nature of the applied voltage. Subsequent behaviour of the vacuum insulated system is determined by the gas evolved, the ability for the gas to disperse and any residual effect on the dielectric spacers used in the construction. Under some circumstances, as discussed in Chapter 2 and elsewhere [1], the insulating properties of the vacuum line may be improved. This phenomenon is utilised in the conditioning process where surface irregularities are destroyed with controlled discharges to produce a high effective breakdown voltage for the system.

Intuitively, the gas pressure in outer space should be low enough that the terrestrial experience on vacuum insulation would apply. However, in practice, the space environment is highly variable and is anything but empty. In the immediate vicinity of a spacecraft, the environment is determined by the combination of the local ambient environment as modified by the presence of the spacecraft and how long it has been in space. The highly dynamic natural space environment has been well characterised [2], and found to consist of a multi-species neutral gas, a local plasma made up of many ion species, solar and cosmic radiation, energetic charged particles, hypervelocity micro-meteoroids, and magnetic and electric fields.

Contributing to the local environment, the spacecraft ejects gases from manoeuvring thrusters, gas evolved from spacecraft materials, and intentional gas/liquid dumps from manned spacecraft. Large spacecraft, therefore, are highly variable sources of contaminating materials which perturb the environment, making it more difficult to specify local breakdown levels in an engineering sense. As a result, for some orbits, it is very difficult to have high confidence that a spacecraft utilising vacuum insulation will operate without failure. Unique to space is the abundance of charged particles and microdust which could, due to a number of processes, produce breakdown even though the local pressure is well within the regime normally associated with laboratory practice.

Within vacuum insulated systems, breakdown often occurs along the insulating materials used as spacers, etc. The Long Duration Exposure Facility (LDEF) has shown that exposure to the space environment produces material degradation due to local chemistry, radiation, and impact due to cosmic/man-made debris, all of which may effect the flashover characteristics of materials [3].

13. 2   Characterisation of the Space Environment

The space environment has been extensively studied and documented [2]. The most serious factors influencing electrical phenomena are the local radiation environment, thermal cycling, local plasma density, neutral particle density, outgassing/effluents, and the meteoroid flux. The earth's magnetic lines of force trap charged species producing distinct "belts" which determine, to a great degree, the local flux and energy spectrum. The radiation environment is fuelled by cosmic sources and the flux generated by the sun, which can vary by factors of one hundred or more depending upon solar activity. As a result, the nature of the particles and the electromagnetic spectrum encountered in space are complex and depend upon such factors as orbital altitude, inclination and current solar activity, and can result in exposures which vary over several orders of magnitude in a given orbital pass. The effects of these particles and electromagnetic radiation can cause major changes in the properties of dielectrics and insulators by ionisation, atomic displacements or local changes due to chemical reactions [4]. The severity of a radiation-induced change under multifactor stressing can depend upon the total dose, intensity, particle species, impingement angle, presence of shielding, mechanical stress, local "system modified environment", temperature, water vapour, and the presence of system generated electromagnetic fields. In some orbits local microdebris and chemical reactions seriously degrade surfaces. The severity of microdebris and space environment-materials interactions has been documented in the published reports of the analysis of LDEF [3].

The sum total of these space environmental factors can have catastrophic effects on exposed high voltage systems and, in general, tend to place constraints on the insulation technology employed. Table 13.1 is a brief summary of the space environment with comments detailing known effects on materials which could be used in high voltage vacuum insulated systems.

13. 3   Technological Influence of the Space Environment

The ability to design a spacecraft to survive in the local environment is key to long life and cost effective performance. The spacecraft as a system may use materials which span the complete range from gases to plastics to metallic solids. Each material used in construction interacts differently with the space environment placing engineering constraints on the materials used and the techniques used to protect them from catastrophic damage.

13.3.1   Residual Gas Environment

Table 13.1 lists the space environmental factors which can influence the electrical behaviour of a spacecraft. The first of these, local pressure, is of significance since the spacecraft and the materials used in its construction are an influence. At altitudes above 150 kilometres, the local pressure is of the order of that used in devices employing vacuum insulation in a controlled terrestrial setting. Unfortunately, the presence of the spacecraft modifies the local pressure in a largely unpredictable way. In this context, Paschen breakdown has been studied for gases known to be constituents in the outgassed species from spacecraft [5], and has confirmed that our ability to employ the local vacuum as an insulating media is obviously a function of the gas species present, the absolute pressure, and the specific spacecraft design. Outgassing rates are determined by a number of processes. Early in the flight history of the space shuttle, there was concern for the level of contamination due to the operation of the shuttle and station keeping manoeuvres [6]. As a result, there was a concerted effort to monitor both the plasma and neutral species around the shuttle as a function of orientation, while using the thrusters for attitude control. Green, Calodonia and Wilkerson [7] have examined the gases and particulates near the space shuttle as a function of mission time. Measured pressures at selected sensor locations within the bay vary from 10^{-7} torr to 10^{-4} torr. The higher pressures correlated with thruster firings which uses monomethylhydrazine (MMH) as the propellant and N_2O_4 as the oxidiser. The measured neutral species measured within the shuttle bay were: H_2O, He, NO, Ar, Freon 12 and 21, trichloroethylene, N_2/CO, O_2, CO_2, and other heavy molecules identifiable with cleaning agents. The principal contaminants were water and helium. The initial decay time for water after launch was approximately 10 hrs and is due to outgassing from the surfaces which had adsorbed gas before launch. The induced ionic component is equally varied. Narcisi et al. [8] have observed O^+, H_2O^+, H_3O^+, N^+, N_2^+, NO^+, O_2^+, and OH^+ within the shuttle bay. All observations correlated with

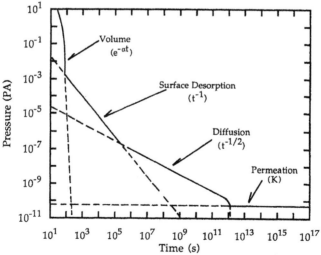

Fig 13.1   Outgassing regimes as a function of time for materials in space. (Reprinted with permission from the ASME, reference [9].)

thruster activation and lasted an order of a second. Thruster events are common, usually several seconds apart, which places uncertainty in the local pressure.

Fig 13.1 illustrates the physical processes which occur when a "gas soaked" spacecraft is thrust into space [9]. Initially the local gas species and pressure are governed by volume diffusion as the gas from voids expand into the "vacuum" of space. This process is rapid with a time scale determined by leakage pathways throughout the spacecraft and existing pressure differential. This region of the outgassing process is most significant in sub-orbital flights such as SPEAR 1 or during the first few minutes of a mission [10]. As time increases, surface desorption begins to dominate the process as adsorbed gases, fractals, etc., are released from the surface. In the case of SPEAR 1, the main source of outgassing was conjectured to be the body of the spacecraft which was about 3 metres from the high voltage spheres used in the experiment.

The purpose of the SPEAR series of experiments was to establish the capability of using high voltages in the space environment without catastrophic arcing. Fig 13.2 illustrates the geometry of the SPEAR 1 spacecraft. The spheres were pulse charged to a voltage of 46 kV with respect to the spacecraft body and the spacecraft monitored to determine whether breakdown occurred.

The pulse applied exponentially decayed with a time constant of 1 second. In order to prevent flashover along the boom structure, resistive grading and field shaping rings were used. The rings were shaped to shield the resistive net connections from direct exposure to the space plasma. The effective resistance of the grading structure was 1.1 megaohms. The telemetry indicated the expected conduction to the spacecraft body via the space plasma but no catastrophic breakdown occurred along the sheath boundary. Further, there was corona along the boom structure but again, the grading scheme effectively eliminated catastrophic failure for the decaying exponential waveform applied to the spheres. The results of the SPEAR 1 experiment are complicated by the fact that the plasma contactor which was to provide good body contact to the local plasma failed to function. As a result, the spacecraft body potential was raised due to the accumulation of charge. In order to explain the results, it is assumed that the pressure at the spheres was close to local ambient and not influenced by the outgassing from the main spacecraft body. Fig 13.3 is a plot of the theoretical local pressure which could be expected superimposed on the experimental measurement of pressure near the body of the spacecraft [11,12]. This process is highly temperature sensitive and can vary over orders-of-magnitude; it also tends to dominate on a time scale of days to weeks which is the typical duration of shuttle missions.

In the time scale from weeks to months, highly temperature sensitive diffusion processes are active. Finally, there are low level contributions to the local environment due to non-zero vapour pressure of the constituents, permeation of species from the interior of structural materials and from sputtering as a result of the constant bombardment of the surfaces by particulates ranging from atomic dimensions to particles several micrometres in diameter.

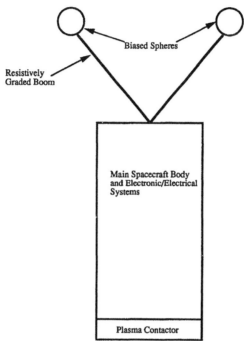

Fig 13.2   Schematic diagram of the SPEAR 1 spacecraft.

The obvious worry associated with the local gas environment is the difficulty in determining exactly what the electrical breakdown characteristics are for specific combinations of gas species and the pressure conditions. The breakdown hold-off strength is a function of local pressure and gap spacing (see Chapter 2). Thus, referring to Fig 13.1, and noting that the long term dominant processes are thermally sensitive, a solution to potential breakdown problems would appear to be thermal conditioning or waiting to turn on high voltage devices until outgassing has proceeded to an acceptable level.

Since the time scale is both volume and total surface area sensitive, these may not be viable alternatives for large systems or short mission times. The fact that there are "local replenishment" sources which can keep a large quantity of gas trapped on surfaces complicates any estimate of the time required before high voltage actuation. Effluents such as water release, propellant residue from station keeping manoeuvres, and cross-contamination can, in fact, produce large uncertainties in the value of the local pressure over an entire mission [6]. Breakdown and surface tracking occur on the

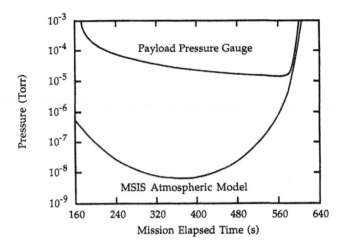

Fig 13.3 SPEAR 1 neutral pressure data. (Reprinted with permission from reference [11], © 1989 IEEE.)

microsecond time scale. Small localised emissions, triggered by a number of mechanisms, may also be catastrophic. Previously dormant systems, drawing full power, may produce local heating effects which promote outgassing. The LDEF experiments show that debris impacts are quite common and large enough to generate a significant local plasma for many microseconds [3]. For many satellites, the surface adsorptivity and emissivity are adjusted to assist in maintaining a relatively constant temperature within the spacecraft body.

An instrumentation package to monitor local pressure changes was included on some of the Apollo missions and on the Viking lander missions to Mars. The data collected confirm that local gas pressure can vary by several orders of magnitude over that assumed for pristine space [5]. These local conditions decay slowly with time, after the diffusion of the gas constituents is no longer collision dominated, and in the case of the Apollo mission, never reached ambient conditions.

13.3.2 Environmentally Induced Electric Breakdown in the Space Vacuum

There are two modes in which the local environment can influence the electrical behaviour of a spacecraft. Passively, the local environment interacts with all spacecraft surfaces, coating, degrading, eroding or storing charge within the dielectric materials. Actively, the local environment can trigger breakdown to the space plasma, or between two surfaces on the spacecraft at differing potentials, and produce internal upsets due to the extremely energetic cosmic rays colliding deep within the spacecraft electronics.

There are numerous circumstances for which a breakdown can occur in a spacecraft. For example, the spacecraft wiring harness contains numerous cables with varied configurations. Breakdown can occur in these wiring bundles and their connectors due to outgassing [13]. Exposed cable traces from the power source to the body of the satellite are also subject to space induced breakdown between conductors at differing potentials and between the conductor bundle and the space plasma [5].

13.3.3   Free Space Breakdown

Fig 13.4 is a plot of the approximate electronic component of the local plasma as a function of altitude which has been adapted from data of Jursa [2]. Note that there is roughly an order of magnitude difference between the minimum and maximum curves. The maximum is associated with maximum solar activity and varies with the eleven year sun spot cycle. The electronic component of the plasma density peaks near 300 km, becoming relatively constant at an altitude of 3000 km. At about 12,000 kilometres, the local electronic component of the plasma density begins to decrease rapidly and is less than 100/cm^3 at altitudes of 50,000 km. In this context Katz et al. [14] give a compilation of breakdown data as a function of the local plasma density from selected experiments from both the laboratory and space. These data are shown in Fig 13.5.

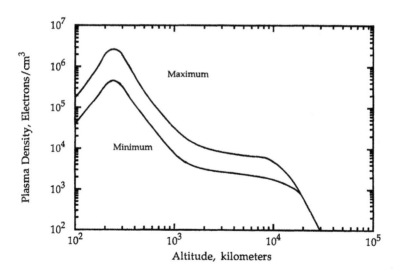

Fig 13.4   Electronic contribution to the local plasma as a function of altitude and solar conditions.

Table 13.1 Summary of space environmental factors relevant to materials degradation. (From Jursa [2], with permission.)

Environmental Parameter	Approximate Parameter Range	Comments
Vacuum	10^4-10^{-8} NT/m^2	Vacuum outgassing results in loss of adsorbed species, solvents, host material. Some condensables emitted.
Electromagnetic Spectrum	Intensity ~ 1.4 kW/m^2 1 mm to 10^{-9} nanometer	Broad spectrum electromagnetic flux peaking near 0.5 mm. Variable over 100 ×. Degrades surfaces, enhances chemistry, produces thermal gradients.
Protons	Energy 0. -4 MeV (400 MeV Max) Flux > 10^8/sec cm^2	Degradation of coatings, cross linking, polymerisation of plastics, etc. Lattice defects produced. Flux and energy variable due to orbit and solar activity. Van Allen belts produce significantly higher intensity.
Electrons	Energy 0.1-4 MeV Flux > 10^8/cm^2/sec	Surface & bulk damage, space craft charging. Defects, polymerisation, scission, cross linking in dielectric materials. Variable due to orbits and solar activity. Van Allen belts produce significantly larger flux. Auroral radiation produces surface doses as high as 10^7rad/hr.
Cosmic Rays	Energy > 10^9 eV ~ 1/cm^2/sec	Limited damage due to low flux. Mechanical tracks, upsets, etc.
Ions		~ 10^2-10^6/cm^2/sec. Primarily O$^+$, N$^+$, H$^+$, He$^+$, NO$^+$. Important in surface chemistry.
Atomic/molecular species	10^6-10^{18}/m^3	Chemical interaction with surfaces producing corrosion. Atomic oxygen most important. H, He, O, O$_2$, N$_2$, A are primary components.
Effluents	Wide density range	Spacecraft specific. Station possible keeping, orbital changes, etc. H$_2$O, H$_2$, N$_2$O$_5$, etc. Surface chemistry affected.
Thermal cycling	T$_{MAX}$ >400 K T$_{MIN}$ ~ 80 K	Microcracking, fatigue, thermal warping, critical surface deterioration.
Meteoroid damage	Grams to micrograms	Micro-abrasion to large puncture. Source of "space dust" on surfaces. Man-made debris becoming more important.

Taking the electronic component of the plasma density as a function of orbital altitude from Fig 13.4, and the breakdown threshold as a function of plasma density from Fig 13.5, it is possible to estimate the threshold for breakdown as a function of orbital altitude. The data from Figs 13.4 and 13.5 suggest that, at low orbital altitudes, there is a high probability of discharge to the space plasma for breakdown between exposed conductors at voltages below 300 V. This is consistent with the results of Hastings et al. [15] and others [16] both from space experiments and laboratory simulation. At high orbital altitudes in excess of 20,000 km the data suggest that the threshold for breakdown is several kilovolts.

The data in Figs 13.4 and 13.5, along with the environmental factors listed in Table 13.1, suggest that an effective limit for the maximum DC voltage between bare conductors is less than 1000 volts for all "near space" applications.

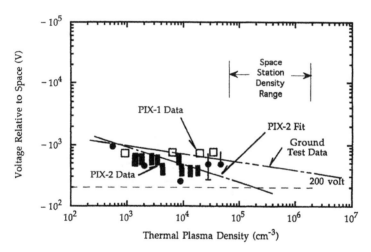

Fig 13.5   Experimental satellite data and laboratory data for voltage breakdown threshold as a function of local plasma density. (Reprinted with permission from reference [14], © 1989 IEEE.)

At the highly populated orbits in the 200-500 kilometre range, the limit appears to be less than 300 volts. For geosynchronous orbits, the breakdown voltage may be on the order of kilovolts.

13.3.4   Breakdown due to Spacecraft Charging

The electromagnetic spectrum (Table 13.1) interacting with local gas species, electrons, protons, cosmic rays, ions and the earth's magnetic field produces a space plasma which is a function of orbital parameters and solar activity [2].

There are numerous mechanisms, related to this environment, by which a spacecraft can accumulate charge. Absolute charging occurs when the spacecraft body potential changes relative to the space plasma as a result of the spacecraft presence. All surfaces/coatings etc. are at the spacecraft reference potential as designed. The spacecraft will accumulate charge from the local plasma environment acquiring a potential determined by the space environmental parameters, the nature of the materials employed in the spacecraft construction and the "free space" capacitance of the spacecraft. Fig 13.6 illustrates the mechanisms responsible, namely:

The flow of environmental ions and electrons to the surface.
Photoemission from the surface.
The emission of secondary electrons due to ion impact.
The emission of secondary and backscattered electrons due to electron impact.

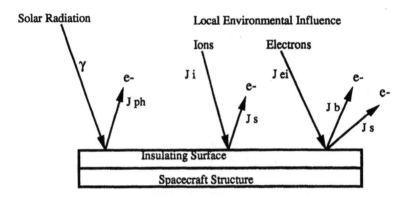

Equilibrium Condition: $\Sigma J = 0$

Fig 13.6   Mechanisms responsible for spacecraft charging.

Equilibrium is reached when the spacecraft is charged to a potential sufficient to maintain a net current of zero [17]. This process is very rapid. During eclipse the plasma current is not partially balanced by photocurrents. In the sunlight, photocurrents tend to make charging take place at a somewhat slower rate. For large structures such as the space station, the capacitance of the structure can result in substantial energy storage when charged to potentials typical of this charging mode.

Cho and Hastings [18] and others [19] have studied the arcing rate for high voltage solar arrays and find that arcing begins at a threshold of about -200 V for orbits typical of Space Station Freedom. Three mechanisms are proposed in the above references.

470

(i)   In a system with a negative return, if there is a thin dielectric layer, intended or as a result of contamination, it can accumulate charge from the space plasma. When sufficient charge has accumulated, the layer breaks down and produces copious rf noise in the system.

(ii)   A second mechanism proposed for arcing on high voltage solar arraysassumes that the pre-breakdown currents, typical of that discussed inprevious chapters of this book, produces sufficient molecular desorption to cause a gas buildup over interconnects which result in surface flashover in the space vacuum.

(iii) Cho and Hastings [18] combined some aspects of both of thesemechanisms into a model based upon charging in the vicinity of a triple junction made up of the space plasma, dielectric and conductorinterface. Charging could take place as mentioned above or due to enhanced field emission.

All of these mechanisms predict to some degree what is observed in simulation facilities [15]. At the moment, there are several flight experiments scheduled over the next few years to clearly study this phenomena in space [20].
Differential charging due to asymmetric interaction with the environment is also a possible charging mechanism. The basic mechanism is illustrated in Fig 13.7. The spacecrafts response to the surroundings is a function of parameters such as which side is exposed to the solar radiation, ram direction, and specific design of the spacecraft. From this mechanism, charge can be attracted to one side of the spacecraft while being ejected from the other. As a result, this charging mechanism can also alter the absolute charging of the spacecraft by its influence on the electronic trajectories in the vicinity of the spacecraft. Any "shaded" dielectric can become highly charged producing large differential voltages. As a result, large potential differences could occur between sections of the spacecraft.
Charge build-up can also occur by deposition into the interior of a dielectric material by energetic particles [4]. Table 13.1 lists an abundance of high energy electrons and protons, especially in times of heavy sunspot activity, which are responsible for this phenomenon. Fig 13.8 illustrates this process, where, the depth of penetration into the material is a function of material parameters and the energy of the incoming particle. If the resistivity of the material is sufficiently high, charge will accumulate. It then follows that if the electric field associated with the embedded charge exceeds some critical field, breakdown from the charge site to regions of lower potential will occur. These discharges may travel directly to the spacecraft common or to the surface of the dielectric material in which the charge was embedded and continue as a surface discharge to some region at spacecraft common.

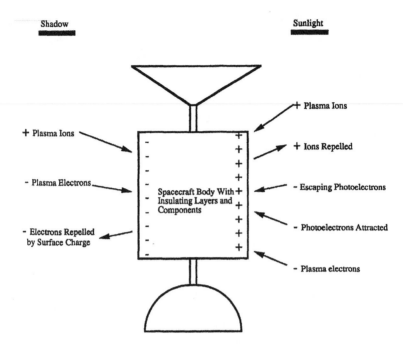

Fig 13.7 Spacecraft charging mechanism illustrating differential charging between surfaces in the sunlight and surfaces in the shadows.

The breakdown field strength for a good insulator is greater than 10^6 V/cm. Hence, the fields associated with this failure mode require substantial charge trapping and are exceedingly high. When breakdown occurs, high amplitude transients are produced which can upset or damage sensitive electronics.

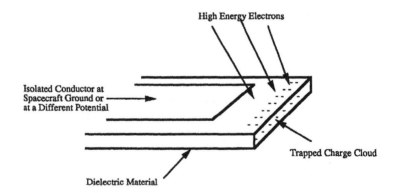

Fig 13.8 Spacecraft charging mechanism illustrating charge deposition beneath the surfaces of dielectric materials.

472

Several spacecraft have been flown to verify charging phenomena. ATS-5 and ATS-6 spacecraft were in geosynchronous satellite orbit [2] and measured potentials as high as 20,000 volts. In fact, P78-2 SCATHA was flown to confirm the mechanisms proposed for spacecraft charging [21]. A summary of the SCATHA results are:

Arcs were observed under a variety of conditions such as eclipse,sunlight, and during beam operations.
A surface potential monitor was used to determine the response of anumber of materials and potentials greater than -1000 V relative to thespacecraft common were observed as these materials charged due to interactions with the space plasma.
One event corresponded to charging producing a potential exceeding 8000 V.

The level of concern for spacecraft charging and the subsequent discharges is a function of the mission of the satellite. If the spacecraft is relatively small, the capacitance to the plasma sheath is small and very little energy can be stored in this capacitance. Since the threshold is low for discharges of this type, it is common also to make the power system bus operate at low voltages, usually less than 50 volts. As a result, if discharges do occur, the electromagnetic interference may be negligible. There have been a number of studies of this breakdown mechanism for large structures such as the proposed space station freedom [19]. It is estimated that the capacitance of a space station module with respect to the space plasma is on the order of 1 millifarad. The operating voltage of the solar array is designed to be approximately 160 volts. With the negative return, balance of leakage currents will drive the structure to about 140 volts. This could lead to arcing through the dielectric coatings on the structure with currents on the order of 1000 amperes. This phenomenon is worst at low earth orbits where there is a substantial plasma density.

At geosynchronous orbits, differential charging and charge deposition within the materials are the most prevalent. This mode is less dependent on the size of the spacecraft. Any of the charging mechanisms capable of producing high power transients must be controlled since high amplitude transients could couple into the spacecraft and cause upset or damage.

13.3.5 Influence of Micrometeoroid/Space Debris

Chemical reactions and radiation strongly influence surface properties of dielectrics and insulators [4]. There is an additional source of surface debris which can contribute to the vacuum breakdown mechanisms. Microparticles ejected from the surface of vacuum insulated lines in the terrestrial environment are postulated to trigger catastrophic breakdown (see Chapter 7).

In space, microparticles with velocities as high as 40 kilometres/sec and with flux densities as high as several particles/sec are prevalent at all altitudes. Mang and Lindewr [22] describe a distribution function which estimates the probability of impact per meter square of exposed spacecraft surface per year as a function of impacting debris size. In the micron region of the distribution, many impacts/day are possible over the surface of a large spacecraft such as the space station. Some of these may be large enough to trigger breakdown at substantially less voltage than that associated with the breakdown strength of the dielectric coatings. This local flux decreases as mass increases resulting in little likelihood of collisions with major fragments of large mass. Based upon the rate at which man-made debris is accumulating, the probability of having catastrophic impacts in many orbits is becoming great enough to warrant considering the redesign of spacecraft to ensure survivability [23].

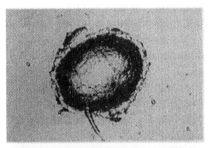

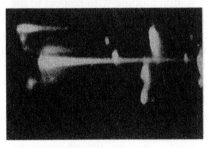

Fig 13.9 Hypervelocity impact at 10 km/s and associated streak camera record showing time span of the luminous plasma generated by the impact. The long luminous trace is approximately 0.5 miliseconds in duration.

Fig 13.9 illustrates the most important processes which occur in hypervelocity impact which might influence breakdown. What is shown is a crater produced by a 50 μm silicon carbide projectile travelling at a velocity of approximately 10 km/s, along with a streak camera record of the plasma produced on impact. Optical emission from the plasma is sufficiently luminous to leave a record for tens to hundreds of microseconds. Hydrodynamic calculations indicate that the temperature associated with impacts at this velocity are greater than

474

10,000 K. The impact event also produces multiphase ejecta, which in turn provide the requisite particulate/contaminant material known to influence vacuum breakdown (see Chapter 7). Such events could also be an important trigger mechanism for discharge through dielectric layers as discussed above, as a means of surface charging, and as a trigger for arcing to the space plasma. In addition, the more massive particles can cause physical damage to interconnecting lines, solar cells, etc.

Fig 13.10 illustrates schematically an experiment to determine the effect of hypervelocity particles on the discharge phenomena associated with charging such as that discussed by Carruth et al. [19] and Weishaupt et al. [24].An argon plasma source produces a plasma number density typical of the space environment while the capacitor C simulates the capacitance of a large space structure such as the proposed space station. The simulated spacecraft coated surface is biased to negative levels while being impacted with hypervelocity particles travelling at speeds up to 14 km/s. Preliminary data [25] indicate that breakdown does occur for voltages as low as 50 volts. A trigger mechanism such as this could aid all the breakdown mechanisms postulated to occur in the space environment.

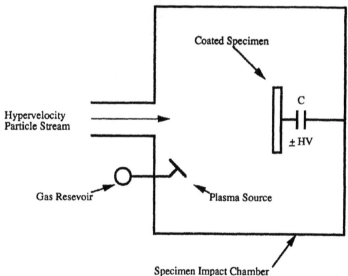

Fig 13.10   Experimental diagram of apparatus designed to measure the effects of hypervelocity impact on discharging of a space structure due to accumulated charge.

In a recent series of experiments, Hill and Rose [26], have conducted hypervelocity triggered breakdown studies for the geometry shown in Fig 13.11. The geometry for the experiment is that of a long wire, biased to some voltage V over a ground plane which is also tied into the facility ground.

The hypervelocity particle generator is apertured to produce a high velocity, low flux particle stream which can impact the ground plane in the immediate vicinity of the wire. With a spacing of 0.5 cm and a wire bias of -1kV, breakdown occurred readily for impacts on the ground plane immediately

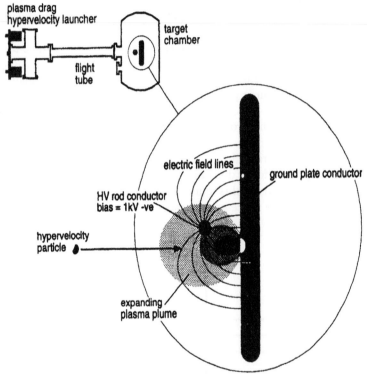

Fig 13.11   Experimental arrangement of apparatus to measure the effect of hypervelocity impact to trigger a discharge between bare conductors at differing potentials.

beneath the biased wire. The field strength is a function of position with respect to the axis of the experiment and varied from 10^5-10^6 V/m within the impact zone. The constant bombardment from particles of all sizes tends to have the following effects which would be detrimental to exposed components of power systems within the space vacuum:

Trigger mechanism for local breakdown [25,26]

Surface puncturing [24]

Spalling/void formation [3]

Debris accumulation [3]

Reduced dielectric insulation effectiveness [25]

Changes in the absorptivity-emissivity ratio [25]

Surface puncture and spalling/void formation lead to shorting and open

476

circuits in laboratory simulation and has been conjectured to be a possible cause for failure of a satellite [27]. For high reliability, a severe debris environment produces the necessity for armouring of components, automated fault isolation, and system redundancy. While tractable at lower power levels, it is unclear whether these mitigation techniques are viable alternatives as the power levels increase.

13. 4   Spacecraft Anomalies Attributed to Breakdown

There have been numerous anomalies related to power for spacecraft in orbit. In many instances, there was sufficient telemetry to allow an assessment of the most probable cause of the failure. Table 13.2 is a compilation of several recorded instances taken from various literature sources [13,27,28].

Table 13.2 spacecraft anomalies related to breakdown

System	Anomaly	Probable Cause
Voyager 1	Power-on reset	Due to ID
SCATHA	34 events	4 due to ID; 30 due to SD
DPS	Star sensor; false flag	Possible ID
DSCS II	Power switching anomalies	Correlated to geo activity
GPS	False command; clock shifts	Total failure 1980; Correlated with ID
INTELSAT	Inadvertent I7, II & IV spin-up	
Skynet 2B	Telemetry	Correlated with eclipse. Possible surface discharge
ANTK	Power Failure	
CTS	Short circuit noise. Power inverter shutdown	Substorm activity
Meteostat	Status changes	Geomagnetic activity
GEOS 4&5	Upsets and failures	Geos 4; possible ID
Solar Max	Upsets, 10/year	
TV-SAT	TWT drop-out	Paschen breakdown in high voltage cable. Gas source unknown
MARECS and ECS	Power loss/failure	Electrostatic discharge with possible trigger mechanism due to micrometeoroid impact.

ID - internal discharge, SD - surface discharge

13. 5   Insulation Mitigation Techniques

The first approach to eliminating unwanted discharges in the space environment is to eliminate as many of the mechanisms as possible due to sound engineering practice. Those due to spacecraft charging may or may not affect the power train. It is of equal interest to determine how effectively the properties of the space vacuum can be used as an insulating scheme. If it can be successfully employed, there may be significant mass savings as well as simplifying of the construction and testing procedures.

13.5.1   Insulation Techniques to Prevent Spacecraft Charging and Breakdown

If the spacecraft solar array is able to operate at low voltage, there is little problem with ambient charging. As the power demands increase, the need for higher voltages and the size of the spacecraft produce a substantial stored energy which can discharge and cause disruption. One solution to this problem, and that proposed for the space station, is to provide a plasma contactor which effectively reduces the charging to an acceptable level [25].

The most obvious solution to the problem of charge deposition and differential charging is to make all "materials conducting" and tied to the spacecraft common. In this case, conducting means that the electrical conductivity is sufficiently high to allow charge bleed off at a rate higher than the charge collection rate for a worst case scenario. In general materials such as kapton, fibreglass, mylar, etc., are too resistive to meet this criteria and a thin metallisation layer is sometimes used. Purvis et al. [29] have formulated a series of guidelines to be used to minimise spacecraft charging phenomena, of which the following are the most important.

Conductive materials must be tied to the spacecraft common with the smallest resistance possible a suggested minimum is $R < 10^9$ ohms divided by the exposed surface area in square centimetres.

Any partially conducting surfaces applied over a substrate must have a resistivity(r) thickness(t) product which satisfies the relation, $rt \leq 2 \times 10^9$ ohm-cm^2.

Partially conducting surfaces applied over a dielectric and tied to the spacecraft common must have an edge resistivity determined by:
$rh^2/t \leq 4 \times 10^9$ ohm-cm^2 where h is the distance from the tie point to the farthest edge of the material and the other units have been defined.

NASA, other space agencies, and the industry have experience with materials which have been found to perform satisfactorily in space. Thus, it is standard practice to use a computer code such as the NASA Charging Analyser Program (NASCAP LEO and GEO) to model the environmental, geometric, and materials aspects of the satellite to determine if it will be susceptible to charging [29]. If the code predicts problems, the geometry and materials to be used can be changed and iterated until acceptable charging is achieved. All critical materials/components are then extensively tested for performance as predicted before reducing to flight hardware.

13.5.2   Insulation of Power Components

Distinct and different from the problems discussed above is the question of how best to insulate the conductors which must distribute power to various parts of the spacecraft. For the most part, these elements are within the spacecraft body or at transitions from one segment to another. As the power levels increase, there is a need to operate the spacecraft power train at higher voltages. Within the SDIO weapons concepts, the advanced concepts for the Space Exploration Initiative of NASA, and such devices as high power space radar, there is a definite need to operate high voltage-high power systems beyond our current practice with as yet unspecified exposure to the full space environment.

It is current practice to insulate cables, transmission lines and power components with sufficient insulation to withstand the design stresses [5]. In that sense, the same amount of insulation is required as for a terrestrial application. The space environment as discussed above make it necessary to ensure that materials are selected with low outgassing rates and if exposed directly to the space environment, minimal degradation as a result of radiation damage. If the average power levels are low, it is possible to encapsulate high voltage power supplies, circuit boards, and critical components with conformal coatings [30]. If fields are allowed to fringe excessively outside the coatings, there is the possibility that charge will be attracted to this area which over time can lead to breakdown or surface tracking [31]. This places the restriction on the coating process that there be no flaws such as voids or cracks which can promote corona in high fields with subsequent degradation of the insulating properties of the coating. In addition, it would be desirous that the coating material has sufficient conductivity to bleed off any accumulated charge. Conducting coatings can easily be employed if the power bus is operated in a direct current mode. All solder joints must be shaped to minimise field enhancement and wherever necessary, surfaces must be shaped to provide field grading to acceptable levels. If the fields are allowed to fringe and promote particle impact from the local plasma, there is a reduction in the flashover strength of the insulator. Maffei et al. [32] have investigated the magnitude of this effect both for positive and negative polarities. They find that at a neutral argon background pressure of 1.5×10^{-5} torr, a 1 cm gap was capable of sustaining a potential of 50 kV. For positive applied potential, and a plasma number density in the range of 10^5-10^6/cm^3, there was a leakage current which was a function of the applied voltage and agreed well with the results of a NASCAP LEO simulation. The currents were electron collection currents and did not constitute catastrophic failure. For negative polarity in the same plasma concentration, the surfaces flashed over at moderate voltages of 5 kV. This is a dramatic reduction in the hold-off strength of the gap by an order of magnitude. These data are consistent with charging from the local plasma with subsequent potential buildup sufficient to produce the necessary charge to sustain flashover. To minimise this possibility, strict field engineering to accurately control fringe fields is necessary.

Gollor and Rogalla [33] in a recent paper have reviewed the field and discussed the ERS-1 satellite which is operating with some of the high voltage components "vacuum insulated". The unit operates at a negative voltage of -15 kV with 15 kW peak power. The author also states the projected need for power levels up to 200 kW pulsed at voltage levels on the order of 150 kV for future systems. The successful operation of this system is predicated upon two factors. First, there must be a "clean" vacuum and second, the system must be designed to be fault tolerant to a an occasional flashover or internal discharge. The ERS-1 radar utilises a travelling wave tube which is somewhat fault tolerant to internal arcing. Further, the high voltage section was designed to provide current limiting and sensing circuits were employed to produce automatic shut-off and restart in the event of a flashover. The high voltage open construction was effectively shielded from the space plasma environment but allowed the "space vacuum" to control the local pressure.

The techniques which must be employed, given that you can guarantee the prerequisites unique to space, are no different than those needed for a terrestrial vacuum insulated system. Extreme care must be taken to maintain gas pressure below an acceptable level by providing the proper venting to the space vacuum. Spacers must be chosen which minimise surface flashover and made from materials with low outgassing rates. The system must be "aged" to remove local surface anomalies which can initiate catastrophic failure, and field fringing/concentration must be controlled. Shielding must also be employed to prevent the constituents of the space environment and effluents, such as those generated by station keeping, from impinging on the part of the distribution system which is at high voltage.

13. 6   Discussion

It is possible to contrast the laboratory vacuum breakdown experience with breakdown in the space environment. Many of the space trigger mechanisms appear common with the terrestrial experience. From the point of view of controlling the process, in a large spacecraft in low earth orbit, there is little in common if the space vacuum insulated power train must be exposed. The Table 13.3 compares vacuum insulation techniques with the space experience to date.

Table 13.3 - Factors Effecting Breakdown in Vacuum and Space

Stimulus	Terrestrial	Space
Gas pressure	Controllable	Largely uncontrollable in the short term and for large tended structures with station keeping.
Residual species	Controllable	Somewhat controllable due to materials selection. Easier for small untended satellites.
Electrode material	Major influence	Negligible if exposed to space environment, important if shielded.
Radiation	Controllable	Uncontrollable if exposed. Requires mass to shield and control.
Particulate debris	Controllable	Uncontrollable if exposed. Requires mass to shield and control.
System bakeout	Easy to do	Very difficult
Temperature	Controllable	Very difficult to control at or near the surface of the spacecraft.
Conditioning	Easy	Very difficult
Dielectric spacers	Important	Important
Field emission	Important	Important

It is possible to use the spacecraft body to shield the high voltage sections of the system [33], and selectively perforated coaxial geometries [34] to allow systems which have been baked out to perform at high voltage in the space environment. The utility of these methods are dependent on the function of the spacecraft and its orbit, and whether it is manned or not. Any cycling effluent release, station keeping manoeuvres, and the normal outgassing from insulators and spacers will, over time, contaminate and degrade the property of the vacuum to act as an effective insulator unless effectively shielded from these influences. While experiments such as SPEAR I indicate that it is possible to design systems which can operate as high as 40 kV for short times in the space environment without catastrophic arcing [10], no long term data exist for exposed conductors at high voltage. The existing data favour low voltage operation for the spacecraft "bus" distribution system, and, if high voltage is necessary, local insulation techniques are to be applied, with vacuum insulation as an option.

The space experience, the results of laboratory simulations, analysis of spacecraft failures and theoretical modelling have been compiled in a single document entitled "Spacecraft Environmental Anomalies Handbook" which clearly illustrates the magnitude of the problem and what can be done to mitigate some of the more serious space environmental factors [28].

All of the issues arising from power-environment interactions will become more serious as we proceed with Space Station Freedom and the Space Exploration Initiative. Mitigation techniques applicable for units operating at levels less than a kilowatt may not be viable options as power levels increase. Since the power levels contemplated for the future are so far beyond our experience profile, it is of extreme importance that spacecraft such as the LDEF, SCHATA, etc., fly and the results of the analysis be used to determine the environmental factors which limit design. The trend to higher power must be accompanied by higher efficiency, higher voltage, and more effective thermal management systems. The following general comments apply to the development of large spacecraft with high power demands such as those envisioned for the SDIO, Space Station Freedom, and the Space Exploration Initiative.

Bus voltages will be as low as possible consistent with sound engineering practice. The magnitude will be determined by system peak and average power requirements, by the power source (solar,nuclear, etc.) and a trade off between thermal management system,joule heating of conductors, and potential interaction with the environment.

Electric and magnetic fields will be precisely controlled to prevent interaction with the space environment which could lead to breakdown.

Transformation to load characteristics will be done at the load with localised insulation techniques applied (pressurise, encapsulate, vacuum) as needed.

Materials exposed to the space environment must be used which can effectively dissipate collected charge while providing the necessary dielectric behaviour.

For large, high power spacecraft in low earth orbit, all key components of the power train will be armoured and must be long-term stable toradiation dose and thermal cycling and may require a plasma contactorto control the spacecraft potential with respect to the plasma.

Effluents must be strictly controlled.

13. 7 References

1. Latham, RV., *In* "High Voltage Vacuum Insulation", Academic Press, London/New York, 1981.
2. Jursa, AS., *In* "Handbook of Geophysics and the Space Environment", Air Force Geophysics Laboratory, Hanscomb AFB, MA, 1985.
3. Levine, AS., Ed. *In* "LDEF - 69 Months in Space, Second Post-Retrieval Symposium", NASA Conf. Pub. 3194, Parts 1,2,3, 1992.
4. Frederickson, AR., Cotts, DB., Wall, JA. and Bouquet, F.L., *In* "Spacecraft Dielectric Material Properties and Spacecraft Charging, Progress in Austronautics and Aeronautics", **107**, AIAA, New York, 1986.
5. Dunbar, WG., *In*, "Design Guide: Designing and Building High Voltage Power Supplies", **11**, AF WAL-TR-88-4143, ll, 1988.
6. Roux, JA. and McCay, TD., *In*, "Spacecraft Contamination: Sources and Prevention", Progress in Astronautics and Aeronautics, **91**, AIAA, New York, 1984.
7. Green, BD, Calodonia, GE. and Wilkerson, TD., *J. Spacecraft*, **22**, No. 5, 500-511, 1985.
8. Narcisi, R., Trzcinski, E., Frederico, G., Woldyka, L. and Delorey, D., *In* "The Gaseous and Plasma Environment Around Space Shuttle", Proc. of the AIAA Shuttle Environment and Operations Meeting, Washington, DC, 47-51, 1983.
9. Gordon, LB., Proceedings of the 23rd IECEC, **1**, 1988.
10. Benson, JD., Allred, DB. and Cohen, HA. *In* "Nuclear Instruments and Methods in Physics Research", **B40/41**, 1071-1075, Elsevier Science Publishers, North-Holland, Amsterdam, 1989.
11. Allred, DB., *In*, "Proceedings of the 7th IEEE Pulsed Power Conference", 312-317, *IEEE* , 89CH2678-2, 1989.
12. Allred, DB., Benson, JD., Cohen, H.A., Raitt, WJ., Burt, DA., Katz, I., Jongeward, GA., Antoniades, J., Alport, M. Boyd, D., Nunnally, WC., Dillon, W., Pickett, J. and Torbert, RB., *IEEE Trans. on Nuclear Science*, **35**, 6, 1386-1393,1988.
13. Geissler, KH. *Proc. of the European Space Power Conference*, ESA SP-294, 451-455, 1989.

14. Katz, I., Jongeward, GA., Mandell, MJ., Maffei, KC. and Cooper, J.R., *Proc. 24th IECEC*, **1**, 447-451., 1989.
15. Hastings, DE., Weyl, G. and Kaufman, D., *J. Spacecraft*, **27**, 5, 539-544, 1990.
16. Thiemann, H., Schunk, RW. and Bogus, K., *J. Spacecraft*, **27**, 5, 563-565, 1990.
17. Garrett, HB., *Reviews of Geophysics and Space Physics*, **19**, 4, 57616, 1981
18. Cho, M. and Hastings, D.E. *J. Spacecraft*, **28**, 6, 698-706, 1991.
19. Carruth, Jr., MR., Vaughn, JA., Bechtel, RT. and Gray, RT., *Proc. 30th Aerospace Sciences Meeting and Exhibits*, AIAA, 92-0820, 1992.
20. Hastings, DE. and Cho, M., *Proc. 30th Aerospace Sciences Meeting and Exhibits*, AIAA 92-0576, 1992.
21. Adamo, RC. and Matarrese, JR., *J. Spacecraft*, **20**, 5, 432-437, 1982.
22. Mang, CR. and Lindewr, WK. *In*, "Hypervelocity Impact", McDonnell, JAM. Ed., 257-261, University of Kent, Canterbury, 1992.
23. Kessler, DJ. and Su, SY., *In* "Orbital Debris, NASA Conference Publication", 236, 1982.
24. Weishaupt, U., Kuczera, H. and Rott, M., *In*, "Proc. 5th European Symposium: Photovoltaic Generators in Space", 175-180, ESA SP 267, 1986.
25. Carruth, Jr., MR., Vaughn, JA., Holt, JM., Werp, R. and Sudduth, RD. *In*, "AIAA Space Programs and Technologies Conference", AIAA 92-1685, 1992.
26. Hill, DC. and Rose, MF., *In* "Proc. 3rd European Space Power Conference", Graz, Austria, 1993.
27. Levy, L., Reulet, R., Sarrail, D., Siguier, JM. and Lechte, H., *In*, "Proc. 5th European Symposium: Photovoltaic Generators in Space", 161-169, ESA SP-267, 1986.
28. Robinson, Jr., PA., *In*, "Spacecraft Environmental Anomalies Handbook", GL TR-89-0222, Geo. Laboratory, Hanscomb AFB, MA, 1989.
29. Purvis, CK., Garrett, HB., Whittlesey, AC. and Stevens, NJ., "Design Guidelines for Assessing and Controlling Spacecraft Changing Effects", NASA Tech. paper, 2361, 1984.
30. Sutton, JF. and Stern, JE., "Spacecraft High-Voltage Power Supply Construction", NASA TN-D 7948, 1975.
31. Kunhardt, EE., Lederman, S. and Levy, E., *In* "Proc. of the 8th Int. Symp. on Discharges and Electrical Insulation in Vacuum", 246-249, Paris, France, 1988.
32. Maffei, KC., Jongeward, G., Katz, I. and Mandell, M., *In*, "Proceedings of the 7th IEEE Pulsed Power Conference", 318-321, IEEE 89CH2678-2, 1989.

33. Gollor, M. and Rogalla, K. *In* "Proceedings of the 15th International Symposium on Discharges and Electrical Insulation in Vacuum", 559-568, IEEE 92CH3192-2, 1992.
34. Taylor, GA. and Bell, WR., *In* "Proc. 7th IEEE Pulsed Power Conference," 325-327, Monterey, CA, 1989.

14

High Voltage Vacuum Insulation at Cryogenic Temperatures

B Mazurek

14. 1   Introduction

Vacuum is a unique insulating medium. It successfully performs the role of thermal insulation, particularly in the range of the lowest temperatures, and at the same time, on account of its dielectric properties, it may be used for electric insulation. Owing to this, it seems to be an ideal medium to be used in cryoelectric equipment, where it could perform both of these functions, i.e. electric and thermal insulation simultaneously. This should therefore simplify the construction of HV cryogenic facilities and make them less expensive. A thorough study of the operating conditions of vacuum insulation in strong electric fields, and under widely fluctuating temperatures conditions, including the lowest temperatures, is also becoming an urgent priority, particularly in view of the development of space technology. This discussion of vacuum insulation under conditions involving the simultaneous interaction of a strong electric field and a low temperature is therefore very opportune. When speaking of low temperature we mean here the temperature of the elements of the solid structures that constitute the insulating system. These will be electrodes, spacers and bushing insulators, as well as other constructional elements. In addition to the practical dimension, investigations carried out at low temperatures also provide information which makes it possible to test the existing mechanisms thought to be responsible for the development of vacuum discharges. Thus, by decreasing the temperature of the various elements of the insulating system (electrodes and insulators), we significantly change their physical properties, which makes it possible to verify certain aspects of the vacuum discharge development mechanism.

In cryoelectrotechnical devices, two physical phenomena are generally put to use, namely the phenomena of increased thermal/electrical conductivity, and that of superconductivity. The highest increase in thermal/electrical conductivity at low temperatures is displayed by commonly used pure conductor materials copper and aluminium. As regards superconducting materials, most experiments have been carried out with niobium and tantalum, and recently with so called high-T_c superconductors, with YBaCuO being the most popular of these.

A very good constructional material that is used for building both vacuum and low-temperature devices is austenitic stainless steel. Furthermore, in some applications there will be HV electrodes made of stainless steel that have to support the high electric fields, and hence participate in the development of discharges: thus, it is a material whose properties should be fully analysed. The second group of materials that should be subjected to testing are dielectric materials which are used to make bushing and spacer insulators. In practice, use is made most frequently of teflon, epoxy resins and ceramics, particularly alumina. This is the reason why the majority of all the publications on the behaviour of vacuum insulation in a strong electric field, and at low temperatures, are concerned with the above mentioned group of materials.

If we intentionally reduce the temperature of an element of a cryoelectrotechnical appliance, then we will be concerned to ensure that it is maintained in this state during its operation. If it is a superconductor, then its temperature should at all times be lower than its critical temperature T_C since, by definition, the flow of current in a superconductor does not cause losses, and so it cannot generate heat and cause the temperature to increase. This, however, is pertinent only when strictly defined conditions are satisfied. Thus, the question arises as to whether the current associated with electron field emission from a superconducting cathode can be emitted without electric losses in the emitter. This is interesting both from the viewpoint of superconductivity theory and the mechanism of vacuum discharge development, as well as from the viewpoint of applicability; i.e. the reliability of operating cryoelectrotechnical appliances. This topic will be discussed in a later section of the chapter, together with the effect on the voltage hold-off capability of a deep pre-cooling of a vacuum insulation system consisting of electrodes made of normal conducting metals in the presence of a spacing insulator.

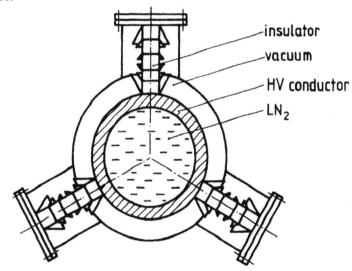

Fig 14.1   Cross-section of a nitrogen cryo-cable taken through the insulating support structure. (After [3].)

In publications on this subject, a number of cryelectrotechnical devices have been described that make use of high vacuum. As an example, we can cite here the construction of cryogenic cables for electric energy transmission. These were developed as long ago as in the early 1970s, and were based on both superconducting systems [1], as well as conductors made of normal metals, such as aluminium [2,3]. The principle of the cryoconducting cable is presented in Fig 14.1. From this, it will be seen that the central canal of the cable is an aluminium pipe, which serves both as the HV conductor and to contain the circulating liquid nitrogen (LN_2) coolant. Correspondingly, the outer surface of this cylindrical conductor is both thermally and electrically insulated by a high vacuum envelope. The high-voltage pipeline is mounted on cyclically repeated segments of three symmetrically distributed supporting insulators. In this construction, the insulators themselves are placed in special chambers welded onto an external shield made of stainless steel. As a result, the external shield can have a relatively small diameter, determined by the vacuum strength of the concentric cylindrical envelope.

This type of electric power line is characterised by having a high transmission power capability with relatively low losses of electric energy. It may be used in electric power stations to join generators with block transformers, or in industrial plants producing aluminium etc. A similar construction may also be used in space for both orbiting and outer space stations complexes, where it is necessary to provide a power link between the central and out-lying modules In the construction of Fig 14.1, the high voltage rating (110 kV) results in a high electric field intensity on the surface of electrodes, with the consequent possibility of prebreakdown field emission currents and a discharge across the vacuum insulation. These emission currents will also be the cause of additional electrical losses, localised heating at both the emission centre and the corresponding region of the anode bombarded by electrons. As a result, prebreakdown emission currents have to be seen as a potentially damaging phenomenon, and that attempts should be made to restrict it.

As discussed in Chapter 12, a similar situation exists with superconducting RF particle accelerators. At present, intensive research efforts are being made to evaluate the possibility of applying deeply cooled pure metals or superconductors (both conventional Nb [4] and high T_c materials [5]) to the development of low-loss particle accelerator systems. It should also be noted that strong electric fields frequently occur in low-voltage, devices, e.g. microelectrics systems, where there are appropriately small distances between electrodes. Indeed, the "switching off" of superconductivity by the flow of field emission currents could be put to practical use in the construction of new devices, such as microswitches [6].

14. 2   Prebreakdown Phenomena with Precooled Electrodes

In this section we shall compare the prebreakdown emissive properties of electrode assemblies made of normal conducting metals (e.g. Cu, Al, Sn, etc.), with those made of superconducting materials (e.g. Pb, Nb, YBaCuO).

14.2.1   Electrodes Made of Normal Conducting Metals

One of the apparent advantages of using vacuum as an electrical insulating medium is its low, or close-to-zero loss; this is a particularly attractive feature for applications operating with high frequencies and at cryogenic temperatures. However, this low loss is only available up to a certain level of the applied electric field intensity: namely, when a current of tunnelling electrons appears from the cathode. This emission follows the well known Fowler-Nordheim (F-N) law, so that, as discussed in Chapter 3, the dependence of the measured emission current I on the electric field intensity E will give a straight line if they are plotted as $\ln(I/E^2)$ versus $1/E$. It follows from the temperature-dependence of the F-N equation that the field emission current is effectively insensitive to the emitter temperature in the range from 0 to 800K. It follows therefore, that the cooling of electrodes from room temperature to cryogenic temperatures should not affect the characteristics of the F-N mechanism. This conclusion is probably only valid for ideally pure and clean surfaces such as can be achieved with sharp pointed emitters. In practice, however, a real emitter surface is covered by an adsorbed layer which affects the emission mechanism. This layer will tend to undergo changes in the course of cooling, and hence it could be expected that a cooled emitter will have different emission characteristics to those obtained at room temperature. Such an anticipation has in fact been confirmed by experiment. By way of example, Fig 14.2 presents the dependence of log I on (E) for a copper cathode with a large surface at 300K and 80K [7].

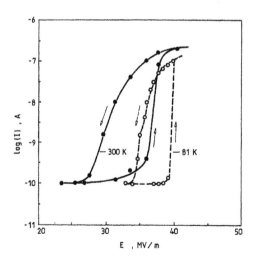

Fig 14.2   Typical I-V characteristics of field electron emission from Cu cathode surface at room temperature and at 80 K. (From Mazurek et al. [7], with permission.)

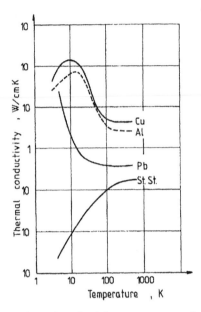

Fig 14.3   Dependence of thermal conductivity on temperature for a selection of metals. (From Berteaux [40], with permission.)

From the law of current continuity, it follows that the emission current must flow through the cathode and supply energy to the anode surface. It has also been shown in Chapter 5 that field emission takes place from one or several sites on the cathode, which are referred to as emission centres (ECs); features that are particularly clearly visible on cathodes with large areas. As a consequence, there are high current densities in the ECs, which can be experimentally observed both in the case of emission from micropoint emitters, and well polished flat cathodes. We are not able to precisely determine the current densities associated with the processes, but with some caution, can assume the data presented in publications [8,9], viz. 10^5-10^9 A/cm^2, to be reasonably accurate. Associated with a high emission current density there will be power losses in the emitter microvolume which will be proportional to $j^2\rho$, where j is the current density and ρ emitter resistivity. When the emitter temperature is reduced, these losses will decrease, mostly for the reason that its resistivity ρ will be reduced according to the Gruneisen-Bloch dependence, viz.

$$\rho = \frac{KT^5}{M\theta^6} \int_0^{\theta/T} \frac{x^5 dx}{(e^x - 1)(1 - e^{-x})}$$

14.1

492

where M is the atomic weight, θ the Debye temperature, and K is a characteristic constant for metals.

At the same time, the temperature reduction results in an increase of thermal conductivity, which makes it easy for the heat to be carried away from the emitter volume to the cathode volume. For completeness, the dependence of thermal conductivity on temperatures for a selected group of metals is presented in Fig 14.3. On the other hand, account has to be taken of the opposite effect resulting from a decrease of the specific heat c at temperatures lower than the Debye temperature, where

$$C = 9R\left(\frac{T}{\theta}\right)^3 \int_0^{\theta/T} \frac{e^x x^4}{\left(e^x - 1\right)^2} dx \qquad 14.2$$

with R being the gas constant. Thus it has to be assumed that the energy dissipated in the emitter volume will cause a significant increase in the emitter temperature, so that it may have a much higher local temperature than that of the remaining part of the cathode. In an appropriately strong electric field, the increase in the emitter temperature may well be high enough to significantly affect the development of an electric discharge. In this context, two cases are distinguished:

a) the generation of a cathode plasma through the violent dissipation of energy in the emission centre (the so called explosive emission process [10,11] discussed in Chapter 6), and

b) the generation of plasma from the pockets of adsorbed gases and impurities located in the vicinity of the emission site [12,13,14].

With small electrode separations (< 0.5 mm), the generation of cathode plasma will fill the interelectrode space and be equivalent to a breakdown. However, with larger electrode separations, it is the generation of anode plasma that becomes the necessary condition for breakdown; here, an essential role is played by the violent increase of anode current accompanying the generation of cathode plasma. It is assumed in the explosive emission theory that the cathode surface is atomically clean, i.e. devoid of adsorbed gases, impurities and alien inclusions, and that the micropoint emitter explodes as a result of the violent energy dissipation of electrons accelerated by the electric field in the emitter structure. An atomically clean surface is, in practice, extremely difficult to obtain: the real surface of an emitter will be covered by adsorption products in which water plays the most important role. That is why the second of the above mentioned possibilities, i.e. the generation of plasma through the ionisation of foreign particles desorbed from the emission site, seems to be very likely. On the other hand, the partial melting observed on the cathode surface may result from the interaction of the positively charged ions of the cathode plasma with the cathode surface. In fact, a model of such an interaction has been presented by Schwirzke [15].

Whilst a detailed discussion of the mechanisms mentioned above is not appropriate for this chapter, it should however once again be pointed out that, in the explosive emission model, as well as in the model involving the ionisation of desorption products, an essential role is played by the dissipation of energy by conducting electrons in the emitter material to produce local emitter heating. However, a direct measurement of the excess temperature of emission centres is extremely difficult, and so we have to estimate it indirectly from observations of the partial melting of cathode emission centres [13].

It follows from the above discussion that the process of generating cathode plasma could be controlled, or even arrested, if it were possible to limit the energy dissipation in the emitter material and, in consequence, limit the increase of the emitter temperature. In fact, such a possibility should, in principle, be realisable by the use of superconducting emitters having zero resistance. The practicability of this assumption will be discussed in the following section.

14.2.2   Superconducting Electrodes

Although superconducting materials are characterised by having zero resistance, a lossless flow of current is only possible under precisely determined conditions. Failure to satisfy these conditions will cause the loss of superconductivity in any locality where the critical parameters are exceeded, even if the remaining volume of material is in the superconducting state. Let us examine a superconducting broad-area cathode on which the applied electric field has sufficient intensity for generating a total field emission current of $\sim 10^{-5}$ A. It is known however, that this relatively small current does not flow out uniformly from the whole cathode surface, but only from one or, at the very most, a few favoured emission centres (ECs) [16]. Their sizes are not known exactly but the experimental results indicate that, even in the case of a point emitter, the emission takes place only from these microscopically localised EC regions [11]. If we assume that the dimension of an EC is 10^{-10} cm^2, then the current density in this case will be $\sim 10^5$A/cm^2. What interests us is whether such emission is possible from a superconductor, i.e. without losses in the emitter. In order to answer this question it would be necessary to know, on the microscale, the density distribution of the emitted current; i.e. if the critical current density j_c is exceeded at any point on the emitter. Thus, if there is a local loss of superconductivity, there will be an onset of ohmic losses, and a resultant generation of heat leading to an increase in temperature of the EC volume. In summary, the emitter will lose its superconducting properties when:

- its temperature is higher than the critical temperature T_c of the material
- the material is subjected to a magnetic field whose intensity is higher than its critical value H_c
- the density of the current flowing through the material is higher than the critical value j_c.

To give the reader an idea of the order of magnitude of the critical values of the parameters mentioned above, the properties of a few characteristic superconducting materials are presented in Table 14.1.

Table 14.1

Material	T_c [K]	H_c [A/cm]	J_c [A/cm^2]
Ta	4.5	600	*
Pb	7.2	660	*
Nb	9.46	2400	10^5 (4.2K)
Nb$_3$ Sn	18.45	220000	10^7 (4.2K)
YBaCuO	92	1500000	10^3 (77K)

*If the superconductor is a pure metal (Type 1), I_c can be estimated as the current level that will create surface magnetic fields sufficient to exceed H_c. This estimate is known as the "Silsbee approximation". The current required to develop a field H_c on the superconductor surface is $I_c = 2\pi R H_c$ (where R is a cylindrical conductor radius).

From a consideration of these data, particularly the critical current values, it is clear that, only in the case of Nb and Nb3Sn cathodes, would it be possible for the emission to take place in the superconducting state. The remaining emitters will pass into the normal state due to their j_c being exceeded. We have used here the conditional form "would be possible" since other limitations will also occur in practice, and these may prevent emission in the superconducting state.

Among these limitations will be those associated with a) the penetration depth of the applied magnetic field, defined by the Meissner effect, and b) the consequences of the Nottingham effect. Meissner found that a superconductor is characterised not only by having zero resistance, but also that it behaves like an ideal diamagnetic. This implies that, after passing into the superconducting state, the material expels the magnetic flux from within its bulk (no connection with Lenz law). Thus, the current in the ideal superconductor flows merely in the superficial layer which is characterised by a "penetration depth λ". In pure metals, the penetration depth at $T \approx 0$ is about 50 nm, but increases with both the presence of impurities and increasing temperature [17]. If we imagine the emitter as a micropoint in the form of a cylinder, as depicted in Fig 14.4, then a current of density j will induce around the cylinder a magnetic field with inductance B which will penetrate the inside of the material to a depth λ. If the diameter of the micropoint is comparable to its penetration depth, then the emitting micropoint will pass to the normal state, in spite of the fact that the remaining part of the cathode will be in the superconducting state.

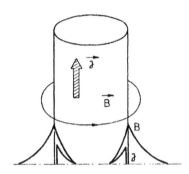

Fig 14.4　Distribution of current density j and induction B in a cylindrical conductor.

A secondary factor, which can threaten the emitter from remaining in the superconducting state, is the Nottingham effect [18] previously discussed in Chapter 5. This effect takes account of the fact that tunnelling electrons are emitted with a mean energy that is approximately equal to Fermi energy. The emitted electrons are replaced by electrons coming from the external circuit, and these convey their energy to a region of the surface layer whose thickness is of the order of the electron mean free path. Thus, the physical consequences of the Nottingham effect are dependent on the temperature of the emitter. If the emitter is at a temperature that is lower than the so called inversion temperature T_i, it will be heated by the Nottingham effect; conversely, if the temperature is lower than T_i, then it will be cooled. This has been confirmed experimentally for both for normal metals and also by work on niobium superconducting emitters [19]. It is very likely that the energy supplied to the EC as a consequence of the Nottingham effect causes it to be heated above the critical temperature T_c, and leads to the emission taking place in the normal state, with the majority of the cathode remaining in the superconducting state. It follows that the factors discussed above are the most likely cause of measurement failures occurring during tests that were designed to establish the differences between emission from the normal and superconducting states [20,21,22,23].

Such differences were anticipated on the basis of the BCS theory [24], from which it follows that the transition to the superconducting state is accompanied by the formation of a new electron order, the so called Cooper pairs, which should reduce the emission current by a value Δi, which for Nb was estimated by Leger to be about 4×10^{-13} A [21]. In fact, Leger's experiment did not confirm those predictions, with similar findings having been reported by other authors [22,23]. On the other hand, Cobourne and Williams [20] have noted that the characteristic $\ln(I/E^2)$ versus $(1/E)$ for Nb electrodes precooled to 5K consists of two straight lines with decidedly different slopes, corresponding to field amplification factors of $\beta_1 \approx 300$ and $\beta_2 \approx 2000$ respectively. However, it was found that conditioning with 18 discharges caused a change of the characteristic into one straight line with $\beta \approx 300$. The authors came to the conclusion that the discontinuity observed on the F-N plot was not connected with a superconducting transition but, most likely, due to a layer of oxide which could have been removed by the discharge. Similar results have been obtained with high-T_c ceramic cathodes, such as YBaCuO [7].

Recent studies by Shkuratov and co-workers, carried out on emitters made of $YBa_2Cu_3O_{6.9}$ monoscrystals [25] have proved that the emitted current is unstable, and that the field enhancing factors, calculated on the basis of the results obtained were within the range $\beta \approx 100\text{-}300$: however, such high values of β do not result from the geometry of the emitter. Next, from the experimentally derived characteristics of log I versus $\text{in}(I/V^2)$ plots, obtained with the use of superconducting broad-area YBaCuO cathodes, a hysteresis phenomenon has been observed, which also occurs with a contaminated copper cathode [7]. Energy distribution measurements of electrons emitted from ceramic superconductors [26] have proved that they are similar to those emitted from metals, and can be described by means of Fermi gas model.

The results given above are evidence that the emission mechanism associated with a superconductor is the same as that for a normal metal. On the other hand, the essential role played by the adsorbate layer on the surface of the emitter remains to be established: it does, however, undoubtedly depend on the temperature of the surface.

14. 3   Vacuum Breakdown with Precooled Electrodes

When speaking of the electric breakdown voltage, we have to define what is meant by this concept. The point is that, whilst the breakdown of a solid or liquid dielectric is an unequivocal notion, in the case of vacuum some misunderstanding may arise. This results from the specificity of this insulating system. Firstly, vacuum belongs to the so called "self-regenerating" class of insulating media, which implies that soon after a breakdown occurs, it intrinsically regains its insulating properties. Secondly, as discussed in the following section, "breakdown" often develops in such a way that it is preceded by a sequence of short-duration, large-amplitude current pulses: however, it frequently happens that, after this phenomenon, the system regains its insulating properties without a full breakdown of the gap. In the third place, the vacuum breakdown voltage will vary according to the state of electrode surfaces. Thus, successive breakdowns at a few minute intervals generally occur at higher and higher voltages; a process which is referred to as the "conditioning" of electrodes. It is extremely difficult to state unequivocally when a given system is conditioned. The number of conditioning breakdowns is dependent on the dimensions of the electrode surfaces, and may range from several dozens for small electrodes to tens of thousands for electrodes with surfaces of a few square metres [27]. It is thus no wonder that, with reference to a similar insulation system, different authors present quite different values of breakdown voltages [3]. It would seem that, in the context where vacuum is used for the high voltage insulation of electrical equipment, then the definition of breakdown given in [28] best captures the essence of the subject. Namely, that "breakdown is such a state of the electric insulation system in which the conditions for the unlimited increase of electric conductivity are satisfied".

In thus defining breakdown, the present author draws upon his observations of a discharge development by using the combination of high speed photography to record optical process, and a fast oscillograph to record of the evolution of the current and voltage. The typical course of the development of a vacuum breakdown with point-to-plane electrodes is presented on the streak photograph of Fig 14.5, as well as in the framing mode photograph of Fig 14.6. We can see from these that at first a faint light appears on the cathode, and after some delay, a much more intensive anode light. The appearance of the anode light is then followed by a sudden drop in voltage across the electrode gap. With an impulse voltage, we are concerned with a typical impulse cut off. Thus, the breakdown voltage is considered to be the value recorded on the storage oscilloscope, which is accompanied by the appearance of anodic plasma observed on the high speed photographs.

14.3.1 Effect of Electrode Cooling on the Development of an Electric Discharge in Vacuum

One of the parameters that characterises an electric discharge is the time-lag of the anodic burning in relation to the cathodic burning. This is readily measurable from the high-speed photographs of Fig 14.5 and 14.6, and leads to the conclusion that anodic phenomena are secondary to, and dependent on cathodic ones. This can be explained by recognising that the cathode is a source of electron current which acts upon the anode as energy and causes it to be heated. Thus, if the anode is heated to a critical temperature, a violent evaporation and ionisation of the anode material will take place and result in the formation of the anodic plasma, as recorded by the high speed camera.

498

Earlier research studies have been able to establish that the time-lag of the appearance of cathodic plasma, in relation to the moment of applying the voltage, is not dependent on the interelectrode distance [11,29]. The measurement of the time difference between the appearance of cathodic burning and anodic burning makes it possible to determine the time-lag of anodic burning in relation to the moment of applying the voltage.

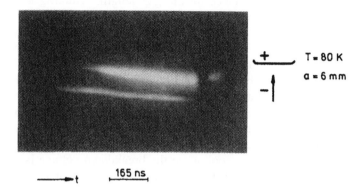

Fig 14.5   Typical streak photograph of the initial phase of discharge development in vacuum for point-to-plane copper electrodes (point cathode). (From Mazurek [3], with permission.).

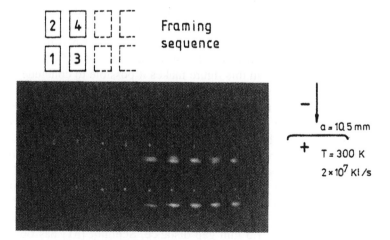

Fig 14.6   Framing photograph of the initial stage of an electric discharge in vacuum. Exposure time 10 ns, time between frames 40 ns, stainless steel electrodes, electrode position and polarity as indicated in the sketch. (From Mazurek [3], with permission.)

499

Fig 14.7 presents the dependence, measured from streak photographs, of the time-lag of anodic burning as a function of the electrode spacing for room temperature 300K and 83K; the latter being obtained after precooling the anode with liquid nitrogen. As can be seen from Fig 14.7, an increase in electrode separation from 1 mm to 10 mm results in an increase in the delay time for anodic plasma to appear; furthermore, this delay has an approximately linear dependence on the gap length. Cooling a flat anode down to 83K causes the time-lag of anodic burning to increase by about 20%. It has not been established, however, whether the cooling of the cathode has an effect on this time lag. According to the concept of electric breakdown described earlier, in which the appearance of anodic plasma plays so significant a role, it may be expected that an increase in the time-lag of the appearance of anodic plasma through the cooling of the anode will be simultaneously accompanied by an increase in the strength of the gap. In fact, investigations have indeed confirmed that such a thesis holds in relation to both DC and impulse voltages.

The effect of electrode cooling on the breakdown voltage is dependent on the polarity of the electrode being cooled. In the case of DC voltages, it will be seen from Fig 14.8 that the cooling of the cathode to about 80K results in only a marginal increase in the breakdown voltage. Although Fig 14.9 indicates that this effect exists for both point and flat cathodes, it is relatively insignificant from the practical point of view. However, the cooling of the anode gives rise to a distinct increase in the breakdown voltage. Thus Fig 14.9 indicates that the cooling of the flat anode causes a 60% increase in the DC breakdown voltage. Correspondingly, Fig 14.10 shows that results of a similar kind are obtained where using impulse voltages, when in both cases, measurements were made with electrode separations of 0.5-12 mm. The family of curves presented in this figure makes it possible to conclude that the effect of cooling becomes weaker when the interelectrode distance decreases. Similar studies have been made by Maitland [30] and Schmidt [31]. For flat copper electrodes having a 0.5 mm electrode gap, with the anode being cooled down to 78K, Maitland has obtained an increase of strength from 5 to 13% compared with room temperature. By considering both the transport of heat from the anode by thermal conduction, as well as the need for locally heating the anode to its boiling point, he was able to come to the conclusion that cooling the anode should cause the breakdown voltage to increase by 20%. On the other hand, for copper electrodes having a rod-plate geometry ($\phi = 30$ mm), and with a 1.5 mm gap distance, Schmidt [31] has obtained a 50% increase in the breakdown voltage when the anode is cooled. It is very likely that the reduction in the breakdown voltage as a result of electrode cooling with small electrode separation is caused by the change of the breakdown mechanism from anode to cathode based. From an analysis of

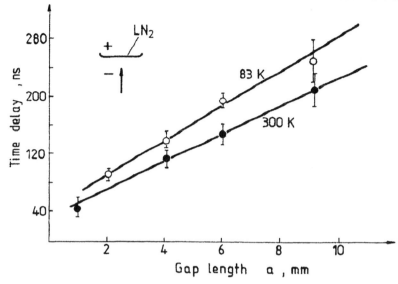

Fig 14.7   Time delay in the appearance of the cathode and anode light as a function of the electrode gap at 83 K and 300 K anode temperatures. (From Mazurek et al. [41], with permission.)

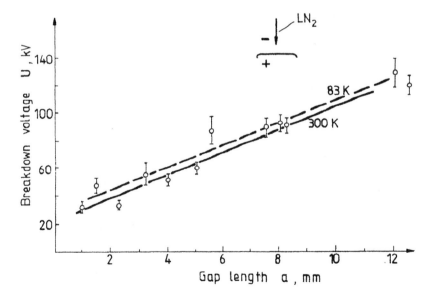

Fig 14.8   DC breakdown voltage as a function of the gap length; point cathode 300 K and 80 K temperatures. (From Mazurek et al. [41], with permission.)

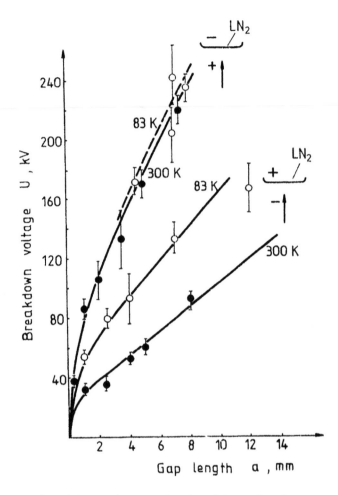

Fig 14.9   DC breakdown voltage as a function of the gap length; parameters of the
plane electrode polarity. (From Mazurek et al. [41], with permission.)

high speed photographs, it is possible to conclude that the mechanism of
vacuum discharge development between electrodes at room temperatures does
not differ from that operating with cooled electrodes. The major difference
that can be recorded centres on the change in the time-lag to anodic burning.

Thus, the cooling of the anode has a similar effect to that of increasing
the electrode separation: i.e. since by reducing the temperature of the anode, it
is necessary to supply an additional quantity of energy to it in order to first
return the temperature of the heated element to room temperature.
Subsequently, the breakdown and associated generation of anodic plasma
proceed in a similar way to the case of electrodes that are at room

temperature. We have a similar phenomenon when increasing the distance between the electrodes. The longer the distance, the lower the density of the current bombarding the anode which, to achieve the same thermal results, should be compensated for by either an appropriately extended time of the action of a flux of electrons upon the anode, or by a higher voltage.

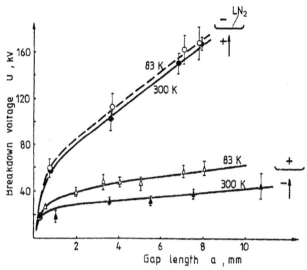

Fig 14.10   Impulse breakdown voltage as a function of the gap length for a pulse slope of 100 kV/μs. (From Mazurek et al. [41], with permission.)

14.3.2   Discussion of the Breakdown Mechanism

As mentioned at the beginning of this chapter, the analysis of electrical breakdown can be reduced to an analysis of the mechanism of plasma generation in the interelectrode gap. If, following Langmuir, we define plasma as an ionised gas, then the last but one stage of breakdown must be the generation of local pockets of gas which will subsequently become ionised. In all kinds of insulation, apart from vacuum, plasma is produced from dielectric materials. In the case of vacuum, plasma must be produced from the electrode material and, more strictly speaking, from the surface layers of the electrodes which will include any particles that are adsorbed on the surface. Plasma can, in principle, therefore be supplied by either the cathode or anode. The formation of cathodic plasma is connected with field emission from the cathode, and may be formed either by violently ionising desorbed gas in the field emission area, or by the so called "explosive" process associated with the ionised vapour in the vicinity of the metal emitter. As can be seen from the high speed photographs, the cathodic plasma in the initial stage of the discharge development increases its size, although with some limitation, but remains connected to the surface. One explanation of this phenomenon may be associated with positive plasma charge which, after throwing out light electrons, maintains positive ions close to the cathode. With small interelectrode distances (<0.5 mm) the generation of cathodic plasma will then fill the interelectrode space and directly create breakdown conditions. With larger electrode gaps, the generation of cathodic plasma creates the necessary conditions for a violent increase of the emission current which, in turn, provides the energy that is indispensable for the melting and evaporation of the superficial anode layer.

The electron current flowing out of the cathode propagates in the inter electrode space along a path that has the shape of a paraboloid of revolution, with its vertex on the cathode and its base on the anode. Of essential importance is the base area on the anode, since this determines the density of anodic current, and thus the density of the heat flux q generated on the anode surface, where $q \sim V j_a$ [W/m^2] in which V is the voltage on electrodes, and j_a is the density of anode current. This heat flux q causes an increase in the temperature of the anode surface being bombarded, which is accompanied by an increase in both the local vapour pressure and in the speed with which anode material is evaporated. It may be assumed [32] that after a temperature of 1546K has been exceeded with a copper electrode, the conditions for producing anodic plasma are satisfied.

The dependence of the anode temperature on the density of the energy flux supplied to its surface by the anodic current can be derived from the basic equations governing thermal conductivity. However, an analytical solution of such a system is not a simple task, particularly if the properties of the materials are assumed to be temperature dependent, i.e. the thermal conductivity $\lambda(T)$, the specific heat $c(T)$ and density $P(T)$; on the other hand it can be solved with the use of numerical methods.

However, by introducing simplifying assumptions, it is possible to estimate the density of the flux q_a which, at a definite time, will cause the anode surface to be heated up to the temperature T_a. (See also the treatment of this topic in Chapter 5.) Thus, assuming a) a constant power flux q_a supplied to the anode and b) constant mean values for λ, c and P, the heat flow q in the direction x in the time interval t will cause a change of temperature T which can be described by the relation

$$q = -\lambda \frac{\partial T}{\partial x} \qquad \qquad 14.3$$

which, for x > 0 and t > 0 can be expressed in the form

$$\alpha \frac{\partial^2 q}{\partial x^2} = \frac{\partial q}{\partial t} \qquad \qquad 14.4$$

The solution to this equation for $q = q_a = const$; at $x = 0$ and $t > 0$ is

$$q = q_a erfc \frac{x}{2\sqrt{\alpha t}} \qquad 14.5$$

where $\frac{\lambda}{Pc}$ is the thermal diffusivity, and

$$erfc = \frac{2}{\sqrt{\pi}} \int_0^\infty e^{-\xi} d\xi \qquad 14.6$$

The above problem has been solved by Carlslaw and Jaeger [33] to obtain the simple dependence

$$T = \frac{2q_a}{\lambda} \left(\frac{\alpha t}{\pi} \right)^{\frac{1}{2}} \qquad 14.7$$

The time t during which the heat flux flows into the cathode can be read out from Fig 14.7 for any given electrode separation. By way of an example, for an electrode gap of a = 6 mm, and for the anode at room temperature $t_{300} = 148 \times 10^{-9}$ s. To achieve local boiling on a Cu anode initially at room temperature 300K, it would be necessary to preheat the electrode by 1546-300 = 1246K, while for an anode precooled down to 83K, the required preheating would be 1546-(-83) = 1463K. By assuming the mean values of $\lambda = 3.76$ W cm^{-1}K^{-1}, C = 0.38 Jg^{-1}K^{-1}, P = 8.94 g cm^{-3} and $\alpha = 1.14$ cm^2 s^{-1}, we can apply the above expressions to calculate

$$q_{300} = \frac{\lambda T}{2\left(\frac{\alpha t}{\pi}\right)^{1/2}} = 1.01 \times 10^7 \text{ Wcm}^{-2} \qquad 14.8$$

In contrast, by identically calculating the heat flux necessary for a precooled anode to reach the same temperature in the same time, one finds,

$$q_{83} = 1.19 \times 10^7 \text{ Wcm}^{-2}$$

These calculations show that the breakdown of a cooled anode occurs at a higher voltage; a conclusion that has been confirmed experimentally, as shown in Fig 14.9 and Fig 14.10. However, the increase in voltage needed to induce a breakdown on the cold surface is slightly smaller than expected, and can be attributed to the adsorbed layers of, mainly, water on the surface of the cathode which can be a source of cathodic plasma. It should also be noted that the power density is proportional to Vj where V is the applied voltage and j is the current density on the anode.

14. 4   Effect of Cooling the Insulator on the Flashover Voltage in Vacuum

A review of numerous publications dealing with electric discharges in vacuum indicates that the least identified phenomenon is the effect of low temperature on the surface strength of solid insulators. The results presented in a relatively small number of papers discussing this subject-matter display considerable differences. By way of example, after cooling a polyethylene insulator to 8K, Kahle and Richter [34] obtained a 70% increase of surface strength. When cooling polyethylene to 233K, Ohki and Yahagi [35] obtained a 10% increase of strength. However, when testing the surface strength of insulators made of different organic materials, Tourreil [36] has found that cooling them to 80K not only fails to improve their strength but may even reduce it. In the papers [37,38], attention has been given to the key fact that a spacer insulator may be heated during the conditioning process, with estimates of the temperature increment amounting to several dozens of degrees (30K according to [37] and 90K according to [38]). The unpredictability of this phenomenon could well be one of the reasons for obtaining different results from the effect of cooling on the flashover voltage. In all the papers quoted above, the insulator was cooled through its contact with the cooled electrode. In a system like this there is a flow of the heat flux towards the heat sink, which is made up of the conduction towards the electrodes being cooled, and the parallel radiant heat flux. The heat flow is accompanied by temperature gradients, which are particularly high at the contacts of the insulator with the electrodes. In this connection, it is not possible for us, on the basis of a one-point measurement of temperature on the electrode or insulator, to determine the temperature of the remaining elements of the system. With the variable microscopic nature of the contact of the insulator with the electrodes, i.e. resulting from the different degrees of surface finish, coupled with the effects of different contact pressures, or material combinations, the local temperature distribution in the vicinity of the electrode-insulator system can be expected to show considerable variations. In the papers cited above, this has not been taken into account and may therefore be a second reason for the discrepancy in the results.

The present author has made measurements of the temperature distribution in an electrode-insulator system using the technique illustrated in Fig 14.11 [39] and, for the case of a 12.5 mm long alumina insulator, obtained the results presented in Fig 14.12. As can be seen, the temperature distribution is dependent on time. After the first hour of cooling, the temperature of the electrode making a thermal contact with the liquid nitrogen tank drops from room temperature to about 110K (curve 5).

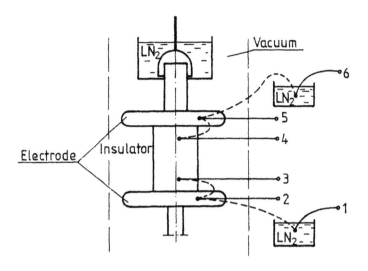

Fig 14.11 Thermocouples for measuring the temperature distribution occurring in the electrode-insulator system; thermocouple endings 1-6 are connected to a voltmeter. (From Cross et al. [39], with permission.)

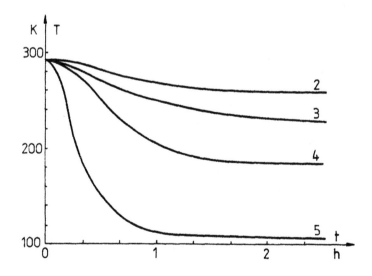

Fig 14.12 Temperature distribution in the electrode-insulator system in vacuum as a function of the cooling time: curves are numbered according to the measuring points shown in Fig 14.11.

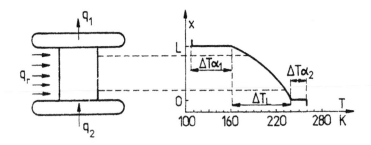

Fig 14.13   Temperature distribution in the electrode-insulator system after two hours of cooling, with allowance made for the radiation heat flux q_r directed towards the side surface of the insulator; q_2 - heat inflow through the electrodes support; q_1 - heat taken out by the cooler electrode. (From Cross et al [39], with permission.)

Later, the temperature of this electrode slowly and asymptematically approaches the temperature of liquid nitrogen. However, it has not been possible to obtain the baseline temperature even after 24 hours cooling. For practical purposes, it was therefore decided to make the measurements of flashover voltages after 2 hours cooling. At this stage, the temperature distribution displayed the form presented in Fig 14.13. This indicates that the temperature drop across the contact between the electrode and the insulator, $\Delta T_{\alpha 1}$ is about 55K, the temperature along the insulator, ΔT_4, is about 75K, while the temperature across on the contact of the bottom electrode with the insulator, $\Delta T_{\alpha 4}$, is about 25K. The overall heat flux q_1 carried away by the copper block to the nitrogen cooler consists of the heat q_2 supplied through the bearer of the bottom electrode, as well as the radiant heat q_r supplied through the side surface of the insulator. For this reason, the temperature distribution is not linear. The example given above indicates how inaccurate is the statement that the "testing was carried out at a definite temperature". When, however, we know the temperature distribution, then we may confine ourselves to giving only one temperature, but we should specify what element of the system it concerns.

Table 14.2

Temperature, T [K]	V_{DC} (kV)		V_{imp} (kV)	
	+	−	+	−
300	98 ± 9	92 ± 9	91 ± 6	82 ± 7
100	101 ± 12	112 ± 12	89 ± 10	122 ± 12

508

The measured results of the flashover voltage of 12.5 mm long alumina insulators are presented in Table 14.2. This shows that the effect of cooling on the surface flashover voltage in vacuum is the reverse to the cooling effect on the breakdown voltage of a pure vacuum gap. Without a solid insulator, the cooling of the cathode does not significantly affect its electric strength,

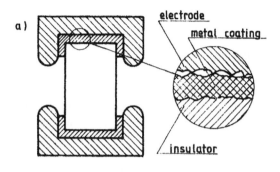

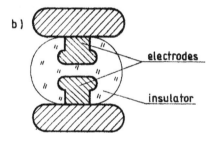

Fig 14.14   (a) Ceramic spacer insulator with metal coated surface of contact, (b) epoxy resin spacer insulator with electrodes controlling the electric field distribution.

whereas the cooling of the anode increases it significantly. Inserting the insulator between the electrodes produces a reverse effect. Cooling from the side of the anode has practically no effect on the value of the flashover voltage, whereas cooling from the side of the cathode causes about a 20% increase of strength with DC voltages, and about 50% with impulse voltages. The above results have been obtained in a system with a simple contact connection of electrodes with the insulator, and emphasise the role of the triple junction in the development of surface discharges. At the triple junction, plasma is generated from the insulator material, which then becomes a source of electron current flowing along the insulator surface towards the anode. The precooling of the triple junction makes it difficult for plasma to be generated, and this is thought to be the reason for the increase in its electrical insulating strength.

The above mechanism can be changed if the triple junction is properly designed in order to eliminate cathode plasma. As illustrated in Fig 14.14(a) and (b), this can be achieved by metallising the contact surface and specially shaping the cathode or the insulator.

14. 5   Conclusions

Vacuum is a very good electrical and thermal insulator. More specifically, its electric strength depends on the temperature of any associated electrodes, where a decrease in temperature generally has an advantageous effect on its insulating properties. As a general rule, there will be an increase in the electric strength whenever the solid surface being cooled is one that can provide a main source of plasma to fill the interelectrode space. For electrode separations larger than 1 mm, an improvement in strength for both DC and pulsed fields can be achieved by cooling the anode: on the other hand, if both electrodes are cooled, significant improvements can be obtained for alternating voltages. However, for electrode separations smaller than 0.5 mm, the cooling effect is negligible. The superconducting state of electrodes has no clear effect on the electric strength of the vacuum gap, and any changes of voltage hold-off strength that are observed should be attributed to the change of electrode temperature.

From a theoretical perspective, a decrease in electrode temperature, ranging from room temperature to that of liquid helium (4.2K), should not however, significantly affect the value of field emission "leakage" currents. This statement also applies to the transition of electrodes from the normal to the superconducting state. Thus, the changes in the emission characteristics that are observed in practice when cooling electrodes should be attributed to the presence of a layer of adsorbate on the cathode surface. In this context, the lower the temperature, the more stable is the layer of the adsorbate, and the better able it is to suppress the onset of "cold" cathode emission currents.

During the early stage of applying the electric field, the emission current is generally very low in value and highly stable. However, when a certain characteristic field value is exceeded, there is a violent increase in the emission current which, as the field continues to increase, comes under the influence of the Fowler-Nordheim law. At this stage, the current is very unstable, and the voltage increment required to produce a given increment in current, increases as the temperature of the cathode is lowered. At the practical level, it is sometimes possible to off-set this undesirable phenomenon by externally limiting the emission currents, particularly when constructing insulation systems having negligible losses and a low noise level. The introduction of an insulating spacer between the electrodes generally causes the electric strength of the vacuum insulation system to be decreased. However, there is some evidence to indicate that the precooling of the insulation system, including the insulator, may lead to an increase in the electric strength, although this will depend on the temperature distribution within the insulation system. The greatest increase in electric strength will occur when either all the elements of the insulation system are cooled to liquid nitrogen temperature, or when the element that is the main source of interelectrode plasma is independently cooled.

To sum up, we may state that at present we have a considerable amount of experimental data concerning the behaviour of high voltage vacuum insulation systems at cryogenic temperatures. Indeed, on the basis of this knowledge, it would be possible right now to construct cryoelectrotechnical and microelectronic devices. Equally this same information would be relevant for the development of high voltage devices designed for operating in outer space. However, we still have not acquired a complete knowledge of physical mechanisms responsible for prebreakdown emission phenomena, particularly the instability of the field emission current. For example, the conventional static methods for measuring emission currents are only of limited diagnostic value since they do not provide the means of dynamically recording the short-term changes that are characteristic of these currents. It is therefore desirable to develop fast-response current measurement systems which will make it possible to discriminate between emission current signals, and disturbances caused by the capacitive current component. Although a challenging problem, it is very attractive from the theoretical point of view, and also from the practical point of view, on account of the many potential applications that will stem from the enormous future exploitation of high temperature superconductivity.

14. 6 References

1. Hadlow, M., Baylis, J.M. and Lindley, B, *Proc. IEE., IEE Review,* **119**, No. 8r, 1003-1032, 1972.

2. Graneau, UP. and Montgromery, DB., *J.Vac. Sci. Tech,* **13**, 1081-1087, 1976.

3. Mazurek, B, "On Development of Electrical Discharges in High Voltage Vacuum Insulation. Selected Problems", Scientific Papers of Electrical Eng. Fundamentals of Worclaw Technical Univesristy, Mongographs 7, 1984.

4. Kirchgessner, J., *Proc. 2nd Annual Conference of Superconductivity and its Applications,* Buffalo, New York, 343-352, 1988.

5.   Dylayen, JR, *Proc. 2nd Annual Conference on Superconductivity and its Applications*, Buffalo, New York, 286-290, 1988.
6.   Shkuratov, SI., *Surface Science*, North Holland, **266**, 88-99, 1992.
7.   Mazurek, B., Latham, RV. and Xu, NS., *J. of Materials Science*, **28**, 2833-2839, 1993.
8.   Rakhovsky, VI., *Trans. Plasma Sci.*, PS-12, 199-203, 1984.
9.   Litvinov, EA., Mesyats, GA. and Proskurovski, DI., *Sov. Phys. Usp.*, **26**, 138-59.
10.   Mesyats, GA. and Proskurovskii, DI., *JETP Lett.*, **13**, 4-6, 1971.
11.   Mesyats, GA. and Proskurovskii, DI., "Pulsed Electrical Discharge Vacuum", Springer Verlag, Berlin, 1989.
12.   Mazurek, B. and Cross, JD., *J. Appl. Phys.*, **63**, No 10, 4899-4904, 1988.
13.   Koval, BA., Proskurovskii, DI., Tregubov, VA. and Yankelevich, E.B., *Sov. Phys. Lett.*, **5**, 246-247, 1979.
14.   Halbritter, J., *IEEE Trans. Elec. Insul*, E1-20, 671-681, 1985.
15.   Schwirzke, FR., *IEEE Trans. on Plasma Sci.*, **19**, No.5, 690, 1991.
16.   Latham, RV., Bayliss, KH. and Bajic, S., *IEEE Trans. Elec. Insul.*, **24**, No.6, 897-900, 1989.
17.   Rose-Innes, AC. and Rhoderick, EH., "Introduction to Superconductivity", Second Edition, Pergamon Press, Oxford, New York, Toronto, Sydney, Paris, 1978.
18.   Nottingham, WB., *Phys. Rev.*, **59**, 907-908, 1941.
19.   Bergeret, H., Septier, A. and Drechsler, D., *Phys. Rev.*, **B-31**, 149-151, 1985.
20.   Cobourne, MH., and Williams, WT., *Physica*, **104**, 50-55, 1981.
21.   Leger, A., *J. de Phys.*, **29**, 646-654, 1969.
22.   Gomer, R. and Hulm, J., *J. Chem. Phys.*, **20**, 1500, 1952.
23.   Klein, R. and Leder, L., *Phys. Rev.*, **124**, 1050-1052, 1961.
24.   Buckel, W., "Superaleitungen, Grundlagen und Anwendung", Physik, Weinheim, 1972.
25.   Shkuratov, SI. and Shilimonov, SN., *XVth ISDEIV*, 127-131, 1992.
26.   Shkuratov, SI., Ivanov, SN. and Shilimanov, S.N. *Surface Science*, **266**, 224-231, 1992.
27.   Heard, HG., Rpt. UCRL., NTIS, 1697, 1952.
28.   Rakhovsky, VI., "Physical Basis of the Communication of Electric Current in Vacuum", Nauka Moscow, 1970.
29.   Chalmers, ID. and Phukan, BD., *Brit. J. Appl. Phys.*, **13**, 122-125, 1962.
30.   Maitland, A., *Brit. J. Appl. Phys.*, **13**, 122-125, 1962.
31.   Schmidt,. BD., *VIIIth ISDEIV*, B2, 1978.
32.   Lewin, G., "Fundamentals of Vacuum Science and Technology", McGraw Hill Book Company, 1969.

33. Carslaw, HS. and Jaeger, IC., "Conduction of Heat in Solids", 2nd Ed., Oxford, Clarendon Press, 1959.
34. Kahle, M. and Richter, KW., Scienctific Paper of the Institute of Elect. Eng. Fundamentals of Wroclaw Technical Univeristy, No. 15, 33-39, 1976.
35. Okhi, Y. and Yahagi, K., *J. Appl. Phys.*, **46**, 3695-3696, 1975.
36. Tourreil, CH., *Adv. in Cryogenic Eng.*, 22, 306-311, 1975.
37. Avdienko, A.A. and Kiselev, A.V., *Sov. Phys. Tech. Phys.,* **12**, 381-384, 1967.
38. Wankowicz, J., PhD Thesis, Technical Univerity of Wroclaw, 1980.
39. Cross, JD., Mazurek, B. Srivastava, KD. and Tyman, A., *Can. Elec. Eng. J.*, **7**, No.4, 19-20, 1982.
40. Berteaux, F., *Revue Generale de L'Electricite*, **79**, No.1, 7-14, 1970.
41. Mazurek, B., Cross, JD., Srivastava, KD., *Physica*, **104C**, 82-87, 1981.

15

Vacuum Arc Initiation and Applications

B Jüttner

15.1 Electrode Spots at Breakdown

15.2 System Arcs

15.3 Applications

15.4 References

15. 1   Electrode Spots at Breakdown

Even with a high series resistor, breakdown transient currents >10 A are possible due to the discharge of electrode and stray capacitances. With powerful networks this stage frequently proceeds into an arc; otherwise the current vanishes when the capacitance is discharged.

In general, such currents are associated with hot spots at the electrode surfaces. With heavily contaminated surfaces, the spots are difficult to detect and perhaps not always present; equally, they may be absent with hot electrodes. However, with relatively clean surfaces at room temperature, they are a decisive component of the discharge, because they deliver ionised matter into the gap, thus creating a conducting connection between the electrodes. Spots at the cathode fulfil a second important task: they emit electrons in order to guarantee current continuity at the cathode surface.

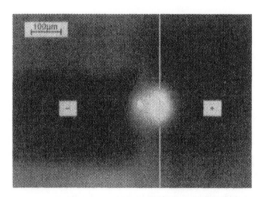

Fig 15.1   Picture of spots at Ti-electrodes during breakdown of a voltage pulse of 15 ns duration. Needle cathode (left) of 300 μm diameter, plane anode (right). (From Anderson et al. [11], with permission.)

In certain circumstances, cathode and anode spots can coexist, where Fig 15.1 gives an impression of how the light output is dominated by the anode spot. This led some authors to the conclusion that anode spots play a more important role for establishing the current conducting channel [1,2]. However, with sufficient sensitivity, Mesyats and co-workers [3] demonstrated that cathode spots occurred first. They also control the discharge in its later stages, because the cathode region is the bottleneck of the discharge: physically, it is much more difficult to pull electrons out of the cathode than drop them down into the anode (most of the discharge current is electronic). Due to its supersonic expansion velocity, the cathode spot plasma becomes almost invisible at distances >0.1 mm, while preserving its ability for current conduction. Thus, in contrast to anode spots, the boundary of the luminous structure of cathode spots is not necessarily identical with the plasma edge.

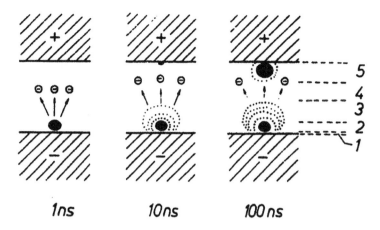

1ns　　　　　**10ns**　　　　　**100 ns**

Fig 15..2　　Discharge zones at breakdown. 1 - space charge zone, 2 - cathode spot plasma, 3 - expanding plasma, 4 - vacuum zone, 5 - anode spot plasma [5].

Therefore, we adopt a scenario established by the pioneering work of the Mesyats group [4] according to which the cathode spots determine the current-time behaviour. Fig 15.2 (from [5]) illustrates schematically a momentary situation. At the cathode there occurs a small ball of dense metal vapour plasma, mostly at the location of the former ignition site. Between this plasma and the cathode surface there exists zone 1, a narrow space charge sheath (with a width of a few nm) due to the charge of the plasma ions. The surface is hot at this location and the combined action of temperature and space charge field produces sufficient electron emission to match the discharge current I. The voltage drop V_c across zone 1 amounts to <100 V. The plasma ball, zone 2, expands towards the anode with velocity v_c, thus forming a plasma region 3 with decreasing density. At the edge of this region there is vacuum that must be bridged by electrons leaving the plasma: however, their negative space charge establishes a limit to the current. Since most of the high electrode voltage is dropped across this zone 4, electrons enter the anode region with high energy and cause evaporation of anode material and eventually the formation of an anode spot plasma, zone 5, expanding with velocity v_a. The voltage drop over the plasma zones 2, 3 and 5 can be neglected, so that the current exhibits a space-charge behaviour, being proportional to $V^{3/2}$, where V is the gap voltage, i.e.

$$I(t) = A \, (V - V_c)^{3/2} F(g) \qquad\qquad 15.1$$

Here A is a constant and g the width of the vacuum zone 4 given by

$$g(t) = d - (v_c + v_a)t \qquad\qquad 15.2$$

where d is the gap spacing. As found in [4,6,7] for point-to plane geometries and confirmed later in [5] for plane-plane geometry, the function F(g) can be approximated by F(g) = (d - g)/g for g > 0. The constant A has been determined to be about 3×10^{-5} AV$^{-3/2}$. For completeness, Table 15.1 presents measured values for the expansion velocity v_c and v_a of cathode and anode spot plasmas respectively.

Table 15.1 Plasma velocities at breakdown (in km/s).

Ref: Material	v_c [4]	d/t_c [4]	v_c [7]	v_a [4]	v_a [1]	v_a [2]	v_a [8]
W		27	19		5.2-5.4		
Ta		35					
Mo	20-22	26	18				
Ni		19	13				
Ti		19					
Nb		30					
Cu	24-26	25	17	7-8		5.5-6.5	5
Al	24	26	18		7.3	8-9.5	
Zn		16	11				
Pb	11	13			5.8		
In		13					
Cd		13					
Bi		11			8.9		
Sn		15			5.4		
C		20					
Stainless Steel					4-5		

In this table, t_c is the commutation time for bridging the gap by the spot plasmas. We have $d/t_c = v_c + \alpha v_a$, where the coefficient $\alpha < 1$ taking account of the fact that anode plasmas appear later than cathode plasmas with a delay time t_d. According to [2], t_d is proportional to the gap length d and at constant velocities $\alpha = 1 - t_d/t_c$. As can be seen from Table 15.1, v_c depends only weakly on the material, being considerably greater than v_a, thus $g \approx d - v_c t$. We have finally therefore

$$I(t) = AV(t)^{3/2}v_c t \, (d - v_c t)^{-1} \qquad\qquad 15.3$$

When discharging a capacitor C into the gap, a second equation is $I(t) = -C^{-1}(dV/dt)$, if the charging current from the voltage source can be neglected at breakdown (high resistor in series). Then integration of equation 15.3 leads to [9]

$$V(t) = V_o f(t) \qquad\qquad 15.4$$
$$I(t) = A V_o^{3/2} v_c t \, (d - v_c t)^{-1} f(t)^{3/2} \qquad\qquad 15.5$$

$$f(t) = \{1 - 0.5 B \, [ln(1 - v_c t/d) + v_c t/d] \}^{-2}$$

$$B = V_o^{1/2} A d/C v_c, \quad V_o = V(0) \gg V_c \qquad\qquad 15.6$$

This solution has proved to be valid for DC and RF voltages [9] and also for pulses with duration down to 10 ns. The latter involves an important fact: the cathode spots do not exhibit a temporal inertia >10 ns (see also next section). Equation 15.5 describes a current pulse as shown in Fig 15.3 [5], whose duration t_p depends on gap length d and capacitance C.

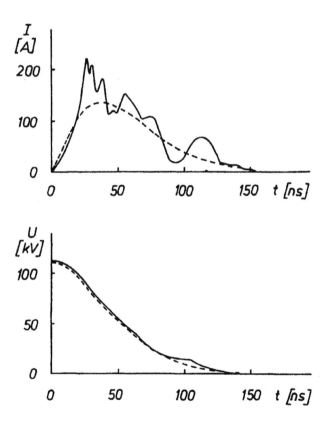

Fig 15.3   Discharge current I and gap voltage V at breakdown. Mo-electrodes, gap distance 4 mm, parallel capacitance 121.5 pF, breakdown voltage 112 kV. Dashed curves according to equations 15.4 and 15.5. (From Jüttner [5], with permission.)

For B >10-20 (i.e. at a sufficiently high ratio d/C in equation 15.6), the capacitor C is discharged before anode spots can become very bright. In this case, open shutter photographs suffice to show that only cathode spots exist. This has been demonstrated in [10] with d ≥ 4 mm and C ≤ 100 pF.

In any case anode spots disappear for $t > t_c$, whereas the cathode spots remain if the circuit delivers sufficient current. Now we are dealing with an arc. The question is, whether the spots change their properties at this moment. To be cautious, we shall treat the spots separately for $t < t_c$ (breakdown) and $t > t_c$ (arc). While at breakdown several zones (Fig 15.2) control the current flow, for the arc the cathode spot is most decisive.

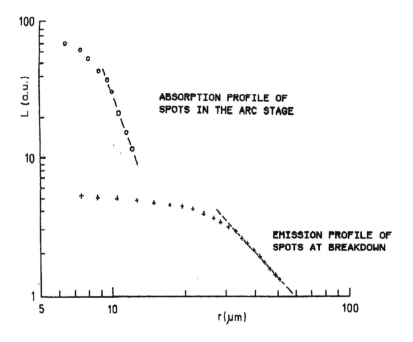

Fig 15.4   Luminance profile of breakdown spots at Ti-cathodes (compare with Fig 15.1). For comparison, a profile of an arc spot is given that was produced by absorption of a laser beam (see Section 15.2.3). The profiles have been processed by Abel inversion.

Spot photographs like those in Fig 15.1 indicate a radius for the luminous plasma cloud of 30-100 μm. Fig 15.4 shows a corresponding radial profile of the light emitted from a spot on Ti during a 15 ns breakdown [11], indicating a radius of the visible plasma of about 30 μm. The material of this cloud stems from the surface, where craters can be found after the discharge. Fig. 15.5 and 15.6 show craters on Mo and Cu produced by the breakdown of nanosecond voltage pulses [12].

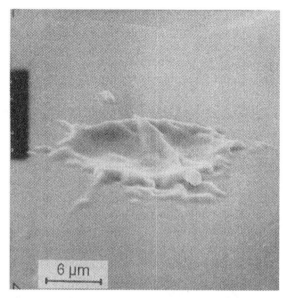

Fig 15.5   Crater picture on a Mo-cathode caused by a high voltage pulse of 10 ns duration. (From Jüttner [12], with permission.)

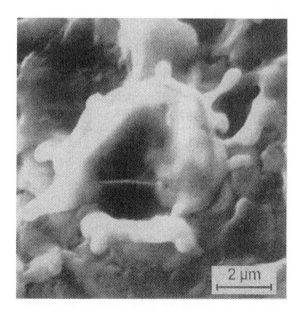

Fig 15.6   Crater picture on a Cu-cathode caused by a high voltage pulse of 10 ns duration. (From Jüttner [12], with permission.)

Similar pictures have been obtained in [13]. The crater radius is ≤10 μm, much smaller than the visible plasma radius. Fig 15.7 shows measured values for Mo and Cu as a function of maximum current in nanosecond discharges [12]. In these experiments, the cathodes consisted of a thin wire, the tip of which was molten prior to the measurements. Cathodes with surface contaminants exhibit a multitude of smaller craters [14], but with the clean surfaces underlying Figs 15.4-15.7, only about one crater was formed during 10 ns.

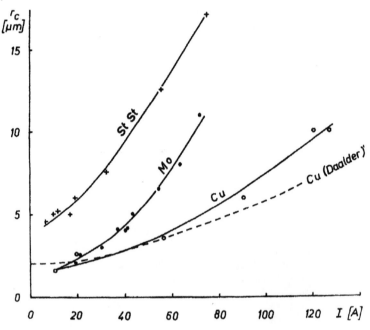

Fig 15.7   Crater radius as a function of breakdown current. For comparison, values for arcs with copper (Daalder, see Fig 15.14) and stainless steel are given. (From Jüttner [12], with permission.)

In [15], the lower bound for crater formation was studied by shortening the pulse length. The result is illustrated in Fig 15.8. Starting with surfaces that had suffered thorough erosion (Fig 15.8(a)), it was found that, below a critical pulse duration, the surface became smooth as shown in Fig 15.8(b); i.e. because the melting process at breakdown could not proceed into crater production. The critical times for this polishing process were 3 ns for Cu and 2 ns for Mo. Probably, in Fig 15.8(b), the liquid metal had somewhat smoothed the surface after the pulse, so small craters could disappear. But in any case these experiments give lower and upper bounds for crater formation of about 1 ns and 10 ns, respectively.

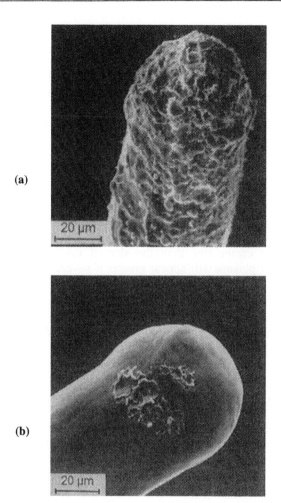

(a)

(b)

Fig 15.8   (a) Eroded Mo-cathode. (b) Eroded Mo-cathode after breakdowns with 2 ns voltage pulses (polishing effect)..(From Hantzsche et al. [15], with permission.)

The spot radius r_s is an important quantity, since both the current and particle densities vary as r_s^{-2}, with both determining the physical conditions of the phenomenon. Therefore, the apparent difference between crater radius $r_c \leq 10$ μm, 30 μm and the radius of the luminous plasma $r_p \geq 30$ μm, needs an explanation. A simple experiment, based on an analysis of the diffused structure of the visible spot edge, can provide an answer [11]. Thus, in Figs. 15.1 and 15.4, the observed light stems mainly from line radiation of the neutral component of the spot plasma. The density n of these atoms should decrease with a power ≥ 2, due to the expansion of the plasma with increasing

523

velocity. If we receive the light from the location where the atoms are excited, the intensity depends not only on n, but on the product of n with the density of the exciting partners (electrons); where the latter will also decrease with the distance. Reducing the temperature with increasing r will also decrease the excitation. Thus, we would expect a variation of the light to be proportional to $r^{-\beta}$, β having a value near 4, or even higher; i.e. in contrast to the experiment which yields $\beta = 2$ (see Fig 15.4). This indicates that many particles excited in the centre leave the spot and radiate outwards. In this case the light intensity outside would only be proportional to the atom density. In fact, the time of flight over a distance of 30 μm is a few ns, comparable to the usual transition times of excited levels.

Thus, the finding that we do not observe a sharp spot edge indicates that the emitted light renders the apparent spot greater than it actually is, i.e. a consequence of plasma expansion. Therefore, we can state $r_c \leq r_s < r_p$, the spot size is nearer to the crater size than the size of the luminous structure. Additionally, the spot plasma will appear greater if several craters are formed during the lifetime of a spot, or if there is noticeable spot displacement during the measurement.

With nanosecond pulse discharges, the spot voltage was measured by Proskurovskii and Puchkarev [16]. These authors employed small probes and found voltages ≤ 50 V, thus confirming that the high gap voltage does not appear between spot plasma and cathode. Alternatively, they were able to show that breakdown and spot formation was possible with voltages ≤ 100 V when using point cathodes and small distances (< 100 μm) without provoking higher voltage transients.

The cathodic erosion is a measure of plasma production in the cathode spot. It was studied with point cathodes in [4] and with needle cathodes in [12]. The erosion rate, defined as the mass removed per current-time integral, was found to depend on the point geometry [4], with an order of magnitude of 1 mg/C. In [12], by observing the decreasing length of needle cathodes, the erosion rate was measured to 136, 100 and 50 μg/C for W, Al and Cu, respectively, independently of pulse duration in the range 10-500 ns. In this case, for discharges > 100 ns, the spots corresponded already to arc spots ($t_p > t_c$). No influence of pulse length on the rate was found.

15. 2   System Arcs

Vacuum arcs can be ignited not only by breakdowns, but also by contact parting, by contacting the cathode with a trigger plasma, by focusing a laser beam onto the cathode surface, or by evaporating a part of the surface by beams of energetic particles. The transient ignition stage may differ among these cases, but after some time (= t_c in Section 15.1) a quasi-stationary arc stage develops which hardly depends on the means of ignition. At very high currents (say $\geq$ 10 kA) the anode can become active; but with moderate currents < 5 kA the most prominent feature of the arc is the cathode spot. As with spots at breakdown, its function is to emit electrons and to deliver ions and atoms into the gap. Therefore, for cathodes at room temperature at moderate currents, dealing with vacuum arcs means dealing with cathode spots.

15.2.1   Main Features

Compared with breakdown spots, much more experimental and theoretical information is available for arc spots, although not yet sufficient for a complete physical picture. In the older literature, a customary distinction between breakdown and arc is the current rise rate. Breakdowns are characterised by values $\geq 10^8$ A/s, whereas arcs are associated with values $<10^7$ A/s [17]. With currents and gap spacings of order 100 A and 5 mm respectively this is equivalent to our definition of the commutation time t_c. A further prominent difference is the duration of the discharge. Breakdowns are nanosecond phenomena whereas arcs can burn for seconds, minutes or even hours. In these longer times the arc spots are very mobile. The spot motion is one of the most typical features of vacuum arcs. In the following we give a condensed description of vacuum arc phenomena. More detailed presentations can be found in [18].

15.2.1(i) Cathode Spot Types

The arc root at the cathode exhibits various forms depending on current and surface state. Table 15.2 shows this schematically, where hot surfaces mean elevated temperatures of areas much greater than the spot area (which is always hot), in contrast to the case where the surface remains cold as a whole.

Most interesting is the difference between Type 1 and Type 2 spots. In the literature, there is some confusion about this nomenclature, we shall therefore add some remarks on the history of the effect. It was introduced by Rakhovsky and co-workers [19] who found a change of the behaviour of arc spots at various currents after a burning time of 10-100 μs: there occurred a transition to slower motion associated with a higher cathode erosion. The effect was explained by heating the surface. Initially, with cold surfaces, there exists a non-stationary spot Type 1 that changes to a stationary Type 2 when the surface becomes sufficiently hot by the action of the arc itself. In Table 15.2, this corresponds to a transition from No. 2 to No. 5. A different view has been presented by another group [20,21]. Here it was found that transitions were accompanied by surface cleaning, i.e. in Table 15.2 from No. 1 to No. 2 or from No. 3 to No. 4, that had already taken place at small currents without gross surface heating, because the running arc spot effectively cleans the surface. Starting with clean surfaces, Type 2 was present from the beginning and starting with contaminated surfaces in a clean vacuum system, the transition was irreversible. In this classification, Type 2 spots are also not stationary.

Table 15.2 Arc spot modes

Running Number	Current (A)	Overall Surface State	Spot Appearance	Spot Name
1	1-1000	Contaminated and cold	Diffused plasma interspersed by small bright points	Type 1
2	1-100	Clean and cold	Concentrated bright plasma point	Type 2
3	>1000	Contaminated and cold	Expanding circle of diffused plasma	Type 1 (high current)
4	>1000	Clean and cold	Expanding ring of bright plasma spots	Type 2 (high current)
5	100-10000	Hot, but no pure thermionic emission	Concentrated bright plasma spots	Type 2, Grouped spot
6	>10000	Molten pool	Concentrated bright plasma area	Contracted
7	<1000	Hot, pure thermionic emission	No spots	Spotless

Indirect evidence for the cleaning effect comes from early works where a change of arc colour was observed [22-24]. This colour was found to be at first blue-white, corresponding to desorbed contaminants, but changed after some arcing to the colour typical for the underlying metal (green for copper). A direct proof was obtained by surface analysis with SIMS (Secondary Ion Mass Spectroscopy) and AES (Auger Electron Spectroscopy) before and after an arc [20,25]. Fig 15.9 gives an example for a Mo-cathode [20]. With Cu, Porto et al. [25] showed that the unarced surface contained only 40 % of Cu, the impurity layer having a depth > 25 nm. However, this was almost completely removed by arcing.

As will be shown below, Type 1 spots have a smaller burning voltage and less noise than Type 2 spots: they also have a higher mobility and a smaller erosion rate. Also, the craters left at the surface have a different appearance, as shown in Fig 15.10 for craters on clean and oxidised Mo-surfaces. Type 1 craters are small and separated from each other, while Type 2 craters often form overlapping strings.

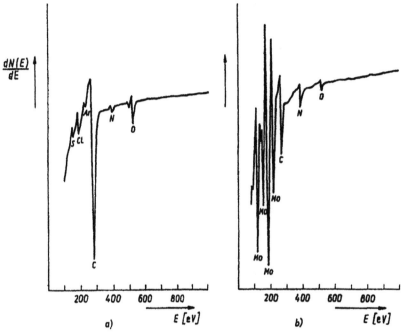

Fig 15.9   Mo-cathode studied by AES: (a) before arcing, (b) after arcing. (Frp, Achtert et al. [20], with permission.)

Another interesting property is indicated in Fig 15.11, which shows a picture of a high current arc taken with a framing time of 16 µs [21]. A multitude of spots were present that formed an expanding ring. One can see the diffused appearance of Type 1 in contrast to Type 2 where relatively sharp spots occurred. In the wake of the Type 1 spots, i.e. in the centre, new spots can be observed that appeared spontaneously with a distance > 1 mm from the existing spots (Fig 15.11(b). This was not found with Type 2 spots (Fig 15.11(c)). Hence spontaneous spot formation is an attribute of Type 1 spots, indicating the ability for spot ignition due to plasma-surface interaction distant from the discharge centre. In Table 15.2, cases No. 3 and 4 differ from 1 and 2 only by the number of spots, while preserving the typical features of Type 1 and Type 2. When the overall temperature of the cathode is high, contaminants are desorbed, so Type 1 disappears. The remaining Type 2 spots become slower, larger craters are formed, and the erosion rate increases. This is the case of No. 5, sometimes referred to as grouped spot [19].

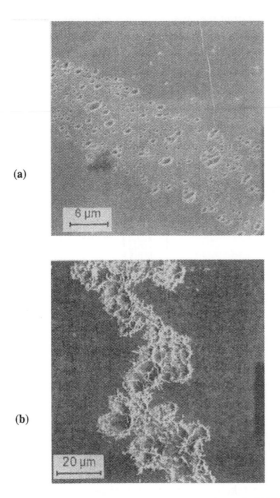

(a)

(b)

Fig 15.10   Crater traces at a Mo-cathode. (a) Type 1, (b) Type 2.. (From Jüttner [107], with permission.

At currents > 10 kA intense anode spots develop leading to a constricted arc column. The numerous cathode spots existing prior to constriction are then concentrated in a molten surface pool. This case, labelled in Table 15.2 as No. 6, is important for high current vacuum interrupters, where one tries to avoid it as far as possible. Here the former statement on the dominance of cathode spots is no longer valid, the behaviour of the discharge being mainly determined by the anode and the arc column. Also, with No. 7 in Table 15.2, cathode spots play no role. In special cases the cathode can reach a temperature sufficient for both thermionic electron emission and stationary evaporation [26]. In this way, a quiet metal vapour arc is possible in a spotless regime. A condensed description of arc spot modes can be found in [27]. In the following we shall mainly discuss arcs with Type 1 and 2 spots.

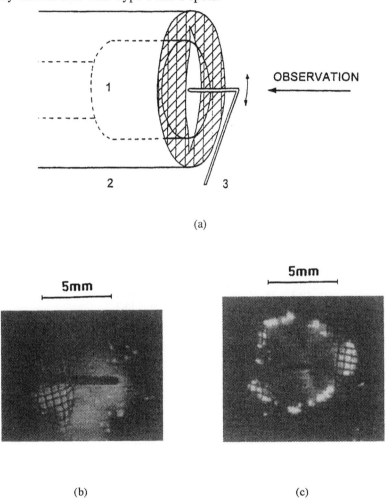

(a)

(b)   (c)

Fig 15.11   High current arc discharge on Cu viewed through a mesh-anode [21].
(a) Coaxial electrode geometry: 1 - cathode, 2 - anode, 3 - trigger pin;
(b) picture of Type 1 discharge; (c) picture of Type 2 discharge. The
discharge expanded from the trigger position. About 300 μs after
ignition the pictures (b) and (c) were taken with a framing time of 16 μs.
(From Bushik et al. [21], with permission.)

15.2.1(ii) Electrical Characteristics

In contrast to gaseous arcs, vacuum arcs have a slightly positive current-voltage characteristic, thus two arcs can burn in parallel without stabilising the resistor. There is virtually no voltage drop outside the electrode regions, therefore the burning voltage does not depend on the gap length, again in contrast to gaseous arcs. With moderate currents, and plane gap geometries, an anode fall is frequently absent. As a consequence, the burning voltage often represents the spot voltage that is concentrated at the narrow space charge zone between plasma and surface (analogous to zone 1 in Fig 15.2).

Table 15.3 Burning voltage (in V) of various spot types for different materials.

Spot Type Reference Current (A)	2 [28] 200	2 [29]	2 [30] 200	2/1 [31] 5-60	1/2 [32]	1 [33]
Material						
Be					17.0	18.6
C (graphite)	16.6			20		
Mg			16	12.5	12.5	11.6
Al	20.8	20	20.5	16.7	15.5	17.2
Ca	13.5					
Ti			20.5			16.8
Cr			20			16.7
Fe			20.5		17.0	17.1
Mild Steel				33		
Stainless Steel				16.2		
Co			20.5		16.0	16.8
Ni	19.3	18	20	15.5	18.0	16.3
Cu	21.5	20	20.5	21.5	16.0	14.7
Zn		12	14	10.7	10.0	9.8
Ga					13.1	
Zr	22.5		22			
Nb			25.5			
Mo	27	26.5	24.5	24		16.6
Rh			24			
Pd			18			
Ag	17	17.5	19	17	13.0	12.1
Cd		11		10	11.0	8.6
In			14		13.0	9.5
Sn		13.5	14	11.3	12.5	10.6
Sb				10		
Te					10.5	11.0
Gd			16.5			
Ho			18			
Hf						16.9
Ta	25.5		24.5			16.8
W		28	28	26		16.2
Pt			20.5			16.0
Au			18		15.0	13.1
Pb		10.5	12.5	9.2		
Bi				8.7		
Th			16			
U			19			

The arc voltage increases weakly with the current, at a rate of about 20 mV/A (Ca, C, Ag, Ni, Al) through about 30 mV/A (Mo, Ta, Zr, Cu) [28]. At currents of 20 A, the burning voltage for Type 1 is smaller than for Type 2 by 1-5 V. In Table 15.3 the classification of Type 1 and Type 2 voltage is somewhat arbitrary because the quoted authors did not consider this possible difference. In work by Reece [31], vacuum and surface conditions were apparently clean enough for Type 2, but the values for Al and Ni seem to belong rather to Type 1. In work by Kesaev [32], many values seem to describe Type 1, but the value for Ni is typical for Type 2. The voltage is relatively small. It is superimposed by spikes with a frequency ≤ 1 MHz [34-37]. Their amplitude is the higher the smaller the current is. With Type 2 spots, at currents ≤ 10 A, the peaks can reach several times the ground level. With Type 1, the noise amplitude remains < 1 V.

There is also a correlation between current and fluctuations of the spot brightness [36,37]. As shown in [13], the peaks are associated with attempts of the plasma system (and to some extent of the external circuit) to restore a dying cathode spot, whereas fluctuations during the lifetime of a spot life have a much smaller amplitude and frequencies > 1 MHz. The latter are more pronounced with the spot brightness [38].

In Table 15.3 the materials are arranged according to the atomic number. Kesaev [17] suggested an influence of the first ionisation potential of the cathode material; other authors however, consider thermal properties, for example, combinations of boiling temperature and thermal conductivity [31], or the evaporation energy [39].

Table 15.4 Spot currents (in A)

Material	Current Spot Type Reference	I_d 2 [34, 41]	I_o 1 [32]
Hg		0.4-0.7	0.07
Cd		8-15	0.19
Zn		9-20	0.30
Bi		3-5	0.27
Pb		5-9	0.5
In		15-18	0.45
Ag		60-100	1.2
Al		30-50	1.0
Cu		75-100	1.6
Cr		30-50	2.5
Fe		60-100	1.5
Co			3.2
Ni			6.0
Ti		70	2.0
C		200	
Mo		150	1.5
W		250-300	1.6

The simplest correlation is with the boiling temperature [40]: the higher this temperature the higher is the burning voltage. However, this is a tendency rather than an absolute law. The weak dependence of the voltage on the current is based on the fact that more spots appear when the current is increased, the current per spot being fixed within certain limits. In order to exist, the spots require a minimum current I_o. Above a maximum current I_d the spots split. Table 15.4 lists values of I_o and I_d published for conditions where Type 2 can be expected [34,41] and those for Type 1 conditions [32]. (The authors themselves did not consider the existence of types 1 and 2.) The materials are ordered according to their boiling points.

With Type 1 spots, Kesaev found $I_d = 2I_o$. The value of I_o depends on the measuring time, because the average burning time decreases with the current (see below); i.e. with limited time resolution, spots cannot be observed below a certain current although they may still exist. This arbitrariness becomes still greater when not considering the existence of spot types, because the critical currents are smaller for Type 1 than for Type 2. Possibly also, Type 2 spots consist of internal fragments not yet resolved in the experiments. With an enhanced time and spatial resolution [8,38,42], Type 2 spots have been found on Cu-cathodes that have currents 20-30 A, smaller than in Table 15.4. Puchkarev and Murzakayev [13] studied nanosecond arcs on clean W-cathodes and they obtained spot currents of 2-5 A: i.e. comparable to those found by Kesaev. As a consequence, the spot is composed of a number of cells, each of them carrying a fixed current. The spot reacts to an increase of current by a growing number of cells, rather than by changing the size of its constituents. The minimum current is reached when only one cell exists. Similar models have been proposed by Sena [43], Chekalin [44], Emtage [45] and Harris [34]. Guile and co-workers developed analogous conceptions for arcs on oxidised cathodes in air [46]. Here, the cell currents could amount to a few mA. Probably with Type 2 spots, the cells are less pronounced than with Type 1, representing local fluctuations rather than separate structures and leading to observable splitting only when they surpass a critical amplitude.

15.2.1(iii) Arc Lifetime

At constant current, the arc spot extinguishes spontaneously after a certain burning time τ_b. These times obey an exponential probability distribution [17,47-49]:

$$W(\tau_b) = <\tau_b>^{-1} exp(-\tau_b/<\tau_b>) \qquad\qquad 15.7$$

$<\tau_b>$ being the mean value. Equation 15.7 is thus a survival law: in a sample of trials, the probability for extinguishing per unit time does not depend on the elapsed time. This holds for Type 2 spots on arced cathodes, where the surface state is not changed by the action of the arc. We have then a most probable value $\tau_b* = 0$. Surface changes during the action of Type 1 spots give rise to a small difference expressed by $0 < \tau_b* \ll <\tau_b>$. The mean burning time is a strongly rising function of the arc current, as shown in Fig 15.12 for different materials. Also, it is higher for rough surfaces [35] and enhanced temperatures [17]. External magnetic fields may greatly stabilise the burning behaviour [50]. Equally, external inductances increase the burning time while external capacitances reduce it [48]. Also the electrode geometry has some influence [48,49,51].

As shown by Smeets [52,53], arc extinction occurs in the form of an accumulation of instabilities. Although it is a transient phenomenon (on a nanosecond time scale, as we shall see below), the cathode spot operates on a large number of repetitive cycles. The probability of failure is small for a single cycle, so the burning time is much higher than the duration of a cycle. The more spots are burning in parallel, the smaller is the probability for extinction. Therefore, the burning time rises with current. In this sense spot fragmentation is advantageous for survival. On the other hand, since the arc voltage is small, every additional voltage-consuming effect reduces arc stability; for example, anodic voltage drops caused by the electrode geometry (long gaps, small anodes.)

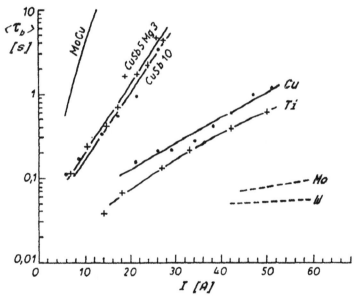

Fig 15.12   Mean arc burning time $<\tau_b>$ as a function of arc current for different cathode materials. (From J [49].

15.2.1(iv) Spot Movement

Arc cathode spots are seldom at rest. This is surprising per se because it seems to be a waste of energy when the spot leaves a surface location where the temperature was high and starts again at a cold place. Several configurations can be distinguished:

(i)   Type 1 spot motion without external magnetic field. A diffused plasma film spread irregularly over the surface with velocities up to 1000 m/s, guiding the displacement of the spot. The latter involves "jumps" over distances of 0.01-1 mm [21, 54].

(ii)   Type 2 spot motion at small currents without external magnetic field. The spot is displaced with elementary displacement steps < 100 µm. The direction is random, resulting in a chaotic motion similar to the diffusion of a Brownian particle (random walk) [12,55]. As discussed below, the apparent displacement velocity depends on the elapsed time, having values in the range 0.1-30 m/s.

(iii)   Type 2 spot motion at high currents without external magnetic field. A ring of spots is formed (compare Fig 15.11(c)), expanding with velocities of 10-100 m/s [21,56]. The ring is the consequence of the mutual interaction of the spots due to their self-magnetic fields.

(iv)   Type 2 spot motion at small currents with an external magnetic field parallel to the surface. The formerly chaotic motion of case (ii) becomes elongated in the direction $\mathbf{B} \times \mathbf{j}$, where $\mathbf{B}$ and $\mathbf{j}$ are respectively the vectors of magnetic field and current density. The direction is retrograde, i.e. opposite to Ampère's rule and perpendicular to $\mathbf{B}$. The velocity increases with the magnetic field, reaching values of about 100 m/s. Fig 15.13 shows a resulting crater trace. It approaches a straight line. However, a chaotic component remains even at fields of about 1 T.

(v)   Type 2 spot motion at high currents with the external magnetic field parallel to the surface. Instead of a ring as in case (iii), the spots are aligned along the field lines, the aligned spots moving as a whole in the retrograde sense [57,58]. The current per spot is increased.

(vi)   Type 2 spot motion at small currents with external magnetic field having a component normal to the surface, i.e. being inclined by an angle Θ to the surface. The motion deviates from the direction $\mathbf{B} \times \mathbf{j}$ by an angle Φ, such that it approaches the direction of $-\mathbf{B}$ (i.e. a deviation from $\mathbf{B} \times \mathbf{j}$ in the direction of the acute angle between $\mathbf{B}$ and the surface) [17,50,59-61]. One finds Φ/Θ = 0.5-1 [59, 60]. When controlling the angle Θ, the effect can be used to steer the arc spot [61].

The retrograde direction has been a puzzling phenomenon for decades. Kesaev [17] established a general rule: the spot moves where the sum of self-magnetic field and external field is maximum. This rule, however, is not yet a physical model. Attempts to explain the retrograde direction are numerous, none of them being unanimously accepted. Among these models we quote: change of a plasma column and/or of the current channel in some distance from the surface with feedback to the ignition of new spots [62-64], confinement of the spot plasma at the retrograde side [65], redistribution of charges in the space charge zone leading to enhanced electrical fields at the

534

retrograde side [66-70], influence of spot ions moving opposite to **j** [71,72], influence of spot electrons moving opposite to **j** [59].

From macroscopic measurements of the chaotic spot motion in case (ii), information on the microscopic spot parameters can be obtained. Assume a two-dimensional random walk with a constant microscopic displacement step R_d covered in a constant time τ. The probability of macroscopic displacements R lying between R and R + dR is then

$$W(R)dR = R\,(2Dt)^{-1}exp\,(-R^2/4Dt)\,dR \qquad 15.8$$

where D is "diffusion constant" containing information about R_d and τ, as defined by

$$D = 0.5\,R_d{}^2/\tau \qquad 15.9$$

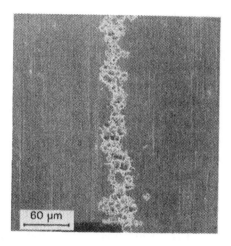

Fig 15.13   Arc trace in a magnetic field. (From Jüttner [107], with permission.).

The mean value of R^2 will be $<R^2> = \int_0^\infty R\,W(R)dR = 4Dt$ so that measurements of $<R^2>$ as a function of arcing time t yield D if t $\gg$ τ. Experiments with Cd, stainless steel, Cu, Mo and Al resulted in D = (5-10) $\times$ 10^{-4} m^2/s [12,55,73];. the values varying only weakly with current and surface temperature.

The apparent spot velocity v_s can be calculated from the mean displacement $<R>$:

$$<R> = \int_0^\infty RW(R)dR = (\pi Dt)^{1/2}$$

$$v_s = d<R>/dt = 0.5(\pi D/t)^{1/2} \qquad 15.10$$

thus v_s decreases as $1/\sqrt{t}$.

With clean surfaces, the arc trace consists of overlapping craters having radii < 10 μm (Fig 15.13). If we assume that the elementary displacement step is near a crater radius, we have $R_d \leq 10$ μm. In this case equation 15.8 yields $t \leq 100$ ns. Therefore, the condition t»t can be fulfilled for t = 1 μs and equation 15.10 then predicts an apparent velocity near to 20 m/s, while for measuring times of milliseconds, the velocity is < 1 m/s.

15.2.2   Material Emitted from Cathode Spots

The vacuum arc emits metal vapour plasma which characterises internal spot processes; this therefore makes it important for technical applications, for example coating devices. The degree of ionisation of the spot plasma is near to 1, the ions having a charge ≥ 1 [28]. Because of the expansion of the plasma,

Table 15.5 Mean ion charge number <Z> and ion charge fraction f_n

Reference <Z> Material	[28] <Z>	[79] <Z>	[80] <Z>	[30] f_1	[30] f_2	[30] f_3	[30] f_4	[30] f_5	[30]
C	1.04			1	1				
Mg		1.50		1.63	0.37	0.63			
Al	1.58	1.42		1.48	0.56	0.39	0.05		
Si				1.46	0.56	0.42	0.02		
Ca	1.47								
Ti			1.79	2.05	0.06	0.82	0.12		
Cr			2.02	1.82	0.25	0.67	0.08		
Fe			1.47	1.73	0.31	0.64	0.05		
Co				1.57	0.47	0.49	0.04		
Ni	1.53	1.37		1.51	0.53	0.44	0.03		
Cu	1.85		1.71	1.70	0.44	0.42	0.14		
Zn				1.14	0.86	0.14			
Zr	2.17			2.33	0.09	0.55	0.30	0.06	
Nb				2.56	0.05	0.46	0.37	0.12	
Mo	1.99		2.82	2.35	0.14	0.47	0.28	0.11	
Rh				1.65	0.46	0.43	0.10	0.01	
Pd				1.64	0.39	0.57	0.04		
Ag	1.36			1.77	0.32	0.59	0.09		
Cd		1.00							
In				1.12	0.88	0.12			
Sn				1.47	0.53	0.47			
Gd				2.07	0.06	0.81	0.13		
Ho				1.93	0.15	0.76	0.09		
Ta	2.69			2.58	0.13	0.39	0.28	0.18	0.02
W				2.74	0.08	0.34	0.36	0.19	0.03
Pt				1.33	0.69	0.29	0.02		
Au				1.58	0.44	0.54	0.02		
Pb				1.36	0.64	0.36			
Th				2.92	0.03	0.15	0.70	0.12	
U				2.62	0.03	0.38	0.54	0.05	

the ions can be analysed outside the spot. Also, neutral atoms are found at some distance, but most of them do not originate from the active spot region, but from flying droplets or from hot surface locations already left by the spot [28,74-78]. Table 15.5 shows the fraction of ion charge numbers f_n and the mean ion charge number $<Z> = \Sigma n f_n$, for different materials. The values in reference [30] are somewhat smaller than those of the other authors because they refer to particle ion currents rather than to electric ion currents (for this difference see [30]). In the case of ionisation equilibrium, the measured ion charge distributions would correspond to rather high plasma temperatures, e.g. about 5 eV for Cu. At the measuring location (outside the spot) the temperature is < 2 eV. Thus, the charge distribution is created in the spot centre and then frozen during expansion of the plasma [81-84]. In spite of the small ion temperature outside the spot (< 1eV) the ion energy is found to be considerable due to a directed velocity component » random velocity. This velocity is associated with the expansion of the spot plasma.

Table 15.6 lists energies W_n for ions of charge n [85]. The average energy is $<W> = \Sigma f_n W_n$ and the average energy per charge number is $<W/Z> = \Sigma f_n W_n / n$. The average velocity is $<v> = (2<W>/m_i)^{1/2}$, m_i being the ion mass.

According to Table 15.6, the energy $<W>$ increases with the charge number, indicating that acceleration by an electric field plays a significant role. The plasma is almost neutral, i.e. any difference in charge densities $\Delta Q = e (n_e - \Sigma Z n_i)$ is small compared to the total charge density of a single component (e.g. $e n_e$). However, this condition does not necessarily mean an absolute smallness of ΔQ, since the ion density can be very large. Therefore, local deviations from total neutrality by a small degree can give rise to high electric fields E, as determined by the Poisson equation div $(\varepsilon E) = \Delta Q$. So, if there is a surplus of ions near the surface, a positive potential hump will develop, accelerating the ions towards the anode [79].

Table 15.6 Ion energies and velocities

Materials	f_1	f_2	f_3	W_1 (eV)	W_2 (eV)	W_3 (eV)	$<W>$ (eV)	$<W/Z>$ (eV)	$<v>$ (km/s)
Cu	0.30	0.55	0.15	37	56	66	51.8	29.8	12.5
Ag	0.65	0.34	0.01	29.5	56	93	39.1	29.0	8.3
Al	0.49	0.44	0.07	28	35	36	31.6	22.3	15.0
Ni	0.48	0.48	0.03	29	24	25	26.2	19.9	9.2
Zr	0.14	0.60	0.21	46	55	69	57.3	28.6	11.0
Ti	0.27	0.67	0.06	46	40	45	41.9	26.7	13.0
Mo	0.09	0.48	0.32	69	122	123	120.2	52.6	15.5
Ta	0.13	0.35	0.28	68	116	126	118.4	48.5	11.1
Ca	0.53	0.47		25	24		24.5	18.9	10.9
C	0.96	0.04		14	12		13.9	13.7	15.0

There is, however, another phenomenon that depends on Z: the arc current is conducted mainly by electrons which undergo friction with the ions on their way to the anode. This leads to an acceleration of the ions in the direction of the electron flow, the effect being nearly proportional to Z [86].

But this is not yet the whole story. As stressed by Miller [87], the ratio <W>/Z decreases with increasing Z (see Table 15.6), so there must be a contribution to <W> that does not depend on Z. The responsible mechanism must be searched for in the hydrodynamics of the plasma expansion (driven by the ion pressure gradient), as elaborated in several theoretical works [81,88-90]. Hantzsche [86,91] computed the relative importance of these processes. He found the electron-ion friction to contribute to about one half of the energy, the ion pressure gradient to about one third and the electric field to the remainder.

Due to the plasma expansion, the ion density n_i decreases with increasing distance r from the surface. With r > 0.1 mm, probe measurements yield $< j_i >r^2/I = C=$ const., for the on-axis ion current density, where I is the arc current [92-94]. Off-axis, at an angle Θ to the surface normal, we have

$$<j_i> = CIG(\Theta)/r^2 \qquad\qquad 15.11$$

According to [74,79,95-98], the function G has the form $G(\Theta) = cos\Theta$, whereas in [99,100] a somewhat different expression is used which is more realistic for $\Theta \approx 0$ (i.e. near the cathode plane), namely

$$G(\Theta) = exp(- \omega^2/k^2),\ \omega = 2\pi (1-cos\Theta),\ k = 4.5 - 4.8 \qquad 15.12$$

Integrating j_i over the half-space ($\Theta = \pi/2-0$) yields the total ion current $I_i \approx \pi CI$. A basic finding by Kimblin [75] is

$$I_i/I = const = 0.07-0.1 \qquad\qquad 15.13$$

for a broad range of materials, which leads to $C \approx 0.03$. With $<j_i> = <Zen_iv> \approx <Z>e<n_i><v>$, we have for the on-axis plasma density (G = 1),

$$<n_i> = CI/(<Z>e<v>r^2) \qquad\qquad 15.14$$

For Cu, with the values for <Z> and <v> taken from Tables 15.5 and 15.6, we have finally

$$<n_i>r^2/I \approx 10^{13}\ m^{-1}A^{-1} \qquad\qquad 15.15$$

Thus, the time averaged mean values of the plasma parameters outside the spots are sufficiently known. Locally and on a time scale <0.1 ms, greater deviations from these values are possible. Ion emission means mass loss of the cathode. The ratio of lost mass to current-time integral is called the erosion rate E_{ri}; this has the form

$$E_{ri} = (m_i/<Z>e)(I_i/I) \qquad\qquad 15.16$$

and can be evaluated using values taken from Table 15.5, in conjunction with equation 15.13, as is done in reference [24,78,101], where m_i is the ion mass.

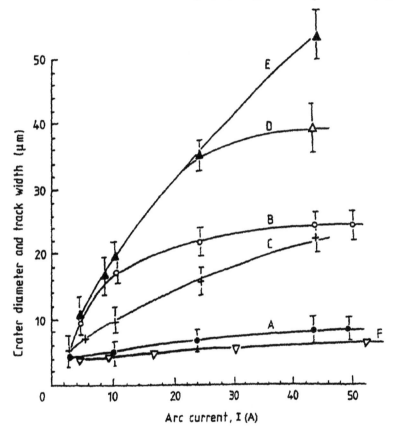

Fig 15.14 Properties of nanosecond arcs on W [13]. A - mean crater diameter, B - mean track width, C - maximum crater diameter (all at room temperature), D - mean crater diameter at 1800 K, E - mean track width at 1800 K, F - mean crater diameter for DC-arcs on Cu according to [24]. (Courtesy of VF. Puchkarev.)

For Type 2 spots (clean surfaces), E_{ri} amounts to 20-200 µg/C, which corresponds to 40 µg/C for Cu. With Type 1 spots, surface contaminants deliver hydrogen and carbon ions, thus m_i is reduced. Furthermore, in equation

15.16 the ratio I_i/I is reduced down to values <0.01 [102]. Therefore, for Type 1 spots the ion erosion rate can drop down to values ≤ 1 µg/C. It should also be noted that equation 15.16 describes the net erosion. In fact, many emitted ions return to the cathode, so that the gross ion erosion can be considerably greater. According to [103], the gross ion erosion rate can reach values of 1 mg/C. Further components of erosion consist of the evaporation of neutral atoms and the ejection of droplets. The former can be neglected, but droplets may enhance the total erosion rate by a factor of 10-100 [104]. The effect increases with the overall temperature of the cathode; this occurs with small cathodes, one has high currents and long arcing times. Droplets are ejected preferentially at small angles to the cathode surface [24,95,104]. Their probability decreases exponentially with increasing size r_d in the range 0.1-100 µm and is proportional to $Q\exp(-ar_d)$, where Q is the current-time integral and a is a constant [101,105]. Although the number of droplets <0.1 µm is highest, the main contribution to the mass loss is due to droplets with a size of 20-60 µm [105]. Small droplets can reach velocities of several 100 m/s [106].

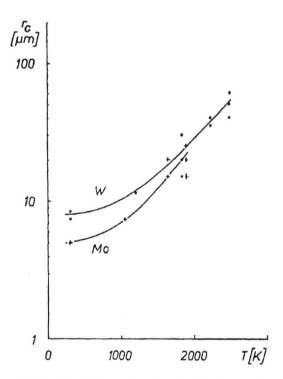

Fig 15.15   Crater radius for W and Mo as a function of surface temperature (From Jüttner [107], with permission.)

The emission of matter leaves craters at the surface as shown in Figs 15.10 and 15.13. Their size is almost equal to the size of craters at breakdown, depicted in Fig 15.7.

Puchkarev and Murzakaev [13] explain the formation of larger craters (at higher currents) by an increase of the number of sub-cells (fragments) whose size remains constant. This is illustrated in Fig 15.14 for nanosecond arcs on clean tungsten: the mean crater diameter varies weakly with the current, whereas the maximum diameter and the track width depend considerably on the current. The figure shows also that the craters become greater at elevated temperatures. The same has been found in [107,108]. Fig 15.15 gives an example [107]. This is in line with increased erosion at higher temperatures, which in turn is associated with enhanced droplet emission.

15.2.3   Cathode Spot Parameters

15.2.3(i) Spot Radius
Fig 15.16 (from [93]) shows the luminance at the edge of Cu-arc spots as a function of distance r. It can be seen that the spot extends over distances $\geq 100\ \mu m$. High speed spot photographs yield a size of about $100\ \mu m$ [19] for

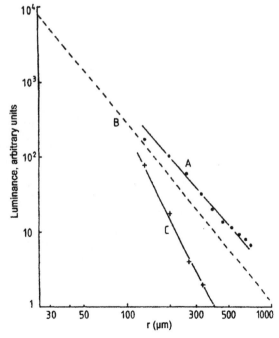

Fig 15.16   Luminance profile of DC-arc spots on Cu. A - normal to the surface (along the axis), C - parallel to the surface. The dashed line B corresponds to a power law of $\alpha\ r^{-2}$. (From Jüttner [93], with permission.)

Type 1 spots, while the minimum value for Type 2 spots is 30-40 μm [41,109]. from the laser experiment (Fig 15.17) in [8], a core density = (3-6) 10^{26} m^{-3} has been estimated for Cu-spots. These high values mean ionisation equilibrium in the centre. Assuming that the charge state distribution measured outside (Table 15.5) reflects the situation within the core, an electron temperature of 5-6 eV has been deduced [83,84]. Such parameters correspond to a very dense, weakly non-ideal plasma. These values are at variance to the much smaller crater radii. Therefore, in the literature, basic spot parameters are sometimes open to animated disputes, as for example, the current density, plasma concentration, energy density, surface field strength. In a sophisticated experiment, pictures of cathode spot plasmas were obtained by absorption of short laser pulses [8]. The duration of the probing laser pulse was 0.4 ns, so the exposure time was extremely short. Fig 15.17 gives an example, indicating a radius of the spot plasma of ≤ 10 μm. A corresponding profile is given in Fig 15.4. The signal decreases with r according to a power ≥4, as it should do, since the absorption coefficient depends on the square of the plasma density [8]. However, in Fig 15.16 the emission profiles indicate a power of about 2 normal to the surface (curve A) and about 3 tangential to the surface (curve C). The latter curve was not processed by Abel inversion as it should be done for the cylindrical symmetry in side-on geometry. As a rule, the inversion reduces the power thus we obtain a value near 2 in both directions, similar to the emission curve in Fig 15.4.

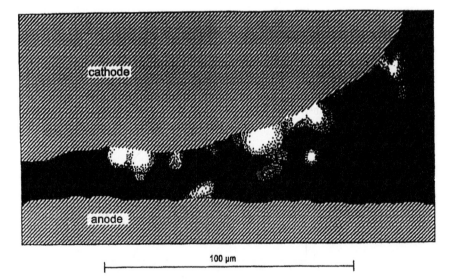

Fig 15.17   Absorption picture of arc spots in a narrow gap between a needle cathode (top) and a plane anode (bottom). The exposure time was 0.4 ns. In the figure the black absorption structures are converted into white structures. (From Anders et al. [8], with permission.)

Now we arrive at the same arguments that apply to breakdown spots (Section 15.1); where the slow decrease of luminance indicates radiation transport from the centre into outer regions due to the finite transition probability of excited particles. As a consequence, the luminous appearance of arc spots cannot be a measure of the size, whereas absorption pictures and surface craters show the right order of magnitude.

Taking a value <10 µm for the spot radius, a spot current of 30 A leads to a current density $>10^{11}$ A/m^2. Evaluating the absorption coefficient deduced from the laser experiment (Fig 15.17) in [8], a core density $= (3\text{-}6)\ 10^{26}$ m^{-3} has been estimated for Cu-spots. These high values mean ionisation equilibrium in the centre. Assuming that the charge state distribution measured outside (Table 15.5) reflects the situation within the core, an electron temperature 5-6 eV has been deduced [83,84]. Such parameters correspond to a very dense, weakly non-ideal plasma.

15.2.3(ii) Spot Time Constants

The extreme concentration of particles and energy in a cathode spot cannot be maintained for long times. With cold electrodes, the residence time of a spot at a given location can amount to few nanoseconds. This has been shown in [8] by imaging the spot by two laser beams, one of them being delayed with respect to the other. As can be seen in Fig 15.18, a time interval between the laser beams of 3 ns can be sufficient to reveal a change of the spot configuration.

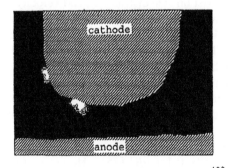
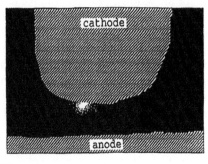

100 µm

Fig 15.18   Two successive laser absorption photographs of arc spots on a Cu-needle cathode. Framing time 0.4 ns, interframe time 3 ns, first picture (left) taken 360 ns after arc ignition, arc current 100 A. The photographs show positional changes of spots that occurred within 3 ns. (From Anders et al. [8], with permission.)

These pictures were taken after about 0.5 µs burning time. At longer times, the surface may become hot, leading to a more stationary situation [110]. The nanosecond time constants in [8] refer to the spot formation time. Still another quantity is the spot lifetime. Some authors define it as the residence time at a given location. From high speed spot photographs, Rakhovsky and co-workers [19] concluded lifetimes of 1-10 µs. However, the definition suffers from being influenced by the available spatial resolution. As discussed above (Fig 15.16), emission pictures yield a size of the order of 100 µm, even if the spot is much smaller. As an example, assume we have a 10 µm spot with nanosecond residence time that displaces at random. Putting R = 100 µm and D = 10^{-3} m^2/s in the relation <R> = $(\pi Dt)^{1/2}$ (equation 15.10), we obtain a smallest discernible time of 3 µs, thus nanosecond residence times could not be observed even if they exist. Taking R = 10 µm the result would be 30 ns, in better agreement with the findings in [8,13,38].

15.2.3(iii) Comparison with Breakdown Spots

With the help of the parameters given above we can compare cathode spots at breakdown with those at arcs. The average crater size is similar in both cases (Fig 15.7), with possible variations being due to a different surface state or a different surface temperature, rather than to different formation mechanisms. As discussed in Sections 15.1 and 15.2.3, the time constants for spot development amount to a few nanoseconds in both cases (see Figs 15.5, 15.6, 15.18). No information is available as to whether, at breakdown, the gap closure by the plasma at t = t_c leads to a change in the existence conditions of the spot. If not, the spot lifetime will also be similar to the pure arc case. Breakdown spots eject droplets [12] as do arc spots. This is shown in Fig 15.19. But for breakdowns, no quantitative data exist on size distribution and velocity.

The expansion velocity of breakdown plasmas (Table 15.1) is higher than for arc spots (Table 15.6) by about 20-40%. Thus, it seems that breakdown plasmas are somewhat more energetic. This has some common features with observations in [110], according to which, the mean ion charge of arc spots is higher by up to 30% during the first microseconds. This effect is explained by an increased surface temperature when the arc has reached steady state conditions. Taking this into account, the difference between arc and breakdown plasmas is reduced.

At breakdown, the erosion rate of point cathodes was found to be 1-4 mg/C [13], whereas arcs have generally 0.1 mg/C [75]. However, this seemingly large difference can be explained by gross surface heating and reduced backflow of matter in the breakdown experiment. Without these effects erosion rates of order 0.1 mg/C have also been found in nanosecond breakdown experiments [12]. On the other hand, DC-arcs may also yield values ≥1 mg/C when the cathode becomes hot [104]. Thus, the erosion rate does not provide simple criteria for distinguishing arcs and breakdowns.

The potential difference between cathode and spot plasma is relatively well known for vacuum arcs because it is near the gap voltage (Table 15.3), having values around 20 V depending on the material. At breakdown, the gap voltage is generally >1 kV. This renders measurements of the spot voltage difficult. A study with floating probes showed values <100 V for tungsten electrodes [16]. Browning et al. [111] found arc spot values at breakdown of field emitter arrays to be: spot voltage about 10 V, the current density 10^{12} A/m^2, directed ion energies about 80 eV and electron energies 6 eV. Therefore, it seems plausible that breakdown spots have similar voltages to arc spots, although still more experimental information is needed for the former, especially with respect to the influence of the cathode material.

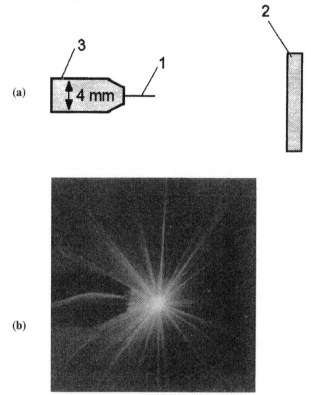

Fig 15.19 Droplets from a tungsten needle at breakdown with 500 ns pulses. (a) Geometry: 1 - cathode, 2 - anode, 3 - cathode holder, (b) open shutter photograph of 20 superimposed shots. (From Jüttner [12], with permission.)

As a result of this comparison, we can state that the essential parameters of arc spots are equal or near to those of breakdown spots. Differences arise when the global surface temperature increases. This occurs mostly with arcs when the burning time surpasses about 10 μs, but with point cathodes, breakdowns can also lead to considerable heating.

As a consequence, cathode spots are non-stationary phenomena with time constants in the nanosecond region. They can form quasi-stationary discharges by a sequence of numerous cycles of formation and extinction. When the cathode is heated as a whole (by the arc itself or by an external source) the processes may be slowed down, but a non-stationary component can exist up to the melting point.

15.2.3(iv) Cathode Spot Theories

A theoretical treatment of arc cathode spots must deal with a number of variables describing electron emission and plasma generation:

(i)   Electron emission density, depending on surface temperature and electric field.

(ii)   Surface temperature, determined by the heat balance at the cathode. The latter depends on the geometry of the heated surface region, on heating by impinging plasma ions and plasma electrons, on Joule heating by the arc current, on energy exchange by the emitted electrons (in most cases cooling), on cooling by evaporation and by heat conduction. Radiation, condensation of neutral atoms, droplet emission and other processes can be neglected.

(iii)   Electric field strength, depending mainly on the ion current density flowing to the cathode. This density is a function of plasma density, ion energy and ion charge, the latter being determined by the temperature of plasma electrons and by acceleration processes.

(iv)   Plasma density, depending mainly on evaporation rate (i.e. on the surface temperature) and expansion.

The corresponding equations are non-linear, rendering exact analytical treatments impossible. Uncertainty also exists about the material parameters at high temperatures and pressures. Finally, in most models, the number of unknown variables is higher than the number of equations. Therefore, simplifications cannot be avoided and experimental values are introduced in order to reduce the number of variables; these include the spot voltage, the ratio of ions to electrons or the ion erosion rate. In general, the theories assume steady state conditions, on the grounds that the time scale for parameter variations is governed by heat conduction within the cathode metal, which can be a slow process. So faster phenomena, e.g. electron emission, ionisation and plasma flow can be treated as having reached equilibrium conditions. The time constant τ for heat conduction can be represented by the thermal diffusivity δ that relates τ to the size X of the region to be heated where $\delta = X^2/\tau$. With most metals, δ is about 10^{-4} m^2/s; thus for X $\geq$ 0.1, mm, τ is $\geq$ 0.1 ms.

Most of the developed stationary theories, for example, those of Kulyapin [112], Djakov and Holmes [113], Ecker [114,115], Hantzsche [116,117], Beilis [118,119], Kubono [120], Harris [89,121], Nemchinskii [122], Benilov [123], trace back to the pioneering work of Lee and Greenwood [124], who achieved a complete description of the arc spot and stressed the importance of thermo-field electron emission. Ecker [114,115] used limiting conditions, instead of exact equations, in order to circumvent the inherent uncertainties. With this very fruitful method, he obtained existence areas where the parameters must be situated, rather than given sharply defined solutions.

A common outcome of these theories is the occurrence of two solutions or two existence areas [113-115,118,119], i.e. as a consequence of the non-linearity of the equations. Ecker called them 0-mode and 1-mode. The 0-mode operates on a relatively low current density $<10^9$ A/m^2 and with surface temperatures <4500 K, whereas the 1-mode exhibits current densities $\geq 10^{12}$ A/m^2 and higher temperatures. It is not easy to say whether these theoretical modes have an experimental analogy (e.g. Nos. 2 and 5 in Table 15.2), because the experiments differ from the theories by a pronounced influence of the time, as expressed by spot motion and parameter fluctuations.

Time dependence comes into play when small surface structures are heated such as thin molten layers within the crater, or liquid micropoints formed by the hydrodynamics of the melt. The plasma exerts a high pressure on the liquid crater surface, leading to rapid variations of surface geometry [125]. With $X \leq 1$ µm, the equation $\delta = X^2/\tau$ yields heating time constants of ≤ 10 ns, already comparable to the time scale of plasma expansion. Theories that consider surface changes during the heating process have been worked out by Daalder [24], Litvinov and Parfyonov [126] and by Prock [127]. Daalder obtained moderate times of order 100 ns from a treatment of crater growth by Joule heating. Prock included the hydrodynamics of the liquid surface, arriving at times of order 10 ns. Values of some nanoseconds have been obtained by Litvinov and Parfyonov who additionally considered plasma heat sources (ion and electron impact) at crater formation.

Still faster phenomena are possible if heating cannot end up in an equilibrium temperature. An example is the thermal run-away of Joule heated volumes, as reported by Hantzsche [128]. Because the electric resistance of most materials grows with temperature, beyond a critical product jI of current density j and arc current I, there is a run-away situation: thermal increase of the resistance leads to enhanced power input, causing further increase of the resistance. The product has the dependency $jI_{crit} = 4\pi\lambda/\kappa_0$, where λ is the heat conductivity and κ_0 the temperature coefficient of the resistivity κ ($\kappa = \kappa_0 T$). For copper $jI_{crit} = 7 \times 10^{13}$ A^2/m^2; thus with current densities $\geq 10^{12}$ A/m^2, no equilibrium temperature can be reached and the result will be an explosion.

Several authors suggested that the spot operates by an uninterrupted sequence of explosions: Rothstein [129] stressed the analogy with thin exploding wires, Nekrashevich and Bakuto [130] explained spot motion by a sequence of explosions, Fursey and Vorontsov-Vel'yaminov [131] and Mesyats and co-workers [4,132] established a spot model on the analogy with vacuum breakdown induced by field electron emission and Mitterauer [133,134] performed thorough calculations on the heating of a field emitter up to explosion, using the results to propose a model of dynamic spot operation.

In connection with these models, the term explosive electron emission has been born [4,131,132]. As already discussed in Chapter 6, it involves the simultaneous emission of electrons and heavy particles (ions) which allows extreme emission current densities. Such an emission takes place at cathode induced breakdown, the current in equation 15.3 being delivered from the cathode in this form. The analogy of spot parameters at arc and breakdown, as stressed above, supports models based on explosive emission. So the investigation of vacuum breakdowns has led principally to new ideas on arc spot mechanisms.

Unfortunately, a closed arc theory will be extremely difficult. The explosive models treat isolated details, or present qualitative descriptions. The advantage of the stationary theories is to provide a quantitative insight into the interrelation of the various complicated processes. Also, there exist situations where the spot dynamics is slowed down. This is the case at enhanced overall surface temperatures, or when the spot is arrested at surface inhomogeneities (inclusions, grain boundaries, accumulated contaminants). With a certain probability, the chaotic motion may guide the spot to locations where it resided some time before, so it may find a pre-heated area. In all of these cases, the active spot area is enhanced, thus the more stationary theories [112-124] may be fully applicable.

On the other hand, have we reached the end of the time scale at values around 1 ns? This question cannot yet be answered. There is a limit to the heating velocity given by electron-phonon relaxation because, within the solid, Joule heat goes to the electrons and must then be dissipated, which is not instantaneous. Theoretical estimates [135] and experimental results [136] indicate values of 10-100 ps, depending on the material. A final limit would be set by Heisenberg's uncertainty relation, as suggested by Liegmann [137].

Finally, we should mention the problem of current conduction in the spot plasma, as raised by Rakhovsky [138]. According to this author, the electric conductivity σ of the plasma is not high enough to sustain current densities $j > 10^{10}$ A/m^2 at a reasonable value of the electric field E caused by the arc voltage ($j = \sigma E$); here σ is the Spitzer-value for an ideal plasma. However, there are contributions to j that do not depend on E. The gradient of electron density n_e may drive an appreciable part of j, even against the external field. This current density has the form $j_{grad} = (\sigma/en_e)\mathrm{grad}(kT_e n_e)$, where T_e is the plasma electron temperature. If plasma expansion already starts at small distances r from the spot centre, say of the order of one micrometre, the value of

n_e^{-1}grad n_e may become sufficiently high. A further component has a convective character: i.e. where the current enters the plasma as a directed beam having an energy corresponding to the cathode fall. This is relatively high and may sustain large current densities until the beam is thermalised inside the plasma. Also, it is not obvious whether the Spitzer-value holds for the dense spot plasma. Thus, current conduction in the spot plasma is a complicated process for which an adequate theoretical description is not yet available.

15. 3   Applications

Arc cathode spots are a simple means for delivering metal vapour plasma, electron and ion beams that can be made for changing surface parameters and for heating and melting large specimens. These properties have led to a broad spectrum of applications of which only a few examples can be presented here. Surface processing by vacuum arcs is still under development, as for example the removal of oxide layers from steel sheets, cleaning of special components, or increasing surface hardness. Another widely applied technique is the use of spot plasmas for switching purposes.

15.3.1   Switching in Vacuum

As a conducting medium, the arc spot plasma plays a decisive role in switching devices and spark gaps. Both of the basic functions of a switch can be fulfilled: interrupting a current, as well as switching on a current. A prominent example is the vacuum interrupter, designed for the interruption of AC-currents up to 70 kA or even higher [139-141]. Its development has promoted intensive research on vacuum arcs and vacuum breakdown [23,31,139,142]. The operational principle is based on the high directional velocity of the metal vapour plasma of v = 10-20 km/s (see Tables 15.1, 15.6). Before interruption, the current passes the closed contacts of the interrupter situated in a vacuum. At interruption, the contacts are separated, thus igniting a vacuum arc. When the alternating current goes to zero (or for DC at artificial current zero), plasma production ceases almost instantaneously, since cathode spots operate on a nanosecond time scale. The remaining plasma leaves the gap and condenses on a vapour shield that surrounds the contacts. With an electrode radius R_e, the gap is cleared in a time of about R_e/v which amounts to a few μs for R_e = 1-10 cm. Thus, no additional means is needed to drive the plasma away, or to cool it, as with gas or oil interrupters. Fig 15.20 shows the principal components of a vacuum interrupter, namely the bellows-activated electrodes the vapour shield and the sealed-off vacuum envelope, consisting in part of an insulator. The size depends on the power level to be handled, but in any case is small compared with other interrupter types.

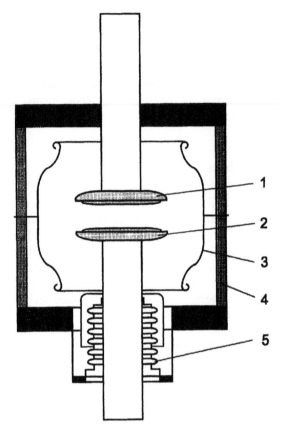

Fig 15.20   Vacuum interrupter. 1, 2 - electrodes, 3 - vapour shield 4 - insulating envelope, 5 - bellows.

Further advantages are: (a) the driving mechanism can be simple, because a vacuum gap is able to hold high voltages at only a few mm gap distance, (b) the dissipated power is relatively small because of the small burning voltage of the arc, (c) the operation is safe (no open flame) and with small noise and (d) there is no need for frequent service operations. Thus, vacuum interrupters are favourable with respect to energy and material consumption and have environmental benefits. A drawback is the cost of manufacturing a sealed-off system that maintains clean high vacuum over many years.

A technical problem is also posed by the voltage hold-off capability. Thus, a general feature of high voltage vacuum insulation is that the voltage strength increases with the gap length more weakly than linearly. It follows therefore that, since all insulating paths of the device should be about one order of magnitude longer than the gap, the overall size of the device grows more than linearly with the required voltage. This consequently implies a rapid increase of volume, weight and costs with increasing hold-off performance. Vacuum interrupters are therefore preferentially designed for medium voltage applications, i.e. 25-50 kV rms.

Interrupting high currents demands precautions to avoid gross melting of the electrodes. Above 1 kA there exist a multitude of arc spots that at currents >10 kA tend to coalesce into an immobile pool because of the constriction of the arc plasma (here, anodic effects also play a role). Magnetic fields, created by the current flow in specially designed electrodes (e.g. spiral types), keep the spots moving in order to maintain the distributed ("diffused") connection of the discharge to the entire area of the electrodes. Severe contraction can lead to failure of the interruption, because in this case vapour production does not cease at current zero. Furthermore, voltage recovery is degraded by an abundance of vapour [143], by flying droplets [105,106,144] and by changes of the hot cathode surface under the influence of electric field and bombarding particles [145].

A problem also exists in the low current regime: this is because of the limited arc lifetime at small currents (see Fig 15.12), an arc extinguishes shortly before current zero. This so-called current chopping is associated with high dI/dt (again because of the nanosecond spot time scale), leading to voltage transients in the external circuit.

The various conflicting demands can be satisfactorily solved by a proper selection of electrode material. An appropriate choice is CrCu all of which is a matrix of chromium filled with copper. Such electrodes have high voltage strength (due to the influence of Cr), small chop currents (due to the combination of two different metals), acceptable contact resistance and little tendency for contact welding. At high currents the arcs also distribute uniformly over the electrodes, provided the material is gas-free and the matrix is sufficiently uniform and has the right size.

Among the switching devices in power networks at technical frequencies (20-200 Hz), we should also mention vacuum contactors and vacuum fuses, where the problems of material, design and manufacturing have many common features with those of interrupters.

As stressed in Section 15.2.3, the formation of cathode spots is characterised by a nanosecond time scale. This allows a broad field of applications in pulsed power technology, mainly to switch - on a current, where Fig 15.21 schematically illustrates some of these cases. The switches operate with a trigger that injects arc plasma into the gap (Fig 15.21(a)). This has two advantages. (i) Without injection, the device can withstand high voltages, but with injection, it is already able to close at low voltages, so the dynamic triggering range is much higher than with high pressure spark gaps. (ii) If the main current begins after the trigger plasma has filled the gap, the current rise rate is not limited by plasma expansion, i.e. in contrast to the case where plasma is produced solely by the switching current.

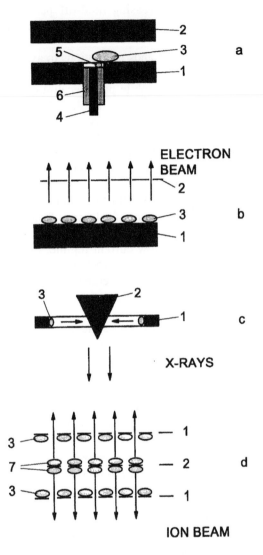

Fig 15.21 Pulsed power applications: (a) triggerable switch, (b) electron beam extraction, (c) X-ray generation, (d) ion extraction with reflecting electrons. 1 - cathode, 2 - anode, 3 - cathodic spot plasma, 4 - trigger, 5 - trigger plasma, 6 - ceramics, 7 - anodic plasma.

The trigger plasma can be generated by an auxiliary discharge as in Fig 15.21(a), or by focusing a powerful laser beam onto the surface. If this plasma is situated at the cathode, the formation of current conducting spots follows immediately. Anodic triggering is less reliable, because the plasma must then expand towards the cathode before the spots can be ignited [5]. Triggered vacuum switches are capable of switching - on high currents (kA-MA) within short times (0.1-100 μs) [146-149]. At high power levels they are often continuously pumped, others are sealed - off.

The opposite action, i.e. switching - off currents, can be found with plasma erosion switches [150-154]. In principle, these are triggered vacuum gaps, too, but after plasma injection, the current reaches a high level >100 kA that "erodes" the plasma near the cathode; i.e. plasma ions are eaten up to such an extent that the resistance of the gap rises to a few ohms. With currents of mega-amperes, this effectively means a nearly switched-off state. The plasma opening switches are used to generate megavolt pulses from inductive energy storage, thus reaching higher power levels than with capacitive storage (because of the higher energy density of the former). Yet another application of the arc plasma as a conducting body, are electromagnetic launchers where the plasma is accelerated in a magnetic field by the Lorentz force in a rail-gun geometry [155,156]. Currents ≥ 100 kA are applied for this purpose, while the voltage remains <10 kV.

15.3.2 Spot Plasma as a Source of Electron and Ion Beams

In pulsed power devices, the cathodic plasma can also serve as a source of intense electron beams [157]. In these cases, illustrated in Fig 15.21(b), the cathode plasma is used during high voltage breakdown (see Section 15.1). As long as the gap is not bridged by the plasma, the extracted electrons gain energy directly in the gap. Thus, in a relatively simple way, relativistic electron beams (REB) can be produced carrying currents ≥ 1 MA at energies of MeVs. The beams are extracted through an anode consisting of a mesh, or a foil, transparent to the energetic electrons. In these devices, the main effort is put into energy storage and pulse forming networks (PFN). The purpose is simulation of nuclear weapons or research on inertia controlled nuclear fusion.

At a moderate power level, the extracted electron beams are a convenient means for generating X-ray pulses. X-rays are often an annoying side-effect of vacuum breakdown, but they can also be cultivated to yield well defined pulses with high doses [6,158-160]. Fig 15.21(c) gives an example.

During the high voltage stage, energetic ions can be extracted from anodic plasma. These powerful ion beams reach mega-amperes and megavolts, with a duration of order 100 ns and are a promising tool for inertial confinement fusion [161,162]. The ions are extracted through transparent cathodes (slotted structures, meshes or foils). Here, one wishes to diminish the electron current fraction in the diode. Thus, Fig 15.21(d) shows the reflex principle, where the anode is also a mesh. A large fraction of the electrons fly through this mesh, being reflected behind the anode by a second cathode. In this way, the electron

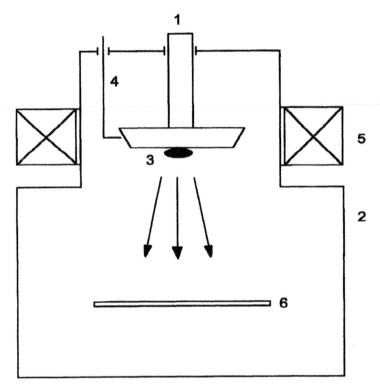

Fig 15.22   Coating principle. 1 - cathode, 2 - vessel (= anode), 3 - cathode spot, 4 - trigger, 5 - magnetic field coil, 6 - substrate.

beam interacts several times with the anode, thus improving the ratio of ions to electrons.

Ions extracted from the cathode plasma must be post-accelerated [30,163-166], but under these circumstances the duration is no longer limited by the plasma expansion, since cathode spot plasmas are also continuously formed after gap closure. So beams of several hundred µs duration have been obtained that have also served for a thorough investigation of the charge state distribution of cathode spot ions (see Table 15.5). In general, these devices are equipped with special ion or plasma optics.

In the pure arc stage, ions can also be extracted as a steady state process. This is widely applied for coating purposes [50,61,167], where Fig 15.22 shows the principle involved. Arc spots are ignited by a trigger spark and steered to the centre of the cathode by a magnetic field, with the anode being the whole recipient. In addition, the magnetic field renders the spot stable at relatively small currents. In contrast to the limited burning time defined by Fig 15.12, with an optimum field of 7-10 mT, arcs with titanium do not extinguish even at currents of 30-50 A [50]. Steering the spot also serves to obtain a homogeneous consumption of the cathode [61]. In this way, coatings of all conducting materials can be easily obtained. The coat rate is high, with up to 400 μm/s having been reported [167]. Further advantages are the high values of ion charge and ion energy as listed in Tables 15.5 and 15.6. A drawback is the droplets emitted from the cathode [168]. However, by deflecting the ions with a magnetic field, their influence can be minimised (magnetic filters), if at the expense of the deposition rate.

15. 4 References

1   Chiles, JA., *J. Appl. Phys.* , **8**, 622-626, 1937.
2   Chalmers, ID. and Phukan, BD., *J. Phys. D: Appl. Phys.* **12**, 1285-1292, 1979.
3.   Bugaev, SP, Iskol'dskii, AM., Mesyats, GA. and Proskurovskii, D.I., *Sov. Phys.-Tech. Phys.* **12**, 1625-1627, 1967.
4.   Mesyats, G.A. and Proskurovsky, D.I., "Pulsed Electrical Discharge in Vacuum", Springer, Berlin, 1989.
5.   Jüttner, B., *Nucl. Instrum. Methods,* **A268**, 390-396, 1988.
6.   Mesyats, G.A., *Sov. Phys.-Tech. Phys.*, **19**, 948-951, 1974.
7.   Mesyats, G.A., *Xth Int. Symp. X-ISDEIV.*, Columbia, 37-42, 1982.
8.   Anders, A., Anders, S., Jüttner, B., Bötticher, W., Lück, H. and Schröder, G., *IEEE Trans. Plasma Sci.*, **20**, 466-472, 1992.
9.   Brutscher, J., "Experimental and Theoretical Investigations on Breakdown in High Vacuum", Thesis, Univ. Frankfurt, 1993.
10.   Jüttner, B. and Siemroth, P., *Beiträge Plasma Phys.*, **21**, 233-245, 1981.
11.   Anders, A., Anders, S., Hantzsche, E., Jüttner, B. and Ziegenhagen, G., *Proc. XXIth Int. Conf. Ion. Phen. in Gases*, Bochum, Pt 1, 3-4, 1993
12.   Jüttner, B. *Beitr. Plasma Phys.*, **19**, 25-48, 1979.
13.   Puchkarev, V.F. and Murzakayev, A.M., *J. Phys. D: Appl. Phys.*, **23**, 26-35, 1990.
14.   Jakubka, K. and Jüttner, B., *J. Nucl. Mater.*, **102**, 259-266, 1981.
15.   Hantzsche, E., Jüttner, B., Puchkarov, V.F., Rohrbeck, W. and Wolff, H.,, *J. Phys. D: Appl. Phys.*, **9**, 1771-1781, 1976.
16.   Proskurovskii, D.I. and Puchkarev, V.F., *Sov. Phys.-Tech. Phys.*, **26**, 1342-1345, 1981.
17.   Kesaev, I.G., "Cathode Processes in Electric Arcs", Nauka- Publishers, Moscow, in Russian, 1968.
18.   Lafferty, J.M, Editor, "Vacuum Arcs, Theory and Application", Wiley, NY, 1980. New edition: Boxman, R.L., Martin, P. and Sanders, D., (editors), "Vacuum Arc Science and Technology", Noyes Publications, New Jersey, USA, in preparation.
19.   Rakhovsky, V.I., *IEEE Trans. Plasma Sci.* **PS-4**, 81-102, 1976.

20. Achtert, J., Altrichter, B., Jüttner, B., Pech, P., Pursch, H., Reiner, D., Rohrbeck, W., Siemroth, P. and Wolff, H., *Beitr. Plasma Phys.*, **17**, 419-431, 1977.
21. Bushik, AI., Jüttner, B. and Pursch, H.,. *Beitr. Plasma Phys.*, **19**, 177-188, 1979.
22. Djakov, BE. and Holmes, R., *Int. Conf. Gas Discharges*, Lond., 468-472, 1970.
23. Farrall, GA., *Proc. IEEE*, **61**, 1113-1136, 1973.
24. Daalder, JE., "Cathode Erosion of Metal Vapour Arcs in Vacuum", Thesis Tech. Univ. Eindhoven, 1978.
25. Porto, DR., Kimblin, CW. and Tuma, DT., *J. Appl. Phys.*, **53**, 4740-4749, 1982.
26. Paranin, SN., Polishchuk, VP., Sychev, PE., Shabashov, VI and Yartsev, IM., *High Temp.* , **24**, 307-313, 1986.
27. Anders, S and Anders, A., *IEEE Trans. Plasma Sci.*, **19**, 20-24, 1991.
28. Davis, WD. and Miller, HC., *J. Appl. Phys.*, **40**, 2212-2221, 1969.
29. Daalder, JE., *J. Phys. D: Appl. Phys.*, **10**, 2225-2234, 1977.
30. Brown, IG., Feinberg, B. and Galvin, JE., *J. Appl. Phys.*, **63**, 4889-4898, 1988.
31. Reece, MP., *Proc. Instn. Electr. Eng.* ,**110**, 793-802, 1963.
32. Kesaev, IG., *Sov. Phys.-Tech Phys.*, **9**, 1146-1154, 1964.
33. Grakov, VE., *Sov. Phys.-Tech. Phys.*, **12**, 1248-1250, 1967.
34. Harris, LP., "Arc Cathode Phenomena", *In* Lafferty, J.M., (Editor), "Vacuum Arcs, Theory and Application", pp.120-168,Wiley, NY, 1980.
35. Fu, YH., *J. Phys. D: Appl. Phys.*, **22**, 94-102, 1989
36. Smeets, RPP and Schulpen, FJH., *J. Phys. D: Appl. Phys.*, **21**, 301-310, 1988.
37. Nazarov, SN., Rakhovsky, VI and Zhurbenko, VG., *IEEE Trans. Plasma Sci.*, **18**, 682-684, 1990.
38. Anders, A, Anders, S., Jüttner, B., Pursch, H, Bötticher, W, Lück, H and Schröder, G., *J. Appl. Phys.*, **71**, 4763-4770, 1992.
39. Vijh, AK., *Sov. Phys.-Tech. Phys.*, **18**, 985-986, 1973.
40. Nemirovskii, AZ. and Puchkarev, VF., *J. Phys. D: Appl. Phys..*, **25**, 798-802, 1992.
41. Djakov, BE. and Holmes, R., *J. Phys. D: Appl. Phys.*, **7**, 569-580, 1974.
42. Anders, S., Jüttner, B, Pursch, H and Siemroth, P, *Contributions Plasma Physics*, **25**, 315-328, 1985.
43. Sena, LA., *Sov. Phys.-Tech. Phys.*, **15**, 1513-1515, 1970.
44. Chekalin, EK., *Proc. VIth Int. Conf. Phen. Ionised Gases*, Düsseldorf, 272-273, 1983.
45. Emtage, PR., *J. Appl. Phys.*, **46**, 3809-3816, 1975.
46. Guile, AE and Hitchcock, AH., *Archiv Elektrotechnik*, **60**, 17-26, 1978.
47. Farrall, GA., Lafferty, JM. and Cobine, JD., *IEEE Trans. Comm. Electronics*, **66**, 253-258, 1963.
48. Farrall, GA., "Current Zero Phenomena" *In* Lafferty, J.M, (Editor), "Vacuum Arcs, Theory and Application", pp. 184-227, Wiley, NY, 1980.

49. Jüttner, B. and Freund, E. *Beitr. Plasma Phys.*, **15**, 47-61, 1975.
50. Karpov, D. and Saksagansky, G., *Contrib. Plasma Phys..*, **30**, 523-545, 1990.
51. Vozdvizhensky, VA. and Kozlov, VB., *Elektrichestvo* No. **6**, 57-60, 1973.
52. Smeets, RP., *J. Phys.D: Appl. Phys.*, **19**, 575-587, 1986.
53. Smeets, RP., *IEEE Trans. Plasma Sci.*, **17**, 303-310, 1989.
54. Slade, PG. and Hoyaux, MF., *IEEE Trans. Parts Mat. Packag.*, **8**, 35-47, 1972.
55. Daalder, JE, *J. Phys. D: Appl. Phys.*, **16**, 17-27, 1983.
56. Sherman, JC., Webster, R., Jenkins, JE. and Holmes, R, *J. Phys. D: Appl. Phys.*, **8**, 696-702, 1975.
57. Perskii, NE, Sysun, VI. and Khromoi, Yu. D, *High Temp.* **27**, 832-839, 1989.
58. Perskii, NE, Sysun, VI and Khromoi, Yu. D, *Sov. Phys.-Tech.Phys.* **30**, 1358-1359, 1985.
59. Litvinov, EA., Mesyats, GA., Parfenov, AG. and Sadovskaya, E., Yu *Sov. Phys.-Tech. Phys. Letters* (1990), *Pis'ma v Zhurn. Tekh. Fiz.* **16**, 92-94, 1990, *XIV-ISDEIV*, Santa Fé, 185-186, 1990.
60. Robson, AE., *Proc. IVth Int. Conf. Phen. Ionised Gases*, Uppsala, IIb, 346-349, 1959.
61. Sanders, DM., Boercker, DB. and Falabella, S., *Proc. IEEE Trans. Plasma Sci ,* **18**, 883-894, 1990.
62. Robson, AE., *J. Phys. D: Appl. Phys.*, **11**, 1917-1923, 1978.
63. Tseskis, AK., *Sov. Phys.-Tech. Phys.*, **23**, 602-604, 1978.
64. Schrade, HO., *IEEE Trans. Plasma Sci.*, **17**, 635-637, 1989.
65. Drouet, MG., *IEEE Trans. Plasma Sci.*, **PS-13**, 235-241, 1985.
66. Longini, RL., *Phys. Rev.* **71**, 642-643, 1947.
67. Ecker, G. and Müller, KG., *J. Appl. Phys.*, **29**, 1606-1608, 1958.
68. Harris, LP, *IEEE Trans. Plasma Sci..*, **PS-11**, 94-102, 1983.
69. St. John, RM and Winans, JG, *Phys. Rev.*, **94**, 1097-1102, 1954.
70. Moizhes, B., Ya and Nemchinsky, VA., *J. Phys. D: Appl. Phys.*, **24**, 2014-2019 , 1991.
71. Tanberg, R., *Phys. Rev.*, **35**, 1080-1089, 1930.
72. Fang, DY., *XI-ISDEIV*, Berlin, 249-252, 1984.
73. Hantzsche, E., Jüttner, B. and Pursch, H. *J. Phys. D: Appl. Phys.* **16**, L173-177, 1983.
74. Eckhardt, G. *J. Appl. Phys..*, **47**, 4448-4450, 1976.
75. Kimblin, CW., *J. Appl. Phys.*, **44**, 3074-3081, 1973.
76. Utsumi, T. and English, JH., *J. Appl. Phys.*, **46**, 126-131, 1975.
77. Tuma, DT., Chen, CL. and Davies, DK., *J. Appl. Phys.*, **49**, 3821-3831, 1978.
78. Daalder, JE., *J. Phys. D: Appl. Phys..*, **8**, 1647-1659, 1975.
79. Plyutto, AA., Ryzhkov, VN. and Kapin, AT., *Sov. Phys. JETP ,* **20**, 328-337, 1965.

80. Lunev, VM., Padalka, VG. and Khoroshikh, VM., *Sov. Phys.-Tech. Phys.* **22**, 858-861, 1977.
81. Wieckert, C. *Phys. Fluids*, **30**, 1810-1813, 1987.
82 Anders, S. and Anders, A *J. Phys. D: Appl. Phys.* **21**, 213-215, 1988.
83. Hantzsche, E., *IEEE Trans. Plasma Sci.* **17**, 657-660, 1989.
84 Radic, N. and Santic, B., *IEEE Trans. Plasma Sci.* **17**, 683-687, 1989.
85. Kutzner, J. and Miller, HC., *J. Phys. D: Appl. Phys.*, **25**, 686-693, 1992.
86. Hantzsche, E, *Contrib. Plasma Phys.,* **30**, 575-585, 1990.
87 Miller, HC., *J. Appl. Phys.*, **52**, 23-30, 1981.
88. Lyubimov, G.A, *Sov. Phys.-Tech. Phys.*, **22**, 173-177, 1977.
89. Harris, LP, 8-ISDEIV, Albuquerque, F1, 1978.
90 Moizhes, B. Ya. and Nemchinskii, V.A, *Sov. Phys.-Tech. Phys.*, **25**, 43-48, 1980.
91. Hantzsche, E., *IEEE Trans. Plasma Sci.*, **20**, 34-41, 1992.
92. Ivanov, VA, Jüttner, B. and Pursch, H., *IEEE Trans. Plasma. Sci.*, **PS-13**, 334-336, 1985.
93. Jüttner, B., *J. Phys. D: Appl. Phys.*, **18**, 2221-2231, 1985.
94. Jüttner, B., *IEEE Trans. Plasma Sci.*, **15**, 474-480, 1987.
95. Daalder, JE and Wielders, PGE., *Proc. XIIth Int. Conf. Phenomena Ionised Gases*, Eindhoven, p. 232, 1975.
96. Drouet, M.G, *Proc. XVth Int. Conf. Phenomena Ionised Gases*, Minsk, p. 481, 1981.
97. Heberlein, JVR. and Porto, DR., *IEEE Trans. Plasma Sci.*, **11**, 152-159, 1983.
98 Meunier, JL. and de Azevedo, MD., *IEEE Trans. Plasma Sci.*, **20**, 1053-1059, 1992.
99. Kutzner, J., *VIII-ISDEIV*, Albuquerque, A1, 1978.
100. Miller, HC. and Kutzner, J *Contrib. Plasma Phys..*, **31**, 261-277, 1991.
101. Daalder, JE., *J. Phys. D: Appl. Phys..*, **9**, 2379-2395, 1976.
102. Puchkarev, VF, *XV-ISDEIV*, Darmstadt, 155-164, 1992.
103. Puchkarev,VF. and Chesnokov, S.M., *J. Phys. D: Appl. Phys.*, **25**, 1760-1766, 1992.
104. Daalder, JE, *Physica*, **104C**, 91-106, 1981.
105. Disatnik, G, Boxman, RL and Goldsmith, S, *IEEE Trans. Plasma Sci.*, **PS-15**, 520-523, 1987.
106. Shalev, S, Boxman, RL and Goldsmith, S *J. Appl. Phys..*, **58**, 2503-2507 1985.
107. Jüttner, B, *Physica ,* **114C**, 255-261, 1982.
108. Fang, DY, Nürnberg, A, Bauder, UH and Behrisch, R, *J. Nucl. Mater.* **111 & 112**, 517-521, 1982.
109. Hantzsche, E and Jüttner, B, *IEEE Trans. Plasma Sci .*, **PS-13**, 230-234, 1985.
110. Anders, A, Anders, S, Jüttner, B and Brown, IG, *IEEE Trans. Plasma Phys.*, **21**, 305-311, 1993.
111. Browning, J, Meassick, S, Xia, Z, Chan, C and McGruer, N, *IEEE Plasma Sci.* **21**, 259-260, 1993.

112. Kulyapin, VM., *Sov. Phys.-Tech. Phys.* **16**, 287-291, 1971.
113. Djakov, BE. and Holmes, R., *J. Phys. D: Appl. Phys.* **4**, 504-509, 1971.
114. Ecker, G., *Beiträge Plasma Phys.* **11**, 405-415, 1971.
115. Ecker, G. "Theoretical Aspects of the Vacuum Arc", *In:* J. M.Lafferty (Editor): "Vacuum Arcs, Theory and Application", pp. 228-320, Wiley, NY, 1980.
116. Hantzsche, E., *Beiträge Plasma Phys.* **14**, 135-138, 1974.
117 Hantzsche, E. *Contrib. Plasma Phys.* **31**, 109-139, 1991.
118 Beilis, I . *Sov. Phys.-Tech. Phys.* **19**, 251-256, 1974.
119 Beilis, I., *High Temp.* **15**, 818-824, 1977.
120 Kubono, T., *J. Appl. Phys.*, **49**, 3863-3869, 1978.
121 Harris, LP and Lau, YY, "Longitudinal Flows Near Arc Cathode Spots", Report General Electric, Schenectady NY, GE TIS REP 74 CRD 154,1974.
122 V.A. Nemchinskii, *Sov. Phys.-Tech. Phys.*, **24**, 764-767, 1979.
123 Benilov, MS., *Phys. Rev .*, E **48**, 506-515, 1993.
124 Lee, TH. and Greenwood, A, *J. Appl. Phys.*, **32**, 916-923, 1961.
125. McClure, GW., *J. Appl. Phys..*, **45**, 2078-2084, 1974.
126 Litvinov, EA. and Parfyonov, AG., *X-ISDEIV*, Columbia, 138-141, 1982.
127 Prock, J., *IEEE Trans. Plasma Sci.,* **14**, 482-490, 1986.
128 Hantzsche, E., *Beiträge Plasmaphysik* ,**12**, 245, 1972.
129 Rothstein, J., *Phys. Rev. ,* **73**, p. 1214, 1948.
130 Nekrashevich, IG. and Bakuto, IA., *Bull. Byelorussian Acad. Sci., Phys. Tech. Institute,* No. 2, 167-177, 1955.
131 Fursey, GN, and Vorontsov-Vel'yaminov, PN,, *Sov. Phys.-Tech. Phys.* **12**, 1370-1382, 1967.
132. Bugaev, SP., Litvinov, EA., Mesyats, GA. and Proskurovsky, DI., *Sov. Phys. Usp.*,**18**, 51-61, 1975.
133 Mitterauer, J., *Acta Phys. Austriaca* ,**37**, 175-192, 1973.
134 Mitterauer, J. and Till, P., *IEEE Trans. Plasma Sci.*, **15**, 488-501, 1987.
135 Litvinov, EA and Shubin, AF, *Sov. Phys.-Tech. Phys.*, **17**, 1131-1132, 1974.
136 Jüttner, B, Rohrbeck, W and Wolff, H, *Proc. 9th Int. Conf. Phen. Ionised Gases*, Bucharest, p. 140, 1969.
137 Liegmann, K, *Beiträge Plasma Phys.*, **23**, 529-550, 1983.
138. Rakhovsky, VI, *IEEE Trans. Plasma Sci.*, **PS-12**, 199-203, 1984.
139 Yanabu, S., Okawa, M., Kaneko, E. and Tamagawa, T., *IEEE Trans. Plasma Sci.*, **PS-15**, 524-532, 1987.
140 Yanabu, S., Tsutsumi, T., Yokosura, K. and Kaneko, E., *IEEE Trans. Plasma Sci.*, **17**, 717-723, 1989.
141 Kaneko, E., Tamagawa, T. Ohasi, H, Okumura, H. and Yanabu, S. *Elektrotechnik und Informationstechnik* ,**107**, 127-133, 1990.
142 Greenwood, AN., "Vacuum Arc Applications", *In*: Lafferty, J.M, (Editor): "Vacuum Arcs, Theory and Application", pp. 321-360, Wiley, NY, 1980.
143 Dullni, E. and Schade, E., *14 ISDEIV,* Santa Fe, pp. 517-521, 1990.

144   Platter, F. and Rieder, W., *13 ISDEIV*, Paris, pp. 335-337, 1988.
145   Jüttner, B., *Elektrotechnik und Informationstechnik*, **107**, 115-117, 1990.
146   Farrall, GA., "The Triggered Vacuum Arc" *In* J. M. Lafferty (Editor) "Vacuum Arcs, Theory and Application", pp. 107-119, Wiley, NY, 1980.
147   Bauville, G., Delmas, A., Haddad, N.. and Rioux, G., *CIEEE Trans. Plasma Sci.*, **177**, 781-785, 1989.
148   Vozdvijensky, VA. and Sidorov, VA., *IEEE Trans. Plasma Sci.*, **19**, 778-781, 1991.
149   Dougal, RA., Morris, G., Jr. and Volakakis, GD., *IEEE Trans. Plasma Sci.*, **19**, 976-988, 1991.
150   Bugaev, SP., Koval'chuk, B.M. and Mesyats, GA., *Proc. 6th Int. Conf. High Power Particle Beams*, Kobe, Japan, p. 878, 1986.
151   Mendel, CW., Jr. and Goldstein, SA., *J. Appl. Phys..*, **48**, 1004-1006, 1977.
152   Stinnett, RW., McDaniel, DH., Rochau, GE., Moore, WB., Gray, E.W, Renk, TJ, Woodall, HN, Hussey, TW, Payne, SS, Commisso, R.J, Grossmann, JM., Hinshelwood, DD., Meger, RA., Neri, JM., Oliphant, WF., Ottinger, PF. and Weber, BV., *IEEE Trans. Plasma Sci.*, **PS-15**, 557-563, 1987.
153.   Hinshelwood, DD., Boller, JR., Commisso, RJ., Cooperstein, J, Meger, RA, Neri, JM,Ottinger, PF. and Weber, BV., *IIEEE Trans. Plasma Sci.*, **PS-15**, 564-570, 1987.
154   Weber, BV., Commisso, RJ., Goodrich, PJ., Grossmann, PJ, Hinshelwood, DD., Kellog, JC. and Ottinger, PF., *IEEE Trans. Sci.*, **19**, 757-766, 1991.
155   Baker, MC, Barrett, BD., Nunnally, WC., Stanford, ER., Headley, C.E and Thompson, JE., *IEEE Trans. Plasma Sci.*, **17**, 786-788 , 1989.
156   Lehr, FM. and Kristiansen, M., *IEEE Trans. Plasma Sci.*, **17**, 811-817 1989.
157   Mesyats, GA., *IEEE Trans. Plasma Sci..*, **19**, 683-689, 1991.
158   Händel, SK., *Brit. J. Appl. Phys.* **14**, 181-185, 1963.
159   Skowronek, M. and Roméas, P., *IEEE Trans. Plasma Sci*, **PS-15**, 589-592, 1987.
160   Skowronek, M., Roméas, P. and Choi, P. *IEEE Trans. Plasma Sci.*, **17**, 744-747, 1989.
161   Yonas, G., Proc., "IAEA Meeting on Advances in Inertial Confinement Fusion Research", 307-323, Kobe, Japan, 1983.
162   Maron, Y., *IEEE Trans. Plasma Sci.*, **PS-15**, 571-577, 1987.
163   Brown, IG., *IEEE Trans. Plasma Sci..*, **PS-15**, 346-350, 1987.
164   Brown, IG., Galvin, J.E, MacGill, R.A and West, M.W, *Nucl. Instrum. Methods* , **B43**, 455-458, 1989.
165   Brown, IG. and Godechot, X, *IEEE Trans. Sci.*, **19**, 713-717, 1991.
166   Rutkowski, HL., Hewett, DW. and Humphries, S, Jr., *IEEE Trans. Plasma Sci.*, **19**, 782-789, 1991.

167   Bababeygy, S, Boxman, R.L and Goldsmith, S, Evaluation of Very Rapid High-Current Vacuum Arc Coatings", *IEEE Trans. Plasma Sci.*, **PS-15**, 599-602, 1987.

168   Boxman, R.L and Goldsmith, S, *Surface and Coatings Technology* , **52**, 39-50, 1992.

Index

Printed and bound by CPI Group (UK) Ltd, Croydon, CR0 4YY

03/10/2024

01040425-0017